Frontiers in Food Biotechnology

Jayachandra S. Yaradoddi •
Bharati S. Meti • Sulochana B. Mudgulkar •
Dayanand Agsar
Editors

Frontiers in Food Biotechnology

Springer

Editors
Jayachandra S. Yaradoddi
Department of Biotechnology
Basaveshwar Engineering College
Bagalkote, Karnataka, India

Sulochana B. Mudgulkar
Department of PG Studies and Research in Biotechnology
Gulbarga University
Kalaburagi, Karnataka, India

Bharati S. Meti
Department of Biotechnology
Basaveshwar Engineering College
Bagalkote, Karnataka, India

Dayanand Agsar
Department of Microbiology
Gulbarga University
Kalaburagi, India

ISBN 978-981-97-3260-9 ISBN 978-981-97-3261-6 (eBook)
https://doi.org/10.1007/978-981-97-3261-6

This Springer imprint is published by the registered company Springer Nature Singapore Pte Ltd.
The registered company address is: 152 Beach Road, #21-01/04 Gateway East, Singapore 189721, Singapore

Foreword

I am happy to put forward the book entitled *Frontiers in Food Biotechnology* edited by Dr. Jayachandra S. Yaradoddi, Dr. Bharati S. Meti, Dr. M. B. Sulochana, and Dr. Dayanand Agsar, which is dedicated to one of the premier institutes BVV Sangha's—Basaveshwar Engineering College, Bagalkote, on the occasion of the Diamond Jubilee year of celebrations (1963–2023).

I have gone through this book and noticed many exciting things in the area of food biotechnology. The book also envisages on several dynamics on one of the most required technological advancements within this specific field.

At present, there is a need for innovative approaches in biotechnology which are inclined towards the cost efficiencies and large-scale production. This book on *Frontiers in Food Biotechnology* intends to provide impetus to the introduction of new technologies thus enhancing the studies on food biotechnology. This book has substantially anticipated the recent developments and trends of biotechnology in food, functional food, food spoilage, food processing and preservation, grain processing and baking technology, food packaging and storage engineering, food safety and toxicology, genetically modified foods, probiotics, nutraceuticals, nutrigenomics, nutri-informatics, food waste management and nanotechnology in food. I am certain that, by successfully converting the fundamental scientific knowledge of food-related biotechnology into industrial and research applications, the book will be able to provide complete up-to-date information on food biotechnology and its wide range of implications for the mankind.

This book offers fast access to a plethora of knowledge for scientists in food processing businesses, biochemists, microbiologists, food technologists, and students from many streams of food science, biotechnology, and food microbiology. The outcome of such a unique book will comprehensively provide and deepen the insight into the pool of upcoming advancements in food biotechnology. The students and research fraternities could benefit in relevant assignments in the food science. In my opinion, this book will serve as a vital resource for all those who want to comprehend the most recent information on food production and processing from a biotechnological perspective in addition to offering a worldwide perspective on the field as a whole.

Basaveshwar Engineering College
Bagalkote, Karnataka, India

S. S. Injaganeri

Preface

Description About the Book

Food biotechnology or food technology uses contemporary biotechnology tools and techniques to benefit the public by excelling the manufacturing and processing of food, for instance, by fermented food (which is conventional biotechnological fermentation process), food additives, plant and animal cell cultures. In bringing revolutionary changes in the field of food biotechnology, there should be sufficient sources of information to understand about the latest outbreaks and innovation occurred during the past few decades. After the Covid-19 disaster, people learnt and understood about how the food acts as a medicine. Nowadays, consumers again started believing the old proverb "Precaution is better than cure." This food biotechnology book could help consumers to recognize their desired foods for better health and well-being.

The global population is quickly increasing, as it has already reached 7.98 billion, which is close to 8 billion, it is massive challenge to feed this population according to their demand. The main objective is not only focused on the production and manufacturing of the food and also food processing; safety and maintaining quality is what highly required from the industries. In this context, this book will cover all types of foods, latest processing, preservation technology, and safety.

Novel fermentation and enzyme technological processes, protein engineering, genetic engineering, metabolic engineering, bioengineering, quorum sensing, and nanobiotechnology have been incorporated to fetch into new dimensions in food biotechnology. A deep insight into all these aspects is emphasized in this work.

The original version of the book has been revised. A correction to this book can be found at https://doi.org/10.1007/978-981-97-3261-6_28

Experience in teaching and research has stimulated us to edit this book. Dissemination of knowledge in basic biotechnology, food Technology, and nutraceuticals is a prerequisite to understanding the significance of the multidisciplinary approach in advancing traditional food production, processing, and safety methods. The current effort will be directed to inspire students, scholars, and scientific communities should take utmost benefit and also effective execution of the present technologies in food biotechnology, simultaneously developing novel technologies in this prominent field.

Bagalkote, Karnataka, India — Jayachandra S. Yaradoddi
Bagalkote, Karnataka, India — Bharati S. Meti
Kalaburagi, Karnataka, India — S. B. Mudgulkar
Kalaburagi, Karnataka, India — Dayanand Agsar

Acknowledgments

My sincere gratitude to Dr. Veeranna C. Charantimath, Honorable Chairman of Basaveshwara Veerashaiva Vidya Vardhaka Sangha, Bagalkote, for his invaluable support and encouragement. We are thankful to Dr. R. N. Herkal, Director of the Technical Education and former Principals Dr. S. S. Injganeri and Dr. D.S. Jangamashetti, Basaveshwar Engineering College, Bagalkote, for their continued support and guidance. We are grateful to the Principal Dr. Veena Soraganvi, for her constant guidance and support throughout the tenure. We are also sincerely thankful to Prof. Dayanand Agsar, Honorable Vice Chancellor, Gulbarga University, Kalaburagi, Dr. M. B. Sulochana, Professor and Head, Department of Biotechnology, Gulbarga University, Kalaburagi, and Dr. Bharati S. Meti, Professor and Head, Department of Biotechnology, Basaveshwar Engineering College, Bagalkote, for their valuable inputs and suggestions in editing the present book. We are also grateful to the editors Dr. Ashok Kumar, Dr. Madhurima Kahali, and production editors Ms. Emmy Lee, Editor, Biomedicine and Life Science Books, Ms. Harshita Bathla, Production Editor, and Dr. Bhavik Sawhnie, for their remarkable suggestions or recommendations, and also thankful to all the authors for their quality chapter contribution towards the book.

Contents

Part V Scope for Commercialization of Foods and Nutraceuticals

Contributors

Ajay S. K. SDM Research Institute for Biomedical Sciences, SDM College of Medical Sciences & Hospital Campus, Shri Dharmasthala Manjunateshwara University, Sattur, Dharwad, Karnataka, India

T. B. Allolli Department of Vegetable Science, University of Horticultural Sciences, Bagalkote, Karnataka, India

Avitha K. M. P.G. Department of Botany, BVVS Akkamahadevi Women's College, Bagalkote, Karnataka, India

Shreya Bhat Department of Biotechnology, Basaveshwar Engineering College, Bagalkote, Karnataka, India

Bhavya D. C. Department of Microbiology, Davangere University, Davangere, Karnataka, India

Bhuvaneshwari G. Department of Postharvest Management, College of Horticulture, University of Horticultural Sciences, Bagalkote, Karnataka, India

Chandan B. M. Plant Molecular Biology Lab (DBT-BIOCARe), Center for Biotechnology Research, College of Horticulture, University of Horticultural Sciences, Bagalkote, Karnataka, India

Chandan K. Department of Post Harvest Management, College of Horticulture, UHS Campus, Sirsi, Karnataka, India

Chetan D. M. Department of Biotechnology Engineering, N.M.A.M. Institute of Technology (Deemed to be University), Nitte, Udupi, Karnataka, India

Sarvamangala S. Cholin Plant Molecular Biology Lab (DBT-BIOCARe), Center for Biotechnology Research, College of Horticulture, University of Horticultural Sciences, Bagalkote, Karnataka, India

Akshay H. Dasalkar Department of Traditional Foods and Applied Nutrition, CSIR-Central Food Technological Research Institute, Mysuru, Karnataka, India

Academy of Scientific and Innovative Research (AcSIR), Ghaziabad, Uttar Pradesh, India

Shivalingsarji V. Desai Department of Biotechnology, KLE Technological University, Hubballi, Karnataka, India

Rashmi G. Gadwal Department of Biotechnology, Gulbarga University, Kalaburagi, Karnataka, India

Vasant M. Ganiger Department of Vegetable Science, Directorate of Extension, University of Horticultural Sciences, Bagalkote, Karnataka, India

Pratiksha Prabhakar Gawali Department of Traditional Foods and Applied Nutrition, CSIR-Central Food Technological Research Institute, Mysuru, Karnataka, India

Academy of Scientific and Innovative Research (AcSIR), Ghaziabad, Uttar Pradesh, India

Rakesh Gehlot Centre of Food Science and Technology, Chaudhary Charan Singh Haryana Agricultural University, Hisar, Haryana, India

Manoj Girish Department of Biotechnology, M S Ramaiah University of Applied Sciences, Bangalore, Karnataka, India

Mallikarjun Goni SDM Research Institute for Biomedical Sciences, SDM College of Medical Sciences & Hospital Campus, Shri Dharmasthala Manjunateshwara University, Sattur, Dharwad, Karnataka, India

Prakash M. Halami Department of Microbiology & Fermentation Technology, CSIR-Central Food Technological, Research Institute, Mysore, Karnataka, India

Jayashree V. Hanchinalmath Department of Life Science, School of Sciences, Jain (Deemed to be University), Bangalore, Karnataka, India

Harshitha T. Department of Traditional Foods and Applied Nutrition, CSIR-Central Food Technological Research Institute, Mysuru, Karnataka, India

Academy of Scientific and Innovative Research (AcSIR), Ghaziabad, Uttar Pradesh, India

Veeranna S. Hombalimath Department of Biotechnology, KLE Technological University, Hubballi, Karnataka, India

Basavaraj S. Hungund Department of Biotechnology, KLE Technological University, Hubballi, Karnataka, India

S. S. Injaganeri Department of Civil Engineering, Basaveshwar Engineering College, Bagalkote, Karnataka, India

Environmental Studies Division, Department of Civil Engineering, Basaveshwar Engineering College, Bagalkote, Karnataka, India

Jagadeesh S. L. Regional Horticultural Research and Extension Centre, UHS Campus, Bengaluru, Karnataka, India

Sanjay Kumar Jain Department of Processing and Food Engineering, University of Maharana Pratap Agriculture and Technology, Udaipur, Rajasthan, India

Palaksha Kanive Javaregowda SDM Research Institute for Biomedical Sciences, SDM College of Medical Sciences & Hospital Campus, Shri Dharmasthala Manjunateshwara University, Sattur, Dharwad, Karnataka, India

Shilpa K. Jigajinni Department of Biotechnology, Basaveshwar Engineering College, Bagalkote, Karnataka, India

Ranjan Kaushik Centre of Food Science and Technology, Chaudhary Charan Singh Haryana Agricultural University, Hisar, Haryana, India

Tanaji Kudre Department Of Meat and Marine Science, CSIR-Central Food Technological Research Institute, Mysore, Karnataka, India

Chaitra C. Kulkarni Kitturu Rani Channamma College of Horticulture, University of Horticultural Sciences, Bagalkote, Karnataka, India

Spoorthi R. Kulkarni Department of Biotechnology, Basaveshwar Engineering College, Bagalkote, Karnataka, India

Ankit Kumar Centre of Food Science and Technology, Chaudhary Charan Singh Haryana Agricultural University, Hisar, Haryana, India

Preeti S. Kumarmath Department of Biotechnology, Basaveshwar Engineering College, Bagalkote, Karnataka, India

Madhumala Y. Department of Biotechnology, Basaveshwar Engineering College, Bagalkote, Karnataka, India

Ghouse Modin N. Mamdapur Department of Library and Information Science, Yenepoya (Deemed to be University), Mangalore, Karnataka, India

Information and Documentation, New Product Development and Research, Synthite Industries Private Limited, Kolenchery, Kerala, India

Renukaradhya K. Math SDM Research Institute for Biomedical Sciences, SDM College of Medical Sciences & Hospital Campus, Shri Dharmasthala Manjunateshwara University, Sattur, Dharwad, Karnataka, India

Gangadhar B. Megeri Environmental Science Division, Department of Civil Engineering, Basaveshwar Engineering College, Bagalkote, Karnataka, India

Bharati S. Meti Department of Biotechnology, Basaveshwar Engineering College, Bagalkote, Karnataka, India

Renuka Meti Department of Food Processing and Nutrition, Karnataka State Akkamahadevi Women University, Vijayapura, Karnataka, India

S. B. Mudgulkar Department of PG Studies and Research in Biotechnology, Gulbarga University, Kalaburagi, Karnataka, India

Vijaykumar B. Narayanpur Department of Plantation, Spices, Medicinal and Aromatics, College of Horticulture, University of Horticultural Sciences, Bagalkote, Karnataka, India

Basavaraj Nainegali Department of Biotechnology, Basaveshwar Engineering College, Bagalkote, Karnataka, India

Parinitha A. Department of Traditional Foods and Applied Nutrition, CSIR-Central Food Technological Research Institute, Mysuru, Karnataka, India

Academy of Scientific and Innovative Research (AcSIR), Ghaziabad, Uttar Pradesh, India

Premjyoti Patil Department of Biotechnology, Basaveshwar Engineering College, Bagalkote, Karnataka, India

Shweta Pattanashetti Department of Food Processing and Nutrition, Karnataka State Akkamahadevi Women University, Vijayapura, Karnataka, India

Ishrat Jahan Peerzade Department of Microbiology & Fermentation Technology, CSIR-Central Food Technological, Research Institute, Mysore, Karnataka, India

Rekha Phogat Centre of Food Science and Technology, Chaudhay Charan Singh Haryana Agricultural University, Hisar, Haryana, India

Ramappa Department of Processing and Food Engineering, University of Maharana Pratap Agriculture and Technology, Udaipur, Rajasthan, India

Neha Rani Centre of Food Science and Technology, Chaudhary Charan Singh Haryana Agricultural University, Hisar, Haryana, India

Sharada P. Department of Biotechnology, Basaveshwar Engineering College, Bagalkote, Karnataka, India

Shivaleela V. B. Pomla Natuals LLP, BEC STEP, Bagalkote, Karnataka, India

Kirankumar Shivasharanappa Department of Life Science, School of Sciences, Garden City University, Bangalore, Karnataka, India

H. Shoba Department of Processing and Food Engineering, University of Maharana Pratap Agriculture and Technology, Udaipur, Rajasthan, India

Agricultural Engineering, College of Horticulture, Munirabad, Karnataka, India

University of Horticulture Sciences, Bagalkote, Karnataka, India

Somya Adusumilli Department of Traditional Foods and Applied Nutrition, CSIR-Central Food Technological Research Institute, Mysuru, Karnataka, India

Academy of Scientific and Innovative Research (AcSIR), Ghaziabad, Uttar Pradesh, India

Veena S. Soraganvi Department of Civil Engineering, Basaveshwar Engineering College, Bagalkote, Karnataka, India

Shefali Srivastava Department of Life Science, School of Sciences, Garden City University, Bangalore, Karnataka, India

Suresha G. J. Department of Post Harvest Management, College of Horticulture, UHS Campus, Bengaluru, Karnataka, India

Manjunatha P. Talwar Department of Life Science, School of Sciences, Garden City University, Bangalore, Karnataka, India

Umme Seema N. KSN University of Agricultural and Horticultural Sciences, Shivamogga, Karnataka, India

Jayashri B. Uppin Department of Biotechnology, Basaveshwar Science College, Bagalkote, Karnataka, India

Gayatri Vaidya Department of Food Biotechnology, Davangere University, Davanagere, Karnataka, India

Meghana Variyar SDM Research Institute for Biomedical Sciences, SDM College of Medical Sciences & Hospital Campus, Shri Dharmasthala Manjunateshwara University, Sattur, Dharwad, Karnataka, India

Virupakshaiah D. B. M. Department of Microbiology, Davangere University, Davangere, Karnataka, India

Yallappa M. Department Agriculture Microbiology, College of Agriculture Bheemarayanagudi, UAS, Raichur, Karnataka, India

Sudheer Kumar Yannam Department of Traditional Foods and Applied Nutrition, CSIR-Central Food Technological Research Institute, Mysuru, Karnataka, India

Academy of Scientific and Innovative Research (AcSIR), Ghaziabad, Uttar Pradesh, India

Jayachandra S. Yaradoddi Department of Biotechnology, Basaveshwar Engineering College, Bagalkote, Karnataka, India

Part I
Introduction to Food Biotechnology

Chapter 1
Prospect of Biotechnology in Foods

Jayachandra S. Yaradoddi, Bharati S. Meti, Shoba H., and S. S. Injaganeri

1 Introduction

The rate and precision of experimentation by scientists can enhance food features, and production methods have increased thanks to modern food biotechnology. Farmers have spent lots of time crossbreeding animals or plants to achieve the precise advantageous qualities they were in the hunt for and circumvent the traits they did not wish for centuries before introducing this technology. In addition to taking a lot of time and effort, the method had no assurance of success. Food biotechnology uses genetic and plant scientific knowledge to carry on this heritage. Modern biotechnology allows researchers to transfer genes for desirable features from one plant to another. The farmer and the consumer profit directly from this procedure as regards the environment and the economy [1].

Agriculture biotechnology is an environment-friendly area where biotechnology greatly impacts environmental protection. Scientists use biotechnology to make food production more environmentally friendly, efficient, and significant. One such

J. S. Yaradoddi (✉) · B. S. Meti
Department of Biotechnology, Basaveshwar Engineering College, Bagalkote, Karnataka, India

Shoba H.
College of Horticulture, Munirabad, Koppal, University of Horticulture Sciences, Bagalkote, Karnataka, India

S. S. Injaganeri
Department of Civil Engineering, Basaveshwar Engineering College, Bagalkote, Karnataka, India
e-mail: principal@becbgk.edu

J. S. Yaradoddi et al. (eds.), *Frontiers in Food Biotechnology*,
https://doi.org/10.1007/978-981-97-3261-6_1

example is the pest and disease resistance of some transgenic foods. This enables farmers to grow a healthy, high-yielding crop using fewer chemicals, mainly herbicides and pesticides. Decreased pesticide use benefits wildlife and the environment and consumers who might be concerned about consuming chemicals when eating fruits and vegetables.

Since many biotech crops have been designed as naturally tolerant to herbicides and used more selectively, they need limited plough to handle weeds. Conservation tillage, which involves leaving crop residue in the agricultural field and reducing or eliminating ploughing, improves water assimilation, reduces soil compaction and erosion, has resulted in the formation of healthier soil. It also helps conserve water from irrigation and rainfall. These advantages help maximize agricultural production and reduce water use [2]. There are various methods (such as PCR, RFLP, RAPD, AFLP, microarrays, and PFGE) that are involved in improving the quality and yield of crop plants (discussed in detail in Table 1.1).

Additionally, crop rotation cultivation practice emits reduced carbon dioxide, or carbon monoxide, than conventional tillage into the environment and supports environments and which is commonly advantageous to insects, birds, and other creatures. Last but not least, biotechnology can aid in reducing deforestation [3]. Transgenic crops yield more and need less land to generate the same goods. Additionally, to help with food production in less arable places, researchers are working on improved growth features, including drought tolerance.

Table 1.1 List of biotechnology techniques involved in various food and agriculture applications

Biotechnology techniques	Applications
PCR (polymerase chain reactions)	Cloning and amplification DNA/RNA
PFGE (pulsed-field gel electrophoresis)	Separating the huge quantity of DNA that can be used for typing of the microbial strains
RFLP (restriction fragment length polymorphism) markers	Helps in marker-based selection
RAPD (random amplified polymorphic DNA)	Widely employed for the identification of herbal medicine due to scarce in availability of DNA
AFLP (amplified fragment length polymorphism)	Can be used to identify the closely related plant species, as well as microbial species
rRNA (ribosomal RNA) gene sequence	Used in identification food and agricultural important microorganisms
DNA microarrays	Useful in analysis of plant defense system, fruit ripening process, seed development, environmental stress responses

A significant field of research is nutritionally enhanced biotech foods and animal feedstock production, which yielded several promising goods. For instance, maize with higher quantity of amino acids, cooking oils with distinctive fatty acid profiles have fewer than 1% trans fats, oils, and minerals that are good for animal feed. Additionally, many items currently being developed are designed to provide nutritional advantages, especially "golden rice,"which includes extra beta-carotene and iron. By removing the offending proteins responsible for most allergic reactions in people, scientists are researching ways to develop foods with fewer allergens, such as soy and peanuts. Fruits and vegetables that contain more vitamins, minerals, and proteins are also being produced [4].

While still possessing the first-generation growth features that result in resilient, high-yield crops, the biotech foods in second-generation promise to offer consumers items that hold freshness for a longer duration, fewer allergies, and higher quantities of beneficial fats, mainly omega-3 fatty acids [5]. Consumers might be inclined to eat transgenic foods. The majority of consumers (53%), according to the International Food Information Council (IFIC's) 2008 Food Biotechnology: A Study from USA Consumer Trends, have unfavorable opinions on plant biotechnology. The majority of people would buy biotech-produced food for several reasons, including increasing the amount of omega-3 fatty acids (78%), lowering trans and saturated fats (76%), and improving the taste and freshness of the food (67%).

1.1 Benefits of Agricultural Biotechnology for Farmers

Farmers' welfare is positively impacted by agricultural biotechnology in both developed and developing nations [6]. Farmers can profit economically from biotech crops while increasing their crop production's sustainability. The higher crop yields from agricultural biotechnology offer significant economic, social, and environmental benefits in light of rising food prices and a growing world population.

The NCFAP (National Centre for Food and Agricultural Policy) published a study report in 2005, accordingly with the advent of biotechnology, there has been tremendous improvement in plants tolerance towards the insects. This herbicide helps US farmers to lower their yearly production expenses from 1.4 billion USD, increasing net earnings by 2 billion USD [7, 8]. Farmers must understand that the decreased crop losses during harsh situations like drought historically help significantly reduce costs. Biotechnology-developed crops varieties are generally designed to sustain their lives even when grown under adverse environmental conditions, such as flood or drought, severe cold or heat, and soil salinity or heavy metals (Fig. 1.1).

Fig. 1.1 Pictures indicating the development pest, disease, drought, and salinity-resistant varieties of plants

1.2 Advantages of Biotechnology in Crop Production and Challenges

Genetically modified plants (GM Plants) produced using genetic engineering technique generally benefit both producers as well as consumers. The gene modification is held so that it defends the crop plant from any biotic and abiotic stress, incorporating the input or desired traits in enhancing the crop yield and benefitting the producer. A variety of profitable genetically modified crops can be placed under this group. Scientists are now focusing extensively on utilizing biotechnology tools and a technique in producing heterogeneous plants, which commonly offer a broad range of benefits to the end users. The varieties of crops are generally manipulated to attain resulting characters or traits. For instance, one of the widely used genetically modified corn for expression of proteins produced by the commonly occurring soil bacteria is *Bacillus thuringiensis* (Bt). Applying Bt as a key insecticidal spray is one of the breakthrough discoveries from the past 40 years. The Bt microbe has been manipulated to express cry proteins, a class of insecticidal proteins. The experimented cry proteins have shown effective results against certain insect pests.

Meanwhile, they are safe for the environment, mammals, humans, and birds. In 1996, the Bt corn plant was introduced and later emerged as a commercial crop. This further led to the most promising advancements in insect pest management during the progression of synthetic insecticides.

One of the severe corn pests *Ostrinia nubilalis* (European corn borer) damages the corn plant by larval feeding injury that will hampers the smooth supply of nutrients with in crop plants [9]. Besides the declining crop yield, infection from the European corn borer clears the way for fusarium, molds spoilage, that initiates the formation of fumonisin, mycotoxin, in corn [10]. Fumonisin is commonly a noxious substance when compared to other substances, causing liver injury in all investigated animals. Therefore, in conclusion, several substantiations support that it can also take part in causing esophageal cancer in humans [11]. Investigation demanding maize hybrids along with European corn borer instituted lower concentrations of fumonisin in expressing specific cry proteins among the transgenic maize varieties, for instance, 2.1 μg/g when compared to 16.5 μg/g on the nontransgenic maize hybrids [10]. Above all the health advantages to the consumers and livestock, infested corn becomes a big economic problem. Corn that exceeds the permissible levels for future utility could be disposed of or utilized for other purposes, resulting in a huge financial loss [12]. In reality, the total loss of about 1 billion dollars/year to farmers as a result of the infected crops (comprising cotton, corn, vegetables, and sorghum) is caused by European corn borer, which is generally largely enhanced by the use of Bt cotton and Bt corn for farm more effectively [13–16].

In recent times, the Department of Agriculture U.S presumed that though the profit and accomplishment of genetically modified crops differ based on several parameters that include level of pest infection and the region, the rapid acquisition of Bt cotton crop plants in the Southwest region and herbicide-resistant soybean has led to the considerable enhancement in overall yield and a major decrease in insecticide applications [17]. On the progression of the similar report conducted by the Economic Research Service investigation held in the same year variation among the regular pesticide used for adoption and nonadoption, farmers indicated that those who were using the genetically modified soybeans, corn, and cotton collectively used about 7.6 million fewer. ERSI (Economic Research Service Investigation) in the same year have found variation among the standard pesticide use of adopted and nonadopted farmers have indicated that those uses GM (genetically modified) soybeans, cotton, and corn collectively uses 7.6 million smaller acre-treatments of pesticides in comparison to nonadopters during 1997. In 1998, adopters amplified this disparity to about 17 million less acre treatment [18]. Besides this, other users have concluded that although there is a significant rise in the Bt corn plantation, the proportion of insecticidal-treated corn has persisted the whole time. The investigation also indicated that pollen from the Bt corn was detrimental to the Monarch butterfly [19]. Under the US and Canada joint collaborative research project, scientists realized the risks of using Bt corn to Monarchs can be lower. This could be because the mass and duration of pollen shed is directly associated with the period of larval activity. Moreover, the number of milkweed plants maturing next to cornfields and the number of fields planted along with Bt corn can be considered another controlling measure reported for their lower risk factor to monarch butterflies [20].

There are other traits were incorporated in various crops to safeguard them from biological damage. Papaya, squash, potato, and tomato are widely used crops that have been manipulated so that they can defend against infections caused by different pests and viruses [21]. Apart from these biological stresses, plant productivity can also be affected by various physical factors (abiotic factors) such as temperature, soil composition, herbicides, and water supply. So, confirmation of the plants with specific genes that will support them to resist a broader range of environmental conditions can enhance productivity. Several plants are being engineered to tolerate the extremities like heat, drought, cold temperatures, and poor soil conditions, especially aluminum contamination and salinity [22–26]. The overall increase in the yield after harvest can also be accomplished by bettering the efficacy of the crop plants' photosynthetic and metabolic pathways.

One instance of such a pathway that facilitated improved crop production includes nitrogen assimilation, modification in photosynthesis, and starch biosynthesis [27–29]. Immediately after the harvest, when the farmer will take crops to the market, is vital in terms of economics because of the perishability among the food crops. The key modification in the delayed ripening process can benefit farmers by minimizing the post-harvest loss and the consumer employing increased shelf life of the foods. To stop the distribution of spoiled fruit, especially matured tomatoes can be harvested; they look green, and vegetables such as tomatoes are commonly ripened artificially by exposing them to ethylene.

The emerging area enzyme biotechnology can also produce specialty oils with desired and favorable fatty acid profiles, mainly higher oleic acid-containing peanut oil [30]. Carotenoids are other nutrients known for their health benefits, such as in macular degeneration and reducing the risk for certain types of cancers. Compared to other plant species, tomatoes (lycopene), papaya (carotene), spinach (lutein), and kale (β-carotene, lutein + zeaxanthin) are rich in carotenoids. Previously, β-carotene has been genetically engineered and expressed in rice cultivars, famously known as golden rice, by inserting genes for three enzymes in the phytoene synthase pathway mainly derived from daffodil (two genes) and another gene from the bacteria *Erwinia uredovora* [31]. This specific strain was reported to be crossed with higher iron-contained rice to give rise to strain with both qualities [32].

Golden rice can be considered one of the most promising and exciting crop plants among all biotechnology crops that yield better profit to the people of developing countries. This variety of rice can potentially lower malnutrition condition and blindness, which is connected with vitamin A deficiency (Fig. 1.2). Conversely, several questions are elevated concerning the efficiency of these rice crops due to many cultural, dietary, and biological barriers that must prevail [33]. Therefore, these basic questions must be addressed or ensured before their introduction to the market. Similarly, phytonutrients that have health benefits generally consist of polysterols, phytoestrogens, and glucosinolates. They can be seen in various food sources; these compounds can be over-expressed selectively to therapeutic levels

Fig. 1.2 Pictures depicting the golden rice, tomato, Bt cotton, mice, maize, sheep, vegetarian egg, pig produced using the transgenic technology of biotechnology

[34]. In another way, biotechnology is on the edge of completely modification of another feature of preventive medicine.

Several techniques are involved in producing vaccines in plants by inserting desired genes. That will express protein antigens in crops, mainly potatoes, bananas, and corn. When consumers eat these antigens-inserted crops, they can trigger an immune response and have been indicated to offer protection against a consequent challenge from the pathogens [35].

The possibility of this method was tested when the mice was allowed to feed with hepatitis B surface antigen expressed in potato tubers, which indicated a main immune response via generating specific antibody to the antigen [36]. Nowadays, companies are showing immense interest in developing into suppliers with a wide variety of biotechnology products, such as animal health products, blood proteins, industrial enzymes, and therapeutic proteins. For instance, a group of scientists from the University of Georgia were found the yellow poplar trees potentiality in absorbing the toxic ion mercury from the soil and modify this toxin into a fairly static form.

They interpreted that the gene was isolated from the mercury-resistant bacteria, which were generally soil-borne and can thrive well at various heavy metal-contaminated regions. Another investigation endorses that the genetically engineered plants have gained the ability to remove mercury ten times more than normal plants. One instance demonstrated how phytoremediation uses plants in cleaning polluted areas when it was associated with biotechnology, which can be an economical, environmentally friendly, and efficient method that can help the soil restore and revegetate contaminated waste sites.

Consumer preference is vital for industries in choosing the right product for its development. New products, mainly seedless melons, squash, peppers, tomatoes, cherries, and strawberries, can be listed as a few likely ways of attracting most consumers. In addition, this fruit has enhanced taste because of the improved total soluble sugar compared to the seeded fruit and could be more profitable for the processing industry [37]. Moreover, the sweet protein present in the *Pentadiplandra brazzeana* an African vine fruit has the thermo stable protein, which is generally 500 times sweeter as sucrose at elevated concentration and also 2000 times sweeter when in 2% (weight) solution. Due to lack of bitterness, that can be considered a potential substitute for lower energy sweeteners [38]. Now people will admit that plant biotechnology has the remarkable potential to impact future society and well-being. About 50 biotechnology crop products have qualified the regulatory review process and are commercialized, with a range of potatoes, corn, squash, and potatoes. It is also important to notice that crops improved by biotechnology are the quickly adopted technologies when considering agricultural history. As in 1996, over 4.3 million acres were farmed with biotech crops; in 2000, the number was increased up to 109.2 million acres [39]. This is due to the various advantages possessed by biotechnology-based crops/foods.

2 Potentials of Food Biotechnology

2.1 Acid Tolerant Plants

In the perseveration of pH of the soil, lime can be employed. This practice can be advanced to be an exceptional mechanism. However, it is a somewhat expensive and short-lived process. Otherwise, it is very much possible to develop an aluminum-tolerant variety of engineered crops or cultivars [40].

Lime could be applied to the soil to preserve the soil pH. This method appears to be exceptional but is also expensive and temporary. On the other hand, it is expected to grow superior cultivars that are tolerant to aluminum.

2.2 *Drought Tolerant Crops*

Among the numerous methods demonstrated, targeted and short gun techniques are the two diverse methods in genetic engineering. These methods have been useful in producing transgenic plants, which encompass the potential to confer drought tolerance [41].

2.3 *Production of Disease-Free Plants*

The method generally employed for producing disease-free plants is through the micropropagation technique. One of the main examples that can be considered under this is banana plants. They are usually grown in nations where they appear to be the main source of revenue/food/employment. Micropropagation is the best way to disease-resistant banana plantlets using tissues of healthy banana plants. This has every possible profit-making revolutionary method that is considerably economical and simple [42].

2.4 *Fortification of Crops*

Both developed and developing countries are suffering due to a shortage of nutritional food, which dictated the emergence of fortified food crops, which are derived from exceptionally quality food sources and are supplemented with additional nutrients for improving the health of malnourished children. One of the main examples of fortified crop is "Protato." Which is GM potato (genetically modified potato), that is being extensively grown and used in India. These will offer approximately one-third to one-half higher proteins than the usual potato. This GM potato can also consist of major amounts of lysine and methionine amino acids (essential amino acids). The product "Protato" can be a good nutrient source in nations using potatoes as the main staple food [43].

Similarly, one more such potential food crop is golden rice. This GM rice has a rich source of beta carotene [44]. The food grains as well as leaves of cowpeas could be used as relish dishes. In various countries, these cowpeas are generally consumed as staple foods. Recently, there is an uptrend in growing the GM varieties of cowpeas in Tanzania [45]. The fortification process can enhance crops' nutritional condition, specially developed using GM organisms (Table 1.1).

2.5 Animal Feeds

GM crops can be practically used as animal feed in several developed countries. Such varieties of crops do have a large potential to provide enough nutrients compared to normal crops [46].

2.6 Aquaculture Development

Advancement in biotechnology has contributed significantly to the realistic applications in aquaculture; biotechnology has stimulated the growth and development of aquatic organisms. Research is ongoing in this area for producing harmless and improved varieties of marine organisms to make them suitable for human consumption [47].

2.7 Pests Tolerant Crops

Throughout the globe, common pests are a major threat to different crops; these crops could range from fodder crops to other crops for deriving food. One of the main variety such crops is Bt cotton. This Bt cotton was developed by introducing *Bacillus thuringiensis* genes to the cotton crops to produce the cry-protein gene. This cry-protein is toxic to Lepidoptera, Diptera, Hymenoptera, Coleoptera, and nematodes. With the effective use of biotechnology, the produced Bt cotton will gain higher resistance against these insect pests, eventually increasing the production yields [48].

2.8 Biofuel

The future prosperity of the market will be based on the supply of secure, equitable, affordable, and safe energy. The production of biofuels is the most promising area of research for the past few decades. Biofuels are now the most dependable alternative to fossil fuels. As per the report of Francisco et al. 2010 [49], six microalgal strains were grown photosynthetically in a photobioreactor. Among these, *Chlorella vulgaris* was mainly used as a feedstock production. Further, the quality of biofuels produced and the quantity of lipid yield could be considered major criteria in selecting biodiesel-producing strains.

2.9 Antifungal Agents

Pigeon pea, groundnut, and chickpea are the main crops of this Kalaburagi region. Previous reports by Sulochana et al. in 2014 [50] endorsed that these crops were routinely infested with different fungal pathogens, causing heavy loss to the farmers. However, in the investigation, they have concluded use of siderophores produced using *Pseudomonas aeruginosa* JAS-25, which was indicated vital antagonistic activity against the *Fusarium udum, Fusarium falciforme,* and *Aspergillus niger* with a significantly high zone of inhibition 3.0 cm, 2.0 cm and 2.0 cm each, respectively.

2.10 Vaccine Production

Vaccines are crucial for human health and well-being; the process of getting the vaccine is known as a vaccination that saves the lives of human beings. Various sources of plants, bacteria, and viruses are being used for vaccine development. Among them, plant-based vaccines can be produced using recombinant DNA technology. This could be introduced into selected plants as antigens to act against particular pathogens. Currently, scientists have made about 200 proteins expressed within plant species. These outcomes are stimulatory for the brighter perspective of plant-based vaccines. The effective use of biotechnology has provided a wonderful platform for researchers and scientists to produce a wide variety of more economical and reliable vaccines with large-scale production [51].

2.11 Fermentation Technology

Fermentation can be considered one of the predominant processes in the production of breweries. Four types are used in breweries: Lager yeast, Belgian yeast, Ale yeast, and Wheat beer yeast., light wine can be produced using the genetic engineering technique. The introduction of a foreign gene, that is, the glucoamylase gene, facilitates the modification of yeast. Glucoamylase enzyme could be expressed using yeast throughout the process via the conversion of starch to glucose [52]. Various yeast strains could be utilized during wine production, initiating malolactic fermentation. The wine-making process generally involves two steps: (1) Primary fermentation that utilizes yeast cells to alter the glucose molecule to alcohol. (2) Secondary fermentation outcome as lactic acid production liberating higher acidity level by the use of bacteria. Expensive substituting strategies were employed in troubleshooting this problem. *Lactobacillus delbrueckii* was introduced with the malolactic gene to counter this issue. This gene suppresses the immediate conversion of malate, which can later minimize the wine acidity level [52].

2.12 *Bioplastic and Biobased Plastics for Environmental Safety and Food Packaging*

As everyone knows advantages and ill effects of using plastics for packaging. After the COVID-19 pandemic, the public is leaning towards better-packaged food rather than open foods. Since the government banned using microsized plastics for packaging, it is necessary to fulfill its gap. So, extensive research was carried out to find the bioplastics (PHA, PHB, PLA, etc.) or biobased plastics substitute materials against the exhaustive use of the synthetic plastics [53–56]. Their investigation found a way to produce biobased/bioplastics using renewable sources and interpreted how they can be effective and safe for human use. In addition, using these alternative packaging materials can reduce waste disposal, helping keep the environment clean. One of fascinating characteristic feature of using such bioplastic in packaging is notably correspond to their complete biodegradability with the simple action of soil or aquatic microbial consortia when they are disposed to such environments with a considerably reduced or null release of green houses gases.

2.13 *Microbial Enzymes*

Enzymes recovered using microbes could play the main role in food industries due to their stability in the most adverse conditions [57–59] compared to plant and animal enzymes. These can be particularly used in the processing and production of various food items on an industrial scale. Food development could be done using different biotechnology tools and techniques, especially genetic modification. These enzymes range from lipases to proteases and carbohydrases. The higher production can be accomplished by cloning genes for the smallest period. These enzymes have a particular role in making cheese, flavoring agents, and curd. It was reported that the USA uses more than 50% of proteases and carbohydrases for industrial food production. Most of them consist of renin and α-amylases [52].

3 Use of Biotechnology to Improve Yield

Milk is a commonly consumed health drink all over the world, which has higher nutritional value. Somatotropin is considered to be the main hormone in the regulation of lactation among cattle. The somatotropin hormone released by the pituitary gland stimulates the enhanced production of milk. Scientists effectively used *Escherichia coli* for inserting bovine somatotropin, which has positively impacted and helped in the production of more amount (10–12%) of milk. By 2050, the worldwide population is expected to reach nine billion. Accordingly, there should be an increase in the production capacity of the food. Fortunately, biotechnology has developed the best solution for all these present and future problems [52].

3.1 Biological Plant Growth Promoters

The use of biofertilizers and biological growth promoters are now considered to be emerging trends. A group of researchers, Sulochana et al., in 2014 [60], were able to isolate different strains of *Pseudomonas* sp. where they were able to produce siderophores that had excellent growth promotion activity on various crops such as *Cajanus cajan* (pigeon pea), *Arachis hypogaea* (groundnut), and *Cicer arietinum* (Chickpea). They treated the seeds of harvests of pigeon pea, chickpea, and groundnut cultivars, and they observed a significant increase in the root and shoot length of these plants compared with their respective controls. They also found excellent seed development of 30, 20, and 20%, corresponding to chickpea, groundnut cultivars, and pigeon peas.

4 Future Prospects

However, there is a huge debate on accepting novel varieties of GM foods due to the safety of human beings and the environment. Since the layperson can raise many questions on the consumption of such GM foods, very few researchers/scientists are giving effective and convincing responses. This might be because they have not done exhaustive research on this field. Moreover, these GM foods can be produced commercially with the utmost assurance from scientists. All the necessary declarations shall provide immense support for including such GM technologies without creating much argument on it.

References

1. International Food Information Council (2004) Food biotechnology: enhancing our food supply. Washington, DC
2. Fawcell R, Towery D (2002) Conservation tillage and plant biotechnology. Conservation Technology Information Center, p 14
3. Ibid, page 1 for benefits to wildlife; pages 9–10 for reduction of CO2. http://www.whybiotech.com/resources/factsheets_drought.asp. Accessed 8 Sep 2008
4. Canadian Produce Marketing Association (2004–2008) Biotechnology and genetically modified foods
5. Sankula S (2006) Quantification of the impacts on U.S. agriculture of biotechnology-derived crops planted in 2005 (Executive summary). National Center for Food and Agricultural Policy. Economics

6. International Food Information Council (2007) Food & Agricultural Biotechnology: health impacts in developing nations. Washington, DC. https://foodinsight.org/Resources/Detail.aspx?topic=Food_Agricultural_Biotechnology_Health_Impacts_in_Developing_Nations_CPE_Program
7. James C (2004) Preview: global status of commercialized transgenic crops: 2004. ISAAA briefs no. 32. ISAAA, Ithaca
8. Brookes G (2006) Global impact of biotech crops: socio-economic and environmental effects in the first ten years of commercial use. PG Economics
9. Obrycki JJ, Losey JE, Taylor OR, Jesse LCH (2001) Transgenic insecticidal corn: beyond insecticidal toxicity to ecological complexity. Bioscience 51:353–361
10. Munkvold GP, Hellmich RL, Rice LG (1999) Comparison of fumonisin concentrations in kernels of transgenic Bt maize hybrids and nontransgenic hybrids. Plant Dis 83:130–138
11. Food and Drug Administration. Center for Food Safety and Applied Nutrition (2000) Draft. Background paper in support of fumonisin levels in corn and corn products intended for human consumption. http://vmcfsan.fda.gov/_dms/fumonbg1.html. Accessed 17 Dec 2001
12. Food and Drug Administration. Center for Food Safety and Applied Nutrition (2000) Guidance for industry. Fumonisin levels in human and animal feeds. http://vm.cfsan.fda.gov/_dms/fumongui.html. Accessed 17 Dec 2001
13. Mason CE, Rice ME, Calvin DD, Van Duyn JW, Showers WB, Hutchison WD, Witkowski JF, Higgins RA, Onstad DW, Dively GP (1996) Introduction. In: European corn borer ecology and management. National Central Regional Extension. Pub. No. 327. Iowa State University, Ames, p 1
14. Falck-Zepeda JB, Traxler G, Nelson RG (2000) Surplus distribution from the introduction of a biotechnology innovation. Am J Agric Econ 82:360–369
15. Gianessi LP, Carpenter JE (1999) Agricultural biotechnology: insect control benefits. National Center for Food and Agricultural Policy, Washington, DC
16. Kalaitzandonakes N (1999) A farm level perspective on agrobiotechnology: how much value and for whom? AgBioforum 2:61–64
17. Fernandez-Cornejo J, McBride WD, Klotz-Ingram C, Jans S, Brooks N (2000) Genetically engineered crops for pest management in U.S. agriculture: farm level effects. Agricultural economics report. No. 786: Resource Economics Division, Economic Research Service, U.S. Department of Agriculture, Washington, DC
18. Heimlich RL, Fernandez-Cornejo J, McBride W, Klotz-Ingram C, Jans S, Brooks N (2000) Genetically engineered crops: has adoption reduced pesticide use? In: Agricultural outlook. Economic Research Service, U.S. Department of Agriculture, pp 13–17
19. Losey JE, Rayor LS, Carter ME (1999) Transgenic pollen harms monarch larvae. Nature 399:214
20. Sears MK, Hellmich RL, Stanley-Horn DE, Oberhauser KS, Pleasants JM, Mattila HR, Siegfried BD, Dively GP (2001) Impact of Bt corn pollen on monarch butterfly populations: a risk assessment. Proc Natl Acad Sci U S A 98:11937–11942
21. Food and Drug Administration. Center for Food Safety and Applied Nutrition (2001) List of completed consultations on bioengineered foods. http://www.cfsan.fda.gov/_lrd/biocon.html. Accessed 25 Dec 2001
22. Thomashow MF (2001) So what's new in the field of plant cold acclimation? Lots! Plant Physiol 125:89–93
23. Alia, Hayashi H, Sakamoto A, Murata N (1998) Enhancement of the tolerance of *Arabidopsis* to high temperatures by genetic engineering of the synthesis of glycinebetaine. Plant J 16:155–161
24. Kasuga M, Liu Q, Miura S, Yamaguchi-Shinozaki K, Shinozaki K (1999) Improving plant drought, salt, and freezing tolerance by gene transfer of a single stress inducible transcription factor. Nat Biotechnol 17:287–291
25. Zhang H-X, Blumwald E (2001) Transgenic salt-tolerant tomato plants accumulate salt in foliage but not in fruit. Nat Biotechnol 19:765–768

26. de la Fuente JM, Ramirez-Rodriguez V, Cabrera-Ponce JL, Herrera-Estrella L (1997) Aluminum tolerance in transgenic plants by alteration of citrate synthesis. Science 276:1566–1568
27. Oscarson P (2000) The strategy of the wheat plant in acclimating growth and grain production to nitrogen availability. J Exp Bot 51:1921–1929
28. Whitney SM, von Caemmerer S, Hudson GS, Andrews TJ (1999) Directed mutation of the rubisco large subunit of tobacco influences photorespiration and growth. Plant Physiol 121:579–588
29. Shewmaker CK, Boyer CD, Wiesenborn DP, Thompson DB, Boersig MR, Oakes JV, Stalker DM (1994) Expression of Escherichia coli glycogen synthase in the tubers of transgenic potatoes (Solanum tuberosum) results in a highly branched starch. Plant Physiol 104:1159–1166
30. McKinney S (2000) Biotech for product developers. http://www.foodproductdesign.com/archive/2000/0500pp.html. Accessed 19 Dec 2001
31. Ye X, Al-Babili S, Kloti A, Zhang J, Lucca P, Beyer P, Potrykus I (2000) Engineering the provitamin A (beta-carotene) biosynthetic pathway into (carotenoid-free) rice endosperm. Science 287:303–305
32. Gura T (1999) New genes boost rice nutrients. Science 285:995
33. Nestle M (2001) Genetically engineered "golden" rice unlikely to overcome vitamin A deficiency. J Am Diet Assoc 101:289–290. (lett.)
34. Beecher GR (1999) Phytonutrients' role in metabolism: effects on resistance to degenerative processes. Nutr Rev 57(II):S3–S6
35. Streatfield SJ, Jilka JM, Hood EE, Turner DD, Bailey MR, Mayor JM, Woodard SL, Beifuss KK, Horn ME, Delaney DE, Tizard IR, Howard JA (2001) Plant-based vaccines: unique advantages. Vaccine 19:2742–2748
36. Richter LJ, Thanavala Y, Arntzen CJ, Mason HS (2000) Production of hepatitis B surface antigen in transgenic plants for oral immunization. Nat Biotechnol 18:1167–1171
37. Rotino GL, Perri E, Zottini M, Sommer H, Spena A (1997) Genetic engineering of parthenocarpic plants. Nat Biotechnol 15:1398–1401
38. Eckles JC (1999) Sweet corn. Liquid Foods Int
39. James C (2001) Global status of commercialized transgenic crops: 2000. No. 23: ISAAA, Ithaca
40. Kole C (2012) Wild crop relatives: genomic and breeding resources: cereals. Springer Science & Business Media
41. Khan S, Khan J (2010) Drought tolerant wheat cultivar (Raj) for rainfed areas of KPK, Pakistan. Pak J Agric Sci 47:355–359
42. Jain A, Singh N, Kumari S, Khan S (2021) Bioentrepreneurship in agricultural biotechnology. In: Agarwal S, Kumari S, Khan S (eds) Bioentrepreneurship and transferring technology into product development. IGI Global, Hershey, pp 183–200
43. Rajasree V, Pugalendhi L (2021) Chapter 7. Breeding vegetables for nutritional security. In: Jovandaric LPE-MZ (ed) Veganism—a fashion trend or food as a medicine. IntechOpen, Rijeka
44. Qamar S, Tantray AY, Bashir SS et al (2020) Golden rice: genetic engineering, promises, present status and future prospects. In: Roychoudhury A (ed) Rice research for quality improvement: genomics and genetic engineering. Springer, Singapore, pp 581–604
45. Mamiro P, Mbwaga A, Mamiro D et al (2011) Nutritional quality and utilization of local and improved cowpea varieties in some regions in Tanzania. Afr J Food Agric Nutr Dev 11:4490–4506
46. Kumar K, Gambhir G, Dass A et al (2020) Genetically modified crops: current status and future prospects. Planta 251:91. https://doi.org/10.1007/s00425-020-03372-8
47. Jin YH, Robledo D, Hickey JM et al (2021) Surrogate broodstock to enhance biotechnology research and applications in aquaculture. Biotechnol Adv 49:107756. https://doi.org/10.1016/j.biotechadv.2021.107756
48. Qaim M (2020) Bt cotton, yields and farmers' benefits. Nat Plants 6:1318–1319. https://doi.org/10.1038/s41477-020-00788-8

49. Francisco ÉC, Neves DB, Jacob-Lopes E, Franco TT (2010) Microalgae as feedstock for biodiesel production: carbon dioxide sequestration, lipid production and biofuel quality. J Chem Technol Biotechnol 85:395–403. https://doi.org/10.1002/jctb.2338
50. Sulochana MB, Jayachandra SY, Anil Kumar SK, Dayanand A (2014) Antifungal attributes of siderophore produced by the *Pseudomonas aeruginosa* JAS-25. J Basic Microbiol 54(5):418–424
51. de Sá Magalhães S, Keshavarz-Moore E (2021) *Pichia pastoris* (Komagataella phaffii) as a cost-effective tool for vaccine production for low- and middle-income countries (LMICs). Bioengineering (Basel) 8:119
52. Haroon F, Ghazanfar M (2016) Applications of food biotechnology. J Ecosyst Ecography 6. https://doi.org/10.4172/2157-7625.1000215
53. Yaradoddi JS, Banapurmath NR, Ganachari SV et al (2022) Bio-based material from fruit waste of orange peel for industrial applications. J Mater Res Technol 17:3186–3197. https://doi.org/10.1016/j.jmrt.2021.09.016
54. Yaradoddi JS, Banapurmath NR, Ganachari SV et al (2020) Biodegradable carboxymethyl cellulose-based material for sustainable packaging application. Sci Rep 10:21960. https://doi.org/10.1038/s41598-020-78912-z
55. Shellikeri A, Kaulgud V, Yaradoddi J, Ganachari S, Banapurmath N, Shettar A (2018) Development of neem based bioplastic for food packaging application. IOP Conf Ser Mater Sci Eng 376:012052. https://doi.org/10.1088/1757899X/376/1/012052
56. Yaradoddi JS, Hugar S, Banapurmath NR, Hunashyal AM, Sulochana MB, Shettar AS, Ganachari SV (2018) Alternative and renewable bio-based and biodegradable plastics. In: Martínez LMT et al (ed) Handbook of ecomaterials. Springer International Publishing AG. https://doi.org/10.1007/978-3-319-68255-6_150
57. Yaradoddi JS, Sulochana MB, Merja KH, Parameshwar AB, Dayanand A (2020) The occurrence of potential and novel isolates of *Oceanobacillus* sp. JAS12 and *Salinicoccus* sp. JS20 recovered from west coast of Arabian Sea, India. Indore Res J Biotechnol 15(9):133–140
58. Yaradoddi JS, Sulochana MB (2020) Screening and characterization of bioactive compounds produced by the moderate halophile *Halobacillus* sp. JS6. Res J Biotechnol 15(12):131–136
59. Jayachandra SY, Parameshwar AB, Mohan Reddy K, Sulochana MB (2012) Characterization of extracellular hydrolytic enzymes producing extremely halophilic bacterium *Virgibacillus* sp. JS5. World J Sci Technol 2(2):23–26
60. Sulochana MB, Jayachandra SY, Anil Kumar S, Dayanand A (2014) Siderophore as a potential plant growth-promoting agent produced by *Pseudomonas aeruginosa* JAS-25. Appl Biochem Biotechnol 174(1):297–308

Chapter 2
Bioactive Components of Foods, Activities and Their Source

Renuka Meti and Shweta Pattanashetti

1 Introduction

In general, bioactive ingredients are prone to deterioration during food processing or storage. Degradation and transformation can result from unstable physical, chemical, or enzymatic processes. Consequently, these bioactive chemicals may lose some of their bioactivity. Because of their complexity, it might be challenging to comprehend the mechanism underlying the breakdown of bioactive substances. Utilizing the microencapsulation technique is one way to safeguard bioactive ingredients and natural extracts. By encasing active chemicals (core material) in a surrounding material (coating agent, encapsulating agent, or wall material), microencapsulation is a technique for executing, safeguarding and transporting these substances in their exact field of activity [1]. Microcapsules of bioactive compounds are generally used to protect and stabilize them from oxygen, heat, light and humidity [2]. Though many more are yet unknown, it has been shown that cereals, vegetables and fruits contain about 5000 distinct phytochemical substances [3]. The highly beneficial qualities of fruits, vegetables and whole grains have been linked to bioactive, non-nutritional chemical components called phytochemicals. Whole foods may include anywhere between 5000 and 25,000 distinct phytochemicals, each of which may have bioactive qualities [4, 5]. It has been established that cereals and pulses are essential providers of dietary fibre, minerals, proteins, calories, vitamins and antioxidants [6].

Glycoalkaloids, antioxidant components and vitamins are among the many types of secondary metabolites found in plant sources. These compounds have been shown to have a wide range of positive health effects on humans, including antioxidative, anti-inflammatory and cardioprotective qualities as well as the ability to

R. Meti (✉) · S. Pattanashetti
Department of Food Processing and Nutrition, Karnataka State Akkamahadevi Women University, Vijayapura, Karnataka, India

J. S. Yaradoddi et al. (eds.), *Frontiers in Food Biotechnology*,
https://doi.org/10.1007/978-981-97-3261-6_2

control diabetes and prevent obesity [7, 8]. Fruit is the fleshy, edible portion of plants that can be either sweet or sour when uncooked. It offers plenty of supplies of vitamins, carbohydrates and bioactive components including fibre and phenols, which have been shown to have biological activity and may slow the onset of several degenerative diseases [9].

In addition to carotenoids (carotenes and xanthophylls), vitamins C and E, and polyphenols including stilbenes, anthocyanins, flavonoids, lignans and phenolic acids are the main components of plant-derived natural antioxidants. These naturally occurring antioxidants, especially polyphenols and carotenoids, have several biological characteristics, including anti-inflammatory, anti-viral, anti-aging, anti-microbial and anti-cancer effects [10–12]. Due to their bioactive qualities, which have been demonstrated to depend on their amino acid sequences, proteins or peptides generated from food have recently garnered attention [13].

Regardless of their nutritional significance, certain amino acid sequences that demonstrate a beneficial biological activity or have a positive impact on human health and bodily functions are referred to as bioactive peptides [14]. It has been documented that when bioactive peptides (BAPs) are present in the original protein, they function as inactive amino acid sequences. Nonetheless, a number of procedures, such as fermentation, food preparation, in vitro or human gastrointestinal tract enzyme-catalysed proteolysis, can easily release BAPs [15–17].

Legumes have been shown to have BAPs that may help prevent chronic illnesses, while there is still a dearth of study on cereals [18]. The bioactive elements of plant foods, such as fruits, vegetables, cereals and grains, are the main topic of this review because of their possible health advantages for humans.

2 Bioactive Components Derived from Plants

Secondary metabolites are plant-derived bioactive substances that are produced during metabolism. They have a number of potential medical uses, particularly in the area of antioxidation. The two main phytochemicals that are thought to be bioactive and can support improved human health are phenolics and carotenoids. Large concentrations of lipophilic chemicals called carotenoids are found in the majority of orange and yellow-coloured fruits and vegetables [19]. These substances are highly valuable for usage in the food industry as colourings and nutritional additives that promote health. For instance, research suggests that zeaxanthin, lutein and cryptoxanthin can reduce the risk of age-related macular degeneration, shield against sunburn-related conditions, lessen cardiovascular problems and avoid cataract development [20–23]. Furthermore, because of their significant antioxidative activity that has been demonstrated, carotenoids have garnered a lot of attention recently and may help lower the chance of developing certain chronic diseases.

Natural antioxidants called polyphenols are mostly found in food and medicinal plants, such as fruits, vegetables, grains, drinks, spices and mushrooms. Polyphenols include, among other things, anthocyanins, flavonoids and phenolic acids. It has been shown that natural antioxidants, particularly carotenoids and polyphenols, exhibit a number of biological characteristics, including anti-inflammatory, anti-aging and anti-cancer effects. In medicinal and culinary plants, polyphenols such as phenolic acids, stilbenes, flavonoids and lignans are widely distributed. Phenolic acids include derivatives of benzoic acid, such as gallic acid and hydroxybenzoic acids (HCA), and derivatives of cinnamic acid, such as ferulic, p-coumaric and caffeic. HCAs are present in food plants in greater amounts than hydroxybenzoic acids. The best foods to consume for dietary catechins are tea and red wine. Usually, catechins are found as aglycones or esterified with gallic acid. The two primary flavones found in the human diet are luteolin and apigenin, which are mostly found in celery and red pepper. Edible plants like berries, plums and eggplants get their violet, crimson or blue hues from anthocyanins like cyanidin, delphinidin and pelargonidin. Legumes are the primary source of isoflavonoids such as genistein and daidzein. The highest dietary sources of these chemicals are soy and items derived from soy [24, 25]. A common flavonoid of the flavonol class, quercetin can be found in a variety of foods, including tea, apples and onions [26]. Natural flavanone naringenin is mostly present in grapes and citrus fruits, and it has been shown to have the ability to preserve insulin signalling in the brain and regulate cognitive functioning [27]. Condensed tannins, or proanthocyanidins, are regarded as the final class of flavonoids. These are polymers or oligomers of catechin (flavan-3-ol), connected by interflavan linkages. They can be present in a number of plant-based meals and drinks, including red wine, grains and fruits [28]. Organosulphur compounds are an additional class of bioactives generated from plants (OSCs).

Rich levels of organic soluble carbohydrates (OSCs) are found in onions (Allium), and because of these biofunctional components, regular Allium consumption is linked to the prevention of a number of chronic illnesses, including diabetes, obesity, cardiovascular disease and metabolic disorders [29]. According to recent research, the main sulphur molecule found in onions is called isoalliin, or organosulphur compounds. These compounds, along with other precursors such S-methylcysteine sulphoxide, or methiin, are the building blocks of a variety of compounds that have physiological and sensory effects [30]. Changing things like food and lifestyle can lead to epigenetic alterations. Finding and creating epigenetic based treatments is currently a vital phase in the development of an anti-cancer drugs [31]. Natural food ingredients or substances, such as those found in Hedgehog, Notch and Wnt/_-catenin pathways, may be able to control cancer stem cell pathways linked to self-renewal [32]. Dietary polyphenol chlorogenic acid (CGA) possesses a number of significant biological properties. It is one of the most representative components of coffee and is produced by a variety of plants. It has been suggested that it functions as an anti-cancer drug by promoting the death of different human cancer cells connected to apoptosis, such as leukaemia and lung cancer cells [33].

Using a human stomach cancer cell line, Afshari et al. [33] assessed the anti-cancer effects of aubergine extract. The researchers came to the conclusion that compared to a normal cell line, aubergine extract had a more lethal effect on human stomach cancer cells. Because of its strong antioxidant qualities and abundance of phenolic components, aubergine may be useful in the detoxification of free radicals. As a result, eggplants may be consumed as a dietary supplement to lower the risk of cancer [34]. Another investigation evaluated the anti-tumour properties of solasodine, which is a naturally occurring aglycone of aubergine glycoalkaloids.

The proper functioning of the metabolic systems is essential for a healthy life. Reactive oxygen species (ROS) are harmful substances that are released during these activities. ROS are mostly produced by incomplete oxygen-containing reactions that release electrons from the mitochondrial electron transport chain. Because of their remarkable physiological effectiveness as pro-vitamins and their ability to participate in antioxidative reactions—specifically, scavenging singlet oxygen—carotenoids have been shown to support good health. Additionally, they can lower the chance of illness, particularly those malignancies that are becoming more and more common place globally [35]. According to a recent article, bioactive compounds like terpenes and phytosterols may be able to treat neuroinflammation and other neurodegenerative diseases [36]. Additionally, there is proof that grape juice and other fruit and vegetable extracts are rich sources of antioxidants that can improve brain function by lessening the effects of oxidative stress related to ageing [33]. Regarding *Caenorhabditis elegans* health, Lu et al. verified the phytoestrogen secoisolariciresinol diglucoside (SDG)'s promising qualities.

The researchers discovered that SDG can increase *C. elegans'* longevity by up to 22.0%, lessen the lethality of heat and oxidative stress, slow the aging-related decline in body movement, relieve the dopamine nervous system's deterioration prompted by 6-hydroxydopamine (6-OHDA) and lessen the toxicity of a protein in *C. elegans* [37]. Red wine's polyphenol resveratrol is a well-studied biomolecule that has been shown to enhance brain function by preventing the aggregation of beta amyloid proteins [38, 39].

3 Microbes in Gut and Marine Microbes as a Source of Bioactive Compounds

Evidence from animal and epidemiological research indicates that 'gut health' may be a potential medical research goal. It has drawn attention as a viable strategy for maintaining human health and resistance to disease, in addition to being a target for the treatment of severe gastrointestinal disorders [34]. The gut's epithelial barrier is a complex structure that separates the internal environment from the host gut lumen. The immune system (e.g. cytokines, immunoglobulin A), the microbiome, the intestinal vascular, neurenteric and endocrine systems, as well as various other components make up the gut barrier in addition to mucus layers and epithelial cells [35, 36].

The gut microbiota may have an impact on the metabolism of lipids and cholesterol by regulating specific metabolic signalling pathways linked to the signalling molecules found in bile acids. Bile acids can act via attaching to a variety of cellular receptors, such as the farnesoid X receptor (FXR) and G-protein-coupled receptors

TGR5. Primary bile acids activated FXR signalling, which controls multiple target genes linked to the metabolism of fats and carbohydrates as well as the gut's innate immunity [40].

Fruits and vegetables, which are rich in fibre and carbs, belong in the diet of the majority of people because they are plant-based foods. Large quantities of other substances (phytochemicals) that have additional effects on the gut flora are also present in them [41]. In addition to polyphenolic compounds, it has been proposed that protein hydrolysates and essential amino acid sequences (peptides) may alter gene expression in a way that reduces cholesterol by binding to bile acids and neutral sterols in the gut and eliminating faecal matter [42].

3.1 Marine Microbes as Bioactive Compounds

Because of this, marine microbes are quite intriguing as possible sources of novel bioactive chemicals that may be beneficial to humans. A multitude of industrial and biotechnological applications can benefit from the production of complex compounds with distinctive biological features by some microbes, as they can thrive in harsh marine settings. As a result, a variety of marine microorganisms, including myxomycetes, fungi, microalgae and bacteria, have previously been identified as producers of chemicals having antiviral, antioxidant, apoptotic and antibacterial properties [43–45].

Carotenoids are the second most common type of natural pigments and are used in food pigments, feed supplements, cosmetics, pharmaceuticals and nutraceuticals. Taking into account of natural and safe properties of microbial carotenoids concerning to their quick, affordable and eco-friendly production, coupled with the paucity of information regarding the health advantages and potential uses of carotenoids synthesized by haloalkaliphilic bacteria [46].

In particular, *Virgibacillus,* which [47, 48, 49] identified to be a promising option for large-scale, low-cost and quick production of carotenoids, warrants further investigation to address the growing body of knowledge regarding the medical and industrial relevance of these compounds. Numerous diseases, including cancer, diabetes mellitus and cardiovascular conditions, have been linked to oxidative stress and have even been shown to grow as a result of it [50].

3.2 Lipids Profile Regulation

According to published scientific research, bioactive substances may be able to preserve lipid homeostasis. Important characteristics of flavonoids and phenolic components can control lipid homeostasis [51]. Rats' antihyperlipidemic response to anthocyanins derived from eggplant peels was studied. According to the study, in a rat model, anthocyanins from aubergine peel significantly decreased the levels of serum total cholesterol, triglycerides and low-density lipoprotein (LDL-C) as well as very low-density lipoprotein (VLDL), while significantly improving high-density lipoprotein cholesterol (HDL-C) [52].

It has been suggested that increasing the intake of high-density lipoprotein cholesterol and lowering blood cholesterol and triglycerides by the use of anthocyanins obtained from eggplants can help prevent obesity [2]. Natural plant sterols, like phytosterols, have a wide range of possible health-promoting properties. These include lowering intestine absorption of cholesterol and lowering blood cholesterol. For these reasons, they are employed as functional dietary components. These substances prevent cholesterol from solubilizing in the intestine because of their structural similarities to cholesterol, which results in reduced absorption. Additionally, they have the ability to compete with low-density lipoprotein (LDL) cholesterol for the purpose of lowering LDL cholesterol levels inside blood vessels [53].

Research published in the literature has established that consuming up to 3 g of phytosterols daily is safe and can have a hypocholesterolemic effect on individuals who need to have a discernible drop in their plasma levels of low-density lipoprotein (LDL) cholesterol [54]. It has been suggested that flavonoids, via neutral sterols and bile acids, can improve the excretion and breakdown of cholesterol.

4 Animal-Derived Bioactive Peptides

Numerous bioactive peptides with antioxidant qualities have been found in marine animals, including prawns, squid, oysters, blue mussels and several fish species. Compared to many other fish sources, puffer fish hydrolysate demonstrated high antioxidant effect [55]. Girgih et al. (2013) [56] documented the salmon protein hydrolysate's antioxidant properties by observing that linoleic acid oxidation was prevented by both the protein hydrolysate and peptide fractions. Strong antioxidant properties have also been discovered for flounder fish muscle hydrolysed with α-chymotrypsin [57]. Antioxidant peptides can also be produced by the enzyme Neutrase hydrolysing the protein of the blue mussel (Mytilus edulis). After this hydrolysate was purified, the active peptide, which had a YPPAK (Tyr-Pro-Pro-Ala-Lys) sequence and improved hydroxyl and superoxide anion radical scavenging properties, was discovered.

Strong antioxidant properties have also been discovered for flounder fish muscle hydrolysed with α-chymotrypsin [58]. Antioxidant peptides can also be produced by the enzyme Neutrase hydrolysing the protein of the blue mussel (Mytilus edulis). The active peptide, YPPAK (Tyr-Pro-Pro-Ala-Lys), with improved hydroxyl and superoxide anion radical scavenging capabilities, was identified upon purification of this hydrolysate [59].

5 Future Prospectus

In order to meet essential nutritional requirements, dietary food supplements are eaten. It is not suggested that they be used to identify, address, mitigate or cure illnesses. These bioactive dietary ingredients include vitamins, fibre, minerals, fatty acids and amino acids in addition to other polyphenolic chemicals. It is estimated that there are around 50,000 different dietary components, most of which are different multivitamins. They are not used in the preparation of regular meals or drinks (such as juices and soups); instead, they are consumed in a medicinal form, such as a powder, tablet or pill [60]. Carotenoids, anthocyanins, leucoanthocyanidin lycopenes and betalains are the primary phenolic chemicals that impart colour to fruits and vegetables. Because they are strong antioxidants and might have therapeutic benefits, they are appealing as 'functional foods' for improving human health. The best strategy to prevent vitamin shortages and preserve good health is to eat a balanced diet. Changes brought on by ageing and the existence of other illnesses may make this effort difficult. Therefore, adding functional foods to the diet—which have a higher concentration of a certain nutrient or bioactive compound—can enhance overall health and well-being and/or lower the risk of disease.

References

1. Ribeiro AM, Shahgol M, Estevinho BN, Rocha F (2020) Microencapsulation of vitamin A by spray-drying, using binary and ternary blends of gum arabic, starch and maltodextrin. Food Hydrocoll 108:106029
2. Joye IJ, Davidov-Pardo G, McClements DJ (2014) Nanotechnology for increased micronutrient bioavailability. Trends Food Sci Technol 40:168–182
3. Liu RH (2013) Dietary bioactive compounds and their health implications. J Food Sci 78:A18–A25
4. Acosta-Estrada BA, Gutiérrez-Uribe JA, Serna-Saldívar SO (2014) Bound phenolics in foods, a review. Food Chem 152:46–55
5. Šaponjac VT, Čanadanović-Brunet J, Ćetković G, Djilas S (2016) Emerging and traditional technologies for safe, healthy and quality food. In: Detection of bioactive compounds in plants and food products. Springer, Cham, pp 81–109
6. Prakash B, Kujur A, Singh PP, Kumar A, Yadav A (2017) Plants-derived bioactive compounds as functional food ingredients and food preservative. J Nutr Food Sci 1:1–7
7. Gürbüz N, Uluişik S, Frary A, Frary A, Doğanlar S (2018) Health benefits and bioactive compounds of eggplant. Food Chem 268:602–610
8. Plazas M, López-Gresa MP, Vilanova S, Torres C, Hurtado M, Gramazio P, Andújar I, Herráiz FJ, Bellés JM, Prohens J (2013) Diversity and relationships in key traits for functional and apparent quality in a collection of eggplant: fruit phenolics content, antioxidant activity, polyphenol oxidase activity, and browning. J Agric Food Chem 61:8871–8879
9. Singh JP, Kaur A, Shevkani K, Singh N (2015) Influence of jambolan (*Syzygium cumini*) and xanthan gum incorporation on the physicochemical, antioxidant and sensory properties of gluten-free eggless rice muffins. Int J Food Sci Technol 50:1190–1197
10. Li Y, Zhang J-J, Xu D-P, Zhou T, Zhou Y, Li S, Li H-B (2016) Bioactivities and health benefits of wild fruits. Int J Mol Sci 17:1258. https://doi.org/10.3390/ijms17081258

11. Zhang J-J, Li Y, Zhou T, Xu D-P, Zhang P, Li S, Li H-B (2016) Bioactivities and health benefits of mushrooms mainly from China. Molecules 21:938. https://doi.org/10.3390/molecules21070938
12. Xu D-P, Li Y, Meng X, Zhou T, Zhou Y, Zheng J, Zhang J-J, Li H-B (2017) Natural antioxidants in foods and medicinal plants: extraction, assessment and resources. Int J Mol Sci 18:96. https://doi.org/10.3390/ijms18010096
13. Miguel M, Alonso MJ, Salaices M, Aleixandre A, López-Fandiño R (2007) Antihypertensive, ACE-inhibitory and vasodilator properties of an egg white hydrolysate: effect of a simulated intestinal digestion. Food Chem 104:163–168
14. Chakrabarti S, Guha S, Majumder K (2018) Food-derived bioactive peptides in human health: challenges and opportunities. Nutrients 10:1738. https://doi.org/10.3390/nu10111738
15. Hartmann R, Meisel H (2007) Food-derived peptides with biological activity: from research to food applications. Curr Opin Biotechnol 18:163–169. https://doi.org/10.1016/j.copbio.2007.01.013
16. Aluko RE (2008) Antihypertensive properties of plant-derived inhibitors of angiotensin I-converting enzyme activity: a review. Recent Prog Med Plants 22:541–561
17. Harnedy PA, FitzGerald RJ (2012) Bioactive peptides from marine processing waste and shellfish: a review. J Funct Foods 4:6–24
18. Cavazos A, De Mejia EG (2013) Identification of bioactive peptides from cereal storage proteins and their potential role in prevention of chronic diseases. Compr Rev Food Sci Food Saf 12:364–380
19. Benke KK, Benke KE (2014) Experimental design issue for assessment of carotenoids lutein and zeaxanthin in age-related eye disease study 2 formulation for age-related macular degeneration. JAMA Ophthalmol 132:904
20. Weikel KA, Garber C, Baburins A, Taylor A (2014) Nutritional modulation of cataract. Nutr Rev 72:30–47
21. Di Pietro N, Di Tomo P, Pandolfi A (2016) Carotenoids in cardiovascular disease prevention. JSM Atheroscler 1:1002. Accessed 2 Nov 2020
22. Cooperstone J, Schwartz S (2016) Handbook on natural pigments in food and beverages. Recent insights into health benefits of carotenoids. Woodhead Publishing, Cambridge, pp 473–497
23. Erlund I (2004) Review of the flavonoids quercetin, hesperetin, and naringenin. Dietary sources, bioactivities, bioavailability, and epidemiology. Nutr Res 24:851–874
24. Liggins J, Bluck LJC, Runswick S, Atkinson C, Coward WA, Bingham SA (2000) Daidzein and genistein contents of vegetables. Br J Nutr 84:717–725
25. D'Andrea G (2015) Quercetin: a flavonol with multifaceted therapeutic applications? Fitoterapia 106:256–271
26. Ghofrani S, Joghataei M-T, Mohseni S, Baluchnejadmojarad T, Bagheri M, Khamse S, Roghani M (2015) Naringenin improves learning and memory in an Alzheimer's disease rat model: insights into the underlying mechanisms. Eur J Pharmacol 764:195–201
27. Pereira-Caro G, Gaillet S, Ordóñez JL, Mena P, Bresciani L, Bindon KA, Del Rio D, Rouanet J-M, Moreno-Rojas JM, Crozier A (2020) Bioavailability of red wine and grape seed proanthocyanidins in rats. Food Funct 11:3986–4001
28. Moreno-Ortega A, Ordóñez JL, Moreno-Rojas R, Moreno-Rojas JM, Pereira-Caro G (2021) Changes in the organosulfur and polyphenol compound profiles of black and fresh onion during simulated gastrointestinal digestion. Foods 10:337
29. Moreno-Ortega A, Pereira-Caro G, Ordóñez JL, Muñoz-Redondo JM, Moreno-Rojas R, Pérez-Aparicio J, Moreno-Rojas JM (2020) Changes in the antioxidant activity and metabolite profile of three onion varieties during the elaboration of 'black onion'. Food Chem 311:125958
30. Pistollato F, Giampieri F, Battino M (2015) The use of plant-derived bioactive compounds to target cancer stem cells and modulate tumor microenvironment. Food Chem Toxicol 75:58–70. https://doi.org/10.1016/j.fct.2014.11.004

31. Li Y, Wicha MS, Schwartz SJ, Sun D (2011) Implications of cancer stem cell theory for cancer chemoprevention by natural dietary compounds. J Nutr Biochem 22:799–806. https://doi.org/10.1016/j.jnutbio.2010.11.001
32. Tajik N, Tajik M, Mack I, Enck P (2017) The potential effects of chlorogenic acid, the main phenolic components in coffee, on health: a comprehensive review of the literature. Eur J Nutr 56:2215–2244. https://doi.org/10.1007/s00394-017-1379-1
33. Afshari F, Seraj H, Hashemi ZS, Timajchi M, Ensiyeh O, Ladan G, Asadi M, Elyasi Z, GanjiBakhsh M (2017) The cytotoxic effects of eggplant peel extract on human gastric adenocarcinoma cells and normal cells. Mod Med Lab J 1:77–83. https://doi.org/10.30699/mmlj17.1.2.77
34. Lu M, Tan L, Zhou X-G, Yang Z-L, Zhu Q, Chen J-N, Luo H-R, Wu G-S (2020) Secoisolariciresinol diglucoside delays the progression of aging-related diseases and extends the lifespan of Caenorhabditis elegans via DAF-16 and HSF-1. Oxidative Med Cell Longev 2020:1293935. https://doi.org/10.1155/2020/1293935
35. Libro R, Giacoppo S, Rajan TS, Bramanti P, Mazzon E (2016) Natural phytochemicals in the treatment and prevention of dementia: an overview. Molecules 21:518. https://doi.org/10.3390/molecules21040518
36. Regitz C, Fitzenberger E, Mahn FL, Dußling LM, Wenzel U (2016) Resveratrol reduces amyloid-beta (Aβ1–42)-induced paralysis through targeting proteostasis in an Alzheimer model of Caenorhabditis elegans. Eur J Nutr 55:741–747. https://doi.org/10.1007/s00394-015-0894-1
37. Singh B, Singh JP, Kaur A, Singh N (2016) Bioactive compounds in banana and their associated health benefits—a review. Food Chem 206:1–11. https://doi.org/10.1016/j.foodchem.2016.03.033
38. Welcome MO (2020) Neuroinflammation in CNS diseases: molecular mechanisms and the therapeutic potential of plant derived bioactive molecules. PharmaNutrition 11:100176. https://doi.org/10.1016/j.phanu.2020.100176
39. Piemontese L (2017) Plant food supplements with antioxidant properties for the treatment of chronic and neurodegenerative diseases: benefits or risks? J Diet Suppl 14:478–484. https://doi.org/10.1080/19390211.2016.1247936
40. Wan MLY, Ling KH, El-Nezami H, Wang MF (2019) Influence of functional food components on gut health. Crit Rev Food Sci Nutr 59:1927–1936
41. Bischoff SC, Barbara G, Buurman W, Ockhuizen T, Schulzke J-D, Serino M, Tilg H, Watson A, Wells JM (2014) Intestinal permeability—A new target for disease prevention and therapy. BMC Gastroenterol 14:1–25
42. Hidalgo-Liberona N, González-Domínguez R, Vegas E, Riso P, Del Bo' C, Bernardi S, Peron G, Guglielmetti S, Gargari G, Kroon PA et al (2020) Increased intestinal permeability in older subjects impacts the beneficial effects of dietary polyphenols by modulating their bioavailability. J Agric Food Chem 68:12476–12484
43. Ameen F, AlNadhari S, Al-Homaidan AA (2021) Marine microorganisms as an untapped source of bioactive compounds. Saudi J Biol Sci 28(1):224–231. https://doi.org/10.1016/j.sjbs.2020.09.052
44. Yaradoddi JS, Sulochana MB, Merja KH, Parameshwar AB, Dayanand A (2020) The occurrence of potential and novel isolates of *Oceanobacillus* sp. JAS12 and *Salinicoccus* sp. JS20 recovered from west coast of Arabian Sea, India. Res J Biotech 15(9):133–140. [IF: 0.49, Q4 Journal]
45. Yaradoddi JS, Sulochana MB (2020) Screening and characterization of bioactive compounds produced by the moderate halophile Halobacillus sp. JS6. Res J Biotech 15(12):131–136
46. Maeda H, Kanno S, Kodate M, Hosokawa M, Miyashita K (2015) Fucoxanthinol, metabolite of fucoxanthin, improves obesity-induced inflammation in adipocyte cells. Mar Drugs 13:4799–4813
47. Fayez D, Youssif A, Sabry S, Ghozlan H, Eltarahony M (2022) Carotegenic *Virgibacillus halodenitrificans* from Wadi El-Natrun salt lakes: isolation, optimization, characterization,

and biological activities of carotenoids. Biology (Basel) 11(10):1407. https://doi.org/10.3390/biology11101407
48. Fayez D, Youssif A, Sabry S, Ghozlan H, El-Sayed F (2023) Some novel bioactivities of *Virgibacillus halodenitrificans* carotenoids, isolated from Wadi El-Natrun lakes. Saudi J Biol Sci 30(11):103825. https://doi.org/10.1016/j.sjbs.2023.103825
49. Jayachandra SY, Parameshwar AB, Mohan Reddy K, Sulochana MB (2012) Characterization of extracellular hydrolytic enzymes producing extremely halophilic bacterium *Virgibacillus* sp. JS5. World J Sci Technol 2(2):23–26
50. Lenaz G (2012) Mitochondria and reactive oxygen species. Which role in physiology and pathology? In: Scatena R, Bottoni P, Giardina B (eds) Advances in experimental medicine and biology, vol 942. Springer, Netherlands, Dordrecht, pp 93–136. isbn:978-94-007-2868-4
51. Loo YT, Howell K, Chan M, Zhang P, Ng K (2020) Modulation of the human gut microbiota by phenolics and phenolic fiber-rich foods. Compr Rev Food Sci Food Saf 19:1268–1298
52. Dey P (2019) Gut microbiota in phytopharmacology: a comprehensive overview of concepts, reciprocal interactions, biotransformations and mode of actions. Pharmacol Res 147:104367
53. Udenigwe CC, Aluko RE (2012) Food protein-derived bioactive peptides: production, processing, and potential health benefits. J Food Sci 77:R11–R24
54. Siasos G, Tousoulis D, Tsigkou V, Kokkou E, Oikonomou E, Vavuranakis M, Basdra E, Papavassiliou A, Stefanadis C (2013) Flavonoids in atherosclerosis: an overview of their mechanisms of action. Curr Med Chem 20:2641–2660
55. Vergara-Jimenez M, Al Matrafi MM, Fernandez ML (2017) Bioactive components in Moringa Oleifera leaves protect against chronic disease. Antioxidants (Basel) 6:91
56. Harada K, Maeda T, Hasegawa Y, Tokunaga T, Tamura Y, Koizumi T (2010) Antioxidant activity of fish sauces including puffer (*Lagocephalus wheeleri*) fish sauce measured by the oxygen radical absorbance capacity method. Mol Med Rep 3(4):663–668. https://doi.org/10.3892/mmr_00000313
57. Girgih AT, Udenigwe CC, Hasan FM, Gill TA, Aluko RE (2013) Antioxidant properties of Salmon (*Salmo salar*) protein hydrolysate and peptide fractions isolated by reverse phase HPLC. Food Res Int 52(1):315–322. https://doi.org/10.1016/j.foodres.2013.03.034
58. Ko JY, Lee JH, Samarakoon K, Kim JS, Jeon YJ (2013) Purification and determination of two novel antioxidant peptides from flounder fish (Paralichthys olivaceus) using digestive proteases. Food Chem Toxicol 52:113–120. https://doi.org/10.1016/j.fct.2012.10.058
59. Wang B, Li L, Chi CF, Ma JH, Luo HY, Xu YF (2013) Purification and characterisation of a novel antioxidant peptide derived from blue mussel (*Mytilus edulis*) protein hydrolysate. Food Chem 138:1713–1719. https://doi.org/10.1016/j.foodchem.2012.12.002
60. Akram NA, Shafiq F, Ashraf M (2018) Peanut (*Arachis hypogaea* L.): a prospective legume crop to offer multiple health benefits under changing climate. Compr Rev Food Sci Food Saf 17:1325–1338

Chapter 3
Nutraceuticals: Classification, Sources and Relation with Medicine

Bhuvaneshwari G., Vasant M. Ganiger, Vijaykumar B. Narayanpur, and T. B. Allolli

1 Introduction

Nutraceuticals are products with biological functions that are acquired only from foods. Nutraceuticals are frequently promoted as having extraordinary health benefits, like the ability to delay the onset of old age. These assertions, however, are controversial and challenging to support. Nutraceuticals and functional foods in the recent years have authenticated a tremendous increase in the interest among the consumers due to their potential of providing health benefits. Food Safety and Standard Act, 2006 streamlined and integrated many regulations concerning dietary supplements, nutraceuticals and also functional foods. Under Clause 22 of the Food Safety and Standards Act [1] nutraceuticals have been defined and from the past 7 years market of nutraceuticals has been growing rapidly. This is mainly due to

Bhuvaneshwari G. (✉)
Department of Postharvest Management, College of Horticulture, University of Horticultural Sciences, Bagalkot, Karnataka, India
e-mail: bhuvaneshwari.g@uhsbagalkot.edu.in

V. M. Ganiger
Department of Vegetable Science, Directorate of Extension, University of Horticultural Sciences, Bagalkot, Karnataka, India

V. B. Narayanpur
Department of Plantation, Spices, Medicinal and Aromatics, College of Horticulture, University of Horticultural Sciences, Bagalkot, Karnataka, India

T. B. Allolli
Department of Vegetable Science, University of Horticultural Sciences, Bagalkot, Karnataka, India
e-mail: registrar@uhsbagalkot.edu.in

J. S. Yaradoddi et al. (eds.), *Frontiers in Food Biotechnology*,
https://doi.org/10.1007/978-981-97-3261-6_3

increasing awareness about the various types of nutraceuticals available in the market, consciousness about health and willingness of consumers to expend on health foods are some of the factors.

2 Definition of Nutraceutical

The term 'Nutraceutical' is rolled out from two words 'nutrition' and 'pharmaceutical', which are the supplements of diet that provide a concentrated form of a feasible bioactive agent from a food, allocate in a non-food matrix, and are used with the purpose of graceful health in dosages that exceed those that could be obtained from normal foods. These are marketed similar to medicine or drugs in the form of pellets, tablets, extracts, solutions, capsules, gels, liquors, powders, granules, etc.

3 Historical Outlook of Nutraceuticals

Almost 2500 years back, the father of modern medicine, Hippocrates (460–377 BC), entrenched the relation of food and its importance for the treatment of various diseases in a very traditional way optimizing various benefits. Even before the development of 'Nutrition' as a distinct scientific discipline, since long time the physicians paid close attention about the role of daily diet in maintenance of health. Scurvy was a common cause of death for crew members on lengthy sea trips during the sixteenth, seventeenth and eighteenth centuries. Under the British Admiralty's guidance, sea captains in the eighteenth century investigated how different meals and eating habits could help keep sailors healthy. While some items made up for the lack of vitamins B and C, others lessened the diet's anti-scorbutic effects. In the past 20 years, biomedical research has shown that nutrition has a significant impact on the development and progression of many of the main chronic diseases, including atherosclerosis and cancer.

Since prehistoric period, pull out of natural materials from plant sources and have been used from mankind for various purposes as medicine. In 1989 Dr. Stephen coined the term 'Nutraceuticals', which is combination of nutrition and pharmaceutical. Since a new trend in accomplice animal care and related changes in the human population arose in the 1990s, nutraceuticals have advanced significantly. The term 'nutraceuticals' has been broadened, to encompass vitamins, minerals, herbs and other botanicals, amino acids and dietary substances for use by humans as dietary supplements with the passage of the dietary supplement health and education act of

1994. While dietary supplements are thought to be more important, the term 'nutraceutical' is not fully accepted by global regulatory systems. There are already more than 470 nutraceutical and functional food items on the market with verified health advantages.

This chapter gives highlights of some important facts regarding therapeutic use of nutraceuticals as the commercial as well as traditional remedies.

4 Classification of Nutraceuticals

There are hundreds of purported nutraceutical ingredients; among of the more well-known ones are isoflavones, tocotrienols, allyl sulphides, fibre and carotenoids. Organizational methods are required in light of the lengthy and expanding list of nutraceutical ingredients in order to make understanding and use simpler. This is especially true for academic training and the creation of food industry products. The best organizational structure for nutraceuticals can vary depending on one's interests and/or educational background. For instance, nutraceuticals that are linked to lowering heart disease risk factors would be of particular interest to cardiologists. In particular, they may be interested in compounds that are said to lower platelet- or free radical-dependent thrombotic activity and to have a favourable effect on hypertension, hypercholesterolaemia and these conditions. They would be especially interested in nutritional components like n-3 fatty acids, phytosterols, quercetin and grape flavonoids. Meanwhile, drugs that target anti-carcinogenic actions would be of greater interest to oncologists. These chemicals may enhance antioxidant defences and microsomal detoxification mechanisms, or they may inhibit the spread of cancer that already exists. Therefore, they may be interested in supplementary therapy or chemo prevention. Contrarily, the nutraceutical interests of food scientists working on the creation of a functional food product will go beyond physiological qualities to encompass stability, sensory and cost-effectiveness concerns. The anti-carcinogenic triterpene limonin serves as a case study for this idea because it is lipid-soluble and extremely bitter, which in some ways restricts its commercial application as a functional food additive. The glucoside derivative of limonin, which shares part of limonin's anti-carcinogenic properties, is practically tasteless and water soluble, which increases its potential usage as an ingredient.

Nutraceuticals can be categorized in many ways like based on the source of the food, another on the basis of the action's mechanism, and a third on the basis of the chemical makeup of the functional ingredients (Table 3.1).

Table 3.1 Classification based on food sources, mechanism of action and chemical nature

a. Classification based on food source

Plants	Animal	Microbial
β-Glucan Ascorbic acid y-Tocotrienol Quercetin Cellulose Lutein Gallic acid Pectin	Conjugated linoleic acid (CLA) Eicosapentaenoic acid (EPA) Docosahexenoic acid (DHA) Sphingolipids Choline Lecithin Calcium Ubiquinone Selenium Zinc	*Lactobacillus acidophilus* *Saccharomyces boulardii* *Bifidobacterium bifidum* *B. longum* *B. infantis* *L. acidophilus* *Streptococcus salivarius*

b. Classification based on mechanism of action

Antioxidant	Positive influence on blood lipid profile	Anticancer	Anti-inflammatory	Bone protective or osteogenetic
LA Ascorbic acid Carotene Tocopherols Tocotrienol Ellagic acid Glutathione Gingerol Hydroxy Tyrosol Polyphenols Luteolin Lycopene Tannins Chlorogenic acid	β-Glucan Tocotrienol Resveratrol Tannins Quercetin Pectin Guar Saponins Sitosterol MUFA Omega-3-fatty acid	Capsaicin Genestein Daidzein Tocotrienol CLA *Lactobacillus acidophilus* Sphingolipids Limonene Diallyl sulphide Ajoene Tocopherol Glycyrrhizin	Linolenic acid EPA DHA GLA γ-linolenic acid Capsaicin Quercetin Curcumin	CLA Soy protein Genestein Daidzein FOS Calcium Casein Phosphopeptides Inulin Fructo - oligosaccharides

c. Classification based on chemical nature

Isoprenoid (Terpenoids)	Phenolic compounds	Fatty acids and structural lipids	Carbohydrate and derivatives	Protein/ amino acid based	Microbes	Minerals
Carotenoids Saponins Tocotrienols Tocopherols Simple terpenes	Coumarins Tannins Lignins Flavonols Flavonoids Isoflavones Anthocyanins	n-3 PUFA CLA MUFA Sphingolipids Lecithin	Ascorbic acid Oligo saccharides Non-starch PS	Amino acids Allyl-S comps Capsaicinoid Isothiocyanates Indoles Folate Choline	Probiotics Prebiotics	Ca Se K Cu Zn

4.1 Classification Based on Food Source

Based on their ability to serve as a source of nourishment for humans, nutraceuticals are organized according to certain more general organizational frameworks. Here, nutraceuticals can be divided into classes of plants, animals and microbes (such as bacteria and yeast) as depicted in Table 3.1

4.2 Classification Based on Mechanism of Action

By their mode of action, nutraceuticals can also be categorized. This method categorizes nutraceutical ingredients according to their physiological qualities, regardless of the source of the ingredients' diet.

Antioxidants, antibacterials, antihypertensive, anti-hypercholesterolemic, anti-aggregate, anti-inflammatory, anti-carcinogenic, osteoprotective and other substances are included in these classes. A person who is genetically inclined to a certain medical issue or researchers working to create potent functional meals for exactly such a person might benefit from this paradigm. This product's creator may take into account the components specified in a number of categories to create a solution to lower blood pressure, low density lipoprotein cholesterol levels and burning sensation.

4.3 Classification Based on Chemical Nature

Predicated on their chemical makeup, nutraceuticals can also be categorized in another way. This method enables the classification of nutraceuticals into molecular/elemental categories. Several major groups are included in this initial model, which provides a foundation for further classification or grouping, etc.

5 Sources of Nutraceuticals

Consumers who are concerned about their health are becoming more and more interested in health-promoting products, and as a result, a wide variety of nutraceuticals that contain phytochemicals from food are now offered on the market. The main classes of plant components known as terpenoids, phenolic metabolites, alkaloids and other nitrogen-based plant elements are where the majority of phytochemicals found in the nutraceutical industry come from.

5.1 Terpenoids (Terpenes)

The greatest family of phytonutrients found in cereals, soy plants and green vegetables are terpenes, sometimes referred to as isoprenoids. Terpenes interact with free radicals to produce an unusual antioxidant function. Terpenes called tocotrienols and tocopherols are found in grains naturally. It has been discovered that the tocotrienols and RRR-tocopherol are efficient cell death inducers for human breast cancer cells. Terpenes may play a role in the impact of a diet rich in fruits, vegetables and grains on lowering the chance of developing tumours.

The antioxidant properties of the carotenoid terpenes have received substantial research. Fruits and vegetables contain these highly coloured chemicals, which come in shades of yellow, orange and red. Carotenes as well as xanthophylls are the two types of fragments that make up carotenoids. Carotenes' biological activity is tissue-specific. All three forms of beta-carotene have vitamin A action. According to Bendich's review, carotenes such as carotene, lycopene and lutein guard against malignancies of the uterus, breast, colon, prostate and lung. They might also guard against the danger of gastrointestinal tumours. Terpenes called limonoids are found in citrus fruit [2] Limonoids appear to have chemotherapeutic effects. The most prevalent monocyclic monoterpene, D-limonene, which is present in orange peel oil, prevents the development of pancreatic cancer. Lung tissue may also be protected by limonoids [3].

An additional significant terpene category is phytosterols. Plants produce sitosterol and its glycoside, which are both sterol compounds. Phytosterols can compete with cholesterol for absorption in the intestine and help the body eliminate cholesterol.

5.2 Phenolic Constituents

The nutraceutical business currently uses a number of polyphenol categories, including anthocyanins, proanthocyanidins, flavones, isoflavones, resveratrol and ellagic acid. According to reports, anthocyanins, as well as berries that are high in them or extracts made from them, have a variety of defensive properties that could be advantageous for both human and animal health. It has been acknowledged that the benefits perhaps connected with the chemical properties along with innate antioxidant capacity of these substances.

Proanthocyanidins are the second most common phenolic in nature. They are present across the entire plant kingdom and are consumed by humans together with fruits (such as grapes, apples and strawberries), beans, almonds, wine and chocolate [4]. The aglycones naringenin and hesperetin, as well as their corresponding glycosides naringin and hesperidin, which are both prevalent in grapefruits and oranges, respectively, are the most representative and extensively studied flavanones.

The important sources of catechins include grapes, berries, cocoa and inexperienced tea. Tea contains gallic acid esters, such as epicatechin, epicatechin gallate and epigallocatechin gallate. Numerous research has recommended that these elements provide protecting benefits via their antioxidant activity and their inhibition of eicosanoid synthesis [5] and platelet aggregation [6]. Experimental proof in telephone way of life systems [7] supports the viable use of tea, especially inexperienced tea, for the prevention of prostate tumour. Catechin is one of the predominant phenolic constituents in grapes and crimson wines, and it is viewed to be reliable for phase of the protective impact of purple wine against atherosclerotic cardiovascular ailment [8–11].

6 Alkaloids and Other Nitrogen-Containing Metabolites

Cruciferous vegetables include glucosinolates, which activate the liver's detoxifying enzymes. Intake of cruciferous vegetables offers a phytochemical defence against mutagenesis, carcinogenesis and other types of electrophile and reactive oxygen toxicity [12, 13]. Compared to the same number of mature vegetables of the same variety, crucifer sprouts may reduce the menace of cancer more efficiently [14].

To sum up, the majority of nutraceuticals are produced from plants and animals. While some are typical dietary ingredients that show up in human metabolism, others are endogenous human metabolites, for example as in lycopene. A number of substances are found in higher plants and are commercially extracted from them, though some, like methylsulfonylmethane (MSM) and dehydroepiandrosterone (DHEA), are not present in sufficient quantities to be exploited for commercial purposes. The n-3 fatty acids, including docosahexaenoic acid (DHA), eicosapentaenoic acid (EPA) and linolenic acid, are typically sold as complex mixtures that contain supradietary quantities of the active ingredients. Partially purifying these fatty acids from the other fatty acids is frequently skipped. On the other hand, more and more complex methods are being used to enhance these entities. DHA/EPA generation by microalgal fermentation is an additional option.

7 Nutraceuticals in Medical Science

All kinds of diseases are having a field day thanks to the contemporary, competitive lifestyle and the continuously demanding conditions. For the majority of diseases, there are allopathic treatments accessible, but they are expensive. First, like all other goods, the price of medicine is rising daily. Second, a number of negative effects are connected to allopathic medications. As a result, a growing number of people are gravitating towards changing their way of life and using nutraceuticals. This promotes a person's general health and aids in warding off numerous ailments.

People's basic eating habits have changed because of the new lifestyle they've adopted. The consumption of unhealthy foods has increased, leading to a wide range of diseases caused by poor nutrition. Obesity is now recognized as a major global issue.

8 Cardiovascular Disease Supplements

In the modern era, cardiovascular disease (CVD) is the prime cause of death worldwide. They have the capacity to drastically lower the risk of chemotherapeutic complications while also lowering the expense of healthcare worldwide. It has been suggested that dietary adjustments and other lifestyle modifications can prevent CVD. Certain nutraceuticals are said to have a positive effect on cardiovascular disorders such as deep vein thrombosis, atherosclerosis, cor pulmonale, coronary artery disease, stroke and chronic disease like heart attack/ischaemia. By regulating the liver's production of cholesterol, binding cholesterol in the intestines and/or enhancing LDL-c receptor uptake in the liver (e.g. berberine, phytosterols), nutraceuticals tend to lower levels of LDL cholesterol in the blood [15]. Excessive consumption of calorie-dense, nutritionally deficient and highly processed foods are digested, easily absorbed and can cause systemic inflammation, decreased insulin sensitivity and a variety of metabolic disorders include obesity, hypertension, dyslipidaemia and glucose intolerance. Tea, cocoa, cocoa derivatives and grapes all contain polyphenols that may be useful in preventing CVD.

9 Supplements for Alzheimer's Disease

The most prevalent and dreaded type of dementia, Alzheimer's disease (AD), accounts for around 70% of all dementia cases and is experiencing a dramatic epidemic as a result of the large rise in the world's elderly population. A wide range of cognitive faculties, including as language, logic, memory and perceptive quickness, frequently deteriorate with advancing age. Green tea catechins, turmeric, Salvia miltiorrhiza, berry anthocyanins and Panax ginseng are a few botanical extracts with antiamyloidogenic activity that have shown to be highly effective in treating Alzheimer's disease [16]. It has been established that bacopa monnieri is a traditional Indian nerve tonic, suggesting an aspirant role in assisting in the prevention of dementia and acting as a unique memory booster [17]. Additionally, astaxanthin-rich algal biomass, fish oil containing omega-3 fatty acids and krill oil have shown potential roles in maintaining remembrance, supporting perception processes and preventing neuroinflammatory, neuromotor and neurodegenerative illnesses in both people and animals [18].

10 Supplements for Diabetes

Diabetes mellitus (DM) is a long-term metabolic illness in which the body cannot use carbohydrates because of an absolute or relative absence of the hormone insulin, which is made by the islets of Langerhans in the pancreas. Isoflavones, which are phytoestrogens with properties comparable to those of human oestrogen, have been consumed by people all over the world. People who have diabetes can benefit from drinking green tea and cinnamon tea [19]. Psyllium-derived dietary fibres have been used to lower lipid levels in hyperlipidaemia and manage blood sugar in diabetic individuals. Supplementing with omega-3 fatty acids can help people with type 2 DM lower their triglycerides and very-low-density lipoprotein (VLDL) cholesterol as well as their arterial pressure and swelling indicators [20].

11 Supplements for Parkinson's Disease

Parkinson's disease (PD), which causes motor impairment, cognitive impairment and dementia, is characterized by the gradual death of dopaminergic neurons of the brain. Numerous hypothesized methods, including the removal of free radicals and reactive oxygen species (ROS), the chelation of iron, the alteration of cell-signalling pathways and the reduction of inflammatory responses are some of the examples for how nutraceuticals can offer neuroprotection. According to studies, curcumin reduces synuclein toxicity and the formation of reactive oxygen species, which are believed to be responsible for programmed cell death. Curcumin also enhances the survival of neurons in the substantia nigra of the midbrain, which is involved in learning and memory [21].

12 Nutraceuticals in Hypertension

In hypertension, the blood pushes against the arterial walls with such power that it may potentially result in health issues, such as heart disease. Numerous naturally occurring substances found in diet, as well as certain nutraceutical adjuncts, vitamins, antioxidants, or minerals, can act like pharmaceuticals and act similarly to a particular class of antihypertensive treatments. Pomegranate juice, grape seed distillate, hesperidin and melatonin have all been found to lower blood pressure [22]. The natural compounds known to lower the risk of stroke and facilitate smooth heart functions are called flavonoids. The carotenoid family member lycopene lowers blood pressure and oxidative stress [23]. Garlic is helpful in treating people who are currently taking medication and lowers blood pressure [24]. Clinical studies demonstrate that 4 weeks after ingestion, seaweed lowers blood pressure [25]. Sesame lowers blood pressure, eases vascular stiffness and boosts antioxidant defence. Drinks: cocoa, coffee and tea aids in lowering blood pressure and enhancing endothelial health [26].

13 Supplements with Anti-inflammatory Properties

White blood cells and the compounds they create in the body use inflammation as a defence mechanism against bacteria, viruses and other foreign invaders. It is the body's reaction to damage or irritation and is characterized by discomfort, oedema, heat and redness. Upright oxidative stress and inflammatory biomarkers are hallmarks of the chronic inflammatory illness rheumatoid arthritis (RA). Many researchers have conducted numerous studies to determine the anti-inflammatory properties of fish oil, primrose oil, curcumin, fenugreek, liquorice, coriander, tomato, carrot, sweet potato, broccoli, green tea, rosemary, hazelnut, walnut, wheat germ, beetroot roots, cucumber fruits, spinach leaves and dates [27]. According to studies, most nutraceuticals have favourable effects on chronic inflammatory disorders, which may be because they contain one or more of the aforementioned phytochemicals.

14 Supplements and Obesity

Obesity causes chronic, excessive adipose tissue enlargement, which raises the risk of type 2 DM, heart-related disease and other amalgamation disorders. Ephedrine, caffeine, Ma Huang-Guarana, chitosan and green tea are examples of herbal stimulants that aid in weight loss [28]. White adipose tissue and skeletal muscle proteins involved in agitation and lipid metabolism are altered by capsaicin, a physiologically active compound found in red chilli peppers. As a result, it promotes fat oxidation and thermogenesis [29]. The dietary supplement's combination of chitosan, fenugreek and vitamin C dramatically lowered body weight.

15 Supplements and Cancer

Nutraceuticals, mostly phytochemicals produced from nutritional or medicinal plants such soybeans, garlic, ginger, tea, honey and others, may have disease preventive properties, according to the findings of several researchers [30]. The isoflavones found in soy foods, curry's curcumin and soya isoflavones all have anti-cancer potential. The lowest incidence of intestinal tumour was observed in people who consumed high amounts of lutein-rich foods, such as chicken eggs, spinach, tomatoes, oranges and green leafy vegetables [31]. Cancer, heart disease and gastrointestinal disorders can all be avoided because of lycopene. Where it concenttrates in the skin, testes, adrenal and prostate where it protects against cancer [32]. Flavonoids that curb the enzymes that make oestrogen lessen the risk of fatality brought on by oestrogen. Isoflavones are a class of phytochemicals that make up the majority of a diet based on soy. Genistein, daidzein and glycitein are the three main isoflavone components demonstrated to have notable effects in cancer prevention and therapy [33].

16 Gene Interactions and Nutraceuticals

Given that nutrients and nutraceuticals interact with genes, it is possible that a genetic explanation may account for the population's persistent prevalence of nutritional disease through nutrient-silencing of phenotype expression. In numerous experimental research, it has been discovered that PUFA (ω-6 and ω-3), milk, calcium, vitamin, iron, ascorbate and saturated fat can influence gene expression.

17 Nutrition, Food Processing Technology and Nutraceuticals

The food business has been keen on distinguishing food fixings that have well-being benefits for a year to use them in the production of utilitarian food varieties and nutraceuticals [34]. There is still a need for platform technologies that are available and close to the market for processing these constituents into marketable value-added functional food and nutraceutical products, despite significant advancements in the identification, extraction and characterization of these ingredients.

18 Nutraceutical Product Development Using Biotechnology

In the past, the food industry applied biotechnology techniques primarily to the main sources of energy-giving meals. Interest in biologically active, non-nutritive elements (or components) from natural goods or foods has grown more lately. Over the past 10 years, significant advancements in every area of genetic engineering and biotechnology have been produced. The extraction of bioactive components using genetic engineering and enzyme/fermentation technology has received great importance from researchers. The employment of this method in the synthesis of enzymes and recombinant microorganisms has become appealing to the food business because of shifting attitudes in society, such as those regarding recombinant DNA and the rising need to investigate all alternative food sources [35].

Enzyme technology used in the food business is significantly impacted by recombinant DNA techniques. The capacity to endure wide pH ranges, improved thermostability, and enhanced enzyme production yields are a few advantages [36]. It is now possible to identify specific genes that code for enzymes from species whose genetic makeup is unknown via genetic engineering. These genes can be inserted into microbes which have been used for preparing food for hundred years using in vitro recombination. Utilizing genetic engineering techniques offers the chance to boost gene expression and thus impact product production. It is also desirable to have information on how biotechnology may impact conventional techniques of manufacturing foods, beverages and nutraceuticals.

19 Enhancing the Bioavailability and Delivery of Nutraceuticals Using Nanotechnology

An empowering innovation that can possibly change the food and agrarian ventures is nanotechnology. Scientists have been involving strategies and information in nanotechnology to address the hardships connected with food and sustenance, driven by developing client interest for quality food things. Because nanoparticles have more surface area per mass unit, they are predicted to have higher biological activity than bigger particles with the same chemical make-up. This presents several food application vistas. For example, nanoparticles can be used as bioactive ingredients in functional foods [37].

20 Production of Nutraceutical Ingredients by Spray Drying

In comparison to other traditional drying techniques, spray drying is making an increasing contribution to food processing applications. In addition to helping to dry fluids, spray drying has been a key factor in the encapsulation and microencapsulation of valuable foods and functional-nutraceutical substances [38]. When compared to other encapsulation techniques, spray drying for microencapsulation is a one-step, cost-effective approach. In the food industry, spray drying encapsulation is mostly used to shield bioactive ingredients or functional foods from environmental factors like light, heat, oxidation. In order to effectively encapsulate vitamins, minerals, flavouring agents, antioxidant chemicals and fatty acids, spray drying is crucial.

21 Nutraceutical Supplements from Plant Sources

The vast majority of plant extracts, along with a few items derived from animal sources, are used as 'traditional' or 'herbal' remedies, and many of these products are also sold as nutritional adjuncts. There are several benefits for businesses to sell these goods as dietary supplements so that they fall under rules for foods rather than drugs.

A plant-based diet can lower the risk of chronic diseases, particularly cancer, according to a large body of research. A 1992 assessment of 200 epidemiological research revealed that persons who ingest diets rich in fruits and vegetables have a lower chance of developing cancer [39]. A vegetable was just half of what it was in people who ate fewer of these meals. More than a handful kinds of these physiologically active plant chemicals—now referred to as 'phytochemicals'—were discovered by Steinmetz and Potter.

The Nutrition Labelling and Education Act of 1990 (NLEA) has helped certain health professionals gradually recognize the function of phytochemicals in health promotion (ADA, 1995). The NLEA required nutrition labels for the majority of foods and permitted disease-related or health-related messages on food labels, allowing customers to know about the product's health advantages.

21.1 Spirulina and Kelp

Both spirulina and kelp have been referred to as miracle cures due to their abilities to ease arthritis pain, boost energy, improve liver function, protect against cardiac disease and cancer, suppress AIDS, regulate appetite and protect cells from X-ray or heavy metal-induced cell damage.

These seaweeds, spirulina and kelp, are beneficial components of a vegetarian or macrobiotic diet since they both include necessary nutrients like carotenoids, protein and minerals that have been linked to health benefits [40].

Spirulina is an excellent treatment for foul breath if the problem isn't brought on by chronic sinusitis or gum disease due to its high chlorophyll content [41]. In chlorophyll breath fresheners, it is frequently the main component.

Iodine-rich kelp is the ideal supplement for an underactive thyroid disease brought on by a deficiency in the iodine mineral. Iodized salt usage is become so common. Before using kelp or any other dietary supplements, go to your doctor if you think you have a thyroid problem because they can be successfully treated with medicine and can be detected with blood testing [42].

21.2 Caffeine

A member of the methylxanthine class of central nervous system (CNS) stimulants is caffeine. The effects of caffeine can be explained by a number of recognized mechanisms of action. The most important is that it prevents the onset of sleepiness brought on by adenosine by reversibly blocking adenosine's action on its receptor [43]. Additionally, caffeine stimulates a few areas of the autonomic nervous system (ANS). Since caffeine primarily functions as a CNS stimulant, when it enters the brain, alertness is the most obvious impact. It is a typical element in drugs to cure or control sleepiness, headaches and migraines since the user will feel more awake and less sleepy [44]. Additionally, numerous studies have shown that habitual coffee drinkers had a lower incidence of Alzheimer's disease and dementia. Those who solely consume high-octane coffee and not decaf can take advantage of these advantages. It also has negative effects. Headaches can result from consuming too much caffeine, which is primarily associated with caffeine withdrawal. On its own, caffeine has no nutritional value. Additionally, caffeine may be included in some drugs.

21.3 Green Tea

Green tea was first produced in China, but it has since expanded to many other Asian nations. There are many different kinds of green tea, and they vary greatly depending on the *Camellia sinensis* type utilized, the growing environment, the horticulture techniques used, the production process, and the time of harvest. The dried leaves of the *Camellia sinensis* bush are used to make all forms of tea and one of the less processed varieties of tea is green tea, which is produced from unoxidized leaves. As a result, it has the highest concentration of polyphenols and antioxidants [44].

Regular green tea is 99.9% water, provides 1 Calorie per 100 mL serving, is lack of significant nutrient content and contains phytochemicals, such as polyphenols and caffeine. It also lacks considerable nutritional value. Despite the numerous health benefits of green tea that have been touted, no solid evidence of any effects has been presented by human clinical studies. At the European Commission's request, a panel of scientists produced a report on the health effects claims in 2011. In general, they concluded that the claims made for green tea were not adequately supported by the available scientific data. Green tea has been a component of traditional Chinese and Indian treatments.

21.4 Grape Tea

The fruit of grapevines is called a grape. Grapevine species Vitis vinifera and Vitis labrusca are both widely distributed. Concord grapes are the common name for Vitis labrusca. The entire grapefruit, skin, leaves and seed are utilized medicinally. Patients with chronic venous insufficiency, which causes swelling in the legs as a result of inadequate blood flow, may benefit from extracts from grape leaves and grape seeds [45]. In addition to treating other problems like high blood pressure, excessive cholesterol and ailments of the heart and blood vessels, grape products are frequently used. The health advantages of grape tea are numerous.

Flavonoids, which are found in grape, may have antioxidant benefits, lower LDL levels (bad cholesterol), relax blood vessels and keep down the risk of chronic heart disease (CHD). The antioxidants in grapes may have other possible health benefits in addition to helping to avoid heart disease [45]. More antioxidants are present in red grape types than in white or blush grape varieties. Grape leaf may have astringent and anti-inflammatory properties. To put it another way, grape leaves appear to have the ability to bring tissue together, which may be able to reduce bleeding and diarrhoea. The crimson leaves seem to have the most of these qualities.

21.5 Wheat Grass

The earliest leaves of the normal wheat plant that have recently grown, or wheatgrass, are eaten as a food, drink or wholesome enhancement. Wheatgrass holds chlorophyll, amino acids, minerals, vitamins and enzymes, just like the majority of plants. Wheatgrass has a variety of health advantages, such as extra nourishment and special healing capabilities.

Juice shops frequently sell wheatgrass juice, but some people also grow their own at home to juice. It is offered in tablets, frozen juice, powder and fresh as produce. Commercially available forms of wheatgrass include sprays, creams, gels, massage lotions and liquid herbal supplements, because it is taken from wheatgrass sprouts, which are the first stages of the 'wheat seed' or 'berry' that is used to make flour. There is no gluten in the wheatgrass juice [46].

The *Triticum aestivum* plant is used to make the meal known as wheatgrass. It is recognized as a highly effective health food that offers incredible advantages. Typically ingested as a fresh juice, it is also available in powdered form.

Wheatgrass is incredibly beneficial for the body since it has a potent combination of nutrients. Wheatgrass may be a fantastic choice to improve your health due to its broad blend of vitamins and nutrients. Antioxidant, bacterial and anti-inflammatory effects are all present in wheatgrass. It includes proteins, proteins, iron, calcium, enzymes, magnesium, phytonutrients, 17 amino acids, vitamins A, C, E, K and B complex [47].

21.6 Mushroom Extract

A fungus known as the 'Reishi Mushroom' has a harsh taste and has been characterized by some as being 'tough' and 'woody'. Mycelium, which connects a clump of mushrooms, and the fruiting body, which is found above ground, are both used medicinally.

Reishi mushroom is specifically used for boosting the resistance against viral infections like the flu, swine flu and avian flu, lung ailments like asthma and respiratory illness, heart diseases and associated conditions like high blood pressure and hyperlipidaemia, kidney disease, cancer and liver diseases [48]. HIV/AIDS, altitude sickness (dizziness, headache, loss of appetite, etc.), chronic fatigue syndrome (CFS), sleeplessness, stomach ulcers, poisoning and herpes discomfort are additional uses for it. Other use includes lowering stress and avoiding weariness. Prostate cancer is treated with Reishi mushroom in combination with other herbs [49].

21.7 Chlorophyll

A green pigment called chlorophyll is present in plants. Chlorophyll is utilized in human medicine. Alfalfa (*Medicago sativa*) and silkworm droppings are two typical sources of chlorophyll used in medicine. In order to treat poor breath and lessen colostomy odour, chlorophyll is utilized [50]. An abdominal incision known as a colostomy allows for the collection of intestinal waste in a bag. Additionally, 'detoxification', constipation and wound healing are all treated with chlorophyll. Almost every photosynthetic creature contains chlorophyll, including eukaryotic algae, prokaryotic blue-green algae and green plants. Chlorophyll is administered intravenously to treat chronic relapsing pancreatitis, a pancreas condition.

21.8 Garcinia Cambogia

The flowering plant genus Garcinia is indigenous to Asia, America, Australia, tropical and southern Africa and Polynesia. It belongs to the family Clusiaceae. The Garcinia Cambogia fruit has drawn the attention of numerous people looking for all-natural weight-loss methods. The little fruit, which is native to India and Southeast Asia and is distributed worldwide, looks like a hybrid between a pumpkin and a tomato.

Hydroxycitric acid (HCA), a component of garcinia cambogia, has been used to promote weight loss. The HCA extract can be acquired in health stores or online in the form of pills or powder [51].

21.9 Blue Tea

A caffeine-free herbal brew known as Blue Tea is created by steeping fresh or dried leaves of the Clitoria ternatea plant [52]. The butterfly pea, also known as blue pea or pigeon wings, is a plant that grows widely throughout most of South East Asia. The flower's rich blue petals have historically been used to manufacture dye. The 'Blue Tea' is a favourite in Thailand and Vietnam, where it is typically served after dinner and sweetened with honey and lemon. When lemon juice is added to tea, the pH of the beverage changes, turning it from deep blue to purple. Today, blue tea leaves can be found in shops and supermarkets all over the world.

21.10 Aloe Vera

The thick leaves of the plant with cactus-like appearance that produces aloe vera are what give the gelatinous material its name. The most common and well-known application of this extract is as a topical ingredient in after-sun lotions and cosmetic preparations. Because it is said to have anti-inflammatory, itch-relieving, pain-killing and healing effects, aloe vera is also a component of ointments used to treat skin disorders. Recently, it has also been marketed as an oral supplement with the promise of lowering blood cholesterol levels and blood sugar in type 2 DM. The leaves of the plant also secrete an unpleasant yellow liquid known as aloe latex or aloe juice, which includes the harsh laxative anthraquinone. When the juice dries, it produces dark brown granules that are recognized by the Food and Drug Administration (FDA) as a counter laxative. Unsurprisingly given the bitter taste and laxative effects of aloe latex, give no dietary use for this plant. As a nutritional supplement, aloe vera gel shouldn't contain any juice and shouldn't cause constipation [53].

There is little evidence in the scientific literature to support the claims that these oral supplements have beneficial effects upon blood lipoprotein profiles or are helpful in the management of type 2 DM, despite the exaggerated claims made by suppliers and advocates of aloe vera supplements, because there were so few controlled clinical trials of oral aloe vera at the time. One systematic review from 1999 stated that neither the clinical effectiveness of oral or topical aloe vera was sufficiently defined at that time [54]. Aloe vera is typically consumed in the UK as concentrated aloe vera gel extract in the form of tablets or capsules, one or two per day. Although there is no evidence of any negative effects, there does not appear to be much confirming to support the oral intake of aloe vera supplements.

Popular medical plant aloe vera has been around for a very long time. The majority of the plant's bioactive substances, including vitamins, minerals, amino acids and antioxidants, are present in the gel.

21.11 Oats

The cholesterol-lowering soluble fibre beta-glucan is found in oat products. The eating of this food can lower total and LDL cholesterol, which lowers the risk of CHD, according to scientific evidence [55].

21.12 Soybean

Soy is not only a high-quality protein, as determined by the FDA's 'Protein Digestibility Corrected Amino Acid Score' technique, but it is also now believed to have preventive and medicinal effects on osteoporosis, cancer, cardiovascular disease (CVD) and menopausal symptoms.

Since soybeans are the sole substantial dietary source of isoflavones (genistein and daidzein), they stand out among the rest. Isoflavones are heterocyclic phenols that resemble oestrogenic steroids structurally. Isoflavones may operate as anti-oestrogens by competing with stronger, naturally occurring endogenous oestrogens (such 17-beta-estradiol) for binding to the oestrogen receptor. This is because they are weak oestrogens. This may help to explain why populations that consume a lot of soy, like those in Southeast Asia, have a lower chance of developing oestrogen-dependent cancer. Bone health may also be enhanced by soy.

21.13 Flaxseed

A-linolenic acid, an omega-3 fatty acid, is present in flaxseed oil in the highest concentration (57%) of all the major seed oils. However, more recent studies have concentrated particularly on lignans, a class of chemicals related to fibre. Enterodiol and enterolactone, the two main lignans found in mammals, are produced in the intestinal system by bacteria acting on plant lignan precursors [56]. The most abundant source of mammalian lignan precursors is flaxseed. As slightly oestrogenic and antioestrogenic substances with structural similarities to both naturally occurring and synthesized oestrogens, enterodiol and enterolactone may help prevent malignancies that are oestrogen-dependent.

Fewer researches have examined how eating flaxseed affects human cancer risk indicators. The use of 10 g of flaxseed per day caused multiple hormonal changes that were linked to a lower risk of breast cancer, according to Phipps et al. Additionally, flaxseed consumption has been demonstrated to lower total and LDL cholesterol as well as platelet aggregation [57].

21.14 Tomatoes

Lycopene, the main carotenoid present in tomato fruit, plays a key effect in lowering the chance of developing cancer.

In a research including more than 47,000 males, those who ingested tomato products 10 or more times per week had a risk of advanced prostate cancer that was less than half that of non-consumers. Surprisingly, lycopene is the carotenoid that is most prevalent in the prostate gland. Breast, digestive tract, cervix, bladder and skin cancer risks have also been shown to be negatively correlated with blood or tissue levels of lycopene, as well as perhaps lung cancer risks [58]. According to theories, lycopene's antioxidant properties may have an impact on cancer risk. In biological systems, lycopene is the singlet oxygen quencher that is most effective.

21.15 Spices

Spices are a class of esoteric food additives that have been used for hundreds of years to improve the sensory quality of dishes. Tropical countries use a disproportionate amount of spices due to their climate. These spice elements give cuisine its distinctive flavour, fragrance or piquancy as well as colouring. Some spices, such as fenugreek, can also change the way food feels. It is a well-known fact that their distinctive perfume increases appetite. Many spices are used in toiletries, cosmetics and perfumes in addition to being used as flavourings and seasonings.

21.16 Broccoli and Other Cruciferous Vegetables

Epidemiological research has linked regular intake of cruciferous vegetables to a down surge the risk of developing cancer. All cruciferous vegetables include glucosinolates, a class of glycosides that are kept in their cell vacuoles. These substances are hydrolysed by the enzyme myrosinase, which is present in plant cells. Hydrolysis products produced by this process include isothiocyanates and indoles. I3C, in particular its mammary gland-specific cancer-chemopreventive characteristics, is now being researched [59]. In addition to initiating Phase I and II detoxification processes, I3C may lower the risk of cancer through regulating the metabolism of oestrogen. I3C considerably increased the extent of oestradiol 2-hydroxylation in women when given to them in humans at 500 mg per day (equal to 350–500 g of cabbage per day) for one week, indicating that this substance may be a novel strategy.

The prevention of cancer in animals has been demonstrated for an abundant range of naturally existing and manufactured isothiocyanates, however, attention has been focused on a particular isothiocyanates isolated from broccoli, known as Sulforaphane which are shown to be the principal inducer of a particular type of Phase II enzyme, quinone reductase.

21.17 Citrus Fruits

Citrus fruits have been demonstrated to be protective against a number of human malignancies in epidemiological investigations. Although vital elements like vitamin C, folate and fibre are primarily found in citrus fruits like oranges, lemons, limes and grapefruit. Fruits like citrus are particularly rich in a group of phytochemicals called limonoids.

Evidence for Limonene's ability to prevent cancer has been mounting over the past 10 years [60, 61]. The effectiveness of this substance against a range of spontaneous and chemically induced mouse cancers was demonstrated by Crowell [62].

21.18 Cranberry

Recent research has concentrated on how cranberry juice can prevent *Escherichia coli* from adhering to uroepithelial cells.

Fructose and a non-dialyzable polymeric compound have both been implicated in this behaviour. It was discovered that the latter substance, which was later extracted from cranberry and blueberry juices, hampers adhesins located on the pili of the surface of some pathogenic *Escherichia coli* [63].

21.19 Wine and Grapes

Research evidences shows that wine, especially red wine, can lower the risk of CVD. When St. Leger et al. [64] discovered a high negative connection between wine intake and death from ischemic heart disease in both men and women from 18 nations, the relationship between wine consumption and CVD initially came to light. Despite diets high in dairy fat, France in particular has a comparatively low rate of CVD [65]. Although the capacity of alcohol to raise HDL cholesterol can help to explain this 'French Paradox', latest studies have concentrated on the non-alcohol components of wine, particularly the flavonoids [66].

Due to the inclusion of the grape skins into the fermenting grape juice during manufacturing, red wine has a high phenolic concentration that is 20–50 times greater than white wine. Phenolic levels in red wine and black seedless grapes are high (920, 1800 and 3200 mg/L, respectively), while phenolic levels in Green Thomson grapes are much lower (260 mg/kg). According to da Luz and Coimbra [67, 68] the positive benefits of red wine to the ability of phenolic substances to stop the oxidation of LDL, a critical step in the process of Atherogenesis, a disorder of the artery wall that involves adhesion of monocytes and lymphocytes to the endothelial cell surface, migration of monocytes into the subendothelial space, ingestion of low density lipoproteins, etc. is what causes red wine's positive effects.

Wine without alcohol has been proven to boost total plasma antioxidant capacity [64]; thus those who want the health benefits of wine without the potential risk may want to take that into consideration. Trans-resveratrol, a phytoalexin present in grape skins, may also be found in substantial amounts in red wine. Resveratrol has been proven to suppress carcinogenesis in vivo and contains oestrogenic effects that may partially account for the cardiovascular advantages of wine consumption [69–71].

21.20 Bee Products

Bees and the pollination services they provide enable a variety of animals and plant species, including humans, to thrive in the natural world.

Bees produce bee products, which are a completely natural source of nourishment. Since the beginning of time, people have used these bee products, and they continue to be a largely suitable source of food now.

In addition to honey, which is unquestionably the most popular bee product, bees also provide us wax, pollen, propolis and royal jelly. Another item that can be useful to individuals is bee venom. The most popular beekeeping-derived product and naturally occurring food is honey. Royal jelly, bee pollen and bee propolis are three by-products of beekeeping that are utilized as dietary supplements [72, 73].

22 Nutraceutical Components from Animal Sources

Functional food can be divided into two categories: plant-based foods and animal-based foods.

Plant-based meals have several ingredients that fight deadly diseases like cancer. Numerous elements, like dairy products and fish oil that are naturally found in animal sources, may be good for your health [74]. To improve health, nutraceuticals give a condensed form of a food's biologically active ingredient. The use of probiotics along with prebiotics has proven to be very successful in treating a number of chronic conditions. Prebiotics and probiotics offer a different source for treating various intestinal diseases. Proper intestinal function and enhanced metabolism are the two main effects that probiotics have on human health.

22.1 Fish

Known as omega-3s, these fatty acids have a double link between carbon and carbon, three carbons away from the chain's methyl terminus. In essence, omega-3 (n-3) fatty acids are a crucial group of polyunsaturated fatty acids (PUFAs) that are largely found in fish oil. The importance of n-3 fatty acids in CVD was initially recognized in the 1970s, when Bang and Dyerberg found that Eskimos had low rates of the condition while eating a diet heavy in fat. Consuming fish has been found to have cardio-protective effects in some future studies [75].

Consuming 35 g or a greater amount of fish day to day has been displayed to bring down the gamble of death from non-unexpected myocardial localized necrosis in the Chicago Western Electric Review, and just one serving of fish each week was related to an essentially lower hazard of complete cardiovascular mortality following 11 years in excess of 20,000 guys [76]. Fish, especially oily fish like salmon and tuna, has significant levels of minerals and protein. Fish is a mainstay of the wholesome Mediterranean diet and frequently has less cholesterol and saturated fat than meat. Iron, zinc and B vitamins are among the vitamins and minerals found in fish.

Omega-3 fatty acids, which are healthy fats that people must consume, are found in oily fish. Potential health advantages of omega-3 fatty acids include their ability to lower the risk of cardiovascular disease. Fish is a low-fat protein source; so including it in your diet may also help you lose weight. One of the most popular dietary supplements is fish oil. It contains a lot of omega-3 fatty acids, which are crucial to good health.

22.2 Dairy Products

Dairy products are unquestionably healthy foods. They are one of the best sources of calcium, a necessary nutrient that helps to fight colon cancer and osteoporosis. Recent studies have concentrated primarily on additional dairy product ingredients, particularly the fermented dairy products known as 'Probiotics', in addition to calcium. According to Fuller [77], probiotics are 'live microbial feed supplements that advantageously affect the host animal by improving its intestinal microbial balance'. Although probiotics have been linked to a number of health advantages, their antagonistic effects against enteric pathogens and other intestinal organisms, as well as their anti-carcinogenic, hypocholesterolemic and other health benefits, have drawn the greatest interest.

23 Marine Sources as Supplements Rich in Nutraceuticals

Compared to terrestrial ecosystems, marine ecosystems have a higher diversity of living things, which offer a variety of supplies for human nutrition and health. Diverse taxonomic groups, including Porifera (sponges), Cnidaria (corals, sea anemones, hydrozoans, jellyfish), Annelida (polychaetes, marine worms), Bryozoa (moss animals or sea mats), Mollusca (oysters, abalone, clams, mussels, squid, cuttlefish, octopuses), Arthropoda (L Macroalgae, microalgae, bacteria, cyanobacteria, certain fish species and Crustaceans are also included in this diverse category since they create secondary metabolites to survive in the harsh marine environment. A wide variety of marine natural products with nutritional components are good to human health or marine nutraceuticals. Numerous marine species contain significant amounts of carbohydrates, which have a number of medicinal and nutraceutical properties that protect against a variety of ailments. Additionally, chitosan and its derivatives, as well as other shellfish waste-derived polysaccharides, have significant uses in biology and biomedicine. The creation of medicines and nutraceuticals uses the marine carbohydrates from varied marine flora and animals.

References

1. FSSA (2006) Food Safety and Standards Act
2. Hertog MGL et al (1993) Dietary antioxidant flavonoids and risk of coronary heart disease: the Zutphen elderly study. Lancet 342:1007–1011
3. Dorow P et al (1987) Effect of a secretolytic and a combination of pinene, limonene and cineole on mucociliary clearance in patients with chronic obstructive pulmonary disease. Arzneimittelforschung 37:1378–1381
4. Santos-Buelga C, Scalbert A (2000) Proanthocyanidins and tannin-like compounds—nature, occurrence, dietary intake and effects on nutrition and health. J Sci Food Agric 80:1094–1117. https://doi.org/10.1002/(SICI)1097-0010(20000515)80:7<1094::AID-SFA569>3.0.CO;2-1
5. Moroney SP, Schultz AB, James AA et al (1988) Load-displacement properties of lower cervical spine motion segments. J Biomechanics 21(9):769–779
6. Pace-Asciak CR, Susan H, Eleftherios PD et al (1995) The red wine phenolics trans-resveratrol and quercetin block human platelet aggregation and eicosanoid synthesis: implications for protection against coronary heart disease. Clin Chim Acta 235(2):207–219
7. Shirakami Y, Shimizu M (2018) Possible mechanisms of green tea and its constituents against cancer. Molecules 23(9):2284. https://doi.org/10.3390/molecules23092284
8. Yang J, Xiao Y (2013) Grape phytochemicals and associated health benefits. Crit Rev Food Sci Nutr 53(11):1202–1225. https://doi.org/10.1080/10408398.2012.692408

9. Ramalho SA, Gualberto NC, Santos MT et al (2014) Catechin and epicatechin contents in wines obtained from Brazilian exotic tropical fruits. Food Nutri Sci 5(5):43249. https://doi.org/10.4236/fns.2014.55053
10. Yilmaz Y, Toledo RT (2004) Major flavonoids in grape seeds and skins: antioxidant capacity of catechin, epicatechin, and gallic acid. Agric Food Chem 52(2):255–260. https://doi.org/10.1021/jf030117h
11. Dohadwala MM, Vita JA (2009) Grapes and cardiovascular disease. J Nutr 139:1788S–1793S
12. Manchali S et al (2012) Crucial facts about health benefits of popular cruciferous vegetables. J Funct Foods 4(1):94–106. https://doi.org/10.1016/j.jff.2011.08.004
13. Miekus N, Marszalek K, Podlacha M et al (2020) Health benefits of plant-derived sulfur compounds, glucosinolates, and organosulfur compounds. Molecules 25(17):3804. https://doi.org/10.3390/molecules25173804
14. Fahey JW et al (1997) Broccoli sprouts: an exceptionally rich source of inducers of enzymes that protect against chemical carcinogens. Proc Natl Acad Sci U S A 94(19):10367–10372
15. Patti AM, Katsiki N, Nikolic D, Al-Rasadi K, Rizzo M (2015) Nutraceuticals in lipid-lowering treatment: a narrative review on the role of chitosan. Angiology 66(5):416–421. [PubMed ID: 25037700]. https://doi.org/10.1177/0003319714542999
16. Afzal O, Dalhat MH, Altamimi ASA, Rasool R, Alzarea SI, Almalki WH, Murtaza BN, Iftikhar S, Nadeem S, Nadeem MS, Kazmi I (2022) Green tea catechins attenuate neurodegenerative diseases and cognitive deficits. Molecules 27(21):7604
17. Fatima U, Roy S, Ahmad S, Ali S, Elkady WM, Khan I, Alsaffar RM, Adnan M, Islam A, Hassan MI (2022) Pharmacological attributes of *Bacopa monnieri* extract: current updates and clinical manifestation. Front Nutr 9:972379. https://doi.org/10.3389/fnut.2022.972379
18. Barros M, Poppe S, Bondan E (2014) Neuroprotective properties of the marine carotenoid astaxanthin and Omega-3 fatty acids, and perspectives for the natural combination of both in krill oil. Nutrients 6(3):1293–1317. https://doi.org/10.3390/nu6031293
19. Warwick KW (2023) CDE. In: Schaefer A (ed) Nutrition
20. Anderson JW, Allgood LD, Turner J, Oeltgen PR, Daggy BP (1999) Effects of psyllium on glucose and serum lipid responses in men with type 2 diabetes and hypercholesterolemia. Am J Clin Nutr 70(4):466–473. https://doi.org/10.1093/ajcn/70.4.466. PMID: 10500014.
21. Dos Santos-Neto LL, de Vilhena Toledo MA, Medeiros-Souza P, de Souza GA (2006) The use of herbal medicine in Alzheimer's disease—a systematic review. Evid Based Complement Alternat Med 3(4):441–445. https://doi.org/10.1093/ecam/nel071
22. Houston M (2014) The role of nutrition and nutraceutical supplements in the treatment of hypertension. World J Cardiol 6(2):38–66. https://doi.org/10.4330/wjc.v6.i2.38
23. Bin-Jumah MN, Nadeem MS, Gilani SJ, Mubeen B, Ullah I, Alzarea SI, Ghoneim MM, Alshehri S, Al-Abbasi FA, Kazmi I (2022) Lycopene: a natural arsenal in the war against oxidative stress and cardiovascular diseases. Antioxidants (Basel) 11(2):232. https://doi.org/10.3390/antiox11020232
24. Ried K (2020) Garlic lowers blood pressure in hypertensive subjects, improves arterial stiffness and gut microbiota: a review and meta-analysis. Exp Ther Med 19(2):1472–1478. https://doi.org/10.3892/etm.2019.8374
25. Wada K, Nakamura K, Tamai Y, Tsuji M, Sahashi Y, Watanabe K, Ohtsuchi S, Yamamoto K, Ando K, Nagata C (2011) Seaweed intake and blood pressure levels in healthy pre-school Japanese children. Nutr J 10:83. https://doi.org/10.1186/1475-2891-10-83
26. Dalibalta S, Majdalawieh AF, Manjikian H (2020) Health benefits of sesamin on cardiovascular disease and its associated risk factors. Saudi Pharm J 28(10):1276–1289. https://doi.org/10.1016/j.jsps.2020.08.018
27. Oyeniran Taiwo O, Adegbenro O, Chioma ED (2017) New era of medicine: role of nutraceuticals in treatment and prevention of various diseases. Univ J Pharm Res 2(5):60–63
28. Valizadeh E, Ghalichi F, Ostadrahimi A (2016) Traditional herbal medicine for weight management: a review. Int J Med Res Health Sci 5(11):393–399

29. Zheng J, Zheng S, Feng Q, Zhang Q, Xiao X (2017) Dietary capsaicin and its anti-obesity potency: from mechanism to clinical implications. Biosci Rep 7(3):BSR20170286. https://doi.org/10.1042/BSR20170286
30. Salami A, Seydi E, Pourahmad J (2013) Use of nutraceuticals for prevention and treatment of cancer. Iran J Pharm Res 12(3):219–220
31. Yu Y, Jing X, Li H, Zhao X, Wang D (2016) Soy isoflavone consumption and colorectal cancer risk: a systematic review and meta-analysis. Sci Rep 6:25939. https://doi.org/10.1038/srep25939
32. Sehgal S, Saji H, Banik SP (2022) Chapter 11—Role of food structure in digestion and health. In: Bagchi D, Ohia SE (eds) Nutrition and functional foods in boosting digestion, metabolism and immune health. Academic, pp 151–165
33. Kim I-S (2021) Current perspectives on the beneficial effects of soybean isoflavones and their metabolites for humans. Antioxidants (Basel) 10:1064
34. Alkhatib A (2020) Antiviral functional foods and exercise lifestyle prevention of coronavirus. Nutrients 12:2633. https://doi.org/10.3390/nu12092633
35. Olempska-Beer ZS, Merker RI, Ditto MD, DiNovi MJ (2006) Food-processing enzymes from recombinant microorganisms—a review. Regul Toxicol Pharmacol 45(2):144–158
36. Patel AK, Singhania RR, Pandey A (2016) Novel enzymatic processes applied to the food industry. Curr Opin Food Sci 7:64–72
37. Singh H (2016) Nanotechnology applications in functional foods; opportunities and challenges. Prev Nutr Food Sci 21(1):1–8
38. Murugesan R, Orsat V (2011) Spray drying for the production of nutraceutical ingredients—A review. Food Bioprocess Technol 5(1):3–14
39. Molina-Montes E, Salamanca-Fernández E, Garcia-Villanova B, Sánchez MJ (2020) The impact of plant-based dietary patterns on cancer-related outcomes: a rapid review and meta-analysis. Nutrients 12(7):2010
40. Ampofo J, Abbey L (2022) Microalgae: bioactive composition, health benefits, safety and prospects as potential high-value ingredients for the functional food industry. Foods 11(12):1744
41. Kyle (2016) Spirulina for bad breath—Does it work?
42. Clark CD, Bassett B, Burge MR (2003) Effects of kelp supplementation on thyroid function in euthyroid subjects. Endocr Pract 9(5):363–369
43. Nehlig A, Daval JL, Debry G (1992) Caffeine and the central nervous system: mechanisms of action, biochemical, metabolic and psychostimulant effects. Brain Res Brain Res Rev 17(2):139–170
44. Bhutia PH, Sharangi AB, Lepcha R, Tamang D (2015) Bioactive compounds and antioxidant properties of tea: status, global research and potentialities. J Tea Sci Res 5(11):1–13
45. Sano A, Tokutake S, Seo A (2013) Proanthocyanidin-rich grape seed extract reduces leg swelling in healthy women during prolonged sitting. J Sci Food Agric 93:457–462
46. All about wheatgrass and wheatgrass therapy. http://www.drwheatgrass.com/wheatgrass/Default.htm
47. Reynolds C Wheatgrass is an important and effective natural therapeutic agent. http://www.wheatgrassprofessional.info/introduction.htm
48. Alhajj MJ, Montero N, Yarce CJ, Salamanca CH (2020) Lecithins from vegetable, land, and marine animal sources and their potential applications for cosmetic, food, and pharmaceutical sectors. Cosmetics 7(4):87. https://doi.org/10.3390/cosmetics7040087
49. Reishi Mushroom (2020), Supplemented.co.UK
50. Pareek S, Sagar NA, Sharma S, Kumar V, Agarwal T, González-Aguilar GA, Yahia EM (2017) Chlorophylls: chemistry and biological functions. In: Fruit and vegetable phytochemicals: chemistry and human health, vol 1, 2nd edn. pp 269–284
51. Sripradha R, Magadi SG (2015) Efficacy of Garcinia cambogia on body weight, inflammation and glucose tolerance in high fat fed male wistar rats. J Clin Diagn Res 9(2):BF01–BF04

52. Goncalves BD, Windson Isidoro HC, Cristina PA, de Andrade Arruda FI, Maria MG (2021) Processing, chemical signature and food industry applications of *Camellia sinensis* teas: an overview. Food Chem 12(12):100160
53. Kiple KF, Ornelas KC (2000) The Cambridge world history of food. Food Service Technology 1(1):2 v. (xlii, 2153). https://doi.org/10.1046/j.1471-5740.2001.00006.x
54. Vogler BK, Ernst E (1999) Aloe vera: a systematic review of its clinical effectiveness. Br J Gen Pract 49:823–828
55. DHHS/FDA (1997) Food labeling: health claims; oats and coronary heart disease. Department of Health and Human Services/Food and Drug Administration. Federal Register. 62, pp 3584–3601
56. Morris D (2001) Essential nutrients and other functional compounds in flaxseed. Nutr Today 36:159–162
57. Chang VC, Cotterchio M, Boucher BA, Jenkins DJA, Mirea L, McCann SE, Thompson LU (2019) Effect of dietary flaxseed intake on circulating sex hormone levels among postmenopausal women: a randomized controlled intervention trial. Nutr Cancer 71(3):385–398
58. Giovannucci E, Rimm EB, Liu Y, Stampfer MJ, Willett WC (2002) A prospective study of tomato products, lycopene, and prostate cancer risk. J Natl Cancer Inst 94(5):391–398. https://doi.org/10.1093/jnci/94.5.391. PMID: 11880478.
59. Bradlow HL et al (1991) Effects of dietary indole-3-carbinol on estradiol metabolism and spontaneous mammary tumors in mice. Carcinogenesis 12(9):1571–1574. https://doi.org/10.1093/carcin/12.9.1571
60. Gould MN (1997) Cancer chemoprevention and therapy by monoterpenes. Environ Health Perspect 105:977–979. https://doi.org/10.1289/ehp.97105s4977
61. Zhou J, Maria AM, Kong L (2021) Effect of limonene on cancer development in rodent models: a systematic review. Front Sustain Food Syst 5:725077. https://doi.org/10.3389/fsufs.2021.725077
62. Crowell PL (1999) Prevention and therapy of cancer by dietary monoterpenes. J Nutr 129:775S–778S. https://doi.org/10.1093/jn/129.3.775S
63. Schmidt DR, Sobota AE (1988) An examination of the anti-adherence activity of cranberry juice on urinary and nonurinary bacterial isolates. Microbios 55(224–225):173–181
64. Bishayee A (2009) Cancer prevention and treatment with resveratrol: from rodent studies to clinical trials. Cancer Prev Res (Phila) 2(5):409–418. https://doi.org/10.1158/1940-6207.CAPR-08-0160
65. St. Leger AS, Cochrane AL, Moore F (1979) Factors associated with cardiac mortality in developed countries with particular reference to the consumption of wine. Lancet 1(8124):1017–1020
66. Renaud S, de Lorgeril M (1992) Wine, alcohol, platelets, and the French paradox for coronary heart disease. Lancet 339(8808):1523–1526
67. da Luz PL, Coimbra SR (2004) Wine, alcohol and atherosclerosis: clinical evidences and mechanisms. Braz J Med Biol Res 37:1275–1295
68. Serafini MG, Maiani G, Ferro-Luzzi A (1998) Alcohol-free red wine enhances plasma antioxidant capacity in humans. J Nutr 128(6):1003–1007. https://doi.org/10.1093/jn/128.6.1003
69. Ko JH et al (2017) The role of resveratrol in cancer therapy. Int J Mol Sci 18(12):2589. https://doi.org/10.3390/ijms18122589
70. Haunschild R, Marx W (2022) On health effects of resveratrol in wine. Int J Environ Res Public Health 19(5):3110. https://doi.org/10.3390/ijerph19053110
71. LiverTox: Clinical and research information on drug-induced liver injury [Internet]. Bethesda (2022): National Institute of Diabetes and Digestive and Kidney Diseases; 2012. Bee products: beeswax, Bee Pollen, Propolis
72. Schmus B (2022) Royal Jelly: benefits, uses, and how it's made. Best Bees
73. Bang HO, Dyerberg J (1972) Plasma lipids and lipoproteins in greenlandic west coast Eskimos. Acta Med Scand 192:85–94. https://doi.org/10.1111/j.0954-6820

74. Albert CM et al (1998) Fish consumption and risk of sudden cardiac death. JAMA 279(1):23–28. https://doi.org/10.1001/jama.279.1.23
75. Fuller R (1999) Probiotics a critical review. In: Probiotics for farm animals. Horizon Scientific, Wymondham, pp 15–22
76. Mitall BK, Garg SK (2008) Anticarcinogenic, hypocholesterolemic, and antagonistic activities of *Lactobacillus acidophilus*. Crit Rev Microbiol 21(3):175–214
77. Mann GV, Schaffer RD, Anderson RD et al (1964) Cardiovascular disease in the Maasai. J Atheroscler Res 4:289–312

Chapter 4
Functional Foods in Brief

Premjyoti Patil

Abbreviations

ADA	The American Dietetic Association
CVD	Cardiovascular diseases
DHA	Docosahexaenoic acid
DSHEA	Dietary Supplement Health Education Act
EPA	Eicosapentaenoic acid
FF	Functional foods
FOSHU	Foods for Specified Health Uses
FSSAI	The Food Safety and Standard Authority of India
FUFOSE	Functional Food Science in Europe
WHO	World Health Organization

1 Introduction

The concept of food during ancient times was to gather and meet the hunger requirements, the way to meet the body's calorie requirement. As the days advanced, nomadic life settled and agriculture evolved, crop species were also developed, food was the center of civilization. There was a dramatic change when the food was abundant after the Industrial Revolution. Humans started to consume more food with unhealthy choices, which led to eating disorders. Then the concept of food

P. Patil (✉)
Department of Biotechnology, Basaveshwar Engineering College, Bagalkot, Karnataka, India

J. S. Yaradoddi et al. (eds.), *Frontiers in Food Biotechnology*,
https://doi.org/10.1007/978-981-97-3261-6_4

from hunger changed to satiety, which led to the concept of "junk food." Every food is functional, because all the food that is consumed has its own nutrients. Anything in excess can cause health disorders, the body requires example fat but if we consume in excess can cause obesity. Similarly excess carbohydrates will be deposited in the body as glycogen converted to fat, leading to obesity. This resulted in complex disorders like obesity, which causes hypertension and more visceral fat deposition in the body, one of a risk factor for cardiovascular diseases [1]. Undernourishment and malnourishment also affect growth and development, especially in children and aged population. So, whether the body is undernourished, malnourished, or overnourished each has its own consequences. Other lifestyle disorders, namely diabetes type II and cancer., are due to unhealthy eating habits and lack of awareness regarding food.

Ayurveda, an ancient medicinal practice in India, signifies the importance of food from natural resources, seasonal food, and place of origin. According to Ayurveda, food has properties have specific functions to perform which have its own effect on physiology and pathology of human health [2]. There was strong urge worldwide to know about food's role in health. Although the concept of food science is old, it got its importance in 1980s.

In 1989, Stephen DeFelice coined the term "nutraceutical" derived from nutrition and pharmaceutical, which is commercially used but is not defined by any regulatory [3].

The concept of functional food got attention in the 1980s when the Japanese government started funding the research programs related to food's health benefits on human physiology. For the first time, Japanese government in 1991 made a policy on FOSHU (Food for Specified Health Uses) for commercial purpose. In 1994, the United States expressed functional foods under the Dietary Supplement Health Education Act (DSHEA). In 1996, Europe established "Functional Food Science in Europe" (FUFOSE), a scientific approach to food [4].

India covered functional food as health supplements, nutraceuticals for special dietary usage, medical foods for special purposes, and functional and novel food under 2016 regulations through the Food Safety and Standard Authority of India [5].

By 2019, FOSHU had approved 1063 products, the total sales increased from 6.2 billion dollars in 2007 to 8.2 billion dollars in 2018, indicating the significance of functional foods in the market [6].

Functional foods are not well defined; each organization has its own definition. Clearly indicating knowledge on functional foods are not addressed across the globe. But still some of the definitions are given below.

FOSHU, Japan, has defined functional foods as "the foods that are processed containing ingredients with functions for health and officially approved to claim its physiological effects on the human body."

The American Dietetic Association (ADA) defines functional food as "whole foods along with fortified, enriched, or enhanced foods that have a potentially beneficial effect on health when consumed as part of a varied diet regularly at effective levels."

European Union, FUFOSE, defines functional food that beneficially affects one or more target functions in the body beyond adequate nutritional effects in a way that is relevant to either an improved state of health and well-being and/or reduction of risk disease. It is consumed as part of a normal food pattern. It is not a pill, capsule, or dietary supplement [7–11].

2 Need of Functional Food

Food is basic need of the body for growth and development. As the food industrialization evolved, accessibility for food became easier leading to complex of overeating. The food industries manufacture highly palatable food which is calorie-rich and has high appetite. Overeating leads to overweight and obesity. Obesity may lead to health disorders like cardiovascular disease, diabetes, high blood pressure, and colorectal cancer. Majority of the population are obese, approximately 650 million people, in the world [11]. To stay healthy and for well being, there is a need to understand the food in better way and its function in the physiology. The food should be designed in a such way that there is a need to reduce the appetite and intake of food. There should not be overnourishment by consuming high calorie food, leading to overweight or obesity; meanwhile, undernourishment of the body should also be taken care of. Functional foods nourish the body and help minimize the risk of diseases related to lifestyle disorder. There is a need to design functional foods to meet the satiety and reduce overeating, providing nourishment for well-being [12].

3 Popular Functional Foods and Potential Health Benefits

Adding certain foods along with basic foods can improve health. Foods rich in fibers, antioxidants, and immunomodulators, for gut, brain health, and muscle health have health benefits on the body. Listing out all functional foods is beyond the scope, but an effort is made to provide brief list of the popular foods [13–15].

Cereals: They provide fibers, both digestible and nondigestible. Wheat germ, oats, corn, and millets provide calories, improve gut health, and reduce appetite.

Vegetables: Tomato, garlic, onions, beetroot, carrots, green leafy vegetables, mushrooms, and crucifers are rich in vitamins, minerals, fibers, and antioxidants, which provide nutrition, are less calorific, and protect the cells from free radicals and reduce the rate of cancer. Lycopene also protects from cardiovascular diseases. Garlic acts as immunomodulators and is also good for cardiac health.

Fig. 4.1 Fruits are rich source of antioxidants

Fruits: Blueberry, papaya, pomegranate, black currant, grapefruit, citrus fruits, and strawberry are rich in vitamins A, B, C, D, E, minerals, flavonoids, polyphenols, and carotenoid pigments which have antioxidants. They also provide fibers and calories to the body. Some of the fruits which are bright colored are rich source of antioxidants shown in Fig. 4.1.

Fermented foods: Kimchi, pickles, and yogurt are rich in probiotics like *Lactobacillus and Bifidobacteria* which improve gut health.

Dairy products: Fortified milk and yogurt are rich in, casein, whey protein, vitamin B2, A, D and E, and calcium which improve bone health. Diary products are also rich sources of probiotics, which increases the gut microbiota and help to develop the immune system.

Algae: *Arthrospira platensis* (Spirulina) and *chlorella* are widely consumed across the globe which are rich in thiamin, riboflavin, omega-3 fatty acids (-DHA, EPA), omega -6 fatty acids (gamma-linoleic, linoleic arachidonic acids, vitamin B12, minerals).

Fish Fat: Fish are rich in omega-3 fatty acids (docosahexaenoic acid—DHA, eicosapentaenoic acid—EPA) and omega-6 fatty acids (gamma-linoleic, linoleic arachidonic acids which improve brain health.

Animal Fat: Fats are rich in conjugated linolenic acid, peptides which improve muscle health.

Probiotics: Widely consumed probiotics are *Lactobacillus and Bifidobacterium* which provide lactic acid, thiamine, vitamin B6, and riboflavin.

4 Challenges with Functional Foods

Every food is functional, as they nourishes the body but the quantity of intake should the concern. The challenges are been listed and shown in Fig. 4.2. Functional food may not be challenging, but the way they are formulated, regulated, prescribed, and marketed are the key factors in deciding consumption of functional foods [12, 15, 16]. Basically, food is consumed for hunger satiety and should be palatable, but here are some of the challenges faced by functional foods:

1. Functional foods that are available in the market are palatable.
2. Food regulations lack clarity in making policies, especially on health claims.
3. Long-term studies on health claims like foods claiming to reduce appetite obesity need to be assessed on strong scientific and appropriate clinical assays.
4. Lack of clinical studies of all the functional foods claiming specific health benefits.
5. There is a gap between nutritionists and medical professionals in creating consumer awareness.
6. The interaction between different bioactive components within the body must be assessed based on the functionality test to consume single or multiple functional foods.
7. Still to explore bioactive ingredients specific to health claim.
8. Dosage and quantity fixation of bioactive compounds of functional food to treat specific disease still need to be properly assessed.
9. Functional foods in the form of pills and capsules lack satiety, which are unsavovry.
10. Manufacturing functional food by pharmaceuticals or the food industry is not clear.
11. Improper advertising and marketing of food products by commercial sector.
12. Proper labeling on functional products regarding dosage, age group, side effects, allergies, and other information needs to be provided.

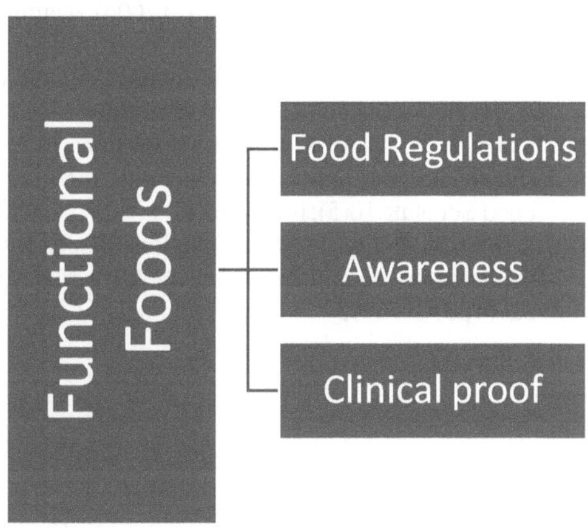

Fig. 4.2 Schematic diagram depicting challenges associated with functional foods

5 Summary

With the lifestyle change, unhealthy food consumption led to health disorders like obesity, diabetes, and cardiovascular diseases. Then, the concept of nutrition, health, and overall well-being became prevalent. Here, an attempt is made to provide a brief concept on functional foods. As food culture differs depending on locality and region, clinical studies regarding functional foods need to be carried out. Functional foods are not well defined across the globe and most of the countries are not aware of and need to be addressed. However, death due to cardiovascular diseases and diabetes-II is still increasing, which are caused because of lifestyle disorders. There is a need to create an awareness worldwide on a scientific basis.

References

1. Litwin M, Kułaga Z (2021) Obesity, metabolic syndrome, and primary hypertension. Paediatr Nephrol 36:825–837
2. Payyappallimana U, Venkatasubramanian P (2016) Exploring Ayurvedic knowledge on food and health for providing innovative solutions to contemporary healthcare. Front Public Health 4:57
3. Kalra EK (2003) Nutraceutical—definition and introduction. AAPS PharmSciTech 5(3):E25
4. Arai S, Osawa T, Ohigashi H, Yoshikawa M, Kaminogawa S, Watanabe M, Ogawa T, Okubo K, Watanabe S, Nishino H, Shinohara K, Esashi T, Hirahara T (2001) A mainstay of functional food science in Japan—history, present status, and future outlook. Biosci Biotechnol Biochem 65(1):1–13
5. FSSAI. https://www.fssai.gov.in/cms/health-supplements.php
6. Iwatani S, Yamamoto N (2019) Functional food products in Japan: a review. Food Sci Human Wellness 8(2):96–101
7. Temple NJ (2022) A rational definition for functional foods: a perspective. Front Nutr 9:1–4
8. Shimizu T (2003) Health claims on functional foods: the Japanese regulations and an international comparison. Nutr Res Rev 16(2):241–252
9. Andlauer W, Furst P (2002) Nutraceuticals: a piece of history, present status and outlook. Food Res Int 35(2–3):171–176
10. Ohama H, Ikeda H, Moriyama H (2006) Health foods and foods with health claims in Japan. Toxicology 221(1):95–111
11. WHO. https://www.who.int/docs/default-source/gho-documents/world-health-statistic-reports/world-heatlth-statistics-2016.pdf
12. Esmaeili M, Ajami M, Barati M, Javanmardi F, Houshiarrad A, Khaneghah AM (2022) The significance and potential of functional food ingredients for control appetite and food intake. Food Sci Nutr 10(5):1602–1612
13. Serafini M, Peluso I (2016) Functional Foods for health: the interrelated antioxidant and anti-inflammatory role of fruits, vegetables, herbs, spices and cocoa in humans. Curr Pharm Des 22(44):6701–6715

14. Carslen MH, Halvorsen BL, Holte K, Bohn SK, Dragland S, Sampson L, Willey C, Senoo H, Umezono Y, Sanada C, Barikmo I, Berhe N, Willet WC, Phillips KM, Jacobs DR Jr, Blomhoff R (2010) The total antioxidant content of more than 3100 foods, beverages, spices, herbs and supplements used worldwide. Nutr J 9:3
15. Goldberg I (1994) Functional foods-designer foods, pharma foods, nutraceuticals. Springer Science + Business Media Dordrecht, Originally Published by Chapman and Han in 1994
16. Jones PJ (2002) Clinical nutrition: 7. Functional foods—more than just nutrition. Can Med Assoc J 166(12):1555–1563

115. [illegible]

Part II
Food Safety and Supply

Chapter 5
Microbiology of Food Spoilage

Madhumala Y. and Veena S. Soraganvi

1 Introduction

Food is one of the necessary elements for sustaining life, providing energy, as well as supporting growth and development. It also helps us maintain health and prevent several diseases. Eating healthy, nutritious, and balanced food is essential for overall well-being and optimal functioning of the human body.

Food is categorized as follows:

1. Stable or nonperishable foods: These have a long shelf life and do not get spoiled easily because of lesser water activity, e.g., dry grains, flour of grains, nuts, pulses, and sugar.
2. Semi-perishable foods: These can stay in a good stage for a week to a month or until they are properly handled. These remain unspoiled for a longer period lesser than stable foods, e.g., fruits and vegetables [1].
3. Perishable foods: This class consists of foods that have a very short shelf life, e.g., milk, meat and poultry, fish, some fruits and vegetables, and eggs.

Food spoilage can occur in various foods because of a combination of factors, including microbial activity, enzymatic reactions, oxidation, and physical changes. Understanding the signs and causes of food spoilage is essential for food safety and quality. Here are some common examples of food spoilage in various types of foods [2].

Madhumala Y. (✉)
Department of Biotechnology, Basaveshwar Engineering College,
Bagalkote, Karnataka, India

V. S. Soraganvi
Department of Civil Engineering, Basaveshwar Engineering College,
Bagalkote, Karnataka, India

J. S. Yaradoddi et al. (eds.), *Frontiers in Food Biotechnology*,
https://doi.org/10.1007/978-981-97-3261-6_5

2 Food Spoilage

Spoiled food is food that has been damaged or, rendering it unsuitable for human consumption. This damage can be caused by various factors, including insect damage, physical injuries such as bruising and freezing, enzyme activity, or microorganisms. Among these factors, microorganisms—specifically bacteria, yeast, and molds—are the primary causative agents of food spoilage. While viruses have the potential to damage plant and animal tissues, they, along with mycoplasma, are generally not considered significant contributors to food spoilage as recognized today [3].

Among yeasts, molds, and bacteria, the latter two are the most important etiologic agents of food spoilage. Vegetables, with their higher water content, provide an ideal environment for the growth of bacteria which cause spoilage. Additionally, their comparatively low carbohydrate and lipid contents indicate that more water content is in an available form for microbial growth. Most vegetables have a pH range that falls within the optimal growth range of many bacteria, making bacterial contamination a common cause of vegetable spoilage. This spoilage is often categorized by a pattern known as bacterial soft rot [4].

2.1 Microbial Spoilage

Bacteria: Bacterial growth can lead to spoilage in various foods, causing off-flavors, odors, and visible changes in texture. For example, sour milk is a result of bacterial spoilage.

Yeasts: Yeasts can ferment sugars in food, leading to the production of alcohol and carbon dioxide. This can cause bread to become stale and develop an undesirable taste.

Mold spoilage: Mold can grow on the surface of various foods, such as bread, cheese, and fruits, resulting in discoloration, off-flavors, and the production of mycotoxins, which can be harmful if consumed [2].

2.2 Enzymatic Spoilage

Enzymes naturally present in food can break down proteins, fats, and carbohydrates, leading to undesirable changes in texture, flavor, and color. For example, fruits such as apples and avocados turn brown because of enzymatic spoilage [1].

2.3 Chemical Spoilage

Oxidation: Exposure to air can lead to the oxidation of fats and oils, resulting in rancid flavors and odors. This is common in products such as nuts and cooking oils.

Hydrolysis: Hydrolytic reactions can cause fats and oils to break down into free fatty acids, leading to off-flavors in products such as mayonnaise and salad dressings [1].

2.4 Physical Spoilage

Freezing and thawing: Repeated freezing and thawing can cause the formation of ice crystals in food, leading to changes in texture and the loss of moisture. This is common in some frozen fruits and vegetables [1].

Dehydration: Moisture loss can result in the hardening or toughening of food products such as dried fruits.

2.5 Contamination

Cross-contamination: Transfer of harmful microorganisms from one food to another can lead to spoilage and foodborne illnesses. It can occur through improper handling and storage of foods.

2.6 Inadequate Packaging

Improper packaging can allow air, moisture, or light to enter, leading to spoilage. Proper packaging, such as vacuum-sealing or airtight containers, can help extend the shelf life of many foods.

2.7 Temperature and Humidity

Incorrect storage conditions, such as storing perishable foods at the wrong temperature or humidity level, can accelerate spoilage. Refrigeration and freezing can help slow down microbial growth and enzymatic reactions [2].

3 Different Types of Spoilage by Various Agents

3.1 Bacterial Agents

Bacterial soft rot caused by *Erwinia carotovora* and *Pseudomonads* such as *Pseudomonas marginalis*. These microbes break pectins, leading to a soft, mushy consistency, unpleasant odor and a water-soaked appearance in various vegetables. Vegetables susceptible to this disease include onions, turnips, garlic, beans (green, lima, and wax), parsnips, cucumbers, beets, lettuce, carrots, rhubarb, spinach, asparagus, potatoes, cabbage, celery, brussels sprouts, cauliflower, broccoli, parsley, watermelons, tomatoes, cantaloupes, peppers, and radishes.

The cementing material in the vegetable body initiates the pectinase production, which hydrolyzes pectin, resulting in the soggy consistency. Due to the rapid growth of these bacteria, molds are often displaced and have less impact on the vegetable spoilage by bacterial agents. Simple nitrogenous compounds, vitamins, and minerals in vegetables are essentially destroyed during this process [2].

Malodors that uphold the growth of contaminating organisms are likely because of unstable compounds such as NH_3 and volatile acids produced by different flora. Complex carbohydrates such as cellulose get degraded at last. *Escherichia coli* strains containing cloned plasmids have been observed to cause the softening of potato tuber slices because of enzymes such as endo-pectate lyases, endo-polygalacturonase, or exo-pectate lyase. These enzymes, along with phosphatidase C and phospholipase A, contribute to soft rot. Carrots contaminated with *Agrobacterium tumefaciens* suffer faster senescence because of increased ethylene synthesis [4].

The *Erwinia* spp., belonging to the *Enterobacteriaceae* family, consists of gram-negative rods related to genera such as *Proteus, Serratia, Escherichia,* and *Salmonella. Erwinia* spp. do not typically need nitrogenous compounds for their growth. The comparatively low levels of proteins in vegetables destroy the plant materials. These grow at around 37 °C; however, some strains can grow even at low temperatures, with a few found to grow at 1 °C. *E. carotovora, pv. atroseptica*, and *E. chrysanthemum* cause potatoes to rot and this symptom is known as "black leg." [2].

3.2 Fungal Agents

Fungal agents are another group of culprits responsible for food spoilage. These fungi can invade various fruits and vegetables, causing different types of rot and decay. Let's explore some of these fungal agents and the produce they affect.

Gray Mold Rot: It is primarily caused by *B. cinerea*, which results in the growth of gray mycelia. This spoilage is ideal at high humidity and warm temperatures. The vegetables susceptible to this condition include onions, beans, carrots, peppers, broccoli, celery, tomatoes, lettuce, rhubarb, Brussels, cabbage, garlic, and potatoes.

Sour and Watery Soft Rot: This circumstance affects various vegetables and is produced by *G. candidum.* The affected vegetables include asparagus lettuce, onions, garlic, carrots, cabbage, sprouts, cauliflower, broccoli, turnips, and tomatoes. Causative fungi are largely found in soils and on decomposing fruits and vegetables. *Drosophila melanogaster* can carry mycelia strands and spores on its body from decomposing produce to cause cracks and damage healthy fruits and vegetables, facilitating the spread of the rot.

Rhizopus Soft Rot: This spoilage is produced by *R. stolonifer* and their related species, resulting in the softening and mushiness of vegetables. Cottony growth of the mold with small black dots of sporangia often covers the affected produce.

Anthracnose: It is a plant disease found as spotting on leaf, fruit, and pods of seed. It is caused by *Colletotrichum coccodes* and other species. While the fungi are normally considered weak pathogens, they can persist on plant debris in the soil and seeds of several plants such as tomatoes. The contamination is favored by warm, moist weather, and vegetables such as beans, watermelons, pumpkins, tomatoes, cucumbers, and peppers are susceptible to this condition [2].

Understanding the impact of these fungal agents and the conditions that promote their growth is crucial for preserving the quality and well-being of fruits and vegetables, particularly after harvesting. Proper storage and handling practices can help mitigate the risk of spoilage [5].

4 Spoilage of Various Specific Foods

4.1 Spoilage of Fruits and Vegetables

Fruit spoilage presents its unique characteristics when compared to vegetables. Fruits typically contain slightly less water but more carbohydrates. On average, fruits have lower protein, fat, and ash content, with protein, fat, and ash levels at around 0.9%, 0.5%, and 0.5%, respectively, which is somewhat lower than that found in vegetables. Like vegetables, fruits also contain essential vitamins and various organic compounds. Depending on their nutrient content, fruits appear to provide an environment conducive to support the bacterial and fungal growth. The pH of fruits tends to be lesser than the level that normally favors the bacterial growth (Fig. 5.1) indicates the different types of microbial spoilage in fruits and vegetables. As a result, fruit spoilage is often driven by various yeast genera commonly found on fruits, and these microorganisms are frequently responsible for causing the deterioration of fruit products [2, 4].

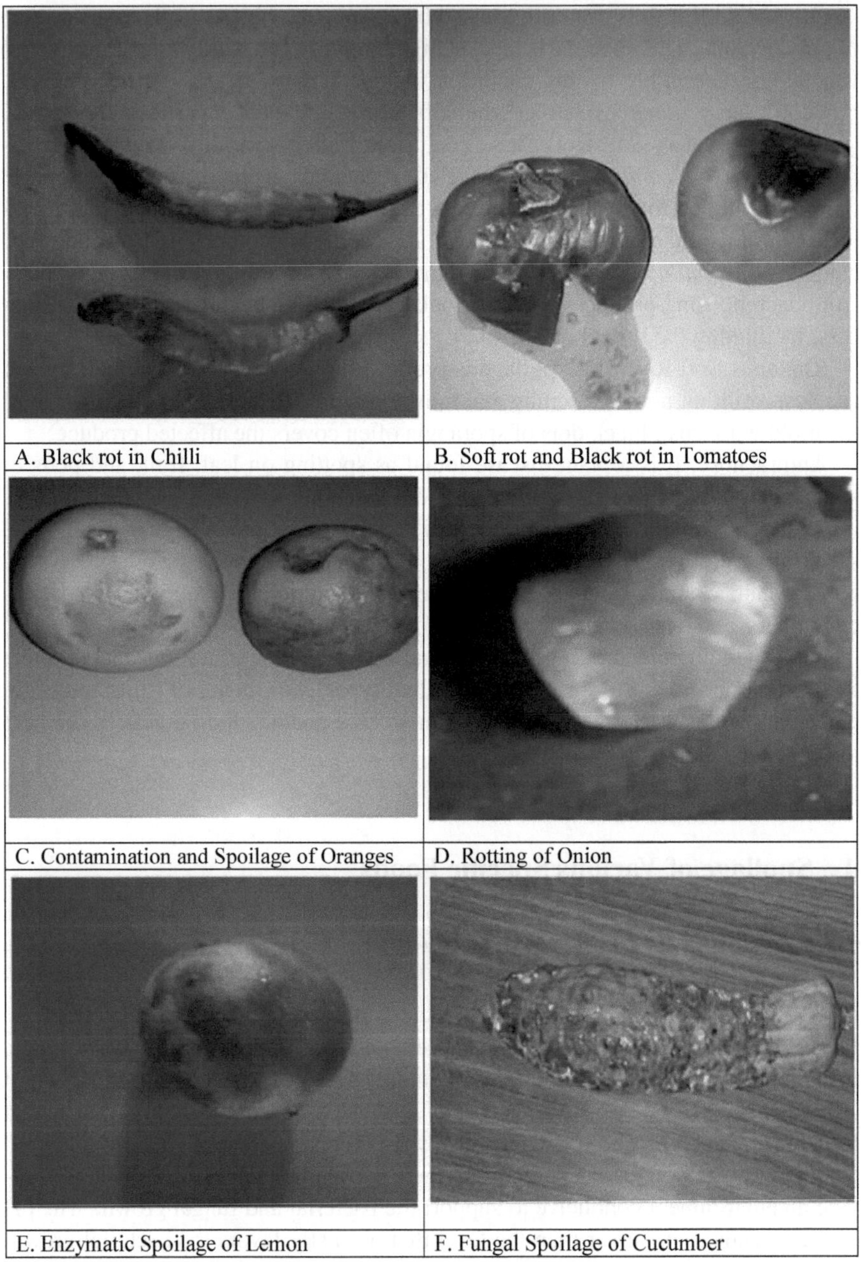

Fig. 5.1 Different types of microbial spoilage in fruits and vegetables

4.2 Spoilage of Eggs

Hen's egg is protected by its inherent parameters. Egg has mainly three structures that act as a barrier to the entrance of microbes—the outer waxy shell layer, inner shell, and lysozyme present in egg white. Lysozyme is efficient to inhibit gram-positive bacteria, egg white has cobalamine that protects egg from spoilage. It has a high pH but the yolk has pH 6.8 and is highly nutritive which forms an excellent supply of nutrition for microbes. Fresh eggs are sterile but after laying the surrounding environmental factors and unhygienic conditions lead to spoilage. Among bacteria, *Proteus, Alcaligenes, Escherichia, Acinetobacter Micrococcus, Salmonella, Aeromonas, Serratia, Enterobacter, Flavobacterium, Staphylococcus,* etc. cause rotting resulting in green rots, black rots, pink rots, red rots, and custard rots based on the type and colored pigmentation produced by bacteria; the condition of spoilage is also known as mustiness. Among molds and yeast, *Mucor, Rhizopus, Penicillium, Cladosporium,* and *Torula* are common spoilage referred as pinspots which cause rotting of egg [6].

The entrance of microbes within the entire egg is favored by high humidity and insanitary environment; most of the organisms are present in the yolk rather than the egg white. Once inside, they grow abundantly producing H_2S and foul-smelling chemical compounds by breaking protein compounds; molds are found as mycelia growth on the outer shell of eggs. The physical changes are moisture loss as well as loss of weight during long-term storage; it includes thinning of egg white, which breaks up the yolk membrane, and thus the yolk gets flattened and is homogenously mixed in the egg white [3].

4.3 Spoilage of Cereals and Dough Products

Cereal grains such as wheat, corn, jowar, oat, and paddy are vital nutrients and energy sources for humans. These are the most regularly eaten food group all over the world and are grown on about 70% of the land. They are consumed in different forms in the food industry. Some of them are cooked and eaten directly (rice and oats). Flours made by grinding cereals and the products are obtained by processing them, e.g., biscuits and cookies. Cereals generally contain 60–74% carbohydrates, 9–16% protein, fat, fiber, vitamins, and minerals with a pH of around 6.5, and is vulnerable to microbial growth leading to spoilage. Microbial growth is on the whole prevented due to very low water activity [2, 5].

Mold is found as the primary organism causing cereal spoilage, along with bacteria and a few yeasts. There are some toxic products produced in cereals by some fungi known as mycotoxins found in cereal crops. There is a significant incidence of mycotoxin infections in cereals found worldwide. Molds which normally produce mycotoxins are *Aspergillus, Penicillium,* and *Fusarium* that cause spoilage of flour; their products are *Rhizopus* (known as bread mold), *Alternaria, Rhizopus,*

Mucor, Cladosporium, Fusarium, Helminthosporium, and *Cladosporium.* The bacteria mostly associated with cereals are *Bacillus, Micrococcus, Lactobacillus,* and *Pseudomonads.* Yeast found in cereals are *Candida, Cryptococcus, Pichia, Sporobolomyces, Rhodotorula,* and *Trichosporon* [7].

Bread and bakery products are normally made using wheat and rye flours by fermentation. Cereals are also necessary to manufacture dough, batter, pasta, noodles, pastries, cake, etc. These cereal products are found to cause physical, chemical, and microbiological spoilage that varies the taste, aroma, look, and overall feature of the end product [1, 7].

4.4 Spoilage of Milk and Dairy Products

Milk is a wholesome food and is considered highly nutritious as it is rich in proteins, fats, carbohydrates, amino acids, minerals, and vitamins. Milk serves as an exceptional source of media for growth of all microorganisms (i.e., bacteria, viruses, fungi, and protozoans). Milk and its products such as butter, yoghurt, curd, cream, and cheese are vulnerable to microbial invasion and spoilage. Contamination of milk happens at different modes. Fresh milk possesses moderately less number of bacteria though *Micrococcus* and *Streptococcus* normally appear in aseptically drawn fresh milk. In the regular milking practice, milk may undergo contamination from udder and adjoining areas of the cattle. Potential sources of contamination are soil, water, cow dung, hands of milkman, and milk transporting utensils [8].

Spoilage of Milk

Souring and curdling of milk to form curd is one of the characteristic spoilage. Bacteria such as *Lactobacillus* produce lactic acid while other *coliforms* produce ethanol, acetic acid, butyric acid, formic acid, CO_2, and H_2. The production of gas in milk shows its evidence by producing foam or gas bubble trapped in curd. More amount of gas is produced by *Clostridium perfringens,* which leads to cracking of curd and stormy fermentation of milk. Proteolytic bacteria cause proteolysis leading to curdling of milk [2].

Ropy structure of milk is found by bacterial and non-bacterial growth. Ropiness is due to a slimy layer of bacteria that develops during storage at lower temperature; another type of ropiness is due to mastitis, presence of thick cream. Several bacteria, yeast, and mold hydrolyze fat present in milk and leads to rancidity. Milk is bothered with the change in flavor such as acid flavor and aromatic flavor due to the production of acetic, formic, and butyric acid. It is not desirable. Aromatic flavor is produced by *Streptococcus lactis* and *Leuconostoc.* Caramel or burnt flavor, bitter flavor, earthy flavor, fruity flavor, soapiness, etc. are caused by other microbes. Growth of pigmented bacteria produces undesirable color. Blue milk is caused by *Pseudomonas syncyaneum,* yellow milk by *Pseudomonas synxantha* and *flavobacterium,* red milk by *Serratia marcescens* and *Micrococcus roseus,* and brown milk by *Pseudomonas putrefaciens* and by enzymatic oxidation of tyrosin by *Pseudomonas fluorescence* [1, 2].

Dairy products such as milk, cheese, cream, and butter are all vulnerable for microbes. So, care should be taken while milking and cleaning. Raw milk kept at lower temperatures always indicates the presence of various microorganisms such as *Enterococcus, Lactococcus, Microbacterium, Streptococcus, Proteus, Leuconostoc, Pseudomonas, Micrococcus, Propionibacterium, Bacillus,* and *coliforms Lactobacillus*. Those that are not able to multiply at the normal low temperature are found in lower numbers. The pasteurization process reduces the microbial number, leaving the thermoduric strains. Pasteurized milk is often spoiled because of the increase in thermoduric *Streptococcus* which converts lactose into lactic acid that reduces the pH to less than 5 or even lesser that results in curdling.

If fungal spores are present, they start multiplying luxuriantly over the surface of the sour milk and elevate the pH to around 7, and allow the growth of *Pseudomonas*, which are proteolytic bacteria, and progress the coagulation of the milk to curd. A similar process occurs in raw milk, especially when held at refrigerator temperatures. Another condition found in fresh milk is ropiness, due to alkaliphiles, which gives a stringy consistency to the product. Butter has 16% water, 80% fat, about 0.6% carbohydrate, and 0.5% protein. Though butter is not an extremely unpreserved product, spoilage occurs generally because of microbes. The microflora of entire milk may be found in cream [8]. Non-microbial defect of butter leads to pink coloration caused by sulfur-dioxide refrigerant; darkening of the surface is due to water evaporation. Discoloration found on the food varies by the variety of organism. Like *Stemphylium* produces spots black in color, green spot is due to *Penicillium*, *Alternaria* create brown spots and *Pseudomonas nigrificans* produce brownish spots. Cream and butter can take in the moisture in the environment and create odors like fishiness because of *Aeromonas* spp. Ester flavor is produced by *Pseudomonas fragi.* Rancid odor is due to lipolytic organism.

Spoilage of Cheese

Spoilage of cheese is due to physical injury or invasion of microbes. Spoilage through microbes is found in the following three stages:

Spoilage During Production

The production of cheese requires lactic starters that are added to undergo lactic acid production. If they are not efficient or if contamination loads are high, many contaminators bring about disagreeable changes in cheese. For example, if the starter culture is not efficient in fermentation, *Bacillus*, *Clostridium,* and *Leuconostoc* may multiply and create textural change like holes in cheese. Acid proteolytes produce bitter flavor. Many other species produce gas, sliminess, proteolysis, and off-flavor which spoils the texture of cheese.

Spoilage While Ripening

In the period of ripening, spoilage occurs because of the enzymes released by autolyzed bacteria or by microbial development while ripening. Microbial cells produce fruity, yeasty flavor, sweet, and bitter, causing discoloration of brown, grayish brown, pink, yellow colored complex on the surface of the cheese.

Completed Cheese

Soft cheese is more readily contaminated than hard cheese such as cheddar and Swiss cheese which are stable. Most common spoilage organisms of finished cheese are molds which grow on the surface or in holes or cracks causing discoloration and off-flavor. *Oospora lactis*, a dairy mold, grows on the surface of soft cheese. *Monilia nigra* causes black discs on hard cheese.

4.5 Spoilage of Beverages

Beverages

Beverages are categorized into two groups: alcoholic and non-alcoholic beverages. These beverages have water, acidification agents, nutritive sweeteners, coloring agents, as well as emulsifying and flavoring agents. They influence the development of microbes in beverages. It is significant to maintain the necessary qualities for consumer satisfaction [2, 9].

Classification of Beverages

Non-alcoholic beverages consist of water, sweeteners, carbolic acids, flavoring and coloring agents, and preservatives (Fig. 5.2).

Fermented beverages such as beers, ales, and table wines also cause spoilage. Beers and ales spoilage is generally known as infections caused by yeasts and bacteria. The contamination of beers and ales is categorized into four groups: sourness, ropiness, sickness, and turbidity. During ropiness, the product turns thick and forms an "oily" stream. *Acetobacter*, *Lactobacillus*, *Pediococcus cerevisiae*, *Gluconooxydans*, and *Sarcinae* produce spoilage. Sickness is the production of honey-like odor. This is a characteristic feature of diacetyl produced by the microbe which destroys the usual odor of beer. Sourness is produced because of *Acetobacter*

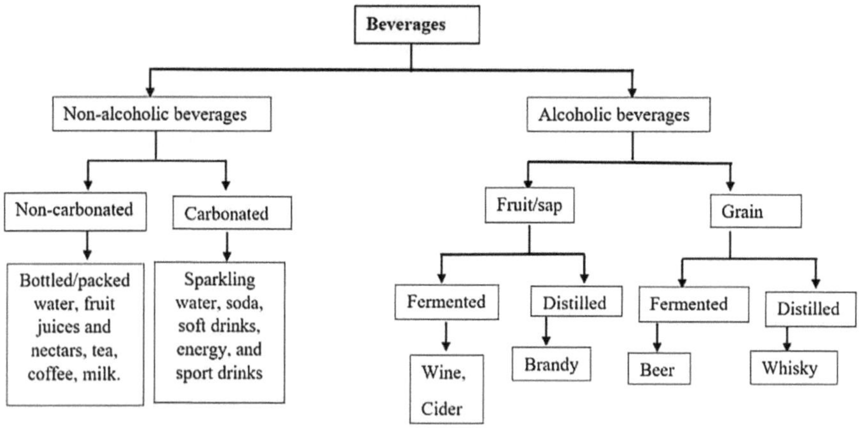

Fig. 5.2 Classification of beverages [10]

spp. Acetic acid-producing bacteria can convert ethanol to acetic acid which results in sourness and it increases the levels of acetic acid in beverages and spoils its taste. Turbidity and off-odors in beers are caused by *Zymomonas* and *Saccharomyces* spp. Growth of microbes is possible in beers due to a normal pH range of 4–5 and a good content of nutrients [1, 2, 9].

4.6 Spoilage of Fish

Fish is an aquatic seafood which is consumed highly and is found to be one of the fast perishable products. It is consumed because of its high nutritive value and the presence of omega-3 fatty acids. The biochemical character of fish makes it to start spoiling process by 12 hours.

The declining of the quality of fish occurs very fast because of the enzymatic and metabolic responses of microbes. The kind and the time of spoilage differ based on factors such as type of the fish, condition of the fish when caught, type and amount of contamination of fish by microbes, and temperature [11].

Spoilage of fish foods refers to the deterioration of fish-based products, which can make them unsafe or unpleasant to eat. Fish, being a highly perishable food, is prone to spoilage because of various factors. Here are some common reasons for the spoilage of fish food. Bacteria, molds, and yeast are microorganisms that can thrive on fish and cause spoilage. These microorganisms break down the fish proteins and lipids, producing off-flavors, odors, and visible signs of spoilage. Fish contains enzymes that, when not properly controlled, can lead to the breakdown of proteins and fats, resulting in off-flavors and odors. Enzymatic spoilage is more common in fresh or minimally processed fish. As time passes, the quality of fish degrades. This is a natural process influenced by factors such as storage conditions, handling, and transportation. Freshness decreases over time, leading to changes in texture, flavor, and aroma. Cross-contamination with other contaminated foods or surfaces can introduce harmful microorganisms to fish, leading to spoilage. Proper hygiene practices during handling and processing are crucial to preventing contamination. Improper processing, such as incomplete cooking or curing, can leave fish susceptible to spoilage. Inadequate preservation methods, such as insufficient salting or smoking, may not effectively prevent spoilage [2, 12].

4.7 *Meat and Their Products*

Meat products are high nutritious food which is usually eaten as food by people. Meat is obtained from a variety of turkeys, chickens, ducks, and animal meat like pork, mutton, and beef. It is processed and consumed in many ways. It is a proteinaceous food and is extremely perishable with a low shelf life. The biochemical characteristics of meat cause a depletion in the quality during slaughtering and till the time it is consumed. Meat products such as ham, sausages, minced meat, cooked meat, dry meat, vacuum-packed meat, smoked meat, and salted meat are all readily spoiled.

Meat spoilage is due to the oxidation of lipids or autolytic enzymes found in the muscle soon after slaughtering [3, 13]. The various reasons for contamination of meat are as follows:

1. Microbial flora of living being
2. Knives, grinders, utensils, hands, clothing, and handlers
3. Pre-slaughter and post-slaughter management of animal meat
4. Handling while slaughtering, evisceration, cooking, processing, and delivery
5. Packaging material used, and means of handling and storage followed

Raw Meat

Raw meat is readily spoiled because of the enzymes and activity of microbes. The autolysis leads to proteolysis and hydrolysis of fats on muscle and connective tissue. The existence and the microbial multiplication is predisposed by the meat composition. Raw meat possesses nutrients such as vitamins, sugars, amino acids, and cofactors, as well as has a pH of 5.4–6.0 and Aw of 0.84 which helps the microbial growth.

Generally, bacteria obtained from raw meat include *Acinetobacter, Micrococci, Flavobacterium, Thermosphacta, Psychrobacter, Moraxella lactic acid bacteria (LAB), Pseudomonas, Brochothrix Staphylococci,* and many other genera of the *Enterobacteriaceae* family. The pathogens isolated in raw and fresh meat are *Listeria monocytogenes Salmonella, E. coli, and Campylobacter.*

The types of defects produced by microorganisms on raw meat by aerobes are Surface slime, Oxidative rancidity, Red, blue, green and brown, black spot

Stickiness, whiskers, Green patches, putrefaction and souring by anaerobes. Sulfide odor, H_2S greening, and blown pack are found in vacuum- packed meat. Putrefaction and Souring is found in Fresh meat, Moldy odor, Greening, Souring Slimy in cured meat. Souring, off-odor in MAP meat, putrefaction, off-flavors, and sliminess are, is found in refrigerated packaged meat. All these lead to textural and change in appearance and quality of the meat [2, 3].

5 Conclusions

Microorganisms influence a wide variety of foods, and spoilage is found everywhere across the world. The intrinsic and extrinsic factors of a particular food product establish the spoilage and the variety of microbes concerned to spoilage. Many different types of organisms spoil and change the physical and chemical characteristics of foods; they also produce toxins such as enterotoxins or mycotoxins in the food product.

The people who handle the food and different equipment used may cause outbreaks of various foodborne diseases and also lead to huge economic loss. To minimize the risk of spoilage, it is essential to follow proper handling, processing, and storage practices. This includes maintaining proper hygiene, controlling temperatures, using appropriate packaging methods, and following the recommended preservation techniques. Additionally, consumers should be aware of storage guidelines and expiry dates for safety and quality maintenance of fish products. Creating awareness among researchers, common men, and industrial producers is the need of the hour. It is also necessary to follow the food-related policies and implement them all over the world.

Effective preservation methods—including refrigeration, freezing, and canning—are essential for mitigating spoilage and extending the shelf life of perishable goods. The economic impact of food spoilage is substantial, with considerable losses incurred throughout the supply chain—from production to distribution and retail. Addressing spoilage-related challenges requires a comprehensive approach, integrating technological advancements, regulatory measures, and consumer education to minimize waste and enhance food security.

References

1. Frazier WC (1995) Food microbiology, 4th edn
2. Jay JM (2000) Modern food microbiology, 6th edn
3. Das AK, Nanda PK, Das A, Biswas S (2019) Hazards and safety issues of meat and meat products. In: Food safety and human health. https://doi.org/10.1016/B978-0-12-816333-7.00006-0
4. Oluawadara A, Olamide A, Mariyana S, Deyan S (2021) Microbial spoilage of fruits vegetables and cereals. Appl Food Res 2:100122. https://doi.org/10.1066/j.afres.2022.100122

5. Gera Y, Odudu JH, Atii RD (2022) Microbial spoilage of fruit and vegetables: implications for human health. J Fam Soc Res 1(2):139–143
6. Robert JR (2004) Factors affecting egg internal quality and egg shell quality in laying hens. J Poult Sci 41(3):161–177. https://doi.org/10.2141/jpsa.41.161
7. Khapre A, Deshpande H, Katke S (2020) A review on microbial contamination of cereal grains. Int J Chem Stud 8(3):1829–1832. https://doi.org/10.22271/chemi.2020.v8.i3y.9474
8. Dhakane RD, Shikshan J, Mandal P (2020) Spoilage and preservation of milk and milk products: a review
9. Kregiel D (2015) Health safety of soft drinks: contents, containers, and microorganisms. https://doi.org/10.1155/2015/128697
10. Esmaeili S, Mogharrabi M, Safi F, Sohrabvandi S, Mortazavian AM, Bagheripoor-Fallah N (2015) The common spoilage microorganisms of beer: occurrence, defects, and determination—a review. Carpathian J Food Sci Technol 7(4):68–73
11. Ikape SI (2018) Fish spoilage in the tropics: a review fish spoilage in the tropics: a review. Octa J Biosci 5(2):34–37. www.sciencebeingjournal.com
12. Zhuang S, Hong H, Zhang L, Luo Y (2021) Spoilage-related microbiota in fish and crustaceans during storage: research progress and future trends. Compr Rev Food Sci Food Saf 20:252–288. https://doi.org/10.1111/1541-4337.12659
13. Spoilage of meat and meat products (2016) Food microbiology: principles into practice, pp 279–295. https://doi.org/10.1002/9781119237860.ch16

Chapter 6
Food Processing and Preservation

S. B. Mudgulkar

1 Introduction

Foods are carbon-based materials that people eat to meet their nutritional needs. Foods come from plants or animals and contain lipids, carbohydrates, proteins and other organic materials together with moisture. Foods can deteriorate due to physical, chemical or microbiological processes. Foods' nutritional content, colour, texture and edibility are all subject to deterioration. Thus, in order to maintain their prominence over an extended period of time, the essential foods need to be kept. Food preservation refers to the method or methods used to address the internal and environmental variables that lead to food decomposition. Food preservation's main goal is to prolong its shelf life while maintaining its distinctive flavour, colour, texture and nutritional content.

As early as 1.5 million years ago, our ancestors began preserving food by simply heating items such as meats, seeds and vegetables. As early as 9600 BC, the most advanced prehistoric societies were found in Mesopotamia and ancient Egypt. The simple methods of food preservation that were used at the time included drying, smoking and salting. The history of 'food preservation' begins in prehistoric times, when a hunt for a large animal that was inedible at one time made the primitive tribe realise they needed to store food. Understanding food preservation methods was the primary and thus most important step towards the emergence of civilisation. Almost identical fundamental methods were employed by several ethics in various eras and places to preserve food products.

S. B. Mudgulkar (✉)
Department of PG Studies and Research in Biotechnology, Gulbarga University, Kalaburagi, Karnataka, India

J. S. Yaradoddi et al. (eds.), *Frontiers in Food Biotechnology*,
https://doi.org/10.1007/978-981-97-3261-6_6

Across the world, conventional methods of food preservation such as freezing, chilling, pasteurisation and chemical preservations were widely used. Irradiation, high-pressure technology, hurdle technology and other emerging technologies are examples of how science is advancing and evolving to support the development of existing ones. Processing food preservation now involves many different disciplines because it involves cultivating, harvesting, processing, packaging and distributing the food. In order to maintain the food products during the food manufacturing and processing phases, an integrated strategy would be helpful.

Most popular techniques for food preserving (Table 6.1) are as follows:

1. Chilling

   By using this method, the food is kept at −18 °C by freezing it below 0 °C. By using this technique, the pace at which germs contaminate food is slowed down while the meal's water solidifies into ice.
2. Ultrafreezing, or deep freezing

   With this type of freezing, food is exposed to extremely low temperatures—below −40 °C—for a brief duration—less than 2 h. Because they don't significantly change the food, these two methods are the greatest for preservation.
3. Air conditioning

   Using this technique, food can be kept at 4 °C, or cold, in refrigerators. In comparison to the earlier options, this one likewise slows down the effects of germs, but it does it more quickly.

Table 6.1 Food preservation methods

Sl. no.	Inhibition	Inactivation	Avoid recontamination
1.	Low-temperature storage	Sterilisation	Packaging
2.	Reduction in water activity	Pasteurisation	Hygienic processing
3.	Decrease of oxygen	Irradiation	Hygienic storage
4.	Increase of carbon dioxide	Pressure treatment	Aseptic processing
5.	Acidification	Blanching	Risk analysis and management
6.	Fermentation	Cooking	Hazard Analysis and Critical Control Point (HACCP)
7.	Adding preservatives	Frying	Good manufacturing practices (GMP)
8.	Control of pH	Extrusion	
9.	Freezing	Light	
10.	Drying	Sound	
11.	Concentration	Magnetic field	
12.	Surface coating	Electrifying	

4. Boiling
   This preservation method is used as a complementary technique to freezing vegetables. It involves boiling the food to eliminate any residue or bacteria that can decompose it.
5. Fermentation
   This method is used in foods such as cheeses and beer. In contrast to what is sought in food preservation, this technique seeks to create microorganisms in food. The difference is that the organisms that are created favour their gastronomic properties and prevent the development of harmful pathogens.
6. Desiccation
   It consists of letting the food dry in the open air and thus eliminating moisture and water needed to develop harmful microorganisms.
7. Dehydration
   Similar to the previous method, but with the variant that machines that completely remove water from the food are often used.
8. Vacuum packed
   This technique seeks to eliminate the air surrounding the food and that bacteria need to contaminate them.
9. Pickling or salted
   These methods entail seasoning meals with salt. Salt draws moisture out of these and stops microbes from growing in the food. In addition to salt, vinegar is added to the pickled food, which balances the salt in the food canning process.
10. Acidification
    It involves lowering the food's pH and halting the growth of bacteria by adding acidic ingredients.
11. Adding sugar
    This technique involves incorporating sugar into food to shield it from the growth of microbes.
12. Additives in food
    It involves incorporating chemicals into the meal to stop the growth of dangerous germs.

Food processing engineering contributes over 10% of India's GDP and employs roughly 15 million people, suggesting that it is a sizable sector of the country's economy. This industry is expanding, and by 2022, prices are predicted to reach up to Rs. 2,58,000 crores. Within the food processing business, there are many opportunities with strong investment returns. The global industry for processed foods is estimated to be worth $7 trillion, and it is steadily expanding. The main drivers of the development of the food processing industries in many nations are rapid industrialisation and globalisation. Based on a review of the UNIDO Industrial Statistics Database (2005), food processing in developing nations has the potential to be a profitable manufacturing sector unit.

India's food processing industry has many obstacles, including low-quality raw materials and infrastructure, a lack of trained labour, high energy costs, improper storage facilities and a lack of market knowledge. India's food processing industry plays a significant part in the country's economic development. As such, the industry must be integrated into the nation's inclusive growth policy by creating a conducive environment that may encourage investments in technology, infrastructure, research and innovation. In order to improve and contribute to India's economic process, the government should also provide incentives for the growth of the food processing sector and raise public knowledge of the state of processed foods.

2 Food Classification

Foods are frequently categorised broadly in accordance with time periods, purposes, nutritional content and processing methods. Foods are grouped according to the time they were produced, and food deterioration is a natural process that causes food to lose its nutritional value, colour, texture, flavour and edibility with time. Eating rotten food can make you sick or, in the worst case scenario, even kill you. They categorise food goods as perishable, semi-perishable or non-perishable based on how long they will last on the shelf.

[6]Perishable: Foods that go bad quickly—from a few days to around 3 weeks—are classified as perishable. Seafood, meats, poultry, eggs and dairy products are a few examples of perishable foods. Should the unique methods of preservation appear to be evaded, the food products spoiled quite sooner (Table 6.2).

Semi-perishable: Foods that, when stored properly, can last up to 6 months or longer are known as semi-perishable foods. Examples of semi-perishable foods include fruits, vegetables, cheeses and potatoes.

**Non-perishable food items are defined as natural and processed foods that can be stored for an extended period of time. You can keep these foods in storage for up to a few years. Non-perishable foods include things like spread, mayonnaise, dry beans, almonds, flour, sugar and canned fruits.

Table 6.2 Storage life of some fresh foods

Storage life of some fresh foods at normal atmospheric conditions			
Sl. no.	Food	Terminology	Storage life
1.	Meat, fish and milk	Perishable	1–2 days
2.	Fruits and vegetables	Semiperishable	1–2 weeks
3.	Root crops	Semiperishable	3–4 weeks

3 Food Deterioration Mechanisms and Types

Food deterioration is the process through which food becomes less edible and compromises food safety. The initial stages of food deterioration are identified by colour, smell, flavour and texture. Food spoiling is explained by a variety of physical, microbiological and chemical processes. Since one mechanism's spoiling can trigger another, these mechanisms don't seem to be equally constrained. The primary factors that contribute to food deterioration are temperature, pH, air, nutrition and different compounds.

1. Material degradation

   Biological spoiling, which is produced by physical variations or ambiguity, is what causes food to become bad. Examples of physical spoiling include moisture gain or loss, moisture movement between components and physical separation of ingredients or components.
2. Content of moisture

   A change in the water content of food products is often the cause of their deprivation. Food moisture transfer is closely related to the food's water activity (Aw), and it would take place within the category of water loss, water gain or water migration. It could be a thermodynamic property described as the ratio of the water pressure in an overflowing system to the pure water pressure level at the same temperature. In place of pure vapour pressure, an equilibrium ratio at the same temperature may also be used, and the water activity in food products reduces with temperature. Foods typically have a water activity of 1.0 at room temperature; however at −20 °C and −40 °C, the water activities are 0.82 and 0.68, respectively.
3. Thermostat

   One of the most important and significant factors in fruit and vegetable putrefaction is temperature; slow growth necessitates an ideal ratio with ideal airborne in fruit and vegetables, and there is an ideal temperature choice for slow ripening to maximise post-harvest life. We refer to these ideal circumstances as modified atmospheres (MA). The temperature typically interferes with the goods' metabolism and modifies the rate of spoiling. Additionally harmful to foods that are susceptible to freeze damage include vasoconstrictor. Lower temperatures cause the food products to partially freeze, which breaks the cells and deteriorates the merchandise. Chilling injury affects the majority of tropical fruits and vegetables, usually occurring before the food products.
4. Pasteurisation

   Pasteurisation is a physical preservation method that involves heating food to a specific temperature in order to kill enzymes and microorganisms that cause spoiling. The majority of harmful bacteria, yeasts and moulds are eliminated during this process, extending the shelf life of the food. The pasteurisation method was developed by Louis Pasteur, a French scientist biologist (1822–1895) who conducted experiments with it in 1862. He treated wine and beer with this method.

The efficiency of the pasteurisation process is dependent on the temperature and time combination; investigations on the thermal death time of heat-resistant microorganisms frequently provide evidence for this claim. Pasteurisation can be divided into three categories based on temperature and warmth exposure: vat (batch), hot temperature short time (HTST), and ultra-high temperature (UHT); HTST and UHT are ongoing processes.

Certain vitamins, minerals and good bacteria may be harmed by the high heat of the pasteurisation process. There is a 20% reduction in antioxidant, a 5% reduction in soluble calcium and phosphorus, and a 10% reduction in thiamine and B. Pasteurisation lowers the amount of vitamin C, carotene and vitamin C in fruit juices. However, in terms of nutrition, these losses will be regarded as minimal.

5. Sterilisation by heat

   An extended amount of time could be achieved via thermal sterilisation, which is a heat treatment procedure that eliminates all living microorganisms, including moulds, yeasts, vegetative bacteria and spore formers. Thermal sterilisation falls into two categories: retorting and aseptic processing.

6. Rebuffing

   Retorting is the process of sterilising food after it has been packaged in a container. Foods that have a pH higher than 4.5 need to be sterilised at a high temperature of 100 °C. This temperature can only be achieved in continuous or batch retorts. Batch retorts are gradually replaced by continuous systems; the most widely used continuous systems in the food industry are rotary cookers and hydrostatic retorts.

4 Eco-friendly and Biobased Packaging

Presently, there is no shortage of funding coming from the public and commercial sectors to support the development of sustainable and ecologically friendly product manufacturing. As a result, biopolymers and biofibres are typically used to create green composites. This has proven to be an intriguing field for scientists, researchers and academicians to spread the word about the significance of these superior green composites as workable alternatives to carbon fibre reinforced composites and non-sustainable glass composites for a variety of food packaging applications [1–5].

7. Aseptic packaging

   Using commercially sterilised food in an extremely sterile container and sealing it in an aseptic atmosphere is known as aseptic packaging. Paper and plastic are used in conventional aseptic packaging. The process of sterilisation will involve either chemical treatment, heat treatment or a combination of the two. It is common practise to preserve fruit slices, dairy goods, juices and seasonings using aseptic packaging. It can unnecessarily prolong the shelf life of food products. For instance, the UHT pasteurisation process can prolong the shelf life of liquid milk from 19 to 90 days, while the combination of UHT processing with aseptic packaging can prolong the shelf life to 6 months or more.

Packages used for aseptic processing are produced from plastics having relatively softening temperatures. Moreover, the aseptic filling can accept a large range of packaging materials, including (a) metal cans sterilised by superheated steam, (b) paper, foil and plastic laminates sterilised by hot oxide and (c) a range of plastic and metal containers sterilised by high-pressure steam. A wide variation in packages thus enhances the proficiency of aseptic packaging and reduces its value. The direct approach of aseptic packaging comprises of steam injunction and steam infusion.

8. Cryogenic

   By causing water below freezing temperature to condense into ice, freezing slows down physiochemical and biochemical reactions and prevents harmful and deteriorating microbes from growing in food. It reduces water activity and the amount of liquid water in the food products. Food items that are frozen undergo a complex process of heat transmission that involves both natural activity and thermal property alteration. Two fundamental sequential processes of freezing are known as nucleation and growth. Nucleation results in the production of frost snow, while growth denotes the subsequent rise in crystal size.

   Freezing time is defined as the time required to lower the initial temperature of a product to a given temperature at its thermal centre. In general, slow freezing of food tissues leads to the formation of larger ice crystals within the extracellular spaces, while rapid freezing produces small ice crystals distributed throughout the tissue.

   A shorter freezing time retards salt diffusion and stops food from breaking down when frozen. With the subsequent cost decrease, IQF (Individual quick freezing) also permits larger capacity for commercial freezing units. On the other hand, setting up a quick-freezing plant requires more funding. Food products are processed using a variety of quick-freezing methods, such as cryogenic freezing, air-blast freezing and contact plate freezing.

9. Chilling

   Foods are chilled to a temperature of between −1 and 8 °C during the procedure. It lowers the items' starting temperature and maintains their final temperature for an extended amount of time. It is used to extending the shelf life of both fresh and processed foods by slowing down the rate of biochemical and microbiological alterations. In real-world applications, freezing occurs most frequently at temperatures below 15 °C. In the modern food industry, partial freezing is used to extend the shelf life of food products. This technique, known as super chilling, lowers the amount of ice that forms on food.

   Because freezing storage is an efficient short-term preservation technique, it has both benefits and drawbacks. The process of chilling inhibits the growth of germs and stops the metabolism of intact plant tissues after harvest and animal tissues after slaughter. Seafood spoilage may be caused by chemical oxidation, autolysis or microbiological activity. More spoiling is caused by microbial activity than by other factors. The majority of spoilage bacteria are Gramme negative, and as a result of their metabolic processes, they give seafood tastes and odours. The conditions of handling, packing and storage influence the proliferation of microorganisms and, in turn, the shelf life of seafood [2, 5–7]

10. Irradiation

    A physical process known as 'irradiation' might include giving a material a strong, quick radiation dosage (IR). There will be two types of infrared radiation (IR): intentionally manufactured IR is made up of accelerated electrons and induced secondary radiation, while natural IR is typically composed of X-rays, gamma rays and high-energy ultraviolet (UV) radiation. IR is used on about 60 additional items in 40 different nations. Disinfestation of grains, fruits and vegetables is one of the outcomes of incubation (IR). Other outcomes include (b) extending the shelf life of fruits and vegetables by preventing sprouting or modifying the rate of their maturation and senescence. Lastly, (c) extending the shelf life of foods by eliminating spoilage organisms and enhancing food safety by eliminating foodborne pathogens.

    Even at high doses, the impacts of radiation have no influence on nutritional characteristics such as lipids, carbs, proteins, minerals and most vitamins. When IR is used in excess, it can lead to the depletion of some micronutrients, including vitamins A, B1, C and E. According to the FDA, IR has the same effects on food's nutritional value as conventional food processing methods.

11. Chemical processes

    One of the conventional techniques for preserving food is to use chemical reagents. The concentration and selectivity of the chemical reagents, organisms that cause food spoiling and the chemical and physical properties of the food products all affect how effective this procedure is. Food additives and preservatives are being consumed and used more widely worldwide. North America led the food preservative industry at this period (2012 data), with Asia-Pacific following closely behind. By the end of 2018, it is anticipated that the food preservative industry would have grown to a value of $2.7 billion.

    However, employing chemical reagents as food additives and preservatives may be a delicate matter due to health risks. Various laws, regulations and governmental bodies oversee and regulate the use of chemical preservatives and food additives in many different nations.

12. Preservatives using chemicals

    Preservatives are described as compounds that have the ability to prevent, delay, or halt the growth of microbes or other degradation. Food preservatives prolong the shelf life of some foods by delaying the deterioration caused by microorganisms, preserving the food's flavour, texture and colour.

    There are two types of food preservatives: natural and artificial. Various compounds found in plants, animals and microorganisms have the ability to preserve food. They serve as flavourings, antimicrobial agents and antioxidants as well. Industrial producers create artificial preservatives, which fall into the categories of antibacterial, antioxidant and antienzymatic.

13. Additives in food

    The major goals of utilising food additives are to increase and preserve nutritional content, uphold standards, reduce waste, uphold customer appropriateness, produce more food that is easily accessible and make food processing easier. Food additives are frequently artificial or natural chemicals that are purposefully added to food during processing, packaging or storage to alter specific food properties. There are two main categories of food additives: purposeful and accidental.

    The two, purposeful additions are tightly regulated by the government; according to the National Academy of Sciences (1973), additives cannot be used to conceal a flawed method, to cover up spoilage, damage or any other inferiority, or even to appear as though they are misleading the customer. Furthermore, even if additives can significantly reduce nutrition, their methods are unrelated.

5 Conclusion

To highlight the methods of preservation and processing, removal of the microbes in food can be done by physical means; such as removing air, water or by lowering and increasing temperatures [8]. It is very true in case of preservation of green leafy vegetables, and it is utmost important to remove moisture from the vegetables since microbes can't survive in such conditions [9]. Similarly in preservation of fish canning, freezing, pickling and smoking can be used, where as in case of milk preservation vast techniques including freezing, refrigiration, UHT treatment, drying, evaporation and fermentation [10–16]. Learning how to preserve food was one of the most important and revolutionary discoveries of human civilisation, as it allowed people to settle down and form societies. Nonetheless, extending the shelf life of food products without sacrificing their original nutritional value remains crucial and difficult. Food is an organic, perishable material that can deteriorate due to physical, chemical or microbiological processes. Various conventional methods, such as fermenting, freezing, chilling and drying, have developed to preserve food while maintaining its nutritional content and texture. Since, higher rate of adaptibility of the bacterial populations in various adverse conditions [17], Modernised and improvement in preservation procedures come with time and with these increasing demands.

The newest developments that are used to preserve food for longer periods of time are radiation, high-pressure food preservation and pulsed field of force effect. Additionally, various chemical reagents have been introduced as preservatives and additives for food. However, there are growing worries regarding the use of chemical additives and preservatives in food items due to potential health risks. The industry for food processing and preservation has been growing quickly to keep up with consumer demand. To ensure food safety and shelf life, it's critical to understand food deterioration mechanisms and preservation methods.

References

1. Shellikeri A, Kaulgud V, Yaradoddi J, Ganachari S, Banapurmath N, Shettar A (2018) Development of neem based bioplastic for food packaging application. IOP Conf Ser: Mater Sci Eng 376:012052. https://doi.org/10.1088/1757899X/376/1/012052
2. Yaradoddi JS, Banapurmath NR, Ganachari SV et al (2020a) Biodegradable carboxymethyl cellulose-based material for sustainable packaging application. Sci Rep 10:21960. https://doi.org/10.1038/s41598-020-78912-z
3. Yaradoddi JS, Banapurmath NR, Ganachari SV et al (2022) Bio-based material from fruit waste of orange peel for industrial applications. J Mater Res Technol 17:3186–3197. https://doi.org/10.1016/j.jmrt.2021.09.016
4. Rangappa SM, Siengchin S, Dhakal HN (2020) Green-composites: ecofriendly and sustainability. Appl Sci Eng Prog 13(3):183–184
5. Yaradoddi JS, Sulochana MB, Merja KH, Parameshwar AB, Dayanand A (2020b) The occurrence of potential and novel isolates of *Oceanobacillus* sp. JAS12 and *Salinicoccus* sp. JS20 recovered from west coast of Arabian Sea, India. Res J Biotech 15(9):133–140
6. Jayachandra SY, Parameshwar AB, Mohan Reddy K, Sulochana MB (2012) Characterization of extracellular hydrolytic enzymes producing extremely halophilic bacterium *Virgibacillus* sp. JS5. World J Sci Technol 2(2):23–26
7. Odeyemi OA, Burke CM, Bolch CCJ, Stanley R (2018) Seafood spoilage microbiota and associated volatile organic compounds at different storage temperatures and packaging conditions. Int J Food Microbiol 280:87–99. https://doi.org/10.1016/j.ijfoodmicro.2017.12.029. Epub. PMID: 29478710.
8. Karel M, Lund D (2003) Physical principles of food preservation, 2nd edn.
9. Handbook of food science, technology, and engineering, 1st edn. 2005 CRC Press
10. Sukumar D (2007) Outlines of dairy technology. Oxford University Press, Oxford
11. Hall GM (1992) Fish processing technology. VCH Publishers, New York
12. Sen DP (2005) Advances in fish processing technology. Allied Publishers
13. Shahidi F, Botta JR (1994) Seafoods: chemistry, processing, technology and quality. Blackie Academic & Professional, London
14. https://intersam.es/en/main-food-preservation-methods-today
15. Food processing and preservation. Scientific Publishers, 2021
16. Shafiur Rahman M (2007) Handbook of food preservation, 2nd edn. CRC Press, Boca Raton
17. Yaradoddi JS, Sulochana MB (2020) Screening and characterization of bioactive compounds produced by the moderate halophile Halobacillus sp. JS6. Res J Biotech 15(12):131–136

Chapter 7
Grain Processing and Baking Technology

Harshitha T., Parinitha A., Pratiksha Prabhakar Gawali, Somya Adusumilli, and Sudheer Kumar Yannam

1 Introduction

Grains are an essential food source for impoverished agricultural communities, as they provide vital energy. Food grains are categorized into cereals, oilseeds, and pulses, as shown in Fig. 7.1. Cereals are staple crops mostly grown from grasses. Pulses are a type of legume seed cultivated for their dry seeds and source of proteins. Oilseeds such as olives, sunflowers, peanuts, mustard, and soybeans are cultivated for their oil content. Botanically, grains are small, dry fruits harvested for consumption by humans or animals. Some grains, such as peas, field beans, and bengal grams, can be consumed when green, but most grains are harvested and dried before use.

The consequences of climate change on crops can have a severe impact, particularly on vulnerable populations. As ecological dynamics and equilibriums shift, crops become more susceptible to diseases, pests, and parasites. These obstacles exacerbate biotic and abiotic stresses, leading to crop failure and storage losses. Enhancing crop yields is a critical research-for-development challenge, primarily due to the limited genetic diversity and low resistance of most grains to various constraints. Maintaining critical moisture content, temperature, and humidity during storage and safer containers reduces storage losses. In this chapter, the authors explore how biotechnological interventions can offer sustainable solutions for improving yields, enhancing processing, and preserving grains to ensure food security.

Harshitha T. · Parinitha A. · P. P. Gawali · Somya A. · S. K. Yannam (✉)
Department of Traditional Foods and Applied Nutrition, CSIR-Central Food Technological Research Institute, Mysuru, India

Academy of Scientific and Innovative Research (AcSIR), Ghaziabad, India
e-mail: syannam@cftri.res.in

J. S. Yaradoddi et al. (eds.), *Frontiers in Food Biotechnology*,
https://doi.org/10.1007/978-981-97-3261-6_7

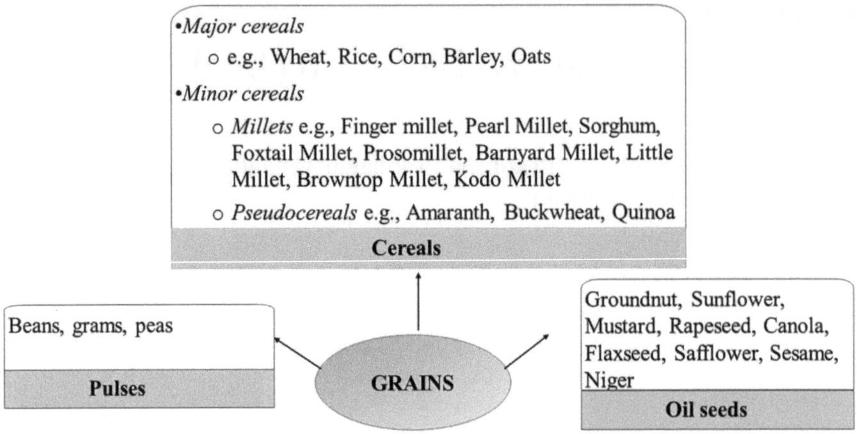

Fig. 7.1 Categories of food grains

2 Processing of Grains

Grain processing is essential for producing safe and quality food, ensuring availability, accessibility, utility, and stability. Grain processing removes inedible parts and preserves nutritional value, making it suitable for various food products. There are three levels of food processing, primary, secondary, and tertiary, based on the degree of transformation of raw ingredients. Primary processing includes cleaning, sorting, and milling and creates edible grains, while secondary processing creates intermediate products, such as flour, dhal, and oil. Tertiary processing produces quick-cooking, ready-to-cook, and ready-to-eat foods. Processing grains enhances their nutritional value and extends their shelf life. Proper storage practices are crucial to prevent spoilage of processed grains.

Grain processing creates diverse food products, enhancing food choices and dietary variety. Cereals undergo processing to create various products, such as flour, semolina, flakes, and various breakfast foods. They can also be used as ingredients in baked goods or extruded into snacks. Pulses, on the other hand, are commonly soaked and sprouted to increase their versatility in cooking. They can be milled into dhal and added to sambar, curry, and snacks. Seed oil can cook fried and sautéed foods, while residual meals are processed into protein isolates and concentrates. Processing offers convenient food options and ensures safe, nutritious, and palatable foods while contributing to economic growth. Enhancing food appreciation involves understanding the various components of its quality. The quality and safety of grains are essential in producing better products.

Quality refers to the degree of excellence of the food. Quality aspects encompass its nutritional value, aesthetic presentation, flavor, ease of cooking, and safety. Individuals make food choices and weigh all these factors to make informed decisions. Consequently, businesses and researchers must have a thorough

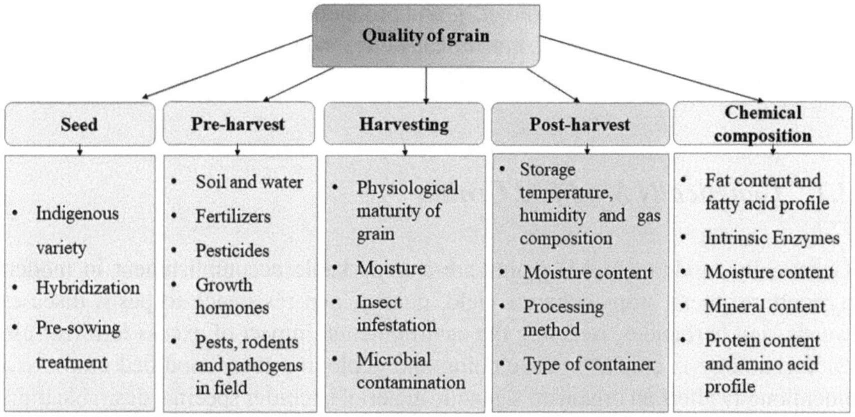

Fig. 7.2 Factors influencing the quality of grain

comprehension of food quality. By comprehensively understanding these factors, they can endeavor to produce superior products. Farmers, processors, and other stakeholders must consider several factors to produce high-quality grain, including seed quality, soil type, irrigation, fertilizer, harvesting time, processing methods, storage conditions, and preservation techniques. Figure 7.2 illustrates some of the factors affecting grain quality. Ensuring the optimal quality of grains depends on implementing sound agricultural practices, precise harvesting methods, effective processing techniques, and correct storage procedures.

Grains are essential to the human diet, providing critical nutrients necessary for overall health and wellness. Due to their significant role in global diets, it is vital to recognize their importance and strive to improve their accessibility. Adopting efficient and sustainable solutions is crucial to meet the increasing demand for grain production and processing. Using cutting-edge agricultural science and technology is vital to achieving this objective. Hybrid plants resistant to pests, diseases, and drought can help boost yields and reduce losses caused by pests, insects, and microbes. Furthermore, preserving crops by minimizing storage losses through thermal and nonthermal processing and using fumigants, pesticides, and rodenticides is essential. The authors suggest referring other books for more details on processing and preservation of grains.

3 Biotechnological Contributions in Grain Production

Biotechnology is a scientific technique to understand and modify an organism's genes to improve well-being. CRISPR/Cas9 gene editing has improved grain crops, such as rice, wheat, maize, and barley, by increasing yield and disease resistance [1]. These modifications make them ideal for future breeding programs and new varieties. Thanks to biotechnology, grain production has increased significantly through improvements in yield, nutritional value, and shelf life and the development of pest-, disease-, and drought-resistant varieties.

3.1 Genetically Modified Grains

Genetically modified (GM) grains are a remarkable accomplishment in modern agriculture: these crops enhance yield, quality, and resistance to pests, diseases, weeds, and herbicides, reducing the environmental impact of excess fertilizer use. Biotechnology is crucial in developing and deploying these modified crops, as it intentionally alters an organism's genetic material to confer specific, desirable traits. They have introduced genes that enable plants to thrive in water-deficient conditions, ensuring food security in drought-prone regions. They also work on GM grains resistant to common fungal pathogens and viruses, enhancing crop protection and reducing the risk of yield loss due to these diseases. Biotechnology enables the rapid detection of disease-causing pathogens in grains, allowing for early intervention and containment and preventing diseases that can devastate grain crops.

GM grains, facilitated by biotechnology, have brought about a new era of improved agricultural productivity, quality, and resilience, offering solutions to critical challenges in modern agriculture, for example, Bt corn exhibits high resistance against corn borers [2]. Moreover, the overexpression of ARGOS genes in *Zea mays* has reduced the crop's sensitivity to ethylene, enhancing its ability to resist drought and increasing grain yield [3]. Crop yields can be significantly improved by overexpressing specific genes, such as those responsible for accumulating cytokinin and abscisic acid. It is achieved by promoting endosperm cell division, increasing carbon resource remobilization to grains, and enhancing rice yield [4]. Additionally, specific genes involved in nitrogen metabolism, such as ammonium transport, nitrate transport, glutamine synthetase, and glutamate synthase, have been shown to play a vital role in increasing biomass, seed number, amino acids, and tissue N levels in crops [5].

Studies have underlined the crucial role of grain number and weight in rice yield improvement. Advanced techniques in molecular biology and methods for mapping quantitative trait loci (QTLs) are employed to uncover the genetic architecture of these characteristics in rice. These methods can potentially identify valuable genes/QTLs associated with grain number and grain weight traits in the rice genome, which could revolutionize rice production. It is crucial to clone and characterize these genes/QTLs to comprehensively understand the molecular mechanisms underlying 2G traits. It is possible to combine yield component traits in rice breeding programs using modern breeding tools, such as marker-assisted selection (MAS), haplotype-based breeding, and allele mining. There is also a possibility of simultaneously introducing GW7 (for grain weight) and DEP2 (for grain number) genes on chromosome 7 to overcome rice's yield ceiling [6].

Recent advancements in in vitro transcription of sgRNAs and CRISPR–Cas/ribonucleoprotein (RNP) complex-mediated editing enable nontransgenic targeting of desirable plant genes while minimizing regulatory concerns. Genome editing breakthroughs offer an excellent opportunity to achieve the goal of oil stability and more efficiently explore valuable rice by-products for additional economic value. Additionally, these advancements have significantly improved the capacity to manipulate specific target genes responsible for lipid rancidity by creating insertions/deletions (indels) in genomes [7]. The cib-3B.1/QGw.cib-3B gene can increase grain weight, boost photosynthesis efficiency, and improve crop yield [8]. Examples in Table 7.1 demonstrate the impact of biotechnology on grain crops. These advancements have made agriculture more sustainable and efficient.

Table 7.1 List of genetically modified grain crops [9]

Crop (trade name)	Developer	GM trait
Triticum aestivum—Wheat HB4 Wheat—ND-ØØ412-7	Bioceres S.A.	Drought stress tolerance
Brassica napus—Argentine canola, rapeseed, turnip MON-941ØØ-2	Bayer Crop Science (including fully and partly owned companies)	Dicamba herbicide tolerance
Vigna unguiculata— Cowpea, Black eyed pea	African Agricultural Technology Foundation (AATF)	Lepidopteran insect resistance
Brassica napus—Argentine canola, canola, oilseed rape, rapeseed, turnip NS-B5ØØ27-4	Nuseed Pty Ltd	Glufosinate herbicide tolerance, modified oil/fatty acid
Brassica napus—Argentine canola, canola, Oilseed rape, Rapeseed, Turnip	BASF (BadischeAnilin und Soda Fabrik)	Modified oil/fatty acid, imazamox herbicide tolerance
Zea mays L.—Mmaize, Corn (GraINzyme Phytase-AGV-PY2Ø3-4)	Agrivida, Inc.	Phytase production, Mannose metabolism
Oryza sativa L.—Rice (Golden Rice—IR-00GR2E-5)	International Rice Research Institute	Mannose metabolism, Enhanced Provitamin A Content
Zea mays L.—Maize, Corn (Enogen—SYN-E3272-5)	Syngenta	Modified alpha amylase, mannose metabolism
Zea mays L.—Maize, Corn (Mavera Maize—REN-ØØØ38-3)	Renessen LLC (Netherlands)	Modified amino acid
Carthamus tinctorius L.—Safflower (GOR-73226-6)	Go Resources Pty Ltd	Modified oil/fatty acid, antibiotic resistance
Glycine max L.—Soybean (DD-Ø26ØØ5-3)	DuPont (Pioneer Hi-Bred International Inc.)	Modified oil/fatty acid, antibiotic resistance, visual marker
Glycine max L.—Soybean (Vistive Gold—MON-877Ø5-6)	Monsanto Company (including fully and partly owned companies)	Glyphosate herbicide tolerance, modified oil/fatty acid

3.2 Biofortification

Biofortification enhances bioavailable vitamins or minerals through agronomic, conventional breeding, or gene editing approaches [10]. It is a promising approach to address hidden hunger and improve global public health by enhancing the nutrient content of staple crops. Over the years, many research studies have utilized methods such as mapping quantitative trait loci (QTLs) and conducting extensive genome-wide association studies to identify precise spots within the genetic structure that can be altered to increase the presence of vital nutrients, such as vitamins and minerals, in crops such as wheat [11]. Approaches to biofortification in rice have enhanced vitamin A [12], thiamine [13], pyridoxine [14], and folates [15]. Genomics research found markers to improve nutrient deficiencies, leading to better crop quality and increased food security globally.

Numerous studies have revealed that genetic alterations in different crops can increase the accumulation levels of certain nutrients or enhance their nutritional value. For instance, overexpression of the OsNAS2 gene in rice grains increases Fe and Zn levels [16], while transgenic *Glycine max* plants that overexpress GmWRI1b exhibit higher oil content [17]. Wheat genetics and associated microbes in the role of iron accumulation were studied to undertake iron biofortification for wheat crops [18]. Microbes that impact iron concentration have hindered the breeding of iron-rich wheat. A promising solution lies in the combination of plant and microbial genetics, which can be an effective biofortification strategy. Conventional breeding methods have yet to yield consistent results in boosting grain mineral concentrations, necessitating exploring alternative approaches. By employing a genetic modification strategy, a study [19] merged the wheat Vacuolar Iron Transporter (VIT) gene TaVIT2-D, expressed exclusively in the endosperm, with the rice Nicotianamine Synthase (NAS) gene OsNAS2, which exhibited constant expression. This method improves the nutritional value of wheat products by increasing the amount of zinc and relocating iron in white flour. In this way, essential minerals are better distributed.

Further studies have investigated the impact of genetic modifications on various aspects of crop quality, such as protein and lipid content, cooking and eating quality, and gluten strength. Genetic modification can enhance Zn and Se and lower Cd levels [20] and modify amylose, lipid, and protein content [21] in rice grains. Genetic modifications also influence β-glucans, protein content, and grain yield in wheat [22].

Genetic modifications have also addressed health concerns in addition to improving nutritional value or quality. The creation of Golden rice, which contains β-carotene, involved incorporating genes from other organisms into the rice genome [23]. It may be a cost-effective approach to alleviate folate and iron deficiencies instead of conventional supplementation programs. Wheat varieties are enhanced with carotenoids to give pasta a more attractive appearance. Phytoene synthase 1 (Psy1) genes are pivotal in determining wheat carotenoid levels. In a study by [24], the Psy1-A1 and Psy1-B1 genes from durum were introduced into bread wheat via interspecific hybridization. This study aimed to synergize gene effects, enhance carotenoid content, and yield carotenoid-enriched bread wheat. Interspecific breeding combined with a marker-assisted selection (MAS) approach using Psy1-A1 and Psy1-B1 alleles resulted in bread wheat prebreeding lines with augmented carotenoid content (16–23% mean). These fortified lines hold promise as new varieties or recurrent parents in wheat breeding initiatives. Similarly, phytase-expressing transgenic maize [25] and RNAi-modified wheat [26] can help reduce the risk of mineral deficiency and celiac disease.

These studies demonstrate the potential for genetic modifications to enhance crop nutritional value, quality, and health benefits.

3.3 Biotechnological Approach to Reduce Preharvest Losses

Preharvest sprouting (PHS) is a condition wherein mature cereal seeds begin germinating, leading to significant losses in harvestable yield, grain quality, storability, and germination capacity. PHS also causes higher moisture content, which increases mycotoxin contamination in the grain [27]. The root cause of PHS is an imbalance between dormancy and germination. Dormancy evolved in wild species to ensure seed germination only when conditions were favorable [27]. However, during domestication, selection was for weak dormancy that allows uniform germination almost independent of the prevailing conditions [28]. If dormancy is too weak, spatiotemporal signals trigger germination malfunction, resulting in PHS.

Regulating dormancy control, thus, PHS, is affected by the relative balance of abscisic (ABA) and gibberellic acid [27]. Researchers have modified ABA-related genes that control dormancy and germination traits and have observed improved PHS resistance in rice [29]. For instance, rice overexpressing OsMFT2, a positive regulator of ABA response genes, has delayed germination and a lower rate of PHS [30]. Glutaredoxin-mediated ABA and reactive oxygen species signaling also control PHS in rice [30].

Scientists have used CRISPR–Cas9 mutations of OsABA8ox and OsABA8ox1 to increase ABA levels and seed dormancy in japonica rice, improving PHS resistance [30]. Novel seed dormancy alleles created by CRISPR–Cas9 editing have reduced PHS in rice and wheat. Genes involved in diverse pathways unrelated to ABA also influence PHS. CRISPR–Cas9-targeted mutagenesis of wheat TaQsd1, which encodes an alanine aminotransferase, prolonged dormancy and germination and prevented PHS [31]. Given the many pathways regulating the dormancy-germination nexus, researchers will likely discover additional genes that impact PHS.

Seed dispersal, or seed shattering, is essential for plant survival in nature. However, in cultivation, nonshattering mutants are preferred [32]. This trait is linked to valuable commercial traits and is selectively retained in some rice cultivars [33]. Researchers have identified genes responsible for seed shattering and have edited them to produce cultivars resistant to shattering. The myeloblastosis transcription factor gene Less Shattering1 (SvLes1) mutation has been associated with reduced shattering in green millet [34]. Using CRISPR–Cas9 editing, researchers produced novel alleles of SvLes1 that lead to a reduced shattering habit.

4 Biotechnology in Grain Processing

The contributions of biotechnologists are vital to the success of the food industry. Through scientific advancements, they enhance food quality, safety, and nutrition while also developing new products, improving processing methods, and ensuring safety. Their expertise in enhancing crop yields, refining baking and brewing techniques, and improving the nutritional value of food products has been essential to the industry's progress. Biotechnologists have optimized fermentation techniques and sensory profiles by utilizing genetically modified enzymes and microbial strains, resulting in superior products. Moreover, they have discovered natural preservatives and flavors in grain processing, which have improved consistency, reduced waste, and increased energy efficiency. Their efforts have paved the way for grain processing, baking science, and brewing technology advancements, resulting in higher quality and more sustainable products.

4.1 Enhanced Nutritional, Cooking, and Eating Quality

The CRISPR/Cas9 method decreased toxic metals in rice grains, specifically cadmium (Cd), by mutating the manganese/metal transporter 5 (Nramp5) gene [35]. Nramp5 transports Cd across the plant membrane. Editing its gene with CRISPR/Cas9 can regulate plant metal ion accumulation [35]. The CRISPR/Cas9 technique has deactivated the GW2 gene in rice, which led to the creation of mutants with elevated levels of proteins and essential nutrients such as iron, zinc, potassium, phosphorus, and calcium in the endosperm of rice grains [36]. Knocking down the orange (Or) gene and replacing it with the calreticulin I (SSUcrtI) and phytoene synthase (ZmPsy) genes has been shown to increase the carotene or provitamin A content in rice. The Or gene produces a protein that hinders the conversion of beta-carotene into vitamin A [37]. By utilizing CRISPR–Cas9 to substitute the Or gene with the SSU-crtI and ZmPsy genes, rice crops could yield higher levels of beta-carotene, which can be transformed into provitamin A when consumed by humans. This innovative approach can be a viable substitute for brown rice [38]. The CRISPR/Cas9 method mutates the Rc gene, which produces red pigment, increasing the production of health-promoting nutrients and creating red in rice [37].

Three physiochemical indicators influencing rice grain quality are amylase level, gel consistency, and gelatinization temperature. The amylase level indicates the amount of starch present in the grain, which affects the texture of cooked rice, improves human health and reduces the risk of serious diseases [39]. The gel consistency affects the rice elasticity when cooked, determining how well the grains stick together. The temperature at which the starch in the grain changes from a solid to a liquid (known as the gelatinization temperature) affects the texture and cooking time of the rice [40]. CRISPR/Cas9 technology decreases the amylose content in rice by modifying the Waxy gene and converting nonglutinous cultivars (XS134 and 9522) to glutinous cultivars [41].

Higher-amylose rice was developed through gene mutations (SBEI and SBEIIb), resulting in altered microstructure and nutritional quality [42]. Through the CRISPR/Cas9 approach, a study discovered that plants with a mutant ISA1 gene had reduced total starch, amylose, and amylopectin contents. In contrast, mutants had higher sugar content and starch gel consistency than wild-type ones [43]. CRISPR/Cas9 has helped produce a new aromatic rice variety by introducing an aroma-related gene into nonaromatic rice. ASD16, a nonaromatic rice variety, has been transformed into an aromatic variety by generating new OsBADH2 alleles [44]. Various genes related to eating and cooking traits were modified using CRISPR/Cas9 in the first decade after its discovery, for example, FAD2-1/FAD2 (increased oleic acid and linoleic acid contents) [45], GS9 (rice grain shape) [46], Hd2, Hd4, and Hd5 (early maturing) [47].

Approximately 1–2% of the population suffers from celiac disease caused by gluten proteins in wheat grains. Specifically, the α-gliadin epitopes found in these proteins trigger the disease. Enhancing the quality of wheat grains can help cater to a more extensive customer base. Using CRISPR/Cas9, an ear of low-gluten wheat was developed with 85% less immunoreactivity by targeting the conserved region of the α-gliadin gene [48]. Deleting two γ-gliadin genes improved gluten quality and reduced celiac epitopes [49]. Editing two α-amylase/trypsin inhibitor proteins (WTAI-CM3 and WTAICM16) resulted in wheat with reduced immunoreactive agents and improved consumer appeal [50]. CRISPR/Cas9 knockout of four wheat quality genes (pinb, waxy, ppo, and psy) improved grain hardness, starch quality, and dough color [51]. Using CRISPR/Cas9, a mutant of the SBEIIa gene was created to develop high amylose wheat with exceptional levels of resistant starch [52]. These new wheat lines might prove healthier than wheat varieties already available on the market. Genetic editing of the waxy gene in *Zea mays* has led to the development of high amylopectin corn [53].

4.2 Fermentation

Fermented batters and doughs of legumes and rice can produce steamed, fried, or baked products. Soybean products, both fermented and unfermented, are available, with soy sauce and soy paste being the fermented products and soymilk and tofu being the nonfermented products. The nutritive value of protein in common beans improved by 8.3% due to fermentation, which increased potassium, amino acids, and soluble fiber content [54]. The use of *Lactiplantibacillus plantarum* FST 1.7 in the fermentation of gluten-free (GF) bread prepared from brown rice flour, soya flour, buckwheat flour, cornstarch, and xanthan gum resulted in an upgrade in the shelf life and quality attributes of GF bread [55]. Certain starter cultures, such as *Saccharomyces cerevisiae* YC1, *Pediococcus pentosaceus* LD7, *Weissella confusa* SD8, and *P. pentosaceus* SA8, can be utilized to prepare sourdough breads from sorghum [56]. As fermentation is time-consuming, freeze-dried sourdoughs can instantly produce fermented gluten-free products [57]. Additionally, fermented drinks made from various GF grains, such as oskikundu from pearl millet flour and sorghum malt [58], katikalla from teff [59], and togwa from maize and finger millet [60], are readily available in different parts of the world. *Saccharomyces cerevisiae* AKP1 can enhance functional composition (nutrients and minerals) and starch digestibility and reduce anti-nutritional factors (phytate and trypsin inhibitor) in multigrain food through fermentation [61]. The yeast species utilized in fermentation exhibit noteworthy biofortifying potential. Further research is warranted to explore its bioactivities in multigrain substrates.

Bakers use microbial starters, such as lactic acid bacteria and sourdough cultures, to control the fermentation process and create diverse flavors, textures, and freshness levels in their bread. The lactic acid in sourdough bread improves its texture and acts as a natural preservative, contributing to the bread's shelf life [62]. Controlled fermentation can produce bread with an exceptional flavor and aroma that consumers will appreciate. Biotechnological interventions empower bakers to create specific textures, shapes, and qualities in their bread, improving quality, nutrition, and customer satisfaction. Moreover, biotechnological methods can enhance bread's nutritional value by adding essential vitamins and minerals to maintain one's health.

4.3 Baking

Baking is a method of cooking using dry heat. Baking, akin to ancient bioprocessing, such as brewing and fermentation, now employs modern biotech for grain enhancement, enzyme aid, and advanced fermentation methods [63]. Biotechnology integration into dough and flour formation has revolutionized the baking industry [64]. Biotechnology enhances dough characteristics, optimizes baking processes, and yields superior-quality bakery products by harnessing biological and genetic advancements. This dynamic fusion of science and baking expertise paves the way for innovative and improved flour and dough processing approaches.

Bread has been a staple in human civilizations and cultures throughout history [65]. The sourdough culture was the primary method of bread production for centuries until the availability of baker's yeast decreased its prevalence [66] but similar favored traits [67]. Recent years, however, have seen a growing interest in understanding the microbiome of sourdough and its contribution to traditional bread-making processes. Bread is also an excellent option for those concerned about blood sugar levels [68] since it contains prebiotics that are beneficial for gut health [69] and have a low glycemic index [68]. The field of biotechnology has provided valuable insights into the microbial ecology of sourdough, paving the way for the development of delicious, traditional-flavored breads that are healthy, gluten-free and have a low glycemic index. Enzymes and microorganisms are genetically modified to achieve this. See the next section in this chapter for more on biotechnology's impact on baking.

4.4 Enzyme-Assisted Grain Processing

Enzymes have played a crucial role in grain processing innovations. Proteolytic enzymes are used to liberate nutrients and fibers from bran without removing the germ and bran layer [70]. Enzyme pectinases are employed to dissolve the pectin sheath encasing cellulose to release essential elements [71]. A conditioning process is introduced using a blend of enzymes such as hemicellulases, pectinases, β-glucanase, and amylase to make grains easier to digest [72]. The enzyme treatment process (WO/2007/051091) to separate layers from cereal grain endosperm cores without causing mechanical damage is patented [73]. Enzyme processing is an eco-friendly approach to processing, leading to the enrichment of grains with bioactive nutrients.

4.5 Mitigating Allergens and Antinutritional Factors

Grains contain substances known as allergens and antinutrients that can have adverse effects on human health. These substances include saponins, tannins, phytic acid, gossypol, lectins, protease, amylase, antivitamin, metal-binding ingredients, goitrogens, and gluten [74]. These antinutrients can cause problems with nutrition and impact the body in harmful ways. Some antinutrients are reduced during cooking, soaking, sprouting, or milling, while others remain. Microorganisms can be beneficial in the fight against them. Utilizing carefully selected sourdough lactobacilli and fungal proteases may be viable in mitigating gluten toxicity. To reduce gluten intolerance in pasta, fermentation of durum wheat semolina was conducted through sourdough lactobacilli [75]. Fermentation significantly enhanced the protein digestibility of black-gram-dhal batter while decreasing its phytic acid content [76]. During cowpea fermentation, a significant reduction in the amount of phytic acid was observed due to the removal of trypsin inhibitors and oligosaccharides [77].

4.6 Mycotoxin Control

Mycotoxins are secondary metabolites produced by different fungi, including *Fusarium*, *Penicillium*, *Aspergillus*, and *Alternaria*. They are transferred along the grain supply chain from raw grains to the final product due to their high stability. The most common mycotoxins detected in grains and grain-derived food products are aflatoxins (AFs), fumonisins (FBs), ochratoxins (OTs), deoxynivalenol (DON), and zearalenone (ZEN).

Mycotoxin contamination in grains and their derived products poses significant health risks to both humans and animals. Trichothecene mycotoxins have been found to disrupt various cellular processes, including protein synthesis, DNA and RNA synthesis, cell membrane integrity, mitochondrial function, and apoptosis [78]. Mycotoxins ZEN, produced by several *Fusarium* species, exhibit estrogenic activity and have been linked to reproductive disorders in animals [79]. Additionally, mycotoxicosis can have toxic effects on various systems, including the hematopoietic, kidney, liver, immune, fetal, and reproductive systems, and may have carcinogenic and mutagenic properties [80]. Mycotoxins such as aflatoxins (G1, B1, G2, B2, and M1) are categorized as Group 1, ochratoxin A as possibly carcinogenic (Group 2B), and zearalenone and deoxynivalenol classified as Group 3 (not classifiable regarding their carcinogenicity) carcinogens to humans [81].

Mycotoxin contamination can occur at both pre- and postharvest stages. To ensure the safety of grains and their products, various strategies have been adopted. Physical methods to remove mycotoxins include sorting of grains (by floatation or based on color), washing, gamma irradiation, ultraviolet rays, microwave irradiation, and use of pulsed light. In chemical methods, structural modifications are induced by the use of chemicals (acids, bases, reducing agents, and oxidizing agents) to reduce the toxicity of mycotoxins [82]. However, some of these conventional methods have a negative impact on grain quality; for example, irradiation can affect the germination of seeds [83], and chemical treatments have been shown to negatively affect the organoleptic properties and nutritional value of grains [82].

Biotechnological methods are considered the best alternative to these conventional approaches. A significant reduction in contamination can be achieved through the development of disease resistance traits during the preharvest stage. Modern transgenic techniques include (1) host-induced gene silencing (HIGS), (2) RNA interference (RNAi), (3) microRNA (miRNA)- or artificial microRNA (amiRNA)-mediated gene silencing, (4) designer transcription activator-like effector(dTALE)-mediated up- or down-regulation of gene expression, (5) Zn-finger nucleases, (6) mega nucleases, (7) transcription activator-like effector nucleases (TALEN), (8) clustered regularly interspaced short palindromic repeats (CRISPR)/Cas9, and (9) Oligonucleotide-directed mutagenesis (ODN)-based gene editing techniques. These techniques can be applied to develop plants with increased resistance to mycotoxin contamination [84].

Furthermore, it has been found that some bacteria, yeast, and molds secrete secondary metabolites and volatile organic compounds that can inhibit mycotoxins

[85]. Lactic acid bacteria (LAB) have demonstrated remarkable antifungal and mycotoxin inhibitory properties among all available biological control agents. Many studies suggest the inhibition of aflatoxin biosynthesis by LAB [86]. However, recent studies have explored the interaction between LAB and *Fusarium* toxins such as ZEN and DON [87]. Many fungal species, such as *Aspergillus*, *Alternaria*, *Absidia*, *Armillaria*, *Candida*, *Dactylium*, *Mucor*, *Penicillium*, *Peniophora*, *Pleurotus*, *Trichosporon*, and *Rhizopus*, have shown mycotoxin inhibition activity [88]. Yeast such as *Yarrowia lipolytica*, *Phaffia rhodozyma*, and *Xanthophyllomyces dendrorhous* exhibited OTA degradation activity [89]. Mycotoxin degradation enzymes such as oxidase, peroxidase, laccase, reductase, esterase, carboxylesterase, aminotransferase, and lactone hydrolase can be extracted from different microbes and employed in mycotoxin control.

4.7 By-product Utilization and Waste Utilization

Efficiently converting the complex carbohydrates found in rice husks into simple sugars for use in biofuels, biochemicals, and other products is known as rice husk saccharification. This process promotes a biobased economy and sustainable waste management by repurposing agricultural waste. Enzymatic hydrolysis breaks down rice husk cellulose and hemicellulose into fermentable sugars. Improving the saccharification process and increasing the yield of fermentable sugars can be achieved by optimizing enzyme cocktails and targeting specific enzyme overexpression. Lytic polysaccharide monooxygenases (LPMOs) enhance rice husk breakdown into simple sugars by cleaving complex carbohydrates more efficiently and quickly [90]. A synergistic combination of LPMO and cellulase enzymes enhances biomass saccharification [91]. The addition of LPMOs to a commercial cellulase cocktail can result in a significant increase in reduced sugars released from rice straw and corn cob [92]. LPMOs oxidatively degrade cellulose, enhancing the saccharification of delignified wheat straw when combined with a cellulase cocktail. Biotechnological methods can optimize the recombinant expression levels of LPMOs to improve their cellulolytic activity.

Biotechnology plays a vital role in effective waste management by transforming organic waste such as rice straw, rice husks, and wheat straw into valuable products, such as biogas, biofuels, and compost. It also contributes to developing eco-friendly packaging materials that help lower greenhouse gas emissions and reduce environmental impact. A Support to Research and Innovation Policy for Biobased Products (BIOSPRI) study reveals that biotechnology has introduced innovative Bio-Based Products such as biopolymers PHA (polyhydroxy alkanoate) and bio-PE (polyethylene) made from natural rubber, lignins, plant fibers, vegetable oils, and animal fats [93]. The production of PHA has resulted in a substantial reduction of up to 200% in GHG emissions, and fossil energy use has fallen by 95%, leading to a significant decrease in waste management requirements.

5 Biotechnology in Baking

Biotechnology in baking encompasses diverse biological, biochemical, genetic, and microbiological techniques. Baking parallel to brewing and fermentation employs modern biotechechnology for grain enhancement, enzyme aid, and advanced fermentation methods [63]. Biotechnology integration into dough and flour formation has revolutionized the baking industry [64]. Biotechnology improves dough and baking using science and genetics for better bakery products. Techniques such as transgenes and RNA interference enhance gluten strength and elasticity [94] while ensuring a balanced starch content with CRISPR–Cas9 [95]. Modifying yeast, enzymes, and proteins can improve baked goods' texture, structure, and other sensorial aspects [96]. Genetic modifications can also boost gluten strength in wheat [97]. Other biotechnology interventions in baking are listed in Table 7.2.

5.1 Genetic Modifications in Flour

Flour plays a crucial role in the composition of baked goods, providing essential structures and substances. The unique properties of wheat can significantly impact the final product's quality, making it more suitable for formulation. Genetic defects in wheat, such as late maturity α-amylase (LMA) and preharvest sprouting (PHS), induce α-amylase expression, lowering the falling number (FN), and grain value. Enhancing α-amylase (TaAmy3) in wheat led to low FN but improved baking quality without harming starch or flour [106]. Increased loaf volume and browning were observed in small-scale baking, challenging the assumption that low FN invariably

Table 7.2 Contributions of biotechnology to baking ingredients and quality parameters

Interventions	Examples	Biotechnological contribution	References
Yeast engineering	Modified yeast strains	Genetic modification for improved fermentation	[98]
Dough rheology control	Xanthan gum, guar gum, etc.	Tailoring dough properties for specific products	[99]
Enzymes	Amylases, proteases, lipases, etc.	Utilized to enhance dough characteristics and quality	[100]
Flavor and aroma	Yeast, lactic acid bacteria	Enhanced through controlled fermentation	[101]
Gluten quality	Modified gluten proteins	Genetic modification to enhance gluten performance	[102]
Acrylamide reduction	Asparaginase enzyme	Enzymatic methods to lower acrylamide levels	[103]
Microbial starters	Lactic acid bacteria, wild yeast	Controlled use of beneficial microorganisms	[104]
Nutrient enhancement	Vitamins, minerals, folic acid, etc.	Fortifying dough with vitamins, minerals, and more	[105]

impairs quality. This highlights the need to better understand the impact of α-amylase on end-product quality. Biotechnological techniques are employed to achieve specific end qualities in wheat flour for optimal baking. Targeted improvements include enhanced gluten strength, balanced starch properties, consistent hydration, and desirable texture. Biotechnology helps tailor wheat varieties with improved protein content, modified gluten composition, and controlled enzyme activity. These advancements ensure better dough elasticity, increased volume, improved crumb structure, and superior baking performance, leading to high-quality bakery products. Genetic modification of wheat can improve its texture and baking properties, resulting in better food quality [107].

CRISPR/Cas9 technology induces targeted mutagenesis of TaSBEIIa in both winter and spring wheat varieties [108]. This has led to the creation of transgene-free wheat with elevated amylose content. Furthermore, the triple-null line flours exhibited notably higher levels of amylose, resistant starch, protein, and soluble pentosan, contributing to consumer health. The baking quality evaluations indicate that high-amylose flours are an excellent choice for crafting cookies and using them as additives.

5.2 *Enzyme Technologies for Dough Improvement*

Enzymes are crucial in the bread-making process, as they play a significant role in improving the dough's texture, volume, and overall quality. Enzymes can also reduce acrylamide levels, which can harm health, by breaking down the substances responsible for their formation. Enzyme technology transforms dough formation.

Amylases, proteases, and transglutaminases enhance extensibility, fermentation, and elasticity [109]. The addition of transglutaminase (TGase) enzyme and a microbial consortium (*Lactobacillus sanfranciscensis* and *Candida milleri*) to dough enhances the rheology, aroma, and shelf life of bakery products [110]. The selected microbial TGase forms isopeptide bonds that enhance bread structure, especially in gluten. Using it with sourdough yields delicious bread with a similar texture to regular bread. Xylanases, lipases, and cellulases improve water absorption and texture [111]. Table 7.3 lists some examples of enzymes used to improve the quality of bakery products. By applying these enzyme technologies, bakers can fine-tune dough properties, leading to improved volume, texture, and overall product quality. The holistic impact of a commercial peptidase on the diverse characteristics of wheat dough and bread quality is well studied [113]. It systematically scrutinizes its influence on dough rheology, microstructure, and gluten characteristics, culminating in discernible effects on the end bread product. Through meticulous analysis of these complex interplays, the study offers invaluable insights for refining bread-making procedures and improving the overall bread quality via targeted peptidase application.

Table 7.3 Examples of enzymes that induce quality improvements in baking [112]

Enzymes	Induced improvement
ά-Amylases, hemicellulases, cellulases, lipases, proteases	Larger volume, high-fiber baking, improving the baking quality of flour
ά-Amylases, hemicellulases	Anti-staling effects and extended shelf life
Hemicellulases, ά-Amylases, proteases, lipases	Softer crumb, fine and uniform pore structure, stability of frozen doughs, better crispness, less hygroscopicity
ά-Amylases, hemicellulases, lipoxygenases	Browning effect, improved crust color, bleaching effect
ά-Amylases, proteases, lipoxygenases, lipases, glucose oxidases	Production of fermentation substrates and aroma precursors
ά-Amylases, hemicellulases, proteases, lipoxygenases, glucose oxidases, lipases	Compensating for recipe changes, bromate replacement, sodium metabisulfite replacement, replacement of vital gluten, reduced-fat baking
Hemicellulases	Increased total and soluble dietary fiber content

In comparative investigations to assess superior bread improvements between lipase enzymes and commercially available emulsifiers, it was observed that both lipase enzymes and DATEM (diacetyl tartaric acid ester of mono- and diglycerides, also known as E472e) yielded favorable outcomes across multiple aspects of bread quality [114]. DATEM, Lipopan F-BG, and LipopanXtra-BG exhibited better effectiveness by reducing fermentation time. However, in instances of prolonged fermentation, in contrast to the third-generation lipase, LipopanXtra-BG, moderate quantities of the second-generation lipase, Lipopan F-BG, significantly amplified bread volume.

Transglutaminase can potentially enhance the quality of wheat-based products by introducing covalent bonds within proteins [115]. Additionally, glucose oxidase, laccase, tyrosinase, and hexose oxidase exhibit activity with the wheat flour arabinoxylan fraction, enabling the cross-linking of proteins through tyrosine and SS covalent bonds within gluten. These findings suggest that enzyme modification can beneficially reshape the properties of damaged gluten, positively impacting wheat-based products.

5.3 Innovations in Leavening Agents

When preparing dough or batter, leavening agents produce air pockets within the mixture, resulting in a lighter, fluffier texture. Three primary types of leavening agents exist: chemical, organic, and mechanical. In baking, chemical leavening agents, including baking powder and baking soda, are frequently utilized to attain the desired texture. Organic leavening agents, such as yeast, are living organisms that convert sugars in the dough or batter into carbon dioxide gas, which causes the mixture to rise. Mechanical leavening agents, such as air or steam, lift the dough or batter by introducing these gases into the mixture. By utilizing leavening agents appropriately, bakers can achieve the desired texture and consistency in their baked goods. Biotechnological advancements are transforming leavening agents. Engineered yeast strains optimize fermentation, elevating dough rise and flavor. Microbial consortia enhance sourdough quality. Bioprocess-derived enzymes, such as amylases and proteases, refine the dough structure and gas retention. These innovations align with industry trends, enhancing bread quality, shelf life, and nutritional value.

Introduction of the *Yarrowia phangngaensis* strain can produce a versatile multi-enzyme complex [116]. This complex holds potential as an enzyme additive for improving the quality of high cassava-wheat composite bread. This research explores an innovative solution to enhance the baking characteristics of composite bread, thereby broadening the scope of cassava-wheat utilization. Sha et al. [117] explored the influence of exopolysaccharide (EPS)-producing lactic acid bacteria (LAB) on sourdough fermentation and steamed bread quality. Comparing *Weissella cibaria* L32 and *Lactobacillus brevis* L17, both high EPS producers, reveals distinct fermentation effects. LAB addition enhances titratable acidity and protease activity, especially *L. brevis* L17. While in situ EPS from LAB improves steamed bread quality, excessive acidification by *L. brevis* L17 degrades the gluten network, resulting in a lower volume and harder texture. EPS-producing *W. cibaria* L32 shows potential for enhanced steamed bread quality, highlighting its application in baking.

To enhance techno-functionality in γ-aminobutyric acid-enriched germinated wheat, xylanase (Xyn) and glucose oxidase (Gox) were integrated to modify key components. The Xyn and Gox combination significantly improved steamed bread quality, optimizing loaf volume and texture. It facilitated a continuous gluten network, enhancing dough viscoelasticity. Xyn increased arabinoxylan (AX) water solubility and homogenous molecular distribution throughout the bread-making process. Enzymes suppressed the polymerization of α-/γ-gliadin and glutenin while promoting AX incorporation into insoluble proteins. Enhanced high molecular weight glycoprotein formation in the dough's liquid lamella improved foam thermal stability, producing superior steamed bread quality. Germinated wheat has potential as a functional ingredient, benefiting from component modification [118].

5.4 Fermentation

Fermentation enhances bread's aroma and flavor, suggesting its potential as an additive-free enhancer for extended shelf life. Driven by the demand for additive-free products, a study explored the use of *Ganoderma lucidum*, a traditional Chinese edible fungus, to enhance bread quality [119]. Fermented broth, rich in polysaccharides, curbs water migration and starch retrogradation in bread crumbs. The addition of the broth, whether sterilized or not, improves bread quality with an optimal dosage. The co-fermentation of *Propionibacterium freudenreichii* with lactic acid bacteria (LAB) enhances vitamin B_{12} content in whole-meal sourdough bread

(WMSB) [105]. It improved bread texture and sensory characteristics, with higher adhesiveness and a more homogeneous texture. This co-fermentation successfully enriched WMSB with vitamin B_{12}, presenting a nutritional improvement opportunity. Microbiological analysis evaluated the biotechnological potential of lactic acid bacteria from Tunisian flours [120]. Isolates have diverse characteristics for improving bread quality and customizing baking processes.

5.5 *Biopreservation*

Among these emerging methods, using microorganisms for biopreservation holds considerable promise. To prolong the shelf life of bakery products, a typeIII sourdough combined with a probiotic mixture (*Lactobacillus* spp., *Bifidobacterium* spp., *Bacillus coagulans*) was assessed [121]. Fresh base pizza and focaccia with bioprotective cultures (BCs) showed stable composition and improved microbiota. Multiomics analysis revealed enhanced carboxylic acids and extended shelf life by 10 days under modified atmosphere packaging. Biotechnology reduces food waste through anti-staling methods, extending shelf life while preserving quality. This holistic approach transforms bread production, aligning it with consumer preferences and sustainability goals.

A study explicitly investigated the application of lactic acid bacteria-fermented whey as a biopreservative ingredient to inhibit the growth of Aspergillus and Penicillium fungi in bread. The study assessed the impact of three different concentrations of this ingredient on various bread properties. The findings reveal that adding 5% lactic acid bacteria-fermented whey has minimal effects on bread quality while simultaneously increasing the presence of antifungal compounds, such as lactic acid and phenyl lactic acid. Consequently, this biopreservation approach contributes to an extended shelf life for bread contaminated with *Aspergillus flavus* and *Penicillium verrucosum*. Importantly, it also significantly reduces the fungal population of *P. verrucosum* in comparison to regular bread after 7 days [122].

6 Quality Control and Safety

Ensuring the quality of grains is a significant aspect that encompasses every stage from breeding and harvesting to processing and eventual consumption of the final product. Traditionally, pesticides and herbicides are utilized to ensure the safety of crops at the cultivation stage, while fertilizers are utilized to improve crop production. However, chemical pesticides and fertilizers cause various environmental hazards [123]. Biotechnological approaches such as genetically modified grains can improve the quality and yield of grains at the crop production (breeding) stage. Genetically modified grains carry traits, such as herbicide tolerance, insect resistance, abiotic stress tolerance, disease resistance, and enhanced nutritional content [124].

The dietary and functional quality of food ingredients determines consumer preferences. To make grains preferable to consumers, grains are refined by different processing methods. Conventional processing includes mechanical and chemical methods that may be associated with a reduction in the nutritional quality of grains because they remove the bran and germ layer of grains. Most grain nutrients (dietary fiber, minerals, vitamins, lignans, phytoestrogens, and phenolic compounds) exist in the bran and germ layer of grains. To address this concern, it is necessary to employ innovative processing methods. Biotechnological methods have emerged as promising tools in the field of food processing. In grain processing, biotechnological concepts such as cell wall-degrading enzymes are utilized to modify grain properties such as taste, texture, and shelf life, with minimal impact on nutritional value [125]. Furthermore, biotechnology facilitates the identification of specific grain varieties, detection of potential contaminants or adulterants, and monitoring of mycotoxins, ensuring that products meet quality and safety standards through genetic markers and molecular tools.

Traceability is the capability of a food chain to monitor and follow the historical journey and movement of food products throughout the entire supply chain, facilitated by various technological advancements such as radio-frequency identification (RFID), blockchain, food sensing technologies, and barcoding. Among them, DNA barcoding has gained more attention for the identification of raw materials and processed foods. The DNA barcode consists of a small portion of the genome, typically less than 1000 base pairs [126]. A specific gene fragment from the target species is affixed to the 5′ end of the barcode, following which the barcode gene is amplified and subjected to sequencing. For the detection and identification of plant-based food, chloroplast and nuclear genomes are utilized for DNA barcoding [127].

It is necessary to identify and authenticate raw materials such as grains used in food products. Various polymerase chain reaction (PCR) techniques are used by-producers to certify and identify crop species as well as by consumers for reliable identification of crop species and their origin [128]. Additionally, with an increase in genetically modified (GM) crops in the food chain, it became necessary to track the origin and transgenes of GM crops [129]. DNA barcoding offers a more accessible option compared to other molecular techniques because it relies on universal markers, eliminating the need for extensive genomic knowledge of each organism, as needed in DNA microarray or fingerprinting techniques [127]. Furthermore, the traceability of food components after industrial processing is difficult by DNA-based methods, as DNA can degenerate during various treatments during processing. However, in barcoding, a small fragment of DNA is used, which limits the risk of fragmentation of DNA, and the mitochondrial or plastid genome used in barcoding has less chance of degradation. Therefore, DNA barcoding could be used for traceability even after industrial processing [130].

The identification of food adulterants is also an important part of food traceability. Grain legumes such as *Phaseolus lunatus* often face issues of adulteration, where high-quality seeds are mixed with lower-quality or even toxic seeds of the same species or seeds from completely different species that look similar in color and shape. Detection of such substitution is impossible through visual inspection [131].

Universal barcoding markers such as the chloroplast *trnL* intron and chloroplast *rpoC1* can distinguish between different legume species using these markers, and we can identify the most distinct and most similar species [132]. This is carried out by using Barcode DNA-High Resolution Melting (Bar-HRM) analysis, which combines DNA barcoding and high-resolution melting analysis (HRMA). Similarly, high-value products such as basmati (known for their fragrance and long grains) are adulterated with nonfragrant varieties. To discriminate between fragrant and nonfragrant varieties, Bar-HRM analysis by combining microsatellites and fragrance typing was carried out [132]. DNA barcoding has been proven to be an important biotechnological tool in grain supply chain traceability compared to other analytical techniques.

7 Sustainable Approach

Sustainable agriculture is crucial for our planet's well-being. It is impossible to overemphasize its significance. It is critical to preserve the environment, conserve natural resources, and optimize nonrenewable resources. Biotechnology is an invaluable tool in achieving these goals, as it enables the creation of crop varieties better suited to the changing environment. Additionally, biotechnology allows for more efficient and cost-effective grain processing methods through enzyme-assisted and microbial-assisted tools.

One of the most significant benefits of biotechnology is its ability to fortify grains with essential nutrients, making it easier to address malnutrition sustainably. A sustainable food supply heavily relies on its promotion and plays a crucial role in ensuring its sustainability. Transgenic crops that are more resistant to pests and environmental stressors can significantly increase food production while reducing food waste. Biotechnology has also led to the development of biodegradable packaging materials that help reduce greenhouse gas emissions and minimize environmental impact.

Moreover, biotechnology is instrumental in efficient waste management, as it can convert organic waste into biogas, biofuels, and compost. This technology has also created innovative biobased products (BBPs), such as biopolymers made from natural rubber, lignins, plant fibers, vegetable oils, and animal fats. These advancements have significantly reduced GHG emissions and fossil energy use, leading to less waste management.

Biotechnology has the potential to impact sustainable agriculture in positive ways significantly. Biotechnology can tackle the diverse challenges of climate change, pest infestations, and malnourishment through the facilitation of eco-friendly and productive food production techniques and waste management practices. It is imperative to pour resources into biotechnological progress to guarantee a sustainable food source for future generations and maintain a thriving planet.

8 Challenges and Ethical Considerations in Biotechnological Interventions

Although genetically modified industrial strains have the potential to offer numerous benefits in terms of nutrition and hygiene, public concerns have hindered their commercial production. It is crucial to differentiate between their presence or absence in different product categories to gain acceptance for genetically modified organisms (GMOs) in food. This calls for developing detection techniques and educating the public about the potential advantages of recombinant DNA technology. By increasing public awareness and understanding, consumers can make well-informed decisions about these products.

Ethical deliberations play a vital role in preserving public confidence and ensuring that these advancements have a constructive impact on the quality and safety of food. While there are advantages, there are also specific challenges and ethical aspects related to grain products, which are elaborated upon below:

- *Allergenic Ingredients:* Utilizing GMOs and novel components in biotechnology raises allergy concerns, necessitating clear labeling to protect sensitive consumers.
- *Transparency and Labeling:* Transparent and accurate labeling of genetically modified grain goods is crucial to inform consumers, enabling them to make informed choices.
- *Preserving Consumer Choice:* Ensuring the availability of nonmodified alternatives becomes an ethical consideration for consumers with objections to biotechnological modifications in food products.
- *Long-term Health Consequences:* Testing baked goods with biotech interventions for long-term health effects is challenging. Safety and thorough testing are ethical priorities.
- *Environmental Impact:* When cultivating GMO crops, biotechnological interventions can have negative environmental consequences. Ethical considerations include assessing potential impacts on ecosystems, biodiversity, and sustainability.
- *Trade and Labor Practices:* Ensuring fair trade practices and ethical treatment of workers in biotech production is crucial, including fair wages and ethical labor practices in the supply chain.
- *Intellectual Property and Accessibility:* Grain biotech innovations should allow access to affordable, nutritious food, especially in underprivileged populations.
- *Regulatory Oversight*: Ethical considerations involve advocating for rigorous regulatory oversight to ensure the safety of biotechnological interventions in grain products and prevent harm to public health.

To move forward, the debates surrounding GMOs should focus on the potential outcomes of integrating this technology into various sectors. It is essential to analyze each industry independently and develop tailored approaches, considering the possibility of the spontaneous release of GMOs. By adopting a constructive approach, we can harness the benefits of genetically modified industrial yeast strains and create safer, higher quality, and more affordable products.

9 Conclusion

The emergence of biotechnological tools has broadened the potential applications of grains through enhanced efficiency, ease of processing, and improved storage capabilities. The cases presented in this chapter catalyze further research, highlighting the need for more dependable hybrid strains that promote sustainable accessibility and availability. Leveraging genomic data, which has already demonstrated significant advantages, can foster growth in the production and processing of grains. Careful monitoring of the regulations and quality control of biotechnologically modified grains and their derivatives is crucial to ensure the success of breeders and processors in devising innovative biotechnological strategies.

References

1. Ahmar S, Hensel G, Gruszka D (2023) CRISPR/Cas9-mediated genome editing techniques and new breeding strategies in cereals—current status, improvements, and perspectives. Biotechnol Adv 69:108248. https://doi.org/10.1016/j.biotechadv.2023.108248
2. Ranjha MMAN, Shafique B, Khalid W, Nadeem HR, Mueen-ud-Din G, Khalid MZ (2022) Applications of biotechnology in food and agriculture: a mini-review. Proc Natl Acad Sci India Sect B Biol Sci 92(1):11–15. https://doi.org/10.1007/s40011-021-01320-4
3. Shi J et al (2015) Overexpression of ARGOS genes modifies plant sensitivity to ethylene, leading to improved drought tolerance in both Arabidopsis and maize. Plant Physiol 169(1):266–282. https://doi.org/10.1104/pp.15.00780
4. Parida AK, Sekhar S, Panda BB, Sahu G, Shaw BP (2022) Effect of panicle morphology on grain filling and rice yield: genetic control and molecular regulation. Front Genet 13:876198. https://doi.org/10.3389/fgene.2022.876198
5. Das S, Ray MK, Panday D, Mishra PK (2023) Role of biotechnology in creating sustainable agriculture. PLoS Sustain Transform 2(7):e0000069. https://doi.org/10.1371/journal.pstr.0000069
6. Singh G et al (2022) 2Gs and plant architecture: breaking grain yield ceiling through breeding approaches for next wave of revolution in rice (*Oryza sativa* L.). Crit Rev Biotechnol 44:1–24. https://doi.org/10.1080/07388551.2022.2112648
7. Bhunia RK, Sinha K, Kaur R, Kaur S, Chawla K (2023) A holistic view of the genetic factors involved in triggering hydrolytic and oxidative rancidity of rice bran lipids. Food Rev Intl 39(1):441–466. https://doi.org/10.1080/87559129.2021.1915328
8. Liao S et al (2023) Genetic dissection and validation of a major QTL for grain weight on chromosome 3B in bread wheat (Triticum aestivum L.). J Integr Agric 23:77. https://doi.org/10.1016/j.jia.2023.04.023

9. ISAAA (2019) Brief 55: global status of commercialized biotech/GM crops: 2019. ISAAA Brief 55
10. Saltzman A et al (2013) Biofortification: Progress toward a more nourishing future. Glob Food Sec 2(1):9–17. https://doi.org/10.1016/j.gfs.2012.12.003
11. Ali MW, Borrill P (2020) Applying genomic resources to accelerate wheat biofortification. Heredity 125(6):386–395. https://doi.org/10.1038/s41437-020-0326-8
12. Bai C et al (2016) Bottlenecks in carotenoid biosynthesis and accumulation in rice endosperm are influenced by the precursor–product balance. Plant Biotechnol J 14(1):195–205. https://doi.org/10.1111/pbi.12373
13. Goyer A (2017) Thiamin biofortification of crops. Curr Opin Biotechnol 44:1–7. https://doi.org/10.1016/j.copbio.2016.09.005
14. Mangel N et al (2019) Enhancement of vitamin B_6 levels in rice expressing Arabidopsis vitamin B_6 biosynthesis *de novo* genes. Plant J 99(6):1047–1065. https://doi.org/10.1111/tpj.14379
15. Blancquaert D et al (2015) Improving folate (vitamin B9) stability in biofortified rice through metabolic engineering. Nat Biotechnol 33(10):1076–1078. https://doi.org/10.1038/nbt.3358
16. Johnson AAT et al (2011) Constitutive overexpression of the OsNAS gene family reveals single-gene strategies for effective iron- and zinc-biofortification of rice endosperm. PLoS One 6(9):e24476. https://doi.org/10.1371/journal.pone.0024476
17. Guo W et al (2020) A Na+/H+ antiporter, K2-NhaD, improves salt and drought tolerance in cotton (Gossypium hirsutum L.). Plant Mol Biol 102(4–5):553–567. https://doi.org/10.1007/s11103-020-00969-1
18. Shi Y, Li J, Sun Z (2020) Success to iron biofortification of wheat grain by combining both plant and microbial genetics. Rhizosphere 15:100218. https://doi.org/10.1016/j.rhisph.2020.100218
19. Harrington SA et al (2023) A two-gene strategy increases iron and zinc concentrations in wheat flour, improving mineral bioaccessibility. Plant Physiol 191(1):528–541. https://doi.org/10.1093/plphys/kiac499
20. Liu C et al (2020) Development of nutritious rice with high zinc/selenium and low cadmium in grains through QTL pyramiding. J Integr Plant Biol 62(3):349–359. https://doi.org/10.1111/jipb.12909
21. Lee G-H, Yun B-W, Kim K-M (2014) Analysis of QTLs associated with the rice quality related gene by double haploid populations. Int J Genomics 2014:1–6. https://doi.org/10.1155/2014/781832
22. Marcotuli I et al (2017) Development of a high-density SNP-based linkage map and detection of QTL for β-glucans, protein content, grain yield per spike and heading time in durum wheat. Int J Mol Sci 18(6):1329. https://doi.org/10.3390/ijms18061329
23. Tang G, Qin J, Dolnikowski GG, Russell RM, Grusak MA (2009) Golden rice is an effective source of vitamin A. Am J Clin Nutr 89(6):1776–1783. https://doi.org/10.3945/ajcn.2008.27119
24. Requena-Ramírez MD, Rodríguez-Suárez C, Ávila CM, Palomino C, Hornero-Méndez D, Atienza SG (2023) Bread wheat biofortification for grain carotenoid content by inter-specific breeding. Foods 12(7):1365. https://doi.org/10.3390/foods12071365
25. Handa V, Sharma D, Kaur A, Arya SK (2020) Biotechnological applications of microbial phytase and phytic acid in food and feed industries. Biocatal Agric Biotechnol 25:101600. https://doi.org/10.1016/j.bcab.2020.101600
26. Gil-Humanes J, Pistón F, Barro F, Rosell CM (2014) The shutdown of celiac disease-related gliadin epitopes in bread wheat by RNAi provides flours with increased stability and better tolerance to over-mixing. PLoS One 9(3):e91931. https://doi.org/10.1371/journal.pone.0091931
27. Tai L et al (2021) Pre-harvest sprouting in cereals: genetic and biochemical mechanisms. J Exp Bot 72(8):2857–2876. https://doi.org/10.1093/jxb/erab024
28. Xu F et al (2022) Antagonistic control of seed dormancy in rice by two bHLH transcription factors. Nat Genet 54(12):1972–1982. https://doi.org/10.1038/s41588-022-01240-7

29. Song S et al (2020) OsMFT2 is involved in the regulation of ABA signaling-mediated seed germination through interacting with OsbZIP23/66/72 in rice. Plant J 103(2):532–546. https://doi.org/10.1111/tpj.14748
30. Xu F, Tang J, Gao S, Cheng X, Du L, Chu C (2019) Control of rice pre-harvest sprouting by glutaredoxin-mediated abscisic acid signaling. Plant J 100(5):1036–1051. https://doi.org/10.1111/tpj.14501
31. Abe F et al (2019) Genome-edited triple-recessive mutation alters seed dormancy in wheat. Cell Rep 28(5):1362–1369.e4. https://doi.org/10.1016/j.celrep.2019.06.090
32. Maity A et al (2021) Seed shattering: a trait of evolutionary importance in plants. Front Plant Sci 12:657773. https://doi.org/10.3389/fpls.2021.657773
33. Konishi S et al (2006) An SNP caused loss of seed shattering during rice domestication. Science 312(5778):1392–1396. https://doi.org/10.1126/science.1126410
34. Mamidi S et al (2020) A genome resource for green millet Setariaviridis enables discovery of agronomically valuable loci. Nat Biotechnol 38(10):1203–1210. https://doi.org/10.1038/s41587-020-0681-2
35. Tang L et al (2017) Knockout of OsNramp5 using the CRISPR/Cas9 system produces low Cd-accumulating indica rice without compromising yield. Sci Rep 7(1):14438. https://doi.org/10.1038/s41598-017-14832-9
36. Achary VMM, Reddy MK (2021) CRISPR-Cas9 mediated mutation in GRAIN WIDTH and WEIGHT2 (GW2) locus improves aleurone layer and grain nutritional quality in rice. Sci Rep 11(1):21941. https://doi.org/10.1038/s41598-021-00828-z
37. Zhu Y et al (2019) CRISPR/Cas9-mediated functional recovery of the recessive *rc* allele to develop red rice. Plant Biotechnol J 17(11):2096–2105. https://doi.org/10.1111/pbi.13125
38. Dong OX et al (2020) Marker-free carotenoid-enriched rice generated through targeted gene insertion using CRISPR-Cas9. Nat Commun 11(1):1178. https://doi.org/10.1038/s41467-020-14981-y
39. Chen L, Magliano DJ, Zimmet PZ (2012) The worldwide epidemiology of type 2 diabetes mellitus—present and future perspectives. Nat Rev Endocrinol 8(4):228–236. https://doi.org/10.1038/nrendo.2011.183
40. Kim MS, Bae E-J, Uhm J-Y (2022) Psychometric properties of the Korean version of ComOn coaching for oncology nurses. Asia Pac J Oncol Nurs 9(4):210–216. https://doi.org/10.1016/j.apjon.2021.12.016
41. Zhang Y et al (2018) Analysis of the functions of *TaGW2* homoeologs in wheat grain weight and protein content traits. Plant J 94(5):857–866. https://doi.org/10.1111/tpj.13903
42. Sun Y et al (2017) Generation of high-amylose Rice through CRISPR/Cas9-mediated targeted mutagenesis of starch branching enzymes. Front Plant Sci 8:298. https://doi.org/10.3389/fpls.2017.00298
43. Shufen C et al (2019) Editing of rice isoamylase gene ISA1 provides insights into its function in starch formation. Rice Sci 26(2):77–87. https://doi.org/10.1016/j.rsci.2018.07.001
44. Ashokkumar S et al (2020) Creation of novel alleles of fragrance gene OsBADH2 in rice through CRISPR/Cas9 mediated gene editing. PLoS One 15(8):e0237018. https://doi.org/10.1371/journal.pone.0237018
45. Bahariah B, Masani MYA, Rasid OA, Parveez GKA (2021) Multiplex CRISPR/Cas9-mediated genome editing of the FAD2 gene in rice: a model genome editing system for oil palm. J Genet Eng Biotechnol 19(1):86. https://doi.org/10.1186/s43141-021-00185-4
46. Zhao D-S et al (2018) GS9 acts as a transcriptional activator to regulate rice grain shape and appearance quality. Nat Commun 9(1):1240. https://doi.org/10.1038/s41467-018-03616-y
47. Liu X et al (2021) Fine-tuning flowering time via genome editing of upstream open Reading frames of heading date 2 in rice. Rice 14(1):59. https://doi.org/10.1186/s12284-021-00504-w
48. Jouanin A et al (2020) CRISPR/Cas9 gene editing of gluten in wheat to reduce gluten content and exposure—reviewing methods to screen for coeliac safety. Front Nutr 7:51. https://doi.org/10.3389/fnut.2020.00051
49. Liu D et al (2023) An elite *γ-gliadin* allele improves end-use quality in wheat. New Phytol 239(1):87–101. https://doi.org/10.1111/nph.18722

50. Camerlengo F et al (2020) CRISPR-Cas9 multiplex editing of the α-amylase/trypsin inhibitor genes to reduce allergen proteins in durum wheat. Front Sustain Food Syst 4:104. https://doi.org/10.3389/fsufs.2020.00104
51. Zhang J, Zhang H, Li S, Li J, Yan L, Xia L (2021) Increasing yield potential through manipulating of an *ARE1* ortholog related to nitrogen use efficiency in wheat by CRISPR/Cas9. J Integr Plant Biol 63(9):1649–1663. https://doi.org/10.1111/jipb.13151
52. Li G et al (2021) A high-quality genome assembly highlights rye genomic characteristics and agronomically important genes. Nat Genet 53(4):574–584. https://doi.org/10.1038/s41588-021-00808-z
53. Waltz E (2016) CRISPR-edited crops free to enter market, skip regulation. Nat Biotechnol 34(6):582–582. https://doi.org/10.1038/nbt0616-582
54. Barampama Z, Simard RE (1995) Effects of soaking, cooking and fermentation on composition, in-vitro starch digestibility and nutritive value of common beans. Plant Foods Hum Nutr 48(4):349–365. https://doi.org/10.1007/BF01088494
55. Moore MM, Bello FD, Arendt EK (2008) Sourdough fermented by Lactobacillus plantarum FST 1.7 improves the quality and shelf life of gluten-free bread. Eur Food Res Technol 226(6):1309–1316. https://doi.org/10.1007/s00217-007-0659-z
56. Ogunsakin OA, Banwo K, Ogunremi OR, Sanni AI (2015) Microbiological and physicochemical properties of sourdough bread from sorghum flour. Int Food Res J 22(6):2610–2618
57. Różyło R, Rudy S, Krzykowski A, Dziki D, Siastała M, Polak R (2016) Gluten-free bread prepared with fresh and freeze-dried Rice sourdough-texture and sensory evaluation. J Texture Stud 47(5):443–453. https://doi.org/10.1111/jtxs.12180
58. Pelembe LAM, Dewar J, Taylor JRN (2004) Effect of germination moisture and time on pearl millet malt quality—with respect to its opaque and lager beer brewing potential. J Inst Brew 110(4):320–325. https://doi.org/10.1002/j.2050-0416.2004.tb00627.x
59. Bultosa G, Taylor JRN (2004) Paste and gel properties and in vitro digestibility of Tef [Eragrostis tef (Zucc.) Trotter] starch. Starch 56(1):20–28. https://doi.org/10.1002/star.200200191
60. Kitabatake N, Gimbi DM, Oi Y (2003) Traditional non-alcoholic beverage, *Togwa*, in East Africa, produced from maize flour and germinated finger millet. Int J Food Sci Nutr 54(6):447–455. https://doi.org/10.1080/09637480120092053
61. Banik A, Ghosh K, Pal S, Halder SK, Ghosh C, Mondal KC (2020) Biofortification of multigrain substrates by probiotic yeast. Food Biotechnol 34(4):283–305. https://doi.org/10.1080/08905436.2020.1833913
62. Capozzi V et al (2011) Biotechnological production of vitamin B2-enriched bread and pasta. J Agric Food Chem 59(14):8013–8020. https://doi.org/10.1021/jf201519h
63. Linko Y-Y, Javanainen P, Linko S (1997) Biotechnology of bread baking. Trends Food Sci Technol 8(10):339–344. https://doi.org/10.1016/S0924-2244(97)01066-2
64. Dahiya S, Bajaj BK, Kumar A, Tiwari SK, Singh B (2020) A review on biotechnological potential of multifarious enzymes in bread making. Process Biochem 99:290–306. https://doi.org/10.1016/j.procbio.2020.09.002
65. Sicard D, Legras J-L (2011) Bread, beer and wine: yeast domestication in the Saccharomyces sensu stricto complex. C R Biol 334(3):229–236. https://doi.org/10.1016/j.crvi.2010.12.016
66. Bigey F et al (2021) Evidence for two main domestication trajectories in Saccharomyces cerevisiae linked to distinct bread-making processes. Curr Biol 31(4):722–732.e5. https://doi.org/10.1016/j.cub.2020.11.016
67. Landis EA et al (2021) The diversity and function of sourdough starter microbiomes. eLife 10:e61644. https://doi.org/10.7554/eLife.61644
68. D'Alessandro A, De Pergola G (2014) Mediterranean diet pyramid: a proposal for Italian people. Nutrients 6(10):4302–4316. https://doi.org/10.3390/nu6104302
69. Poutanen K, Flander L, Katina K (2009) Sourdough and cereal fermentation in a nutritional perspective. Food Microbiol 26(7):693–699. https://doi.org/10.1016/j.fm.2009.07.011
70. Ailea WF, Thompsoa JJ Enzymatic treatment of cereal grains. United States Patent Office

71. Marie E, Blanchon J, De Villiers A (1963) Production of a food product by the enzymatic treatment of cereal grains. United States Patent Office
72. Tobey Jr. JF, McGee SJ, Cobb CW, Cortner W (1994) Enzymatic grain conditioner and methods of using it. US5662901A
73. Robert Jeffery P (2007) Enzymatic treatment process for cereals. WO2007051091A2
74. Samtiya M, Aluko RE, Dhewa T (2020) Plant food anti-nutritional factors and their reduction strategies: an overview. Food Prod Process Nutr 2(1):6. https://doi.org/10.1186/s43014-020-0020-5
75. Ercolini D et al (2013) Microbial ecology dynamics during rye and wheat sourdough preparation. Appl Environ Microbiol 79(24):7827–7836. https://doi.org/10.1128/AEM.02955-13
76. Yadav S, Khetarpaul N (1994) Indigenous legume fermentation: effect on some antinutrients and in-vitro digestibility of starch and protein. Food Chem 50(4):403–406. https://doi.org/10.1016/0308-8146(94)90213-5
77. Ibrahim SS, Habiba RA, Shatta AA, Embaby HE (2002) Effect of soaking, germination, cooking and fermentation on antinutritional factors in cowpeas. Nahrung 46(2):92–95. https://doi.org/10.1002/1521-3803(20020301)46:2<92::AID-FOOD92>3.0.CO;2-P
78. Rocha O, Ansari K, Doohan FM (2005) Effects of trichothecene mycotoxins on eukaryotic cells: a review. Food Addit Contam 22(4):369–378. https://doi.org/10.1080/02652030500058403
79. Tian Y et al (2018) Antagonistic and detoxification potentials of Trichoderma isolates for control of Zearalenone (ZEN) producing Fusarium graminearum. Front Microbiol 8:2710. https://doi.org/10.3389/fmicb.2017.02710
80. Alam S, Nisa S, Daud S (2022) Mycotoxins in environment and its health implications, pp 289–318. https://doi.org/10.1007/978-3-030-96523-5_12
81. Ostry V, Malir F, Toman J, Grosse Y (2017) Mycotoxins as human carcinogens—the IARC monographs classification. Mycotoxin Res 33(1):65–73. https://doi.org/10.1007/s12550-016-0265-7
82. Wan J, Chen B, Rao J (2020) Occurrence and preventive strategies to control mycotoxins in cereal-based food. Compr Rev Food Sci Food Saf 19(3):928–953. https://doi.org/10.1111/1541-4337.12546
83. Kottapalli B, Wolf-Hall CE, Schwarz P (2006) Effect of electron-beam irradiation on the safety and quality of fusarium-infected malting barley. Int J Food Microbiol 110(3):224–231. https://doi.org/10.1016/j.ijfoodmicro.2006.04.007
84. Haque MA, Wang Y, Shen Z, Li X, Saleemi MK, He C (2020) Mycotoxin contamination and control strategy in human, domestic animal and poultry: a review. Microb Pathog 142:104095. https://doi.org/10.1016/j.micpath.2020.104095
85. Peles F, Sipos P, Kovács S, Győri Z, Pócsi I, Pusztahelyi T (2021) Biological control and mitigation of aflatoxin contamination in commodities. Toxins (Basel) 13(2):104. https://doi.org/10.3390/toxins13020104
86. Ahlberg SH, Joutsjoki V, Korhonen HJ (2015) Potential of lactic acid bacteria in aflatoxin risk mitigation. Int J Food Microbiol 207:87–102. https://doi.org/10.1016/j.ijfoodmicro.2015.04.042
87. Li J, Wang W, Chen S, Shao T, Tao X, Yuan X (2021) Effect of lactic acid bacteria on the fermentation quality and mycotoxins concentrations of corn silage infested with Mycotoxigenic fungi. Toxins (Basel) 13(10):699. https://doi.org/10.3390/toxins13100699
88. Adebo OA, Njobeh PB, Gbashi S, Nwinyi OC, Mavumengwana V (2017) Review on microbial degradation of aflatoxins. Crit Rev Food Sci Nutr 57(15):3208–3217. https://doi.org/10.1080/10408398.2015.1106440
89. Yang Q, Wang J, Zhang H, Li C, Zhang X (2016) Ochratoxin A is degraded by Yarrowia lipolytica and generates non-toxic degradation products. World Mycotoxin J 9(2):269–278. https://doi.org/10.3920/WMJ2015.1911
90. Jia L, Zhao L, Qin B, Lu F, Liu D, Liu F (2023) Enhancement of rice husks saccharification through hydrolase preparation assisted by lytic polysaccharide monooxygenase. Enzyme Microb Technol 171:110319. https://doi.org/10.1016/j.enzmictec.2023.110319

91. Ogunyewo OA, Randhawa A, Gupta M, Kaladhar VC, Verma PK, Yazdani SS (2020) Synergistic action of a lytic polysaccharide monooxygenase and a cellobiohydrolase from *Penicillium funiculosum* in cellulose saccharification under high-level substrate loading. Appl Environ Microbiol 86(23):e01769–e01720. https://doi.org/10.1128/AEM.01769-20
92. Semenova MV et al (2019) Enhancement of the enzymatic cellulose saccharification by Penicillium verruculosum multienzyme cocktails containing homologously overexpressed lytic polysaccharide monooxygenase. Mol Biol Rep 46(2):2363–2370. https://doi.org/10.1007/s11033-019-04693-y
93. Degli Esposti M et al (2021) The role of biotechnology in the transition from plastics to bioplastics: an opportunity to reconnect global growth with sustainability. FEBS Open Biol 11(4):967–983. https://doi.org/10.1002/2211-5463.13119
94. Beasley JT et al (2019) Metabolic engineering of bread wheat improves grain iron concentration and bioavailability. Plant Biotechnol J 17(8):1514–1526. https://doi.org/10.1111/pbi.13074
95. Singh A, Yadav A, Roy JK, Bhati KK (2021) Role of genome engineering for the development of resistant starch-rich, allergen-free and processing quality improved cereal crops. In: Genome engineering for crop improvement. Wiley, Hoboken, pp 333–350. https://doi.org/10.1002/9781119672425.ch19
96. Lafiandra D, Sestili F, Sissons M, Kiszonas A, Morris CF (2022) Increasing the versatility of durum wheat through modifications of protein and starch composition and grain hardness. Foods 11(11):1532. https://doi.org/10.3390/foods11111532
97. Ruan Y et al (2020) High density mapping of quantitative trait loci conferring gluten strength in Canadian durum wheat. Front Plant Sci 11:170. https://doi.org/10.3389/fpls.2020.00170
98. Higgins VJ, Bell PJ, Dawes IW, Attfield PV (2001) Generation of a novel Saccharomyces cerevisiae strain that exhibits strong maltose utilization and hyperosmotic resistance using nonrecombinant techniques. Appl Environ Microbiol 67(9):4346–4348. https://doi.org/10.1128/AEM.67.9.4346-4348.2001
99. Rosell CM, Rojas JA, Benedito de Barber C (2001) Influence of hydrocolloids on dough rheology and bread quality. Food Hydrocoll 15(1):75–81. https://doi.org/10.1016/S0268-005X(00)00054-0
100. Sheikholeslami Z, Mahfouzi M, Karimi M, Ghiafehdavoodi M (2021) Modification of dough characteristics and baking quality based on whole wheat flour by enzymes and emulsifiers supplementation. LWT 139:110794. https://doi.org/10.1016/j.lwt.2020.110794
101. Farahmand E, Razavi SH, Mohtasebi SS (2021) Investigating effective variables to produce desirable aroma in sourdough using e-nose and sensory panel. J Food Process Preserv 45(2):e15157. https://doi.org/10.1111/jfpp.15157
102. Zhou Z et al (2022) Promoter DNA hypermethylation of TaGli-γ-2.1 positively regulates gluten strength in bread wheat. J Adv Res 36:163–173. https://doi.org/10.1016/j.jare.2021.06.021
103. Abedi E, Pourmohammadi K, Sayadi M (2022) Synergic effect of phytase, amylase, galactosidase, and asparaginase activity on the mitigation of acrylamide and hydroxymethylfurfural in roll bread by co-culture fermentation. J Food Compos Anal 106:104355. https://doi.org/10.1016/j.jfca.2021.104355
104. Fang L et al (2023) Effects of mixed fermentation of different lactic acid bacteria and yeast on phytic acid degradation and flavor compounds in sourdough. LWT 174:114438. https://doi.org/10.1016/j.lwt.2023.114438
105. Zhang Y, Momoisea P, Lin Q, Liang J, Burrow K, Serventi L (2023) Evaluation of sensory and physicochemical characteristics of vitamin B12 enriched whole-meal sourdough bread fermented with Propionibacterium freudenreichii. Sustainability 15(10):8157. https://doi.org/10.3390/su15108157
106. Ral J et al (2016) Engineering high α-amylase levels in wheat grain lowers Falling Number but improves baking properties. Plant Biotechnol J 14(1):364–376. https://doi.org/10.1111/pbi.12390
107. Zhao X et al (2023) Population genomics unravels the Holocene history of bread wheat and its relatives. Nat Plants 9(3):403–419. https://doi.org/10.1038/s41477-023-01367-3

108. Li J et al (2021) Modification of starch composition, structure and properties through editing of *TaSBEIIa* in both winter and spring wheat varieties by CRISPR/Cas9. Plant Biotechnol J 19(5):937–951. https://doi.org/10.1111/pbi.13519
109. Caballero PA, Gómez M, Rosell CM (2007) Bread quality and dough rheology of enzyme-supplemented wheat flour. Eur Food Res Technol 224(5):525–534. https://doi.org/10.1007/s00217-006-0311-3
110. Scarnato L et al (2017) New bread formulation with improved rheological properties and longer shelf-life by the combined use of transglutaminase and sourdough. LWT Food Sci Technol 81:101–110. https://doi.org/10.1016/j.lwt.2017.03.042
111. Wang X, Pei D, Teng Y, Liang J (2018) Effects of enzymes to improve sensory quality of frozen dough bread and analysis on its mechanism. J Food Sci Technol 55(1):389–398. https://doi.org/10.1007/s13197-017-2950-8
112. Poutanen K (1997) Enzymes: an important tool in the improvement of the quality of cereal foods. Trends Food Sci Technol 8(9):300–306. https://doi.org/10.1016/S0924-2244(97)01063-7
113. Gu M et al (2022) Effects of a commercial peptidase on rheology, microstructure, gluten properties of wheat dough and bread quality. LWT 160:113266. https://doi.org/10.1016/j.lwt.2022.113266
114. Moayedallaie S, Mirzaei M, Paterson J (2010) Bread improvers: comparison of a range of lipases with a traditional emulsifier. Food Chem 122(3):495–499. https://doi.org/10.1016/j.foodchem.2009.10.033
115. Pourmohammadi K, Abedi E (2021) Enzymatic modifications of gluten protein: oxidative enzymes. Food Chem 356:129679. https://doi.org/10.1016/j.foodchem.2021.129679
116. Nwagu TNT et al (2021) A novel strain of *Yarrowiaphangngaensis* producing a multienzyme complex; a source of enzyme additives for baking high cassava-wheat composite bread. Food Biotechnol 35(2):158–177. https://doi.org/10.1080/08905436.2021.1910520
117. Sha HY, Wang QQ, Li ZJ (2023) Comparison of the effect of exopolysaccharide-producing lactic acid bacteria from sourdough on dough characteristics and steamed bread quality. Int J Food Sci Technol 58(1):378–386. https://doi.org/10.1111/ijfs.15881
118. Wang G et al (2023) Enhancing the technofunctionality of γ-aminobutyric acid enriched germinated wheat by modification of arabinoxylan, gluten proteins and liquid lamella of dough. Food Chem 404:134523. https://doi.org/10.1016/j.foodchem.2022.134523
119. Xu Z et al (2024) Effect of liquid fermented Chinese edible Ganoderma Lucidum fungus on wheat bread: a quality improver and staling inhibitor. Food Control 155:110060. https://doi.org/10.1016/j.foodcont.2023.110060
120. Nachi I, Fhoula I, Smida I, Ouzari H-I, Hassouna M (2019) Microbiological analysis and assessment of biotechnological potential of lactic acid bacteria isolated from Tunisian flours. Ann Microbiol 69(1):29–40. https://doi.org/10.1007/s13213-018-1365-8
121. Calasso M et al (2023) Shelf-life extension of leavened bakery products by using bio-protective cultures and type-III sourdough. LWT 177:114587. https://doi.org/10.1016/j.lwt.2023.114587
122. Dopazo V et al (2023) Evaluation of shelf life and technological properties of bread elaborated with lactic acid bacteria fermented whey as a bio preservation ingredient. LWT 174:114427. https://doi.org/10.1016/j.lwt.2023.114427
123. Qaim M (2020) Role of new plant breeding technologies for food security and sustainable agricultural development. Appl Econ Perspect Policy 42(2):129–150. https://doi.org/10.1002/aepp.13044
124. Kumar K et al (2020) Genetically modified crops: current status and future prospects. Planta 251(4):91. https://doi.org/10.1007/s00425-020-03372-8
125. Singh A, Karmakar S, Jacob BS, Bhattacharya P, Kumar SPJ, Banerjee R (2015) Enzymatic polishing of cereal grains for improved nutrient retainment. J Food Sci Technol 52(6):3147–3157. https://doi.org/10.1007/s13197-014-1405-8
126. Mishra P et al (2016) DNA barcoding: an efficient tool to overcome authentication challenges in the herbal market. Plant Biotechnol J 14(1):8–21. https://doi.org/10.1111/pbi.12419

127. Hollingsworth PM, Graham SW, Little DP (2011) Choosing and using a plant DNA barcode. PLoS One 6(5):e19254. https://doi.org/10.1371/journal.pone.0019254
128. Salem HH, Ali BA, Huang T, Qin D, Wang X, Xie Q (2007) Use of random amplified polymorphic DNA analysis for economically important food crops. J Integr Plant Biol 49(12):1670–1680. https://doi.org/10.1111/j.1744-7909.2007.00582.x
129. Auer CA (2003) Tracking genes from seed to supermarket: techniques and trends. Trends Plant Sci 8(12):591–597. https://doi.org/10.1016/j.tplants.2003.10.010
130. Galimberti A et al (2013) DNA barcoding as a new tool for food traceability. Food Res Int 50(1):55–63. https://doi.org/10.1016/j.foodres.2012.09.036
131. Hou X, Liu J-E, Zhao Y-Z (2008) Molecular phylogeny of Caragana (Fabaceae) in China. J Syst Evol 46(4):600–600
132. Nehal N, Choudhary B, Nagpure A, Gupta RK (2021) DNA barcoding: a modern age tool for detection of adulteration in food. Crit Rev Biotechnol 41(5):767–791. https://doi.org/10.1080/07388551.2021.1874279

Chapter 8
Food Packaging Technology

Shoba H., Ramappa, and S. K. Jain

1 Introduction

The ancient people long ago simply consumed whatever they could acquire from their natural surroundings. But slowly they understood the importance of processing and preservation of the foods, when they were shifted to a living or sheltered regions and ever since people discovered how crucial it was to preserve food for the longer periods with the aid of development of food packing and storage technologies. People used plant leaves to hold or cover food in the prehistoric era, which may have prompted the emergence of food packing.

The idea of safeguarding food goods emerged upon the civilization developed in response to the growing demands. In order to hold food for prolonged period, until the 1800s packaging materials were from naturally existing things like leaves, gourds, shells, bamboo, wood, and grasses have been used to construct baskets. Glass, paper, and earthenware were a few of the earliest materials that were shaped into food containers [1].

Shoba H. (✉)
Department of Processing and Food Engineering, CTAE, Maharana Pratap University of Agriculture and Technology, Udaipur, Rajasthan, India

Agricultural Engineering, College of Horticulture, Munirabad, Karnataka, India

University of Horticulture Sciences, Bagalkote, Karnataka, India

Ramappa · S. K. Jain
Department of Processing and Food Engineering, CTAE, Maharana Pratap University of Agriculture and Technology, Udaipur, Rajasthan, India

J. S. Yaradoddi et al. (eds.), *Frontiers in Food Biotechnology*,
https://doi.org/10.1007/978-981-97-3261-6_8

2 Packaging

Food packaging is done to protect food components since numerous direct and indirect pollutants contact could cause the damage, leakage, and other potential problems. Food packaging supports the supply chain's efforts to preserve and store food. In general, packaging is the final step in which already-packaged goods or products are placed into courier bags, bubble envelopes, larger boxes, cartons, cardboard, etc. But packaging and packing are frequently used in the same sentence. A product's presentation through packaging is an art. It is more important to focus on the end product's appearance, color, design, or presentation when it comes to luring customers. Additionally, packaging is crucial for luring clients and influencing their purchase of products.

In recent years, the packaging sector has undergone a significant upheaval. Approximately a few decades ago, the only packaging options were gunny sacks, metal cans, and glass jars; nevertheless, today's items are typically available in attractive, well-organized packets. In recent years, the production of new food packaging materials has increased customer convenience while simultaneously extending the shelf life of foods and improving their safety and quality.

Various package types are listed as below:

2.1 Types of Packaging

There are three basic types of packaging:

- Primary packaging
- Secondary packaging
- Tertiary packaging

- **Primary packaging:** Such type of package generally has a direct touch with the packed goods. It holds the item and protects it. Typically, this is the smallest distribution or usage unit, for example, a bag of chips.
- **Secondary packaging**: The material existing outside the primary packing is referred to as secondary packaging. Additionally, it is employed to group of primary packages, for example, breakfast cereal boxes made of paperboard, such as cornflakes, and cans of beverages in cartons or trays.
- **Tertiary packing:** The substance used to hold secondary packets is referred to as tertiary packing. It is applied to bulk packing to make handling, distribution, shipping, and storage easier, for example, cardboard boxes and wooden crates (Fig. 8.1).

Fig. 8.1 Indicating various types of the food packaging

2.2 Functions of Food Packaging

Packing serves three main functions:

- Protection
- Preservation
- Promotion

The main purposes of food packaging are to shield food products from external factors including mechanical, chemical, or biological harm, to enclose the food also in maintaining within its packaged state, and provide customers with nutritional information and ingredient. The relevance of packaging's auxiliary purposes, such as convenience, traceability, and tamper or theft indication, is rising. The main aim of the food packaging is to fill and present food in an economical manner that satisfies both industry standards and consumer expectations, needs, and desires [2].

2.3 *Food Packaging Characteristics*

Here are a few key characteristics of an effective package:

- Food was shielded from physical harm including breakage, spillage, and theft during transit or storage as well as microbial contamination.
- Compressing and lowering the volume of particular goods.
- The good packaging inhibits permeability of water vapor, gases (e.g., O_2, CO_2, N_2, and ethylene), and other unstable compounds producing odor, out or into the package.
- Packaging shouldn't be poisonous, also shouldn't allow compounds to move from one part of the food packed to other part inside the packing material.
- Packaging must have effective printing or tagging capabilities.
- The good packaging should make handling process, storing, and transportation easier.
- It ought to assist with product branding, differentiation, and advertising for instance, Maggi noodles, Kissan jam, and Amul butter.
- It must be appealing to entice the observer to give it a try.
- It ought to be inexpensive, reusable, or simply discardable.

Advantages

1. Packaging does not let taste or flavor because of its impermeability to gases, moisture, vapors, odors, and microorganisms. This keeps products fresh for a long time.
2. It does not react with the food that is packaged or travel into food products since it is inert.
3. It is suitable for heat sterilization or heat processing when hermetically sealed since it can tolerate high processing temperatures.
4. It is microwave-transparent.
5. It is translucent, allowing the contents to be seen by the consumer and allowing them to see the product.
6. It can be shaped into a wide variety of hues and shapes.
7. Glass color differences can shield contents that are sensitive to light.
8. Its filling rates are comparable to those of cans.
9. It may be stacked without damaging the container since it is rigid, provides good insulation, and has good vertical strength.
10. It can be recycled, reused, and resealed.

3 Various Classes of Food Packaging Material

A food product's shelf life is significantly influenced by the type of packaging used. In order to preserve product quality during distribution and storage, it is crucial to choose the proper packaging materials and technologies. Paper, paperboards, glass, metals, and plastic are some of the packaging materials that have historically been used in food packing. For imparting practical or aesthetic qualities, modern food packaging materials frequently combine numerous other materials.

3.1 Food Packaging Material

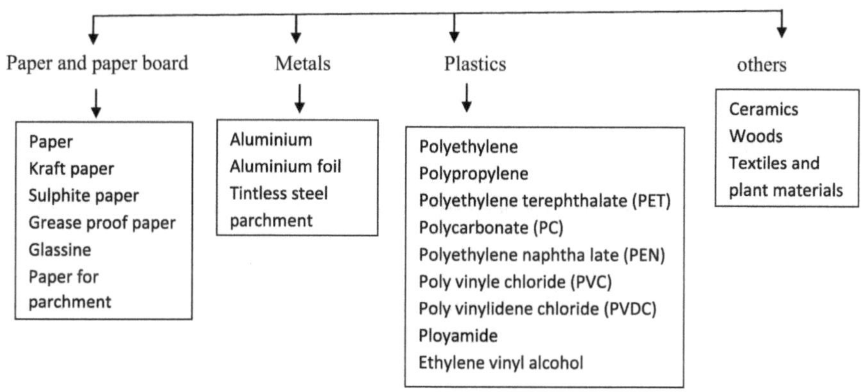

3.2 Paperboard and Paper

Paperboards and paper have been used to package of food since the seventeenth century, and their use for this purpose increased in the latter half of the nineteenth century [3]. The majority of papers used to package food are typically created from cellulose fiber that comes from wood. Pulping is the initial step in the production of paper or paperboard, and it is carried out by mechanically grinding wood. Compared to chemical pulp, paper derived by such type of pulp is comparatively weaker. Chemical pulp can be produced by treating wood chips through an alkali or an acid, followed via washing (sulfite pulp) [4].

Paper

The plain paper is having low barrier qualities, and is not utilized for food packaging for extended periods of time in order to enhance its functional and protective qualities; it is frequently coated, laminated, treated, or lacquered by substances like resins, lacquers, or waxes. The following are some examples of papers that are used as packing materials.

Kraft Paper

To make kraft paper first sulfate pulp is alkali-treated. There are two types of kraft papers: unbleached (brown) and bleached. The strongest paper is kraft that hasn't been treated. It is a versatile paper used for wrapping and bagging. Flour, sugar, dried fruits, and vegetables are also packed in it.

Sulfite Paper

This type of paper generally produced from pulp that has undergone acid digestion. Compared to kraft paper, sulfite paper is weaker and lighter. It is fixed to enhance its aesthetics, boost its wet strength, and also improve its oil barrier properties, which is frequently coated and combined with plastic or foil laminates to enhance print quality. It is used to package biscuits and candy in the form of wrappers, bags, or sachets.

Greaseproof Paper

Greaseproof paper is produced from sulfite pulp, which undergoes a procedure known as beating. This process involves breaking down the fibers and rendering them gelatinous due to an extended hydration period longer than usual. After being densely packed, these finely fragmented fibers are barrier to oils, except wetting agents. Greasyproof paper is employed to cover fatty items like candy bars, pastries, and snack foods.

Glassine

It is a processed version of greaseproof paper that has undergone extra hydration to create extremely tightly packed sheet with a fantastic glossy and smooth surface. It has a little barrier property to moisture penetration compared to greaseproof paper. It serves as a liner for biscuits, fast food, baked goods, and cooking lipids like butter.

Paper for Parchment

Paper for parchment is manufactured from pulp that has been acid-treated (passed through a sulfuric acid bath), then it is rinsed, neutralized, and dried. The acid changes the cellulose, making it smooth and water- and oil-impermeable. It is a poor barrier to moisture and air, though. Butter and other fats are packed using it.

3.3 Metals

The metals are utmost adaptable among all packaging materials, metal is crucial to the "Canning" process, which involves preserving and storing food. In the last century, both industrialized and developing nations have incorporated canned food into their diets. It has strong barrier and physical protective qualities. It is a versatile material due to its ability to be formed and printed, recyclability, and customer acceptance. Aluminum and steel are the metals that are used in food packaging the most frequently. Metal food packaging materials including coated plates (tinplate, tin-free steel, and polymer coated steel) are also frequently used.

Aluminum

Aluminum is a silvery-white, lightweight metal. It has an extremely effective barrier property against air, temperature, moisture, and chemicals. It is frequently used as the primary package in the food sector. Because it can be easily formed into thin sheets and folded or rolled, unlike many other metals, it is more adaptable. It is the perfect material for recycling because it is simple to recover and transform into new goods. Corrosion is very resistant to aluminum and its many alloys. In most cases, aluminum is coated for packaging purposes. Aluminum's main drawbacks are its high price in comparison to other metals like steel and its inability to be welded, which limits its use to the production of seamless containers. Aluminum foil and aluminum cans are the two main types of aluminum used for food packaging.

Aluminum Foil

It is made by rolling extremely thin sheets of pure aluminum metal and then annealing them. There is a large variety of thicknesses available for aluminum foil. Food is wrapped in the thinner foils, and trays or other containers are made out of the thicker foils. It is a great barrier to light, gas, scents, tastes, moisture, and microbes. To preserve food, aluminum foil is used to wrap it. It is frequently offered as rolls on the market.

Tinless Steel

Electrolytic chromium-coated steel (ECCS), generally termed as tin-free steel, is made up of steel plates that have had chromium electrodeposited on both sides. It offers less corrosion protection than tinplate and needs an organic material layer to offer full corrosion protection. Compared to tinplate, it is more resistant to sulfur taint and weak acids. Tin-free steel is unsuitable for welding due to the presence of chrome oxide, but it works incredibly well for coatings like lacquers, paints, and inks.

3.4 Plastics Packaging

However, in the year 1800s the discovery of plastics took place, packaging did not begin to employ them until the twentieth century. It was originally employed during World War II, although it wasn't until after the war that food packaging became commercially viable. Due to their extreme malleability, plastics may be molded into a variety of solid things. Because of its lower price, simple to producibility, and hydrophobicity, it is employed in a broad array of goods on various scales. Approximately one-third of plastics are generally utilized in packaging process by the industrialized nations [5].

Despite the fact that plastic poses numerous safety risks, it is increasingly used for food packaging instead of more expensive materials like glass and metal because of its low cost and practical benefits, such as its microwave-capability, thermo sealability, optical properties, and ability to be made in an infinite number of sizes and shapes [6]. By adding, condensing, or crosslinking monomer units during polymerization, plastics can be created. Thermosets and thermoplastics are the two main subcategories of plastics. Thermoplastics are polymers that melt and soften when heated and then resolidify when cooled. Polyethylene, polypropylene, and polyvinyl chloride are a few examples. Thermosets are polymers that, unlike thermoplastics, solidify upon cooling; once they have taken on their final shape, they cannot be reshaped or changed back to their initial form. They are sturdy and firm.

3.5 Various Plastics

Plastics come in a variety of forms, as will be mentioned below. The two most widely used forms of plastic are polyolefins and polyesters.

Polyolefins

Polyethylene and polypropylene, the two polymers most frequently used in food packaging, are collectively referred to as polyolefins. Similar characteristics shared by polyethylene and polypropylene include flexibility, lightness, strength, stability, resistance to moisture and chemicals, and ease of processing.

1. **Polyethylene**

   The most straightforward, adaptable, and affordable plastic is polyethylene (PE), which is made by polymerizing ethylene. One of the first plastics that was frequently used for food packaging was polyethylene (PE). Low-density (LDPE), high-density (HDPE), linear low-density (LLDPE), and very low density (VLDPE) polyethylene are among the varieties currently in use [7].

   (a) LDPE Imperial Chemical Industries invented the first LDPE in 1933. It is pliable, simple to seal, durable, sturdy, and moisture-resistant. LDPE is utilized in applications that require heat sealing since it is comparatively transparent. Because of its great cold resistance (up to −70 °C), LDPE is utilized in the packaging of frozen foods, for example, squeezable food bottles, flexible lids, and bread and frozen food bags. The HDPE is stiff, resilient, strong, impermeable to gas and moisture, simple to process, and simple to form. It is also resistant to chemicals. In comparison to LDPE, HDPE is a stronger plastic with a higher melting point. It is employed in the production of margarine tubs, juice, milk, and water bottles.

2. Polypropylene (PP)

   Compared to polyethylene, it is denser, tougher, and more transparent. It is an excellent water vapor barrier and has good chemical resistance. The melting and hardness points of polypropylene's various forms vary. Due to its high melting point (160 °C), it is better suited for uses requiring heat resistance, such as hot-filling and microwaveable packaging. It is mostly used in yoghurt and margarine tubs.

3. Polyesters

   When carboxylic acid and alcohol react to form ester monomers, polyester is created through condensation polymerization. Polycarbonate (PC), polyethylene naphthalate (PEN), and polyethylene terephthalate (PET) are available varieties of polyesters. However, PET is the polyester that is most frequently used in packaging the foods.

4. PET (Polyethylene Terephthalate)

   Ethylene glycol and terephthalic acid are condensation polymerized to create PET, a thermoplastic polyester. PET dominates a sizable portion of the market for beverage and mineral water packaging. Its gas barrier capabilities, transparency, shatter resistance, and light weight, make it a top choice.

5. Polycarbonate (PC)
   Carbonyl dichloride (phosgene) and a sodium salt of bisphenol acid are polymerized to create PC. Although its gas barrier characteristics are only fair, it offers a very good barrier against flavors and odors. PC can withstand a broad range of temperatures. It has a 230 °C melting point and a −100 °C brittle temperature. Cleaning polycarbonate requires caution since abrasive cleaners like sodium hypochlorite can trigger the release of bisphenol-A, which is thought to pose a risk to human health.
6. Polyethylene Naphthalate (PEN)
   Ethylene glycol and dimethyl naphthalene dicarboxylate are condensation polymerized to produce PEN. It can withstand temperatures better. Despite having strong barrier qualities for carbon dioxide, oxygen, and water vapor that are superior to those of PET, PEN is chemically comparable to PET. It offers improved heat resistance, making it more ideal for hot refills, rewashing, and reuse. It costs more than PET does.
7. PVC, or Polyvinyl Chloride
   By polymerizing vinyl chloride monomer, PVC is produced. It is a substance that is translucent, amorphous, ductile, rigid, strong, and heavy. Grease and oil, as well as substances like acids and bases, are all effectively withstood by it. It has consistent electrical properties and good flow characteristics. PVC is used to create a variety of flexible items.
   The usage of alternative non-phthalate plasticizers, such as adipates, which have the potential to leach into food but at lower levels than phthalates, is employed in place of phthalates in food packaging materials due to a number of safety and health concerns (HHS 2005). PVC is mostly utilized in non-food applications, but it can be found in bottles and packaging materials for food. For meat product blister packing, PVC sheets are frequently utilized. Burning PVC results in environmental issues because of the chlorine concentration.
8. PVDC or Polyvinylidene Chloride
   Vinylidene chloride is polymerized to produce PVDC. It has a great barrier to fats, oils, gases, and water vapor. It can be heat-sealed. The packaging of cured meats, poultry, cheese, tea, coffee snack foods, and confectionery are among the common uses of PVDC. Retorting, hot filling, low-temperature storage, and modified-atmosphere packaging are further uses for it. PVDC also poses environmental issues when burned, since it contains twice as much chlorine as PVC. It is commonly used in meal trays, disposable plasticware, disposable cups, plates, and egg trays.
9. Polyamide (Nylon)
   Nylon is a popular name for polyamides. It is formed through the condensation polymerization of diamine and diacid. The repeating units in polyamides are bound mutually by amide links in the main chain. The amount of carbons in the starting monomer serves to distinguish various forms of polyamides. Nylon-6, for instance, which is used in packaging, comprises six carbon atoms.

Nylon is strong, low gas permeable, and has good chemical resistance. It has similar utility to PET because of its similar mechanical and thermal characteristics.

10. EVOH (Ethylene Vinyl Alcohol)

    Ethylene and vinyl alcohol are linked together to form EVOH in a single polymer chain. It functions admirably as a barrier against oxygen, fat, and grease. EVOH provides great gas performance in dry weather gas-barrier characteristics. However, it is significantly impacted by moisture, hence multilayer co-extruded films are typically utilized.

3.6 Ceramics

The majority of ceramics are used in kitchenware. Ceramics may include heavy metal contamination since they are created from natural elements like clay, silica, and kaolin. These might contaminate the food within by migrating. The EU Framework Regulation EC 1935/2004 regulates ceramics used as packaging materials. According to the rules, when using ceramics, it must be noted the Specific Maximum Migration Levels (SML) of cadmium and lead into food. Additionally, the ceramics manufacturer must certify on each product that it does not, in fact, go above this migration limit. All the nations with active ceramics regulations are currently revising them.

3.7 Wood

Wood is frequently used to package fresh produce, although it is rarely used to package manufactured items. The most frequent uses of wood in food packaging include barrels for wines, beers, spirits, salted fish, and vegetables in brine; tea chests; small decorative boxes for items aimed at a tourist or gift market; and the construction of pellets. Other typical uses include wooden crates, notably for returnable bottles. Compared to cardboard boxes, wood is more durable and offers superior defense against crushing and impact. However, it is more expensive and heavier. Wooden containers can be designed to be both light- and leak-proof. Wood is a porous substance, so it doesn't make a perfect barrier to air and moisture. Depending on how they are made, wood containers can offer good protection from insects. Wooden barrels are extremely challenging to build, and the necessary training takes years. They are also highly expensive, which leads to their constant reuse; any damage is reported back to Coopers for repair.

3.8 Textiles and Plant Fibers

Among the materials used for food packaging are cotton, jute, sisal, and linen. Instead of being utilized for tiny retail or consumer containers, these materials are more frequently used as shipping containers to convey bigger amounts of food. Foods sold at specialty markets, such as tourist trinkets, or ornamental containers made of locally manufactured jute or cotton material, which may have strong promotion potential, are one application for cloth packaging. Textile packaging does not significantly impede the passage of moisture, smells, or air. It also does not shield food from microorganisms, insects, rodents, birds, or mechanical harm such as crushing or puncturing.

Therefore, they are utilized for foods that are resistant to scent pickup, humidity changes, and food that cannot simply be crushed. They are primarily designed as a portable, lightweight container to keep the food together in a packaging that is simple to carry. They are employed for items that flow freely, including grains, sugar, salt, spices, tea, and coffee beans. Although the protection provided to goods transported in this fashion is minor, they are also frequently used for the short-term transport of a wide variety of other commodities, including fresh fruits and vegetables and dry fish. The ability to create textiles locally using readily available materials and the simplicity of repair are their key benefits. They weigh little but are good at resisting slipping, allowing for the secure stacking of sacks. They may be washed and reused, and when thrown, they decompose naturally.

4 Food Packing Methods

Today, packaging has a significant role in the quality of food products by protecting against physical, chemical, and environmental hazards. Following is a discussion of the many packing methods [8]. The simplest and most popular technique of packaging that has been around for a while is vacuum packaging. The food is packaged in a substance with a low oxygen permeability, and then vacuum is established by removing air from the container before sealing. The package collapses around the product as the air escapes. The primary objective of this packaging method is to increase the shelf life of the food products. Products with a high fat content get rancid when oxygen from the package reacts with them. Vacuum packaging thus appears to be a potential method for reducing the amount of oxygen present inside the box.

1. Aseptic Packing

   It involves filling a sterile packaging material with a commercially sterile product while doing so in a sterile environment. Tight enough seals must be achieved between the packaging and the product to prevent contamination. Prior to filling, the packages are sterilized using a variety of techniques to destroy any bacteria present. Sterilization of the packaging material is a crucial step in the aseptic packaging method. The product is heated at an ultra-high temperature (UHT) for a predetermined period of time, then cooled and flowed via a system; in aseptic packaging a sterile product is poured into a sterile packaging material in a closed system connected to the packaging machine. The equipment and packaging materials can be sterilized using a variety of techniques, including heat, radiation, infrared light, and hydrogen peroxide. Items including milk, juice, mineral water, drinks, wine, tea, nutritional beverages, concentrates, dairy, sauces, and tomato-based items can all be packed. This method of packaging preserves the flavor and nutritional value of the products [9].

2. Modified Atmosphere Packaging (MAP)

   It is a process that replaces the air within food containers with a specific combination of gases before the package is sealed, changing the internal atmosphere. Typically, an inert gas is introduced to do this. Commercially, MAP uses nitrogen, carbon dioxide, and oxygen. Most commonly, nitrogen gas, which is odorless, tasteless, colorless, nontoxic, and non-inflammable, is used to change the atmosphere. In MAP, there are two ways to produce the changed atmosphere.

   (a) Before sealing flexible packages like pouches, a predefined gas mixture is used to replace the air inside the package.
   (b) Before sealing, the predetermined combination of gas is introduced in the case of trays after the air has been removed by a vacuum pump. The change extends the shelf life of the items and offers protection against food spoilage, dehydration, oxidation, weight loss, and freezer burn. It usually refers to freshly prepared or barely processed foods that are still in the process of respiration.

5 Other Contemporary Methods of Food Packaging

In order to preserve and increase product shelf life, active packaging entails incorporating specific additives into the polymer, packaging film, or packing containers. Instead of relying only on traditional food processing methods such as high heat treatments, dehydration, brining, acidity adjustments, and preservative additions, modern approaches increasingly utilize packaging techniques. When packaging serves a specific purpose in the preservation of food in addition to acting as a barrier against the elements, it is said to be active.

5.1 Bioactive Packaging

By producing better packaged foods, bioactive packaging is a cutting-edge packaging technique that directly affects consumer health. Enzyme encapsulation, microencapsulation, nanoencapsulation, and enzyme immobilization, all use biopolymers in this packaging technology to hold preferred bioactive doctrine under ideal conditions till their ultimate discharge into the food products [10]. By taking into account the unique product or functional substance characteristics or requirements, this technique maintains the bioactive substance until their regulated or quick release within the packed food during its storage time or prior to consumption.

5.2 *Biobased and Biodegradable Packaging Materials*

Trends now shifted toward the use of alternative packaging material to the conventional plastic packaging, it was long due as there were lack of availability of the alternative packaging material to plastic, especially for the food packaging sectors [11].

The term "smart" or "intelligent" packaging refers to a cutting-edge packaging technique that combines knowledge from various academic disciplines, including physics, chemistry, biochemistry, electronics, and food science and technology. The quality and safety of food products are monitored using a range of chemical sensors or biosensors in smart packaging, also known as intelligent packaging. According to Bambang et al. [12], these sensors can effectively monitor food quality and safety and provide information about the product's freshness, pH, oxygen, carbon dioxide, pathogens, gas leakage, ripeness regulators and indicators, bioprobes, radio frequency indicators, and toxin indicators.

Despite the discoveries of these biobased/biodegradable materials for the packaging applications, yet there is a problem that needs to be addressed by developing inexpensive methods, and processes for the production of these materials [11].

6 Packaging That Is Edible

Food waste from industrial food production and/or agricultural waste are used to create edible films and coatings, adding value in the process. Such packaging material satisfies the global demand for natural and eco-friendly foods. Biopolymers comprised of polysaccharides including cellulose, starch, chitosan, pectins, gums, and alginates as well as proteins derived from both animal and plant sources are used to create edible films and coatings.

Despite the fact that this technology does not entirely replace conventional packaging materials, it can lower the cost and bulk of conventional packaging [13]. By serving the fundamental purpose of barrier to gases and moisture as well as controlling the release of food additives and nutrients to the packaged food, the use of these packaging materials can enhance the process of food preservation [14, 15].

6.1 The Cycle of Packaging

The field of packaging integrates practically all branches of science, technology, commerce, social science, and engineering. As a result, creating a cohesive road map for the conversation to follow is the most challenging aspect of teaching a subject that spans so many different professions. This chapter will be helpful and continue to be the fundamental road map for packaging. In various packaging cycle components, a brief explanation of each stage of packaging cycle can be assisted to demonstrate by means of how the annular reduction of a worldwide system could be designed, even if the following chapters in the book go into more detail on these topics. These include the raw ingredients that go into the production of packaging materials like metal ores, wood fiber, oil, and energy. Both primary and secondary sources of converters combine the conversion processes, although it can be helpful to understand the distinction between primary and secondary converters in order to comprehend issues with material procurement and production.

First-stage converters: These are the converters that turn raw materials like plastic resin pellets, paper, or other recognizable single materials into a single form.

Operations including processing, filling, and sealing after the product has been prepared, it is combined with the package to create a collective system which must continue to be safe/secured and marketable up to user uses it. Unfortunately, a lot of people (and quite a few businesses) mistakenly believe that this marks the start as well final of the interaction among food processing and packaging, which might not completely understand the ecological impact to package-product system should withstand prior to use.

6.2 Transportation and Distribution

An industrial food production system must take into account distribution, inventory storage and management, and means of transportation as well as the risks associated with each. Through food, end user products, other types of commodities being carried from one side of the planet to the other as the global economy continues to integrate, they will only get worse.

Without these options, the package adds to the strain on landfills and incinerators, which are in certain places expensive and in short supply due to public opinion, government regulations, and the need for extensive environmental engineering. Source-reduction strategies have increased in popularity in developing the many packaged customer goods as well as reduced the per capita production of municipal solid waste (MSW) in various industrialized nations due to the numerous economic incentives for reducing material use and improved engineering practices.

This information, when combined with engineering data, can also lead to higher processing system performance, better product safety, and longer product shelf lives. A very accurate demographic picture can be produced by combining data from customer records with other data. This picture then influences marketing campaigns and promotional products advertised in specific markets, and often it directly influences the product mix that a consumer products company will field.

The production, transport, distribution, and handling process for food contained inside the packaged material make it a crucial component for any food industry. Previously, packaging materials were made to safeguard food products from external agents like germs, water vapor, oxygen, and light, while they were being stored [16–18]. Additionally, packaging aids in preventing the loss of beneficial substances (e.g., volatiles that give a product its aroma). Paper, metal, glass, and plastic are some of the packaging materials employed by the worldwide food business. The latter is commonly used as mono- or multi-materials combined in different ways [19, 20].

Recent decades have seen significant advancements in packaging and coatings, driven by the necessity for circular economy adaption as well as the increased demand for food that is both safe and of a high caliber [21]. In recent years, the idea of a circular economy has proliferated and is receiving growing attention on a global scale. However, a significant section of the packaging industry is built on a linear paradigm where packaging is created, used, and then discarded. As a result, a significant amount of post-consumer trash is produced, which is currently a worry for the entire world. In this regard, a lot of studies try to explain how package design might help a system move from a linear to a circular one [22].

Scientific and technological studies emphasize the following factors being recurrence of packing materials: (1) increases the efficacy of the supply chain of packaging in packaging; (2) decreases in the ecological conflict of packaging; (3) advancements in packaging process development; and (4) the effects of packaging's regulatory compliance [23]. As was mentioned, the primary goal of this field's research is to examine how packaging affects the environment. Within this framework, the three primary possibilities are reduction, reuse, and recycling. Recycle mention to the continued exploration for various applications, whereas reduction refers to minimizing the usage of energy input, raw materials, also the waste creation.

As it was mentioned, the primary goal of this field's research is to examine how packaging affects the environment. Within this framework, the three primary possibilities are reduction, reuse, and recycling. Reuse refers to the continued use of goods or components for their intended function, whereas reduction refers to minimizing the usage of raw materials, energy input, and waste creation [23]. Papers are generally recyclable, and are derived from renewable resources, hence substituting conventional plastics packing by papers packing is a growing research field. Since paper, even with a coating, exhibits lesser barrier qualities than plastics, it is important to note that this change in material type may result in a decrease in the items' shelf life. As a result, the scientific and industrial communities now face serious challenges with the management and processing of materials with little potential for recycling [24]. To address these challenges, methods for an efficient waste cycle are being developed.

The tremendous development of food systems, including packaging, will eventually have a negative influence on the environment. As a result, actions must be taken to lessen the negative effects of the food manufacturing and packaging chain, including the application of LCA approaches to evaluate the consequences of various goods [25, 26].

An estimated 10% of energy is used for packing, compared to 50% for food production [27]. Additionally, when greenhouse gas emissions from the food chain are taken into account, agricultural production accounts for around 40% of CO_2-equivalent emissions while packing accounts for roughly 5%. Therefore, compared to the packaging itself, the environmental effect of food loss caused by bad packaging design is far worse [28]. In order to meet the demand for sustainable packaging, polymers from renewable sources stand out as significant renewable and biodegradable possibilities [29]. Conventional polymers like PE and PET made from renewable resources (e.g., sugarcane) are offered on the market but are not biodegradable. The fact that they can be recycled is a big plus. Additional biological sources show amazing potential for new packaging materials. In order to provide a suitable relationship between performance and production cost, they still require extensive research. For instance, proteins and polysaccharides are frequently used to create films and coatings [29].

However, these basic materials' extreme sensitivity to water vapor and poor mechanical qualities continue to be major obstacles in the development of packaging based on them [19, 20]. The use of nanofillers, bioactive substances (e.g., essential oils, plant extracts, and/or pure active components), and blends with other polymers have all been suggested as solutions to the drawbacks of biopolymer-based films [30, 31]. Functional chemicals can be added to polymeric matrices to create "smart/intelligent" active coatings and films that can prolong the shelf life of food goods. The connection existed among the packaging substances and food products is made possible by modern active and intelligent packaging technologies. Defect detection, quality control, follows packaged food products toward regulating the storage condition between the production stage and the utilization stage. Further, by using different sensors and indicators such as gas and humidity sensors, time-temperature indicators, respectively, are some of the advantages of smart packaging. While using diffusion and adsorption/absorption mechanisms for different components, such as ethylene scavengers, oxygen, also the CO_2 absorbers, flavor adsorption process, antimicrobials, antioxidants, and active packaging aids in increasing the shelf life of food items [32].

Modified atmospheric packaging (MAP) is a recent development in packaging technology that is also incredibly important. After harvest, fresh and minimally processed foods continue to go through physiological and metabolic processes, which make them more prone to quality loss and shorter shelf lives. The mathematical integration of the dynamic physiological qualities of the product, the properties of the packing material, and the ideal equilibrium atmospheric conditions for the product in question result in successful MAP design. The type of product, the materials used for packaging, and the storage temperature will all affect the precise gas combination in the package for each application [33]. A widely used alternate option is to apply a hover to the packing in order to lower the O_2 level. The idea for this technology came from the evacuation of interstitial air, which led to compression as a result of the vacuum [16–18]. It is mostly used on processed packaged meals and items that are oxidation-prone. Many items are processed inside their packaging in order to prevent post-process contamination; as a result, the packaging must be able to survive the processing conditions.

Understanding the connections between packaging, food, and processing is so important. Some packaged food processing methods, like pasteurization and sterilization, have already achieved widespread industry consolidation. The focus of investigations, however, has recently shifted to cutting-edge techniques for processing packaged goods, including high-pressure processing, microwave-assisted sterilization, irradiation, plasma, ultrasound, UV light treatments, and pulsed light, among others [34]. Due to the numerous mechanisms of action used by these technologies, each one's packaging needs must be examined in terms of both their integrity in relation to various processing techniques and their effectiveness over the course of the food's shelf life.

The product or consumer side can be used to determine the shelf life. Between the product side, the deterioration of the product is examined as a function of time and may involve the development of unfavorable bacteria, a reduction in desired components (e.g., vitamins), or an increase in undesirable components (e.g., dark colors or moisture). However, the consumer's side is composed of requesting customer approval or rejection of food that has been in storage for a variety of lengths of time [19, 20].

For the effective use of the created packaging, the safety of food contact materials (FCMs) should be considered with all supplementary elements of packaging technology and food stability. Food protection and preservation are the goals of FCMs. But occasionally, when they come into touch with food, various compounds that were previously present in the packaging may migrate and contaminate the food [16–18]. Polymers with lower molecular weight, solvents, antioxidants, leftover plasticizers, and monomers are a few examples of these compounds. Since they may be dangerous to both humans and the environment, it is necessary to restrict the movement of these compounds.

It is crucial to evaluate how well a package in contact with food complies with the law because it must ensure the product inside is safe, high-quality, and compliant, avoiding the possibility of contamination from physical, chemical, or biological sources [19, 20]. In general, modern advancements in food safety and packaging should center on circular economy-friendly materials with suitable protective qualities. Packaging must be both economically feasible and safe for contact with food. The consumer's equally important preferences, like convenience, functionality, and personalization, shouldn't be overlooked by any of these needs. The combination of these factors may serve as a roadmap for present and future research, yielding exceptional outcomes that will benefit society and the environment.

7 The Packaging of the Future

Today's packaging has many detrimental effects on both human health and the environment. But the packaging industry will continue to develop as long as packaging technology advances.

Packaging will undoubtedly continue to develop and provide:

1. Plastic substitute
2. Packaging made of more environmentally friendly materials
3. Packaging that prevents the transfer of harmful substances to food
4. Better packaging design
5. Reasonably priced indicators of high quality

8 Conclusion

Various packaging materials can be employed; they are commonly in line with the change according to the comfort, lifestyle, and also the increasing the customer interests based on the quality, moreover, with a special focus on supplying the safe meals. This work has been intended to deliver the very important aspects of food packaging which will also take care of the environmental sustainability. Conversely, beside this environmental feasibility, we have also put effort to promote the existing revolutionary food packaging technologies which should be reinforced to obtain the reduced costs without compromising the quality of the packaging material in protecting different foods. This report will definitely encourage researchers and stir them to take part exploring the new challenges in food packaging sector.

References

1. Berger K, Welt B (2005) A brief history of packaging. http://edis.ifas.ufl.edu/pdffiles/AE/AE20600.pdf. Accessed 4 Jan 2009
2. Marsh K, Bugusu B (2007) Food packaging-roles, materials, and environmental issues. J Food Sci 72(3):R39–R55
3. Kirwan MJ (2003) Paper and paperboard packaging. In: Coles R, McDowell D, Kirwan MJ (eds) Food packaging technology. Blackwell Publishing, London, pp 241–281
4. Brennan JG (2006) In: Brennan JG (ed) Food processing handbook, vol 2. 2nd edn. Wiley-VCH Verlag GmbH & Co. KGaA, Weinheim, p 300
5. Andrady AL, Neal MA (2009) Applications and societal benefits of plastics. Philos Trans R Soc Biol Sci 364(1526):1977–1984
6. Lopez-Rubio A, Almenar E, Hernandez-Muñoz P, Lagarón JM, Catalá R, Gavara R (2004) Overview of active polymer-based packaging technologies for food applications. Food Rev Intl 20(4):357–387
7. Risch SJ (2009) Food packaging history and innovations. J Agric Food Chem 57(18):8089–8092

8. Majid I, Nayik GA, Dar SM, Nanda V (2018) Novel food packaging technologies: innovations and future prospective. J Saudi Soc Agric Sci 17(4):454–462
9. Ansari IA, Datta AK (2003) An overview of sterilization methods for packaging materials used in aseptic packaging systems. Food Bioprod Process 81(1):57–65
10. López-De-Dicastillo C, Catalá R, Gavara R, Hernández-Muñoz P (2011) Food applications of active packaging EVOH films containing cyclo dextrins for the preferential scavenging of undesirable compounds. J Food Eng 104(3):380–386
11. Yaradoddi J, Patil V, Ganachari S, Banapurmath N, Hunashyal A, Shettar A (2016) Biodegradable plastic production from fruit waste material and its sustainable use for green applications. Int J Pharma Res Allied Sci 5(4):56–66
12. Bambang K, Wicaksono Y, Jayus AA, Heng LY, Ahmad M (2011) Sensing and instrumentation for food quality and safety smart packaging: sensors for monitoring of food quality and safety. Sens Instrumen Food Qual 5:137–146. https://doi.org/10.1007/s11694-011-9120-x
13. Campos CA, Gerschenson LN, Flores SK (2011) Development of edible films and coatings with antimicrobial activity. Food Bioprocess Technol 4(6):849–875
14. Yaradoddi JS, Banapurmath NR, Ganachari SV et al (2020) Biodegradable carboxymethyl cellulose-based material for sustainable packaging application. Sci Rep 10:21960. https://doi.org/10.1038/s41598-020-78912-z
15. Yaradoddi JS, Sulochana MB, Merja KH, Parameshwar AB, Dayanand A (2020) The occurrence of potential and novel isolates of *Oceanobacillus* sp. JAS12 and *Salinicoccus* sp. JS20 recovered from west coast of Arabian Sea, India. Res J Biotech 15(9):133–140
16. Marangoni Júnior L, Coltro L, Balardin F, Dantas H, Vieira RP (2022) Research on food packaging and storage. Coatings 12(11):1714. https://doi.org/10.3390/coatings12111714
17. Marangoni Júnior L, Jamróz E, de Ávila Gonçalves S, da Silva RG, Alves RMV, Vieira RP (2022) Preparation and characterization of sodium alginate films with propolis extract and nano-SiO2. Food Hydrocoll Health 2:100094
18. Marangoni Júnior L, de Ávila Gonçalves S, da Silva RG, Martins JT, Vicente AA, Vercelino Alves RM, Vieira RP (2022) Effect of green propolis extract on functional properties of active pectin-based films. Food Hydrocoll 2022:107746
19. Marangoni Júnior L, Vieira RP, Anjos CAR (2020) Kefiran-based films: fundamental concepts, formulation strategies and properties. Carbohydr Polym 246:116609
20. Marangoni Júnior L, Cristianini M, Anjos CAR (2020) Packaging aspects for processing and quality of foods treated by pulsed light. J Food Process Preserv 44:e14902
21. Alias AR, Wan MK, Sarbon NM (2022) Emerging materials and technologies of multi-layer film for food packaging application: a review. Food Control 136:108875
22. Robertson G (2013) Food packaging: principles and practice. CRC Press, Boca Raton. isbn:9781439862421
23. Zhu Z, Liu W, Ye S, Batista L (2022) Packaging design for the circular economy: a systematic review. Sustain Prod Consum 32:817–832
24. Geueke B, Groh K, Muncke J (2018) Food packaging in the circular economy: overview of chemical safety aspects for commonly used materials. J Clean Prod 193:491–505
25. Kuclourya T, Monroy R, Cuan Urquizo E, Roman-Flores A, Ahmad R (2022) Scientometric analysis and critical review of fused deposition modeling in the plastic recycling context. Clean Waste Syst 2:100008
26. Sazdovski I, Bala A, Fullana-i-Palmer P (2021) Linking LCA literature with circular economy value creation: a review on beverage packaging. Sci Total Environ 771:145322
27. Dantas FBH, Dantas ST (2020) M*é*todos de Processamento Permitem Maior Conserva*ção* e Disponibilidade Durante o Ano Todo, vol 31. Instituto de Tecnologia de Alimentos, Campinas, p 3
28. Garnett T (2008) Cooking up a storm—food, greenhouse gas emissions and our changing climate. University of Surrey, Guildford, p 155
29. Abbate E, Rovelli D, Andreotti M, Brondi C, Ballarino A (2022) Plastic packaging substitution in industry: variability of LCA due to manufacturing countries. Procedia CIRP 105:392–397

30. Marangoni Júnior L, Vieira RP, Jamróz E, Anjos CAR (2021) Furcellaran: an innovative biopolymer in the production of films and coatings. Carbohydr Polym 252:117221
31. Marangoni Júnior L, da Silva RG, Vieira RP, Alves RMV (2021) Water vapor sorption and permeability of sustainable alginate/collagen/SiO2 composite films. LWT 152:112261
32. Soltani Firouz M, Mohi-Alden K, Omid M (2021) A critical review on intelligent and active packaging in the food industry: research and development. Food Res Int 141:110113
33. Belay ZA, Caleb OJ, Opara UL (2016) Modelling approaches for designing and evaluating the performance of modified atmosphere packaging (MAP) systems for fresh produce: a review. Food Packag Shelf Life 10:100307
34. Oliveira M, Abadias M, Usall J, Torres R, Teixidó N, Viñas I (2015) Application of modified atmosphere packaging as a safety approach to fresh-cut fruits and vegetables—a review. Trends Food Sci Technol 46:13–26

Chapter 9
Understanding Food Allergies and Allergens: A Comprehensive Guide to Diagnosis, Treatment, and Management

Jayashri B. Uppin, Gayatri Vaidya, and Ghouse Modin N. Mamdapur

1 Introduction

Food allergies and allergens constitute significant subjects within public health and nutrition. A grasp of the fundamentals of this topic is vital due to the considerable impact that allergies to specific foods can have on an individual's health and their overall quality of life. Food allergies emerge when the immune system erroneously identifies certain proteins in foods as threats, setting off an allergic response. Although many foods carry this potential, common allergenic culprits encompass nuts, shellfish, milk, eggs, and wheat [1]. These reactions can span from minor discomfort to life-threatening anaphylactic shock, underscoring the gravity of food allergies. The identification and effective management of food allergies are paramount for those directly affected and healthcare providers and food producers. Heightened awareness regarding allergenic components in food items, accurate labeling, and transparent communication regarding potential allergens are pivotal to ensuring the safety of individuals with food allergies [2]. In this context, a comprehensive understanding of the scientific underpinnings of food allergies and allergens, coupled with practical strategies for prevention and management, assumes a pivotal role in safeguarding public health and fostering overall well-being [3].

J. B. Uppin (✉)
Department of Biotechnology, Basaveshwar Science College, Bagalkote, Karnataka, India

G. Vaidya
Department of Food Biotechnology, Davangere University, Davanagere, Karnataka, India

G. M. N. Mamdapur
Department of Library and Information Science, Yenepoya (Deemed to be University), Mangalore, Karnataka, India

Synthite Industries Private Limited, Kolenchery, Kerala, India

J. S. Yaradoddi et al. (eds.), *Frontiers in Food Biotechnology*,
https://doi.org/10.1007/978-981-97-3261-6_9

A food allergy is a condition in which specific foods trigger an abnormal immune response. It occurs when our immune system mistakenly identifies certain food proteins as threats, harming our body [4]. In response, our body employs protective measures, such as releasing chemicals like histamine, which can cause inflammation. Individuals with food allergies, often called allergic individuals, can experience allergic reactions even when exposed to tiny amounts of allergenic food. Allergies are now among the most prevalent immunological diseases, representing one of the most common and continuously growing chronic health issues. According to the World Health Organization, allergies are the fourth most frequent chronic disease worldwide, with over 5% of adults in the United States affected.

Food allergies manifest through a range of symptoms, including hoarseness, coughing, wheezing, chest tightness, nausea, vomiting, diarrhea, dizziness, fainting, and, in severe cases, anaphylaxis [5]. Anaphylaxis, characterized by hives, swelling, throat constriction, plummeting blood pressure, and shock, is a medical emergency. These reactions can be exacerbated by factors like alcohol consumption or physical activity, leading individuals to doubt their diagnosis. Such doubts can be perilous, especially when dealing with life-threatening anaphylactic reactions. Preventing such reactions requires individuals to diligently avoid allergenic foods for a lifetime, including being mindful of cross-contamination risks from shared food processing equipment. The United States mandates certain food allergen labeling to help individuals make informed dietary choices [6].

Inadequate treatment of allergies, particularly in cases of anaphylactic reactions, can be extremely hazardous. Moreover, it can give rise to various long-term health issues like malaise, malabsorption, and inflammation [7]. The psychological and mental health effects of grappling with a chronic health condition, including the burden of avoidance and fear of exposure, can lead to anxiety and a reduced quality of life. Many older adults live on fixed incomes and face substantial medical expenses yearly, making it challenging to afford allergy-friendly, healthy foods [8].

In the early 1980s, food allergies were perceived as exceptionally rare, with limited public awareness and skepticism among clinicians [9]. While the perceived prevalence at the time was similar to today (around 20%), some medical professionals believed it to be less than 1%. It is estimated that 5% of adults and 8% of children have severe food allergies, though these numbers may be underestimated, especially among older adults [10]. Older adults need to recognize that their food allergies and sensitivities may not diminish with age, especially for allergies to peanuts, tree nuts, and seafood [11].

Diagnosing food allergies is often a complex process involving skin tests, blood tests, and food challenges [12]. However, it can be challenging for older adults, who may have other health issues resembling allergy symptoms. Allergy testing can also be less reliable in older individuals, potentially yielding false-negative results [13]. Moreover, the presence of multiple prescription drugs can complicate allergic reactions. For example, certain medications, like beta blockers, can reduce the effectiveness of epinephrine, while others may increase the risk of cardiac issues when combined with epinephrine. This underscores the importance of healthcare providers taking age and comorbidities into account when treating allergies and responding to allergic reactions [14]. Recent data suggest that the prevalence of food allergies among the global population ranges from 1.1 to 10.8%, with indications of a rising trend in food allergy rates [15].

1.1 Definition of Food Allergies

Food allergies refer to the immune system's abnormal response to specific proteins found in certain foods. When an individual with a food allergy consumes the allergenic food, their immune system mistakenly identifies these proteins as harmful invaders, like bacteria or viruses. In response to this perceived threat, the immune system releases chemicals, including histamines, to protect the body. These chemicals trigger allergic symptoms, ranging from mild to severe, and may include skin rashes, digestive issues, respiratory problems, or even life-threatening anaphylaxis. It's essential to note that food allergies are different from food intolerances. Food intolerances, such as lactose intolerance or non-celiac gluten sensitivity, do not involve the immune system and typically result in digestive discomfort rather than an immune response. Food allergies are a serious medical condition, and even trace amounts of the allergenic food can lead to significant health risks for individuals affected by them. Managing food allergies involves strict avoidance of allergenic food, awareness of potential cross-contamination, and preparedness for allergic reactions when they occur (Fig. 9.1).

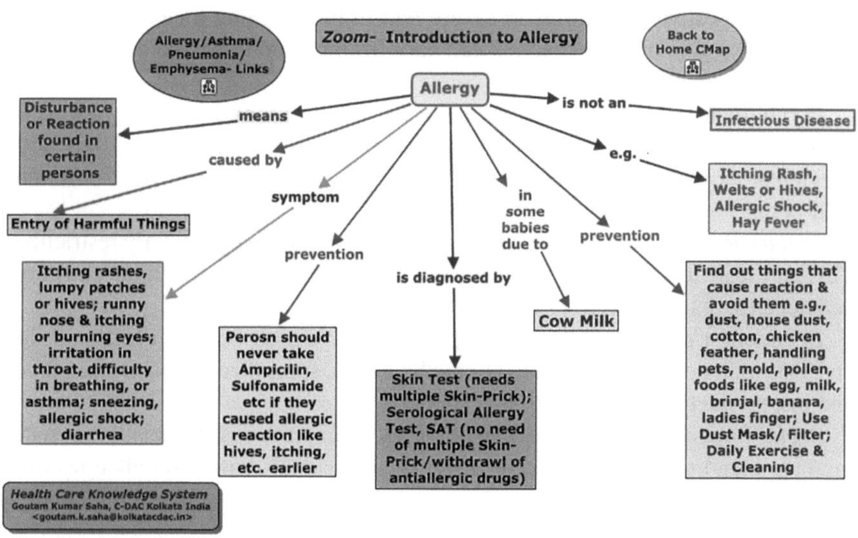

Fig. 9.1 Schematic representation of by means, symptoms, causes, diagnosis, and preventive measures of allergy (Courtesy: Regional Institute of Medical Sciences [16])

1.2 Prevalence and Impact

The prevalence and impact of food allergies have been a growing concern in recent years, both from a public health perspective and on an individual level. Understanding the scale of the issue is crucial for effective management and support for those affected. Here's an overview of the prevalence and impact of food allergies:

a. Prevalence of Food Allergies

Rising Incidence: Food allergies have significantly increased prevalence, particularly in developed countries. While the exact reasons for this rise are unclear, it is a global health issue affecting millions of people.

Common Allergens: Certain foods are more frequently implicated in food allergies. These common allergens include peanuts, tree nuts, milk, eggs, soy, wheat, fish, and shellfish.

Childhood Allergies: Food allergies are often first diagnosed in childhood, with many children eventually outgrowing their allergies. However, some allergies persist into adulthood, and new allergies can develop at any age.

b. Impact of Food Allergies

Health Risks: Food allergies can lead to various health issues, from mild skin reactions and digestive discomfort to severe and life-threatening anaphylactic shock. The severity of an allergic reaction can vary from person to person and even from one exposure to another.

Psychological Impact: Living with food allergies can have a profound psychological impact on individuals and their families. Constant vigilance, anxiety about accidental exposure, and social challenges related to food can affect emotional well-being.

Dietary Restrictions: Managing food allergies requires strict dietary restrictions. This can lead to challenges in maintaining a balanced diet, potential nutritional deficiencies, and limited food choices.

Economic Costs: Food allergies impose economic costs on individuals and society. This includes expenses related to medical care, specialized allergen-free foods, and lost productivity due to illness or anxiety.

Social Impact: Food allergies can affect social interactions, especially when dining out, attending social events, or traveling. Individuals with food allergies often need to communicate their dietary restrictions clearly and ensure the safety of their food.

Quality of Life: Overall, food allergies can significantly impact an individual's quality of life, affecting their physical health, emotional well-being, and social interactions.

Recognizing the increasing prevalence and understanding the multifaceted impact of food allergies underscores the importance of education, awareness, research, and support for affected individuals and their families. Improving food labeling, developing treatments, and raising public awareness are crucial in addressing this complex health issue.

1.3 How to Get Rid of Some Seasonal Allergies?

- **Education**: This guide aims to educate readers about the fundamentals of food allergies and allergens, from the immune response to the practical aspects of managing allergies.
- **Awareness**: We seek to raise awareness about the prevalence and impact of food allergies, emphasizing empathy, understanding, and support for those affected.
- **Safety**: Our objective is to promote safe practices by providing clear guidance on identifying allergenic ingredients, preventing cross-contamination, and responding to allergic reactions.
- **Empowerment**: We aim to empower individuals with food allergies and their caregivers with knowledge and resources to manage their condition and improve their quality of life confidently.
- **To Generate New Knowledge**: These studies aim to contribute new knowledge or insights to a particular field. These may be exploratory or innovative studies seeking to advance understanding.

2 Understanding Allergies

2.1 How Allergies Develop?

The development of allergies is a complex process influenced by genetic, environmental, and immunological factors. While the exact mechanisms are not fully understood, here is a simplified explanation of how allergies develop:

(a) **Genetic Predisposition**: Allergies often have a genetic component. Individuals with a family history of allergies are likelier to develop allergies. Specific genetic factors related to the immune system play a role in determining susceptibility.
(b) **Initial Exposure**: Allergic reactions typically require initial exposure to an allergen. During this first encounter, the immune system usually identifies the allergen as harmless.

(c) **Sensitization**: During the initial exposure, the immune system of a genetically predisposed individual may produce antibodies called immunoglobulin E (IgE) against the specific allergen. This process is known as sensitization. IgE antibodies are designed to recognize and bind to the allergen, tagging it for future encounters.
(d) **Memory Cells**: After sensitization, the immune system creates memory cells, specifically memory B and T cells, which "remember" the allergen. These memory cells are poised to respond rapidly if the same allergen is encountered again.
(e) **Re-Exposure**: When the individual is re-exposed to the same allergen, the memory cells recognize it and trigger an immune response.
(f) **Release of Chemicals**: The immune response includes the release of various chemicals, including histamines, from immune cells called mast cells and basophils. Histamines and other chemicals are responsible for the allergic symptoms experienced, such as itching, sneezing, swelling, and inflammation.
(g) **Allergic Reaction**: The allergic reaction can manifest in various ways, depending on where the immune response occurs. For example, in the skin, it may lead to hives or eczema. In the respiratory system, it can cause sneezing, runny nose, or asthma symptoms. It may result in nausea, vomiting, or diarrhea in the gastrointestinal tract.
(h) **Severity and Sensitization**: The severity of allergic reactions can vary. Some individuals may experience mild symptoms, while others may have severe reactions, including anaphylaxis. Repeated exposures to the allergen can increase the severity of allergic reactions, a phenomenon known as "progressive sensitization."
(i) **Allergenic Substances**: Allergens can be proteins or other molecules found in common substances like foods (e.g., peanuts and shellfish), pollen, dust mites, pet dander, insect venom, medications (e.g., penicillin), or latex.

It's important to note that not everyone who is exposed to allergens will develop allergies, and the exact triggers and mechanisms can vary from person to person.

2.2 Common Allergenic Foods

Common allergenic foods are those that frequently trigger allergic reactions in susceptible individuals. These foods contain proteins that some people's immune system recognizes as harmful, leading to allergic responses. Here is a list of some common allergenic foods:

(a) **Peanuts**: Peanuts are among the most allergenic foods and can cause severe reactions, including anaphylaxis, in some individuals.
(b) **Tree Nuts**: Tree nuts, including almonds, cashews, walnuts, and hazelnuts, are common allergens. Allergic reactions to one type of tree nut often indicate a heightened risk of allergies to others.

(c) **Milk**: Allergies to cow's milk are common in children, but many outgrow this allergy as they get older. Some individuals may also be allergic to milk from other animals, such as goats or sheep.
(d) **Eggs**: Egg allergies are prevalent in children but can persist into adulthood. The proteins in egg whites often trigger allergic reactions.
(e) **Soy**: Soybeans and soy products contain allergenic proteins, and soy allergies can manifest as digestive or skin symptoms.
(f) **Wheat**: Wheat allergies can be challenging to manage, as wheat is a common ingredient in many foods. Individuals with wheat allergies need to avoid wheat-based products.
(g) **Fish**: Allergies to various fish, such as salmon, tuna, and cod, can lead to allergic reactions. Cross-contamination is a concern when cooking fish.
(h) **Shellfish**: Shellfish allergies can be divided into two categories: crustaceans (e.g., shrimp, crab, and lobster) and mollusks (e.g., clams, mussels, and oysters). Allergic reactions can range from mild to severe.
(i) **Sesame**: Sesame allergies are becoming more recognized as a significant allergen. It is found in various foods, including bread, sauces, and salad dressings.
(j) **Mustard**: Mustard allergies are less common but can cause reactions in individuals sensitive to mustard seeds or mustard products.
(k) **Sulfites**: Sulfites are food additives used to preserve the color and flavor of certain foods and beverages. Some individuals may be sensitive to sulfites commonly found in dried fruits, wine, and processed foods.
(l) **Celery**: Celery allergies may manifest as oral allergy syndrome, where individuals experience itching or swelling in the mouth and throat when consuming raw celery.

It's essential to note that the severity of allergic reactions can vary widely among individuals. While some may have mild symptoms like hives or itching, others may experience more severe reactions, including anaphylaxis, a life-threatening condition requiring immediate medical attention. Individuals with known food allergies must carefully read food labels, be aware of potential cross-contamination risks, and carry prescribed medications (such as epinephrine auto-injectors) to manage allergic reactions.

2.3 *The Immune System's Role*

The immune system plays a central role in allergic reactions. Allergies result from the immune system's overreaction to typically harmless substances, known as allergens. Here's how the immune system is involved in allergic responses:

- Recognition of allergens
- Sensitization
- Memory cells formation
- Re-exposure

- Release of chemicals
- Inflammatory response
- Allergic symptoms
- Severity and anaphylaxis
- Repeated exposure

2.4 Types of Allergic Reactions

Allergic reactions can manifest in various ways, depending on the allergen and the individual's immune system. There are several types of allergic reactions, categorized based on their clinical presentation and underlying immune mechanisms,

- Immediate hypersensitivity (Type I reaction)
- Delayed hypersensitivity (Type IV reaction)
- Allergic granulomatous angiitis (Churg–Strauss syndrome)
- Serum sickness (Type III reaction)
- Oral allergy syndrome (OAS)
- Exercise-induced anaphylaxis (EIA)

3 Common Allergens

Common allergens are substances that frequently trigger allergic reactions in susceptible individuals. These substances can provoke various allergic responses when they come into contact with the immune system of sensitive individuals. Here are some common allergens,

- Pollen
- Dust mites
- Pet dander
- Mold spores
- Cockroach allergens
- Insect stings and bites
- Medications
- Foods
- Metal allergens
- Perfumes and fragrances
- Foods that cross-react with pollen
- Inhaled allergens in occupational settings

It's important to note that the severity of allergic reactions can vary widely among individuals, and not everyone exposed to these allergens will develop allergies. Managing allergies often involves allergen avoidance, medications, and, in some cases, allergen-specific immunotherapy to desensitize the immune system. Individuals with known allergies should work with healthcare providers to develop personalized allergy action plans.

3.1 Major Food Allergens

Food allergies can manifest in various forms, and individuals may be allergic to multiple foods. Some of the most common types of food allergies include:

- **Peanut Allergy**: Peanut allergies are among the most prevalent and severe food allergies. Reactions to peanuts can range from mild itching or hives to life-threatening anaphylaxis.
- **Tree Nut Allergy**: Tree nuts, such as almonds, walnuts, and cashews, can also trigger allergic reactions. These allergies tend to be persistent and often lifelong.
- **Shellfish Allergy**: Shellfish allergies are divided into two groups: crustaceans (e.g., shrimp, crab, and lobster) and mollusks (e.g., clams, mussels, and squid). Allergic reactions to shellfish can be severe and sometimes cross-react between the two groups.
- Milk allergy is common in infants and young children but can also affect adults. It involves an adverse reaction to proteins found in cow's milk, which may cause gastrointestinal and skin symptoms.
- **Egg Allergy**: Egg allergies typically affect children and may lead to reactions like hives, digestive issues, or respiratory symptoms. Most children outgrow this allergy as they get older.
- **Wheat Allergy**: Wheat allergies can cause a range of symptoms, including skin reactions, digestive problems, and respiratory distress. People with wheat allergies must avoid wheat-containing products.
- **Soy Allergy**: Allergic reactions to soy can vary in severity, with symptoms such as hives, digestive issues, and, rarely, anaphylaxis. Soy is a common ingredient in many processed foods.
- **Fish Allergy**: Fish allergies can be specific to certain types of fish or involve multiple varieties. Reactions can range from mild to severe, and individuals with this allergy should be cautious when dining out.
- **Sesame Allergy**: Sesame allergies are becoming more recognized, and symptoms can include skin reactions, gastrointestinal distress, and even anaphylaxis. Sesame is found in various foods, including bread and condiments.
- **Mustard Allergy**: Mustard allergies are less common but can lead to a range of symptoms, from mild skin reactions to more severe respiratory issues.

It's important to note that food allergies can vary in severity from person to person. Some individuals may only experience mild symptoms, while others can have severe and life-threatening reactions. Proper diagnosis, avoidance of allergenic foods, and preparedness for managing allergic reactions are essential for individuals with food allergies to lead safe and healthy lives. If someone suspects a food allergy, they should seek medical evaluation and guidance for diagnosis and management.

4 Diagnosis and Testing

Diagnosing allergies involves medical history assessment, physical examination, and specific allergy testing. Proper diagnosis is essential for identifying allergens and developing an effective management plan. Here are the key components of allergy diagnosis and testing,

- Medical history assessment
- Physical examination
- Allergy testing (skin prick test, blood tests, and patch testing)
- Elimination diets and oral food challenges
- Provocation testing
- Nasal or bronchial provocation tests
- Component-resolved diagnostics

Once specific allergens are identified through testing, healthcare providers can work with patients to develop personalized allergy management plans. These plans may include allergen avoidance strategies, medication management, and, in some cases, allergen-specific immunotherapy (e.g., allergy shots or sublingual tablets) to desensitize the immune system to allergens over time. Allergy diagnosis and management should always be conducted under the guidance of a qualified healthcare professional.

4.1 Recognizing Allergic Reactions

Recognizing allergic reactions is crucial for timely intervention and management. Allergic reactions can vary in severity, from mild symptoms to potentially life-threatening anaphylaxis (Fig. 9.2). Here are some common signs and symptoms to help you recognize allergic reactions:

(a) **Skin Reactions**:

  - Hives (urticaria): Raised, itchy, red, or pale welts on the skin.
  - Eczema (atopic dermatitis): Red, itchy, and inflamed skin, often with dry patches.

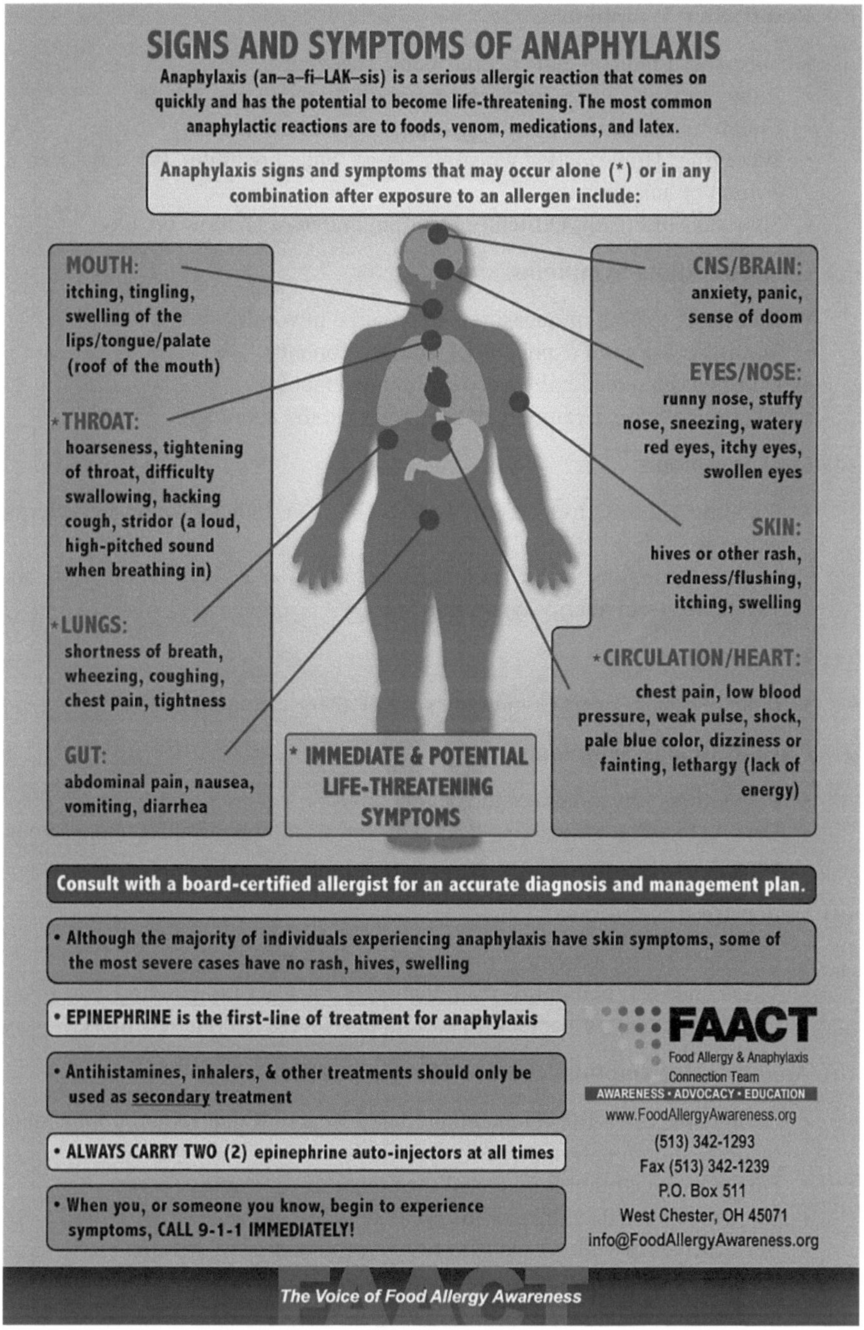

Fig. 9.2 Range of symptoms (mild, moderate, and severe symptoms) of allergy (Courtesy: Food Allergy & Anaphylaxis Connection Team [17])

(b) **Respiratory Symptoms**:

- Sneezing: Frequent and uncontrollable sneezing.
- Runny or stuffy nose: Excessive nasal discharge or congestion.
- Coughing: Persistent coughing, especially with asthma.
- Wheezing: High-pitched whistling sound when breathing, typically seen in asthma or anaphylaxis.
- Shortness of breath: Difficulty breathing, rapid or shallow breaths.

(c) **Gastrointestinal Symptoms**:

- Nausea: A feeling of queasiness or the urge to vomit.
- Vomiting: Forceful expulsion of stomach contents.
- Diarrhea: Frequent and watery bowel movements.
- Abdominal pain: Cramping or discomfort in the abdomen.

(d) **Oral Symptoms**:

- Swelling of lips, tongue, or throat: This can indicate a severe allergic reaction.
- Itching or tingling in the mouth: Especially after eating certain fruits and vegetables (oral allergy syndrome).

(e) **Eye Symptoms**:

- Itchy or watery eyes: Commonly seen in allergic conjunctivitis.

(f) **Cardiovascular Symptoms**:

- Palpitations: Awareness of the heartbeat.
- Drop in blood pressure: Can lead to dizziness, fainting, or loss of consciousness in severe cases.

(g) **Generalized Symptoms:**

- Fatigue: Feeling extremely tired or weak.
- Mood changes: Irritability, restlessness, or a sense of impending doom.
- Sense of anxiety or panic: Can accompany severe allergic reactions.

(h) **Anaphylaxis Symptoms (Severe Allergic Reaction):**

- Anaphylaxis is a life-threatening allergic reaction that requires immediate medical attention. Symptoms may include:
- Difficulty breathing: Rapid, shallow breathing or gasping for air.
- Swelling of the face, lips, or throat: Potentially obstructing the airway.
- Severe hives or widespread skin rash: Accompanied by itching.
- Rapid or weak pulse: Irregular heartbeat.
- Severe drop in blood pressure: Leading to shock.
- Loss of consciousness: Fainting or loss of responsiveness.

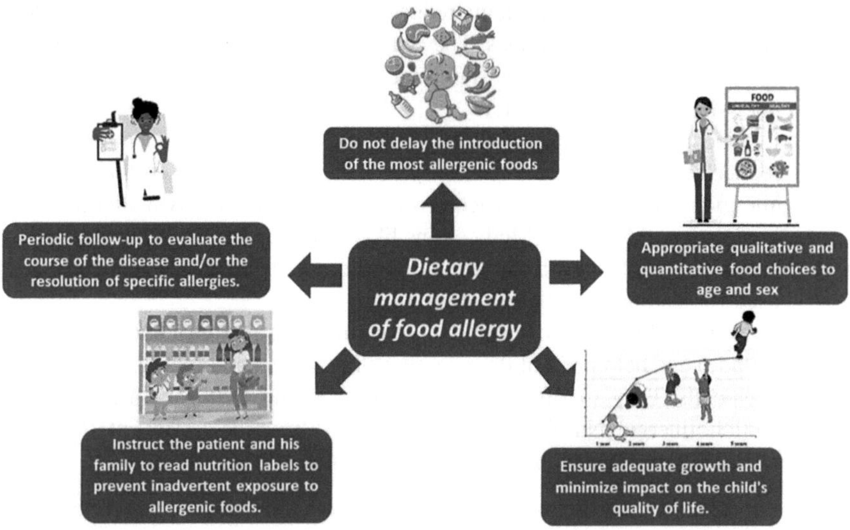

Fig. 9.3 Showing signs and symptoms of anaphylaxis (Courtesy: Leone et al. [18])

It's important to note that the severity and combination of symptoms can vary from person to person and with different allergens. Anaphylaxis is a medical emergency, and epinephrine should be administered immediately if anaphylactic symptoms are suspected (Fig. 9.3). If you or someone else experiences symptoms of anaphylaxis, call 911 or seek emergency medical help immediately. For milder allergic reactions, over-the-counter antihistamines or other medications may be sufficient for symptom relief. However, if symptoms persist or worsen, it's advisable to consult a healthcare professional for proper evaluation and guidance on allergy management. If you know specific allergens that trigger your reactions, allergen avoidance remains a key strategy in preventing allergic symptoms.

5 Treatment and Medications

Allergy treatment aims to manage symptoms and, in some cases, modify the immune system's response to allergens. The choice of treatment depends on the type and severity of allergies. Here's an overview of common allergy treatment options:

(a) **Allergy Medications**:

- Antihistamines: These medications block the effects of histamine, a chemical released during allergic reactions, to relieve symptoms like sneezing, itching, and runny nose. Examples include cetirizine (Zyrtec), loratadine (Claritin), and fexofenadine (Allegra).

- Decongestants: Decongestants reduce nasal congestion by narrowing blood vessels in the nasal passages. They can be used in combination with antihistamines. Common decongestants include pseudoephedrine (Sudafed) and phenylephrine (Sudafed PE).
- Corticosteroids: Nasal corticosteroids (e.g., fluticasone and mometasone) reduce inflammation in the nasal passages and are effective for managing allergic rhinitis. Topical corticosteroids are also used for skin allergies.
- Leukotriene Modifiers: Medications like montelukast (Singulair) block leukotrienes, substances that contribute to allergic reactions. They are primarily used for asthma and allergic rhinitis.
- Mast Cell Stabilizers: Cromolyn sodium (NasalCrom) and nedocromil (Tilade) can help prevent the release of histamine and other chemicals from mast cells. They are often used for allergic conjunctivitis and asthma.
- Anticholinergic Nasal Sprays: Ipratropium bromide (Atrovent) can help relieve severe nasal congestion associated with allergic rhinitis.

Figure 9.4 shows dietary management against different food allergies.

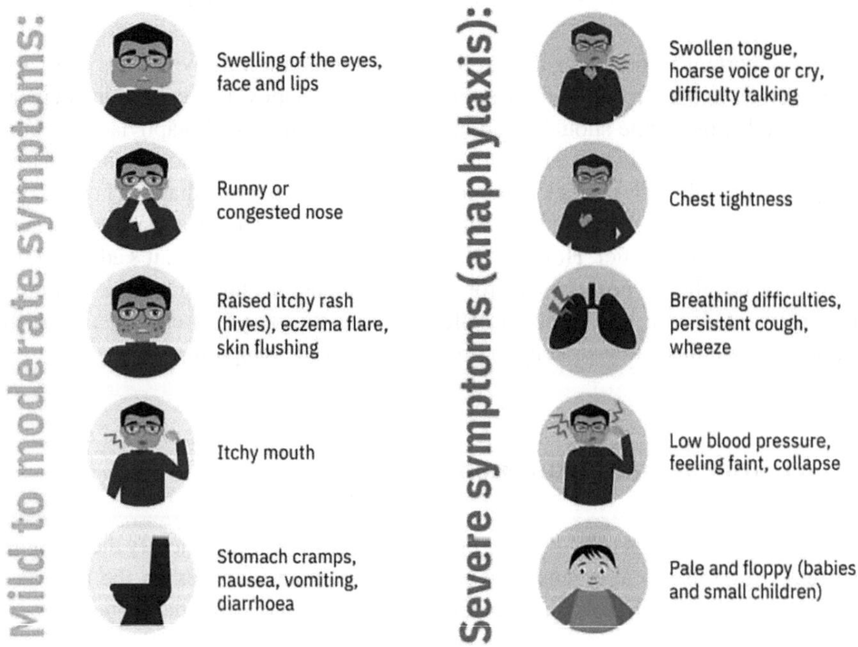

Fig. 9.4 The dietary management against different food allergies (Courtesy: Hogg [19])

(b) **Immunotherapy**:

Immunotherapy, also known as allergy shots or allergen-specific immunotherapy (AIT), is a long-term treatment for allergies that can modify the immune system's response to allergens. It involves receiving regular injections of allergen extracts over several years. The doses start small and gradually increase to build tolerance to the allergen. Immunotherapy is effective for allergies to pollen, dust mites, animal dander, and insect venom. Sublingual immunotherapy (SLIT) is an alternative to injections. It involves placing allergen extracts under the tongue. SLIT is available for allergens, such as grass pollen and dust mites.

(c) **Emerging Treatments**:

Ongoing research is exploring new allergy treatments and approaches, including:

- Biologic Medications: These targeted therapies block specific immune system molecules involved in allergic reactions. Examples include omalizumab (Xolair) for severe asthma and dupilumab (Dupixent) for atopic dermatitis and asthma.
- Vaccines: Researchers are developing allergy vaccines that could offer long-lasting protection against specific allergens.
- Microbial Exposure: Studies suggest that early-life exposure to certain microbes may reduce the risk of developing allergies. Research in this area may lead to preventive strategies.

6 Food Allergies in Children

Food allergies in children can pose significant challenges for the child and their parents or caregivers. Managing food allergies in pediatric populations requires vigilance, education, and effective communication with healthcare providers, schools, and childcare facilities. Here are some key considerations and guidance for parents of children with food allergies.

- Diagnosis and education
- Allergen identification
- Emergency action plan
- Communication
- Allergy-safe school and childcare environment
- Meal planning
- Teach your child
- Supervision
- Be prepared for field trips and special events

Managing food allergies in children requires ongoing vigilance and teamwork between parents, healthcare providers, and educational institutions. Children with food allergies can thrive in a safe and inclusive environment with proper education, communication, and preparation.

7 Legal and Regulatory Aspects

Legal and regulatory aspects play a crucial role in managing food allergies and ensuring the safety of individuals with food allergies. Key areas of focus include food labeling laws and allergen disclosure requirements. Here's an overview of these aspects:

(a) **Food Labeling Laws:**

Many countries have established laws and regulations that govern food labeling to protect consumers with food allergies. These laws often require manufacturers to provide clear and accurate information about the presence of common allergens in packaged foods.

(b) **Allergen Disclosure Requirements:**

Common allergen disclosure requirements typically include the following aspects:

- Allergen Labeling: Manufacturers must clearly label foods that contain common allergens. These labels are often found in the ingredient list, with the allergen name in plain language.
- Precautionary Allergen Labeling: Manufacturers may use precautionary allergen labeling, such as "may contain" or "made in a facility that processes," to indicate potential cross-contamination risks. These statements are voluntary but are used to inform consumers of possible allergen exposure during food production.

(c) **Major Food Allergens**:

In the United States, the Food and Drug Administration (FDA) has identified eight major food allergens that must be disclosed on food labels[20].

- Gluten Labeling: Some countries, such as the United States and Canada, have specific labeling requirements for gluten-containing grains, such as wheat, rye, and barley. Foods labeled as "gluten-free" must meet particular gluten content thresholds.
- Regulation of "Free-From" Claims: Many countries regulate the use of "free-from" claims, such as "dairy-free" or "nut-free," to ensure accuracy and prevent misleading statements on food labels.
- Consumer Education: Regulatory authorities often work to educate consumers about food allergies and label reading. This includes providing guidelines on interpreting food labels and allergen disclosure statements.
- Penalties for Non-Compliance: Regulatory agencies can impose penalties and fines on manufacturers who do not comply with allergen labeling laws, especially when there is a risk to public health.
- International Harmonization: Efforts are ongoing to harmonize allergen labeling requirements internationally to provide consistent information for consumers and facilitate international trade.

8 Global Perspectives on Allergies

Allergies are a global health concern, and their prevalence and trends can vary significantly based on cultural, regional, and environmental factors. Here is an overview of allergy trends worldwide and some insights into cultural and regional differences:

(a) **Allergy Trends Worldwide:**

Allergies, including allergic rhinitis, asthma, and food allergies, have risen globally. The reasons for this increase are complex and likely involve genetic, environmental, and lifestyle factors.

- Common global allergy trends include:
- Urbanization: Allergies tend to be more prevalent in urban areas, possibly due to increased exposure to indoor allergens, pollution, and lifestyle changes.
- Westernization of Diets: Adopting Western-style diets, which may include a higher intake of processed foods and fewer fresh fruits and vegetables, has been associated with an increased risk of allergies.
- Hygiene Hypothesis: Some researchers suggest that reduced exposure to microbes and infections in early childhood may increase the risk of allergies. This theory is known as the hygiene hypothesis.

(b) **Cultural and Regional Differences:**

- Allergy prevalence and patterns can vary significantly by region and culture. Here are some examples of cultural and regional differences:
- Food Allergies:(1) Peanut Allergy: Peanut allergy is more common in Western countries, while it is less prevalent in Asian countries where peanuts are introduced into the diet at an earlier age. (2) Shellfish Allergy: Shellfish allergies are more common in coastal regions, where these foods are often part of the traditional diet.
- Asthma and Allergic Rhinitis: The prevalence of asthma and allergic rhinitis can vary between countries and regions. For example, Nordic countries like Finland and Sweden have high rates of allergies and asthma.
- Environmental Factors: Environmental factors, such as climate, pollen exposure, and pollution levels, can influence the prevalence and severity of allergies. For instance, regions with high pollen counts tend to have more cases of allergic rhinitis.
- Cultural Practices: Cultural practices, including dietary habits, breastfeeding patterns, and early introduction of allergenic foods, can impact the development of food allergies in children.

(c) **Healthcare Infrastructure**: Access to healthcare and allergy diagnosis and treatment can vary significantly between regions. Individuals with allergies may have limited access to allergists and specialized care in some areas.

(d) **Traditional Medicine**: In certain cultures, traditional or alternative medicine practices may be used to manage allergy symptoms alongside or in place of conventional medical treatments.
(e) **Awareness and Education**: Allergy and education initiatives can differ widely by region. In some areas, there may be less awareness of allergies and their management, leading to delayed diagnosis and treatment.

9 Summary and Conclusion

Food allergies are a significant health concern that affects individuals worldwide. These allergies involve adverse immune responses to specific food proteins, leading to symptoms ranging from mild discomfort to life-threatening anaphylaxis. The prevalence of food allergies has been on the rise globally, impacting people of all ages, but children are particularly vulnerable. This study has explored various aspects of food allergies, starting with definitions and an understanding of how allergies develop. We've discussed common allergenic foods and the immune system's role in allergic reactions.

Furthermore, we delved into the types and recognition of allergic reactions, emphasizing the importance of timely intervention. Managing food allergies involves diagnosis through various testing methods and personalized treatment plans. We've outlined the importance of proper diagnosis and the different tests used to identify allergens. Treatment options include medications, allergen avoidance, and immunotherapy, with emerging therapies showing promise. Navigating school and childcare settings can be challenging for children with food allergies. Parents and caregivers must proactively educate themselves, teachers, and school staff about food allergies and emergency procedures.

Finally, legal and regulatory aspects, such as food labeling laws and allergen disclosure requirements, play a crucial role in protecting individuals with food allergies. These regulations vary by region and culture but are essential for ensuring that allergenic ingredients are clearly labeled on packaged foods. Understanding and managing food allergies requires a holistic approach that combines medical knowledge, effective communication, and awareness. By staying informed and taking proactive measures, individuals with food allergies and their caregivers can lead safer and healthier lives while minimizing the risks associated with allergenic foods.

References

1. Park YW, Haenlein GFW (2017) Therapeutic, hypo-allergenic and bioactive potentials of goat milk, and manifestations of food allergy. In: Park YW, Haenlein GFW, Wendorff WL (eds) Handbook of milk of non-bovine mammals, 1st edn. Wiley, Hoboken, pp 151–179. https://doi.org/10.1002/9781119110316.ch2.4

2. Bird JA, Jones S, Burks W (2019) Food allergy. In: Clinical immunology. Elsevier, pp 625–631.e1. https://doi.org/10.1016/B978-0-7020-6896-6.00045-4
3. Groetch M, Sampson HA (2016) Management of food allergy. In: Pediatric allergy: principles and practice. Elsevier, pp 420–429.e1. https://doi.org/10.1016/B978-0-323-29875-9.00048-3
4. Turnbull JL, Adams HN, Gorard DA (2015) Review article: the diagnosis and management of food allergy and food intolerances. Aliment Pharmacol Ther 41(1):3–25. https://doi.org/10.1111/apt.12984
5. DerSarkissian C (2022) Allergies and anaphylaxis. https://www.webmd.com/allergies/anaphylaxis
6. Marchisotto MJ, Harada L, Kamdar O, Smith BM, Waserman S, Sicherer S, Allen K, Muraro A, Taylor S, Gupta RS (2017) Food allergen labeling and purchasing habits in the United States and Canada. J Allergy Clin Immunol 5(2):345–351.e2. https://doi.org/10.1016/j.jaip.2016.09.020
7. Kulthanan K, Chusakul S, Recto MT, Gabriel MT, Aw DC-W, Prepageran N, Wong A, Leong JL, Foong H, Quang VT, Zuberbier T (2018) Economic burden of the inadequate management of allergic rhinitis and urticaria in Asian countries based on the GA^{2}LEN model. Allergy Asthma Immunol Res 10(4):370. https://doi.org/10.4168/aair.2018.10.4.370
8. Minaker LM, Elliott SJ, Clarke A (2014) Exploring low-income families' financial barriers to food allergy management and treatment. J Allergy 2014:1–7. https://doi.org/10.1155/2014/160363
9. Bhatti A (2018) Cognitive bias in clinical practice—nurturing healthy skepticism among medical students. Adv Med Educ Pract 9:235–237. https://doi.org/10.2147/AMEP.S149558
10. Kamdar TA, Peterson S, Lau CH, Saltoun CA, Gupta RS, Bryce PJ (2015) Prevalence and characteristics of adult-onset food allergy. J Allergy Clin Immunol 3(1):114–115.e1. https://doi.org/10.1016/j.jaip.2014.07.007
11. Hadley C (2006) Food allergies on the rise?: determining the prevalence of food allergies, and how quickly it is increasing, is the first step in tackling the problem. EMBO Rep 7(11):1080–1083. https://doi.org/10.1038/sj.embor.7400846
12. Lavine E (2012) Blood testing for sensitivity, allergy or intolerance to food. Can Med Assoc J 184(6):666–668. https://doi.org/10.1503/cmaj.110026
13. Ansotegui IJ, Melioli G, Canonica GW, Caraballo L, Villa E, Ebisawa M, Passalacqua G, Savi E, Ebo D, Gómez RM, Luengo Sánchez O, Oppenheimer JJ, Jensen-Jarolim E, Fischer DA, Haahtela T, Antila M, Bousquet JJ, Cardona V, Chiang WC et al (2020) IgE allergy diagnostics and other relevant tests in allergy, a World Allergy Organization position paper. World Allergy Organ J 13(2):100080. https://doi.org/10.1016/j.waojou.2019.100080
14. Meals on Wheels of Central Maryland (2023) Did you know?: food allergies in older adults. https://www.mealsonwheelsmd.org/foodallergyawareness/
15. Food Insight (2023) Food allergies across the globe. https://foodinsight.org/food-allergies-across-the-globe/
16. Regional Institute of Medical Sciences (2023) Zoom-Introduction to allergy [Internet]. Available from: https://www.rims.edu.in/healthcarecdackolgksaha/Zoom-Introduction%20to%20Allergy.html
17. Food Allergy & Anaphylaxis Connection Team (2023) Food Allergy & anaphylaxis: Signs and symptoms of anaphylaxis [Internet]. Available from: https://www.foodallergyawareness.org/food-allergyand-anaphylaxis/signs-symptoms/signs-symptoms/
18. Leone L, Mazzocchi A, Maffeis L, De Cosmi V, Agostoni C (2023) Nutritional management of food allergies: prevention and treatment. Front Allergy 3:1083669
19. Hogg S (2023) Allergy management in primary care setup: Texas Speciality Clinic [Internet]. Texas Specialty Clinic. Available from: https://texasspecialtyclinic.com/blogs/allergy-management-in-primarycare-setup-texas-speciality-clinic/
20. FDA (2023) Food allergies. https://www.fda.gov/food/food-labeling-nutrition/food-allergies

Chapter 10
Food Toxicology and Safety

Sharada P.

1 Introduction

The condition of malnutrition is not well recognized and ill-treated. The various causes of malnutrition can be attributed to a lack of essential nutrients in our diet, decreased rate of absorption in micro- and macronutrients, lack of maternal care, and poverty. The various percentage of malnutrition includes 35.5% for stunted growth, 19.3% for low weight for height, and 32.1% for underweight, respectively, according to NFHS- (2015–2016) [pib.gov.in]. According to UNICEF- 2016, certain countries have reported a 40% increase in the same issue. This crisis has become worse due to climate change, COVID-19, and conflicts between countries [downtoearth.org]. On the other side, consumption of similar or such kinds of food was from the kids to healthy adults. The majority of the key nutrient classes, such as organic acids, peptides, and essential amino acids, which cannot be produced by the human body, are primarily derived from food [1]. Food can also cause a number of health issues when it is consumed in excess or without taking a balanced diet into account. Food that is improperly consumed might have negative effects on health. The presence of additional hazardous ingredients in food is a key factor in determining food quality. Examining food pollutants and their quantitative analysis is crucial for consumers of food. Therefore, it is crucial to continuously monitor the safety and quality of food in order to protect the public's health [2]. Another area of concern in the shifting circumstances is food dangers. Hazards associated with food are another area of focus in the changing scenario. Heat treatment rapidly eliminates biological and physical hazards that rarely cause serious food poisoning [3]. Several instances of chemical hazards have been explored, which include contamination of melamine in the infant's formula, heavy metal residue remains in the fish products,

Sharada P. (✉)
Department of Biotechnology, Basaveshwar Engineering College,
Bagalkote, Karnataka, India

J. S. Yaradoddi et al. (eds.), *Frontiers in Food Biotechnology*,
https://doi.org/10.1007/978-981-97-3261-6_10

and egg contamination with the pesticide have immediate impacts on consumer health [4]. Chemical analysis of food components allows us to analyze the quality of food in this regard [5]. Chemical hazards and contaminants present in food can be identified by spectroscopic techniques such as UV detection and fluorescence techniques, alone or in combination with other separation techniques, heat treatment, electrophoresis, and specific quantitative assays detected [6, 7]. Analytical methods also use techniques of chromatography such as gas-liquid chromatography and high-performance liquid chromatography [8]. Further, various factors can proportionately influence the chemical properties of foods, leading to changes in their bioactive along with nutritional parameters. Therefore, the implementation of effective, versatile, and reliable analytical techniques for assessing food authenticity has been the focus of interest [9].

The issue related to food safety now poses an alarming situation for the food industry. In addition to the continuously growing world population, the need for food is also substantially increasing creating an even larger and further complicated food chain. However, food can be intoxicated with toxins, parasites, bacteria, and pathogens, as well as contaminants that can cause many deadly diseases, including serious infectious diseases [10]. Food safety is therefore a major domain in public health concern for food industries, retailers, distributors, and consumers, with approximately 48 million people falling ill, approximately 1.28 million hospitalized, and approximately 3000 deaths each year because of consumption of unsafe and adulterated food products [11]. The main objective of the evaluation of food is related to the maintenance of food safety and security. Food technology/quality control labs replace the older technologies with innovative and modern analysis techniques to achieve newer versions [12, 13]. One of the persistent concerns in food product safety is accurately displaying accurate information efficiently [14]. There are still many problems to be solved in the area of food authenticity and traceability. Many analytical innovations and applications have been observed in food analysis [15].

Usually, the food is more frequently contaminated by various physical factors like moisture, heat, and biological factors like pathogens and bacteria, which reduces its efficacy in terms of quality. Microorganisms put the public's health at risk by increasing the risk of diseases spreading through the food chain. Therefore, the packing material is utilized to prevent contamination of food, minimize wastage of food, and further maintain the effective quality of the food [16]. Since plastic packaging materials cannot degrade in the environment, they pose a threat to the natural system. Thus, the society of today needs packaging materials that are both environmentally friendly and biodegradable. The use of biodegradable plastics for food packaging has captured the attention of scientists and researchers across the globe [17].

2 Food Safety

Food poisoning affects 80 million Americans each year, according to estimates. A food-borne sickness could be moderate (lasting for about 24–48 h) or severe (requiring immediate hospitalization or even can lead to death). People with chronic illnesses, children, pregnant women, and pregnant women are particularly at risk. For living things to survive, grow, and maintain healthy body functions, they must have access to food. Humans eat 30 tons of food on average in their lifetime, and this food is consumed in a variety of fundamental dietary forms on local, national, and international scales. Humans and other species may look for alternate sources of food with excellent and high nutritional value as the demand rises. However, food products might never be entirely safe or contaminant free. Basically, food is a mixture of chemicals. Basically, components of foods are categorized into four respective categories: nutrients, toxins of natural origin, contaminants, and additives. These nutrients accommodate more than 99.0% of the food. The major classes of nutrients are primarily in the form of carbohydrates, proteins, vitamins, fats, and minerals, and almost all of them may lead to drawbacks like toxicological effects to consumers. The physical, chemical, and immunological aspects of food products may be concealed by the metabolism process. The metabolic reaction may hinder the physical, chemical, and immunological properties of food products.

Browning responses have a significant impact on various food quality attributes, which include flavor, appearance, color, texture, nutrient content, digestibility, and functionality. There are numerous types of contaminated foods that pose serious dangers to people. The material used for packing is used to retain the whole quality and natural condition of food in order to maintain the safety of foods. The eating of tainted food exposes people to various microorganisms, increasing the risk of developing food-associated illnesses. Even if there are many analytical methods used to examine food products, contaminated food still has social repercussions and is a growing problem that gravely jeopardizes human health and well-being. Therefore, it is essential for food safety and security to identify pathogenic agents quickly and effectively in order to prevent food-borne illness. The many food-borne illnesses and safety precautions are highlighted in this chapter. This chapter highlights the various food-borne illness and safety measures. Moreover, it also focuses on the various parameters required for food storage and its effect on human health due to the consumption of contaminated food and even focuses on the importance and challenges involved in the safety of food.

3 What Is the Importance of Food Safety?

Before understanding the basic details of safety involved in food, let us know the common terminologies involved in food safety and toxicology.

- Basic hygiene—a condition conducive to the preservation of health
- Microorganisms—any organism too small to be viewed by the unaided eye (bacteria, some fungi, algae, etc.)
- Contamination—a condition of impurity resulting from the transfer of microorganisms like bacteria from one source to another
- Personal protective equipment (PPE) or protective clothing apparel, such as hand gloves and aprons used to minimize contamination and clothing that needs to be worn during the processing and packaging of foods
- Cleanliness—a state of personal neatness

4 What Are the Major Reasons of Food-Borne Illness?

Major reasons can be marked by the pathogenic microorganisms—very tiny living beings that are visible only through a proper microscope. Poor aseptic conditions and improper food handling practices enable harmful microbes to multiply and spread leading to food contamination. The three major hazards that are involved in food-borne illness are caused by eating food that is contaminated.

1. **Biologically hazardous agents**
2. **Physically hazardous substances**
3. **Chemically hazardous agents/substances**

4.1 Biologically Hazardous Agents

- Live things
- Bacterial agents
- Viral agents
- Parasites
- Fungal agents

4.2 Physically Hazardous Substances

- Foreign materials/objects—glass pieces, metals, scraps, rubber products
- Man-made materials
- Toxic metals

4.3 Chemically Hazardous Agents/Substances

- Food additives and preservatives
- Agricultural chemicals
- Chemicals from over-processing or high temperature (e.g., acrylamide and furans)
- Environmental contaminants
- Natural toxins released from microbes and animals (e.g., marine toxins)

1. **Bacterial agents**

   They are usually and abundantly found in various food sources and liquid sources and even found on our skin as well as clothes. Some bacterial populations in very low concentrations might not lead to sickness. It is the type and concentration of bacteria in the food that makes us sick.

   - *Campylobacter jejuni*
   - *Clostridium botulinum*
   - *Escherichia coli O157:H7*
   - *Salmonella (over 1600 types)*
   - *Staphylococcus aureus*

2. **Viral agents**

   These agents infect living host cells and then spread by gaining control over the host and thereby multiply inside the host cells, forming new viral particles. This machinery of viral replication continues.

   - Norwalk virus
   - Hepatitis
   - Rotavirus

3. **Parasites**

   These entities are larger than bacterial agents or viral particles. Few of the parasites can be observed even without the aid of a microscope. These need a suitable host but can reproduce on their own. Examples of parasites include roundworms and tapeworms.

4. **Fungal agents**

   There are many types of fungal agents like mold and yeast. Few fungal molds and yeasts can be used to produce bread and cheese. However, few of the molds are harmful and can cause contamination of food. Many fungal species and strains multiply and reproduce and synthesize by-products that are harmful. Examples of fungal by-products include toxins, gases, and alcohol that might cause illness.

5. **Molds**: Food spoilage from molds can be evident from the deterioration of food or other organic materials, which is essentially reasoned by the growth of mold fungi. Molds are a type of fungi that thrive in moist and warm environments. They diminish the nutrient quality, texture, and flavor of foods.

6. **Yeast**: The growth of yeast can be established by changes in food quality, deterioration, off flavors, texture, and gas production.

5 Most Common Reasons for Food Born Illness

- 21% would be accounted for holding/storage time of food (time between preparation of food and serving of food).
- 40% would be accounted for improper method used for cooking the foods.
- 20% would be accounted for persons with infected hands touching the food.
- 16% would be accounted for inadequate cooking of foods.
- 16% would be accounted for improper storage of food.
- 12% would be accounted for inadequate levels of heating of food
- 11% would be accounted for contamination from microbes.
- 7% would be accounted for cross-contamination by other sources.
- 7% would be accounted for improper cleanser used.
- 4% would be accounted for use of leftovers.

5.1 Cross Contamination

Hands are often the culprit—handling raw chicken and preparing the sandwich without cleaning the hands intermittently. In some cases, 80% of raw chicken was tested positive for Salmonella in the US.

5.2 Hand Washing Experiment

Every individual in the group shall "dirt" the respective hands as mentioned below.

Make a light coat of cooking oil or petroleum jelly on your hands. Then sprinkle one of the hands with a small quantity of cinnamon and further rub both the hands together gently and cover the front and back across, along with the fingers of both the hands.

5.3 Wash Both the Hands as Mentioned Below

- Student assigned as 1: Wash your hands with warm water and rinse (do not scrub or add soap) for 5 s.
- Student assigned as 2: Wash your hands with warm water and scrub for 20 s under the water faucet, but without using the soap.
- Student assigned as 3: Wash your hands with warm water along with the soap, scrub your hands (not under the water faucet) for 5 s, and rinse just until no soap is left on both the hands.

- Student assigned as 4: Wash your hands with warm water along with the soap, scrub the hands (not under the water faucet) for 20 s, and rinse unless no soap is left on both hands.

Correct Method Involved in Washing Both the Hands

- Washing the hands
- Wet the hands
- Apply the soap
- Scrub hands, between fingers, and forearms
- Scrub underneath the fingernails.
- Rinse both the hands along with forearms
- Dry both the hands.
- Turn off the water with the help of towel
- Open the door
- Discard the towel

6 Parameters that Make the Pathogens Grow (FATTOM)

- F—Food
- A—Acidity
- T—Temperature
- T—Time
- O—Oxygen
- M—Moisture

6.1 Parameters for the Growth of Bacteria

1. Food—Food that is high in protein
2. Acid—Food that has pH 4.6 or higher
3. Temperature—Foods that are at a temperature in between 40 and 140 °F
4. Time—Cooked foods left at least 4 h to allow enough bacteria growth to cause the illness.
5. Oxygen—Aerobic or anaerobic (microorganisms that grow in oxygen or in the absence of oxygen)
6. Moisture—Microbes thrive in the wet environments

6.2 *Measures for the Prevention and Reduction of Growth of Bacteria During Cooking*

Procedure to Be Followed for Personal Cleanliness

- People should avoid handling the foods when they have any infection or illness persists or if there are cuts or sores on either of their hands.
- Wash your hands before preparing food. Further wash the hands after coughing, sneezing or while in the restroom when touching hair or face, and while handling eggs or raw meat.
- Keep your hair away from the face.
- Wear proper clean clothing/apron (dirty clothing may carry the bacterial population).
- People should avoid tasting the food with their fingers while cooking—licking of fingers should be prohibited.
- Maintain proper cleanliness of kitchen.
- People should use warm and soapy water on dishes.
- People should not wipe their hands on dish towels use paper towels so dishes don't get bacteria.

Storage of Food

- Danger zone: 41–140 °F
- Store the cold foods at 41 °F or below
- Store the hot foods at 140 °F or higher
- Thaw the frozen foods in the refrigerator
- Maintain the refrigerator temperature at 38–41 °F

Avoid Cross-Contamination

- Do not mix the foods to avoid the microbes from one food to get into another.
- Always maintain proper hygiene and keep the work areas clean.
- Store especially raw products and cooked food products separately during the food preparation.
- After the finish of your work, the cutting boards and other tools used for cutting raw meat or eggs should be washed thoroughly and disinfected (sanitized), to avoid future contamination chances.
- Do not place the cooked meat on the plate that was held with raw meat prior.
- Always store fresh meat products on trays on the lowest shelf in the refrigerator.

Destruction of Organisms

- The purpose of cooking food is primarily not to enhance flavor but it essentially enables to destroy the bacterial population.
- Follow the proper cooking conditions in terms of temperatures.
- Temperature of 165° and above needs to be implemented for chicken, pork, and leftovers.
- Temperature of 155° and above needs to be implemented for ground beef.
- Temperature of 145° and above needs to be implemented for lamb, fish, beef, and eggs.
- Temperature of 135° and above needs to be implemented practiced for rare roast beef and sushi.

7 Various Challenges Involved in Safety of Food

Most of the times foods we like and use for good health are many of the times contaminated with a varied number of germs that can lead to sickness and prove to be dangerous and life-threatening. A great deal of progress is essentially required to safeguard people and decrease food-borne illnesses.

Latest global issues and challenges to food safety will further continue to emerge, largely because of:

- Modification in strategies related to food production and supply.
- Variations in the environment lead to contamination of food.
- Engineered and exotic bacteria, toxins, and manipulation lead to multiple drug resistance.
- Multiple changes and requirements by consumers in terms of preferences.
- Regular improvisations in the tests that detect various food-borne ailments.

8 Future Prospectus of Food Safety and Toxicology

The knowledge and understanding of various parameters related to food safety and toxicology need to be addressed with at most clarity by the scientific community and relevant authorities in the present scenario, as there are lots of threats related to the respective fields. Lots of data remain obscure. Overall, the future of food safety and toxicology will likely involve a multidisciplinary approach, integrating technological innovations, regulatory frameworks, emerging food-borne risks, personalized nutrition, and collaborative efforts to ensure the safety, quality, and sustainability of the global food supply.

References

1. Morisasa M, Sato T, Kimura K, Mori T, Goto-Inoue N (2019) Application of matrix-assisted laser desorption/ionization mass spectrometry imaging for food analysis. Foods 8(12):633. https://doi.org/10.3390/foods8120633
2. Mishra GK, Barfidokht A, Tehrani F, Mishra RK (2018) Food safety analysis using electrochemical biosensors. Foods 7(9):141. https://doi.org/10.3390/foods7090141
3. Ren J, Deng T, Huang W, Chen Y, Ge Y (2017) A digital PCR method for identifying and quantifying adulteration of meat species in raw and processed food. PLoS One 12(3):e0173567. https://doi.org/10.1371/journal.pone.0173567
4. Hua MZ, Li S, Wang S, Lu X (2018) Detecting chemical hazards in foods using microfluidic paper-based analytical devices (μPADs): the real-world application. Micromachines 9(1):32. https://doi.org/10.3390/mi9010032
5. Granato D, Putnik P, Kovačević DB et al (2018) Trends in chemometrics: food authentication, microbiology, and effects of processing. Compr Rev Food Sci Food Saf 17(3):663–677. https://doi.org/10.1111/1541-4337.12341
6. Jha SN, Jaiswal P, Grewal MK, Gupta M, Bhardwaj R (2016) Detection of adulterants and contaminants in liquid foods—a review. Crit Rev Food Sci Nutr 56(10):1662–1684. https://doi.org/10.1080/10408398.2013.798257
7. Rodriguez RS, O'Keefe TL, Froehlich C, Lewis RE, Sheldon TR, Haynes CL (2021) Sensing food contaminants: advances in analytical methods and techniques. Anal Chem 93(1):23–40. https://doi.org/10.1021/acs.analchem.0c04357
8. Nolvachai Y, Kulsing C, Marriott PJ (2017) Multidimensional gas chromatography in food analysis. TrAC Trends Anal Chem 96:124–137. https://doi.org/10.1016/j.trac.2017.05.001
9. González-Domínguez R (2020) Food authentication: techniques, trends and emerging approaches. Foods 9(3):346. https://doi.org/10.3390/foods9030346
10. Fukuda K (2015) Food safety in a globalized world. Bull World Health Organ 93(4):212–212. https://doi.org/10.2471/blt.15.154831
11. Scallan E, Hoekstra RM, Angulo FJ et al (2011) Foodborne illness acquired in the United States—major pathogens. Emerg Infect Dis 17(1):7–15. https://doi.org/10.3201/eid1701.P11101
12. García-Cañas V, Simó C, Herrero M, Ibáñez E, Cifuentes A (2012) Present and future challenges in food analysis: foodomics. ACS Publications
13. Hong E, Lee SY, Jeong JY et al (2017) Modern analytical methods for the detection of food fraud and adulteration by food category. J Sci Food Agric 97(12):3877–3896. https://doi.org/10.1002/jsfa.8364
14. Wrona M, Nerín C (2020) Analytical approaches for analysis of safety of modern food packaging: a review. Molecules 25(3):752. https://doi.org/10.3390/molecules25030752
15. Galvez JF, Mejuto JC, Simal-Gandara J (2018) Future challenges on the use of blockchain for food traceability analysis. TrAC Trends Anal Chem 107:222–232. https://doi.org/10.1016/j.trac.2018.08.011
16. Han J-W, Ruiz-Garcia L, Qian J-P, Yang X-T (2018) Food packaging: a comprehensive review and future trends. Compr Rev Food Sci Food Saf 17(4):860–877. https://doi.org/10.1111/1541-4337.12343
17. Dilkes-Hoffman LS, Lane JL, Grant T, Pratt S, Lant PA, Laycock B (2018) Environmental impact of biodegradable food packaging when considering food waste. J Clean Prod 180:325–334. https://doi.org/10.1016/j.jclepro.2018.01.169

Part III
Applied Technologies and Frontiers

Chapter 11
Nanotechnology in Food Industry

Preeti S. Kumarmath

1 Introduction

The science of manipulating nano-sized particles for their useful qualities for a variety of reasons is known as nanotechnology. Through creative and inventive methods, it contributes significantly to the field of food and agriculture industries, improving safety and food value, increasing harvests, and advancing human health [1]. Globally, nanotechnology has gradually but firmly taken over a number of industries. The developed world is where the fastest-paced technological development is most evident, as nanomarkets have risen to prominence in the previous several years. Since nanotechnology has evolved into a versatile technology, it is not a novel idea. Four generations of nanomaterials—active and passive nanoassemblies, small-scale molecular nanosystems, and general nanosystems—have surfaced and are employed in integrated scientific domains [2]. The evidence for nanoscale manufacturing, which will be applied to practically every branch of science and technology, is found in the field of nanoscience's explosive growth.

This chapter examines how nanotechnology has recently advanced in applications and highlights the significance of nanoscience. Nanomaterials are small particles that can range in size from 1 to 100 nm, are insoluble or bio-persistent in nature, can be synthesized in a variety of ways, and are used in a wide range of industries, primarily the agricultural, pharmaceutical, food, healthcare, chemical, mechanical, oil, and gas sectors [3, 4]. A brief overview of the drawbacks of nanotechnology is also given for each industry to aid in the scientific community's research towards finding answers and raising awareness of both the benefits and drawbacks of the technology. Combining the essential elements of the biological with basic physical and chemical sciences is known as nanotechnology. These

P. S. Kumarmath (✉)
Department of Biotechnology, Basaveshwar Engineering College,
Bagalkote, Karnataka, India

J. S. Yaradoddi et al. (eds.), *Frontiers in Food Biotechnology*,
https://doi.org/10.1007/978-981-97-3261-6_11

operations occur in the nanometer range. Physically, the size is diminished; chemically, new bonds and properties are resolved; biological actions, like drug binding and distribution at a specific spot, are formed at the nanoscale [5, 6]. To fully utilize the enormous promise of this new science, billions of dollars and euros are being invested globally in nanotechnology, particularly in the developed nations of China, Europe, and USA [7].

Due to the desire of both industrialized and emerging nations to increase their investments in nanotechnology, there has been a new industrial revolution [8]. As a result, there are several potentials provided by nanotechnology (Fig. 11.1) for creating and utilizing new materials, structures, or systems in a variety of industries, including medical, nutrition, agriculture, etc. Researchers are trying to figure out how to enhance food quality with the least number of negative effects on the nutritional value of the product in response to growing customer concerns about food quality and health advantages. The food industry has seen a rise in demand for materials based on nanoparticles because many of these compounds are non-toxic especially vital nutrients [9]. At high pressures and temperatures, it also exhibits greater stability [10].

Complete food solutions, from food manufacturing and processing to packaging, are provided by nanotechnology. Complete food solutions, including processing, packaging, and food production, are made possible by nanotechnology. Nanomaterials have a significant impact on food's health benefits in addition to its safety and quality. Novel methodologies, methods, and products that enable the direct application of nanotechnology in food science are being developed by numerous researchers, with the inter organizations corporations [11]. Numerous industries have benefited from advances in nanotechnology, including batteries, electronics, food items, and medical. In the food sector, nanotechnology is utilized to enhance food safety, prolong product shelf life, enhance taste, and better transport food ingredients to desired regions.

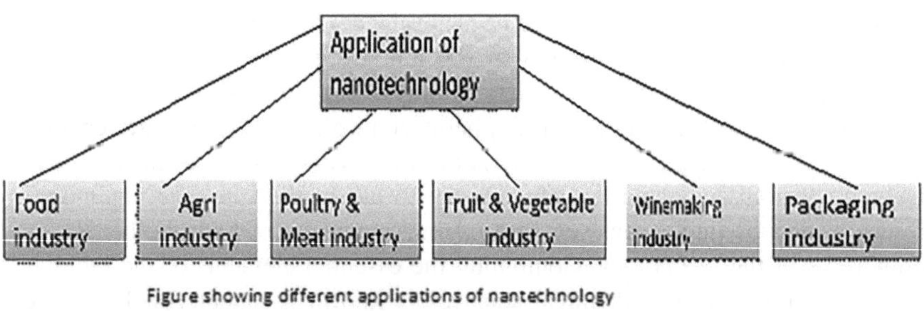

Fig. 11.1 Nanotechnology and its applications

2 Utilizing Nanotechnology in the Bioprocessing Sector

Contributions to the bioprocessing sector are correlated with scientific and technical precision in nanotechnology. Scholars are intrigued by the potential use of nanoparticles in the superior manufacturing of novel biomaterials and food [12, 13]. The research community has focused on a number of nanotechnology applications over the years, including food monitoring, biosensing devices, pathogen detection, and smart packaging materials, especially reusable and biodegradable ones, as well as nanoencapsulation of active food ingredients. Ultimately, the effective introduction of nano-based bioprocessed products into common markets for public use will depend on societal acceptance and management of social, ethical, and cultural concerns [14, 15].

2.1 Nanotechnology Use in the Bioprocessing Sector

The need for food rises along with the global population, raising more questions about how to provide consistent, safe, and healthful food options. With enhanced sensitivity equipment, sensors and diagnostic tools will support food quality monitoring [16]. Furthermore, a system that might support brandshield during bioprocessing could readily conceal the use of counterfeit industrial food uses thanks to nanotechnology [17]. The food business might regulate the amount of electricity used in food production by using comprehensive nanotechnology applications. In the end, the environment would benefit from reduced electricity use. In the future, this could help allay environmental worries by directly addressing the interplay of climate, nutrition, and nanotechnology [18].

Fermentation processes are one of the major effects of nanotechnology in the bioprocessing sector. Unlike the conventional fermentation procedures, these technologies are used for increased industrial demand and expanded production of biomolecules at extremely cheap cost [19]. The production of biocompatible, safe, and non-toxic chemicals and nanostructures with broad applications in the food, bioprocessing, and winemaking areas has been made possible by the successful application and integration of fermentation and nanotechnology [20].

Food supply chain management and surveillance, which are used in a number of subsectors including production, distribution, storage, and toxicity control, are two further significant applications. In an effort to enhance overall analytical performance in terms of response time, selectivity, sensitivity, dependability, and accuracy, nanodevices and nanomaterials are being included in biological and chemical sensing technologies [21]. Modern nano-based materials such nanotubes, nanocomposites, nanowires, nanosheets, nanorods, and other materials utilized for component immobilization and tagging are gradually replacing conventional food control approaches [22].

These techniques can be controlled optically or electrochemically. A number of food monitoring assays based on nanotechnology are suggested and put into practice. Tests related to biology and immunity, as well as electrochemical and optical techniques like colorimetric and surface-enhanced Raman scattering, can be included in this [13]. During commercial processing and industrial bioprocessing, materials ranging from heavy materials to insecticides, allergies, bacteria, and antibiotics can be easily monitored. Additionally, biocomposting materials have seen amazing transformations because of nanotechnology.

The need for sustainable solutions is growing as the world's demand for biodegradable composites to lessen their impact on the environment and boost industrial production efficiency rises [23]. This will create nanocomposites that are more valuable mechanically than conventional polymers, proving their industrial relevance. Mechanical, electrochemical, catalytic, optical, and electrical qualities have all been improved [24].

In addition to the bioprocessing industry, which uses these biodegradable polymers to produce food with pertinent benefits, other applications for these polymers include therapeutics, biomedicine, packaging, drug delivery technology, biotechnological base tissue engineering, water treatment, sensing, and the cosmetics and food industries [3, 5, 13, 25]. These nanocomposites are suitable for usage in products because of their exceptional qualities, which include reduced toxicity, antibacterial activity, thermal stability, and improved biodegradation features [26].

To guarantee that consumer pleasure, public benefit, and safety regulations are satisfied, it is still necessary to do thorough toxicity and safety profiles on these and other nanomaterials [26]. Furthermore, the creation of functional meals has benefited from the continued advancement of nanotechnology. With the world population growing every year, there is an increasing demand for more sustainable, safer, and better food items, which has been made possible by the spread and integration of nanotechnology with the food sector [27]. The broad use of nanotechnology in the food business globally is necessary because of its overall beneficial effects on food packaging, processing, manufacturing, disease detection, monitoring, and production profiles [5, 28]. Recent studies have demonstrated how bioactive substances and necessary elements can be better delivered to food through the use of nanoencapsulation or nanomaterials [29]. These approaches avoid needless interactions with other food additives while improving the protective function and sensitivity of bioactive substances. As a result, the bioactivity and solubility profiles of nanofoods are greatly enhanced, enhancing their advantages to human health. But because these meals come into direct contact with the human body, their safety needs need to be carefully controlled through safety profiles [30].

2.2 Nutritional and Nourishing Value

The majority of bioactive substances, including proteins, vitamins, lipids, and carbohydrates, are sensitive to extremely acidic conditions as well as the duodenum's and stomach's enzyme activity. These bioactive substances can withstand such harsh environments and easily integrate into food products because of their encapsulation. This is difficult to accomplish in a noncapsulated form since these bioactive chemicals are poorly soluble in water. Tiny edible capsules based on nanoparticles are being developed to enhance the way that vitamins, medications, or delicate micronutrients are delivered in everyday meals to offer major health advantages [31, 32].

The various methods used to encapsulate compounds in tiny forms in order to more efficiently deliver nutrients like protein and antioxidants for precisely targeted nutritional and health benefits include nanoemulsification, nanostructuration, and nanocomposite. It has been discovered that polymeric nanoparticles are appropriate for encapsulating bioactive substances (such as vitamins and flavonoids) in order to protect and deliver those substances to their intended functions [33].

2.3 Nanotechnology in the Food Sector

As the graphic illustrates, there are a plethora of uses for nanotechnology in the food sector, including smart diets, point-of-care drug administration, packaging, food manufacturing, and other contemporary preservatives. Because of their excellent barrier qualities against environmental impacts, nanomaterials like polymer clay nanocomposites are used in packaging materials [34]. Similar to this, combinations of nanoparticles are employed as antimicrobial agents, especially in canned goods, to shield food from rapidly degrading due to microbiological activity. Similar to this, a number of assays based on nanosensors and nanoassemblies are employed in food preparation and storage to identify microorganisms [35]. In addition to microscopically small pathogenic organisms, nanoarrays can detect minuscule gases and both chemical and inorganic residues [36]. However, because of the potential for toxicity, the majority of these metallic nanoparticles are not put straight to food. Undertaking specific research to mitigate their toxicity can help boost the adoption of such items in the market [37]. In light of this, it is reasonable to state that the food business is quickly adopting nanotechnology for uses in sensing, antibacterial, storage, and packaging [38].

Global dairy production is likewise changing as a result of nanotechnology [39]. Improved processing techniques, food contact and blending, higher yields, longer shelf life and product safety, better packaging, and antimicrobial resistance are a few potential effects of nanotechnology on the dairy business [40]. Furthermore, medicines, biologically active compounds, and improved flavors, colors, odors, and other food attributes are being added to dairy products via nanocarriers at an increasing rate [41]. These substances have improved absorption, solubility, and delivery properties for the system they are intended for. However, for the wider commercialization of these products, the issue of public acceptance owing to concerns about unknown or potential negative effects linked with dairy and food items based on nanotechnology needs to be addressed [42].

2.4 The Use of Nanotechnology in Agriculture

The foundation of the economies of many countries worldwide is agriculture. It contributes significantly to the global economy as a whole and, by providing for people's nutritional needs, is essential to maintaining the population. Agriculture is predicted to suffer greatly as a result of the drastic changes in global weather patterns brought about by global warming. The impact of nanoparticles on plant growth and development has been the subject of numerous studies. By prolonging the shelf life of agricultural products—especially fruits and vegetables—nanotechnology has been used to enhance quality and lower postharvest loss [43].

It is always preferable in this scheme to take proactive steps to improve farming practices and make them safer and more sustainable than they were previously. As a result, modern technology is utilized everywhere. A significant part of this interaction between sustainable technologies is also played by nanotechnology. It is crucial to the manufacturing, processing, packing, storage, and transportation of agricultural industrial products [44]. Precision farming has benefited from the introduction of various nanotechnology-based approaches that enhance plant nutrient uptake and enhance pathogen detection to prevent agricultural diseases. Nanoclays and zeolites are beneficial additions to fertilizers because they effectively replenish soil nutrient broths and restore soil fertility [44]. The seeds in modern seed banks and intelligent seed concepts are also designed to sprout in environments that are conducive to their survival. Coating is done in this framework using nanopolymer mixes [45]. The use of nanotechnology in pesticides, herbicides, fungicides, and insecticides is also transforming these products. Additionally, the use of nanotechnology in treatment and disinfection procedures has been taken into consideration with the goal of improving the interrelated poultry and animal husbandry regions.

2.5 Nanotechnology in the Poultry and Meat Industry

The poultry industry is a big part of the food industry and donates millions of dollars to the food industry worldwide every year. Various commercial food chains worldwide have their roots in the healthy poultry industry. The emergence of widespread foodborne diseases in poultry, dairy, and meat farms is of great concern to the food industry. Nanobiotechnology is certainly playing a productive role in combating foodborne pathogens, such as those that proliferate through *Salmonella* and

Campylobacter infections, by enabling increased poultry consumption while maintaining the affordability and safety of manufactured chicken products [46]. The authenticity and traceability of chicken products are verified using a variety of nano-based instruments and materials, including surface biocides, packing materials, biosensors, air, nano-based disinfectants, water filters, protective clothes, and detective devices [47]. Furthermore, spoilage organisms and foodborne pathogens are reduced prior to the food entering the supply chain through the application of materials based on nanotechnology [48].

2.6 The Fruit and Vegetable Industry Using Nanotechnology

As previously said, the food business has seen significant advancements in nanotechnology. This cannot be ignored in the medicinal, fruit and vegetable, or agricultural sectors. In order to meet the nutritional needs of an expanding population, scientists are working to increase the organic vegetables and shelf life of fresh. Nanotechnology is vital to the safety and production of fruits and vegetables, from horticulture to packaging, food processing, and pathogen detection [49].

Nanotechnology is currently displacing conventional technologies because it is more cost-effective, produces better results, and has a longer shelf life overall. Nanotechnology has not yet been linked to any high-level toxicity to fresh, green, organic vegetables; however, there may still be some hazards. These innovations lessen postharvest waste, a significant issue in developing nations, while offering consumers reliable and enough food sources [37]. Reduced humidity, oxygen transfer, and ideal water vapor permeability are benefits of nano packing. With the aid of nanotechnology, the long-term durability of such items can be raised to the required level [50].

2.7 The Use of Nanotechnology in the Wine Industry

One significant commercial use of the food business globally is the wine industry. In this industry, nanotechnology is likewise being used more and more. Through the use of amperometric, fluorometric sensors, or biological, as well as nanoelectronics and nanoelectrochemistry, nanotechnology facilitates sensing. These nanoparticles aid in the analysis of wine constituents, such as organic acids, polyphenols, sulfur dioxide, biogenic amines, and guarantee their proper presence during the winemaking, and processing processes [51].

More work is being done to advance sensory nanotechnology in order to process, boost wine production, testing selectivity, accuracy, sensitivity, and quick response times [35]. Carbon nanorods, nanodots, nanotubes, nanodots, and metallic nanoparticles including silver, gold, iron oxide, zinc oxide, and other nanocomposites are among the specific nanoassemblies utilized in the wine industry. The concepts of nanoliquid chromatography, electronic tongues, mesoporous silica, and magnetic nanoparticle applications in winemaking goods have all been presented in recent research investigations [52]. However, a full description of these nanoparticles is outside the purview of this study. An elaborate account of these nanomaterials is outside the purview of this study. All things considered, though, it is accurate to state that enology is benefiting greatly from the application of nanotechnology.

2.8 Use of Nanotechnology in the Packaging Sector

With the global awareness of environmental protection, the packaging sector is always evolving. The packaging sector is related to a number of issues. To provide the consumer with the greatest quality, food safety should be the top priority when it comes to packaging. Additionally, packaging needs to be ecofriendly in order to lessen pollution from food waste and increase the sustainability of industrial processes. An effort is being made to lessen the load by substituting environmentally friendly biopolymer-based organic materials that have been treated at the nanoscale to make use of the advantageous aspects of nanotechnology for non-biodegradable plastic packaging materials [53].

Because it increases food safety and production while addressing environmental challenges, nanomanufacturing of packaging biomaterials has proven effective in the food packaging business [54]. These packaging nanomaterials include nanoadditives, anti-caking agents, among many other things, and nutraceutical delivery systems. To increase the functionality of the nanofillers, packaging, and biopolymers are mixed to create packaging material nanocomposites [55]. To avoid contaminating the packaged substance in these situations, preference is given to antibacterial nanoparticles combined with a polymer. It is crucial to remember that packaging nanotechnology is not just being used in food packaging; it is also being embraced by a number of other industries, leather goods, including textiles, and cosmetics, all of which stand to gain greatly from it [49].

3 Safety Concerns

The food business can benefit greatly from nanotechnology, but it is important to consider the safety concerns around nanomaterials. Numerous researchers addressed the potential for nanoparticles to enter food from packaging materials and their effects on consumer health when discussing safety concerns with nanomaterials [56]. Even though some materials are classified as GRAS (generally recognized as safe), more research is necessary to determine the danger associated with their nanocounterparts because the physicochemical features of nanostates differ greatly from those of macrostates. Furthermore, these nanomaterials' small size may raise the possibility of bioaccumulation in bodily tissues and organs [57]. For instance, when exposed to human lung cells, silica nanoparticles used as anticaking agents may be harmful [58]. Particle surface shape, aggregation, surface energy, concentration, and adsorption are only a few of the numerous variables that influence resolution. Cushen et al. (2014) created a model to investigate the migration of particles from food packaging [59]. They investigated the copper and silver migration from nanocomposites.

They discovered that one of the most significant factors influencing migration—more significant than particle size, temperature, or contact time—was the percentage of nanofillers in the nanocomposites. Because every nanomaterial is unique, toxicity will probably need to be assessed on an individual basis [60]. In order to guarantee product quality, health and safety, and environmental rules, regulators must also create some criteria for commercial products.

4 Conclusion

There is a great focus on attention and activity in this field of research since the usage of structures in the nanoscale range in the food business has grown in popularity recently. In the fields of materials and machine sciences, dynamic and incredibly cheap nanotechnologies are being introduced. Devices and materials based on nanobiotechnology are getting smaller and more sensitive as a field. This study provided a thorough overview of these technologies. Researchers and practitioners from various fields will find this overview useful in exploring the applications of nanotechnology in their own fields of expertise. It is generally recognized that it can be used in the areas of food safety and packaging. Although nanotechnology has a vast range of potential applications, the risks that come with these unbounded possibilities are yet unknown and unclear. Furthermore, food preservation with nanomaterials has shown promise in preventing food from fats, gases, bad tastes, absorbing moisture, and smells. Therefore, in order to find more answers in the field of nanotoxicology, more work needs to be linked and thoroughly recognized.

Moreover, it is advised that professionals in the fields and in education collaborate with researchers, technicians, and industrialists to investigate possibilities and effectively employ nanotechnology in field studies. They offer top-notch delivery mechanisms for bioactive substances to reach the intended tissues. Even though nanotechnology is developing at a rapid pace, there are still a lot of obstacles to overcome, lots of room for improvement, and concerns from the public regarding the potential effects of nanotechnology that need to be addressed.

Furthermore, in order to investigate choices and effectively apply nanotechnology in field trials, it is advised that technicians, researchers, and industrialists collaborate on research area and educational levels. They offer top-notch delivery mechanisms for bioactive substances to specific tissues. Despite the fact that nanotechnology is constantly evolving and setting new standards, there are still a lot of obstacles to overcome, lots of chances to improve existing technology, and concerns from customers regarding the potential effects of nanotechnology still need to be addressed.

References

1. Duncan TV (2011) Applications of nanotechnology in food packaging and food safety: barrier materials, antimicrobials and sensors. Curr Opin Colloid Interface Sci 363:1–24
2. Zelzer M, Ulijn RV (2010) Next-generation peptide nanomaterials: molecular networks, interfaces and supramolecular functionality. Chem Soc Rev 39:3351–3357
3. Anselmo AC, Mitragotri S (2019) Nanoparticles in the clinic: an update. Bioeng Transl Med 4:e10143
4. Rickerby D, Morrison M (2007) Nanotechnology and the environment: a European perspective. Sci Technol Adv Mater 8:19
5. Bhushan B (2017) Introduction to nanotechnology. In: Handbook of nanotechnology. Springer, Berlin, pp 1–19
6. Kumar S (2014) Nanotechnology in computers. Int J Inf Comput Technol 4:974–2239
7. Schulte J (2005) Nanotechnology: global strategies, industry trends and applications. Wiley, Hoboken
8. Qureshi MA, Karthikeyan S, Punita K, Khan PA, Uprit S, Mishra UK (2012) Application of nanotechnology in food and dairy processing: an overview. Pak J Food Sci 22:23–31
9. Roselli M, Finamore A, Garaguso I, Britti MS, Mengheri E (2003) Zinc oxide protects cultured enterocytes from the damage induced by Escherichia coli. J Nutr 133:4077–4082
10. Sawai J (2003) Quantitative evaluation of antibacterial activities of metallic oxide powders (ZnO, MgO and CaO) by conductimetric assay. J Microbiol Method 54:177–182. https://doi.org/10.1016/S0167-7012(03)00037-X
11. Dasgupta N, Ranjan S, Mundekkad D, Ramalingam C, Shanker R, Kumar A (2015) Nanotechnology in agrofood: from field to plate. Food Res Int 69:381–400. https://doi.org/10.1016/j.foodres.2015.01.005
12. Hulla J, Sahu S, Hayes A (2015) Nanotechnology: history and future. Hum Exp Toxicol 34:1318–1321
13. Bajpai VK, Kamle M, Shukla S, Mahato DK, Chandra P, KyuHwang S, Kumar P, Huh YS, Han YK (2018) Prospects of using nanotechnology for food preservation, safety, and security. J Food Drug Anal 26:1201–1214
14. Roco M, Müller B, Wagner E, Borchard G, Di Francesco T, Jurczyk K, Braegger U, Jurczyk M, Bartolucci C, Ijabadeniyi O et al (2018) Nanoscience and nanotechnology: advances and developments in nano-sized materials. De Gruyter, Berlin
15. Yu H, Park JY, Kwon CW, Hong SC, Park KM, Chang PS (2018) An overview of nanotechnology in food science: preparative methods, practical applications, and safety. J Chem 2018:5427978
16. Yadav C (2021) Food and bioprocessing industry. In: Handbook of polymer nanocomposites for industrial applications. Elsevier, Amsterdam, pp 295–324
17. McNeil SE (2005) Nanotechnology for the biologist. J Leukoc Biol 78:585–594
18. Anmol L, Veena G (2018) Nanotechnology approach in food science: a review. Int J Food Sci Nutr 3:183–186
19. Palit S (2020) Recent advances in the application of nanotechnology in food industry and the vast vision for the future. In: Nanoengineering in the beverage industry, pp 1–34
20. Garde-Cerdán T, Costa BS, Rubio-Bretón P, Pérez-Álvarez E (2021) Nanotechnology: recent advances in viticulture and enology. J Sci Food Agric 101:6156–6166
21. Ghosh T, Raj GB, Dash KK (2022) A comprehensive review on nanotechnology based sensors for monitoring quality and shelf life of food products. Measurement: Food 7:100049
22. Thangadurai D, Sangeetha J, Prasad R (2020) Nanotechnology for food, agriculture, and environment. Springer, Berlin
23. Hea X, Deng H, Hwang HM (2019) The current application of nanotechnology in food and agriculture. J Food Drug Anal 27:1–21
24. Kalita D, Baruah S (2019) The impact of nanotechnology on food. In: Nanomaterials applications for environmental matrices. Elsevier, Amsterdam, pp 369–379

25. Bhardwaj A, Bhardwaj A, Misuriya A, Maroli S, Manjula S, Singh AK (2014) Nanotechnology in dentistry: present and future. J Int Oral Health 6:21
26. Fadiji AE, Mthiyane DMN, Onwudiwe DC, Babalola OO (2022) Harnessing the known and unknown impact of nanotechnology on enhancing food security and reducing postharvest losses: constraints and future prospects. Agronomy 12:1657
27. Pandey G (2018) Challenges and future prospects of agri-nanotechnology for sustainable agriculture in India. Environ Technol Innov 11:299–307
28. Bongomin O, Ocen GG, Nganyi EO, Musinguzi A, Omara T (2020) Exponential disruptive technologies and the required skills of industry 4.0. J Eng 2020:4280156
29. Xue J, Luo Y (2021) Sustainable food and agriculture system: a nanotechnology perspective. ES Food Agrofor 5:1–3. [CrossRef]
30. Vishwakarma K, Upadhyay N, Kumar N, Tripathi DK, Chauhan DK, Sharma S, Sahi S (2018) Potential applications and avenues of nanotechnology in sustainable agriculture. In: Nanomaterials in plants, algae, and microorganisms. Academic, Cambridge, pp 473–500
31. Yan SS, Gilbert JM (2004) Antimicrobial drug delivery in food animals and microbial food safety concerns: an overview of in vitro and in vivo factors potentially affecting the animal gut microflora. Adv Drug Deliv Rev 56:1497–1521
32. Koo OM, Rubinstein I, Onyuksel H (2005) Role of nanotechnology in targeted drug delivery and imaging: a concise review. Nanomed Nanotechnol Biol Med 1:193–212
33. Langer R, Peppas NA (2003) Advances in biomaterials, drug delivery, and bionanotechnology. AICHE J 49:2990–3006
34. Huang H, Wang D, Zhu J, Seeger S, Chu Z (2022) Sand-based economical micro/nanocomposite materials for diverse applications. ACS Appl Mater Interfaces 14:43656–43665
35. Biswa R, Alam M, Sarkar A, Haque I, Hasan M, Hoque M (2022) Application of nanotechnology in food: processing, preservation, packaging and safety assessment. Heliyon 8:e11795
36. Rashidi L (2021) Different nano-delivery systems for delivery of nutraceuticals. Food Biosci 43:101258
37. Jagtiani E (2022) Advancements in nanotechnology for food science and industry. Food Front 3:56–82
38. Weiss J, Takhistov P, McClements DJ (2006) Functional materials in food nanotechnology. J Food Sci 71:R107–R116
39. Singh T, Shukla S, Kumar P, Wahla V, Bajpai VK, Rather IA (2017) Application of nanotechnology in food science: perception and overview. Front Microbiol 8:1501
40. Hamad AF, Han JH, Kim BC, Rather I (2018) The intertwine of nanotechnology with the food industry. Saudi J Biol Sci 25:27–30
41. Chellaram C, Murugaboopathi G, John A, Sivakumar R, Ganesan S, Krithika S, Priya G (2014) Significance of nanotechnology in food industry. APCBEE Proc 8:109–113
42. Berekaa MM (2015) Nanotechnology in food industry; advances in food processing, packaging and food safety. Int J Curr Microbiol App Sci 4:345–357
43. Saini J, Bhatt R (2020) Global warming—causes, impacts and mitigation strategies in agriculture. Curr J Appl Sci Technol 39:93–107
44. Usman M, Farooq M, Wakeel A (2020) Nanotechnology in agriculture: current status, challenges and future opportunities. Sci Total Environ 2020:137778
45. Vijayakumar MD, Surendhar GJ, Natrayan L, Patil PP, Ram PMB, Paramasivam P (2022) Evolution and recent scenario of nanotechnology in agriculture and food industries. J Nanomater 2022:1280411
46. Lamri M, Bhattacharya T, Boukid F, Chentir I, Dib AL, Das D, Djenane D, Gagaoua M (2021) Nanotechnology as a processing and packaging tool to improve meat quality and safety. Foods 10:2633
47. King T, Osmond-McLeod MJ, Duffy LL (2018) Nanotechnology in the food sector and potential applications for the poultry industry. Trends Food Sci Technol 72:62–73
48. Sharma PK, Dorlikar S, Rawat P, Malik V, Vats N, Sharma M, Rhyee JS, Kaushik AK (2021) Nanotechnology and its application: a review. Nanotechnol Cancer Manag 1–33

49. Bhusare SK (2021) Applications of nanotechnology in fruits and vegetables. Food Agric Spectr J 2:231–236
50. Enescu D, Cerqueira MA, Fucinos P, Pastrana LM (2019) Recent advances and challenges on applications of nanotechnology in food packaging. A literature review. Food Chem Toxicol 134:110814
51. Loira I, Morata A, Escott C, Del Fresno JM, Tesfaye W, Palomero F, Suárez-Lepe JA (2020) Applications of nanotechnology in the winemaking process. Eur Food Res Technol 246:1533–1541
52. Hosnedlova B, Sochor J, Baron M, Bjorklund G, Kizek R (2020) Application of nanotechnology based-biosensors in analysis of wine compounds and control of wine quality and safety: a critical review. Crit Rev Food Sci Nutr 60:3271–3289
53. Primožič M, Knez Ž, Leitgeb M (2021) (Bio)nanotechnology in food science—food packaging. Nanomaterials (Basel) 11:292
54. Sahani S, Sharma YC (2021) Advancements in applications of nanotechnology in global food industry. Food Chem 342:128318
55. Jaiswal L, Shankar S, Rhim JW (2019) Applications of nanotechnology in food microbiology. Front Microbiol 8:43–60
56. Bradley EL, Castle L, Chaudhry Q (2011) Applications of nanomaterials in food packaging with a consideration of opportunities for developing countries. Trends Food Sci Technol 22:603–610
57. Savolainen K, Pylkkänen L, Norppa H, Falck G, Lindberg H, Tuomi T et al (2010) Nanotechnologies, engineered nanomaterials and occupational health and safety—a review. Saf Sci 6:1–7
58. Athinarayanan J, Periasamy VS, Alsaif MA, Al-Warthan AA, Alshatwi AA (2014) Presence of nanosilica (E551) in commercial food products: TNF-mediated oxidative stress and altered cell cycle progression in human lung fibroblast cells. Cell Biol Toxicol 30:89–100
59. Cushen M, Kerry J, Morris M, Cruz-Romero M, Cummins E (2014) Evaluation and simulation of silver and copper nanoparticle migration from polyethylene nanocomposites to food and an associated exposure assessment. J Agric Food Chem 62:1403–1411
60. Mahler GJ, Esch MB, Tako E, Southard TL, Archer SD, Glahn RP et al (2012) Oral exposure to polystyrene nanoparticles affects iron absorption. Nat Nanotechnol 7:264–271

Chapter 12
Fermented Foods and Their Potential

Manoj Girish, Jayashree V. Hanchinalmath, Shefali Srivastava, and Kirankumar Shivasharanappa

Abbreviations

μg	Micrograms
AD	Atopic dermatitis
BMI	Body mass index
BCE	Before the common era
BSM media	Skerman's basal mineral salt media
CGXIIG growth medium	Corynebacterium glutamicum XII containing glucose growth media
g	Grams
GM17 media	Growth medium 17
HPP	High pressure processing
IBS	Irritable bowel syndrome
mg	Milligrams
M9 agar media	Minimal 9 agar media
N	Number of people participated in the trial
NAFLD	Nonalcoholic fatty liver disease

M. Girish
Department of Biotechnology, M S Ramaiah University of Applied Sciences, Bangalore, Karnataka, India

J. V. Hanchinalmath
Department of Life Science, School of Sciences, Jain (Deemed to be University), Bangalore, Karnataka, India

S. Srivastava · K. Shivasharanappa (✉)
Department of Life Science, School of Sciences, Garden City University, Bangalore, Karnataka, India
e-mail: kirankumar.v@gardencity.university

J. S. Yaradoddi et al. (eds.), *Frontiers in Food Biotechnology*,
https://doi.org/10.1007/978-981-97-3261-6_12

PEF	Pulsed electric field
pH	Potential of hydrogen
SOBsuc media	Sorbitol-based sucrose medium
WBM media	Whey-based liquid medium

1 Introduction

Fermentation is a method of preserving food that has been used for thousands of years [1]. It works by breaking down substances into simpler compounds with the help of microorganisms, which can be found naturally or intentionally added through inoculation. Fermentation is a biochemical metabolic process that entails various hydrolysis (catabolism) and synthesis (anabolism) processes by microorganisms. Fermented utilizes living organisms, like bacteria and yeasts to break down carbohydrates and other nutrients found in food. As a result, it creates substances such, as vitamins, minerals, amino acids and other microbial metabolites that are beneficial for humans [2]. Metabolites produced by microorganisms during fermentation are given in Table 12.1. The biochemical pathway by which the fermented products are formed is shown in Fig. 12.1.

Table 12.1 The metabolic diversity of microorganisms in fermentation

Metabolite	Microorganism	Fermentation substrate	References
Amylase	*Bacillus subtilis*	Sugarcane bagasse hydrolysate	[3]
Beta-glucanase	*Trichoderma viride*	Oatmeal and peptone	[4]
Cellulase	*Aspergillus niger*	Apple pomace	[5]
DL-methionine	*Corynebacterium glutamicum*	Glucose, urea and molasses-based media	[6]
Invertase	*Saccharomyces cerevisiae*	Red carrot residue	[7]
L-arginine	*Corynebacterium glutamicum*	Sucrose	[8]
L-asparagine	*Pseudomonas fluorescens*	M9 agar medium	[9]
L-cysteine	*Pantoea ananatis*	Luria-Bertani broth	[10]
L-histidine	*Corynebacterium glutamicum*	Molasses (15%)	[11]
L-leucine	*Corynebacterium glutamicum*	CGXIIG growth medium	[12]
L-lysine	*Bacillus methanolicus*	SOBsuc medium	[13]
L-phenylalanine	*Aspergillus oryzae*	Potato dextrose agar	[14]
L-tryptophan	*Escherichia coli*	Seed fermenter	[15]
L-tyrosine	*Brevibacterium lactofermentum*	Processed Iranian cane and beet molasses	[16]
L-valine	*Corynebacterium glutamicum*	Brain heart infusion broth	[17]
Lactase	*Lactobacillus acidophilus*	Fermented ragi	[18]
Peptidase	*Debaryomyces hansenii*	Fermented sausage	[19]

Table 12.1 (continued)

Metabolite	Microorganism	Fermentation substrate	References
Protease	*Aspergillus niger*	Skerman's basal mineral salt media (BSM)	[20]
Vitamin B1 Vitamin B3	*Bacillus subtilis, Lactobacillus* sp., *Leuconostoc mesenteroides, Bifidobacterium longum*	Cashew apple, cashew apple juice	[21]
Vitamin B12	*Propionic acid bacteria*	Glycerol and glucose	[22]
Vitamin B6	*Bacillus subtilis, Escherichia coli*	Minimal medium with yeast extract	[23]
Vitamin B2	*Propionibacteria*	Whey-based liquid medium (WBM)	[24]
Vitamin K	*Lactococcus, Lactobacillus, Enterococcus*	GM17 media	[25]

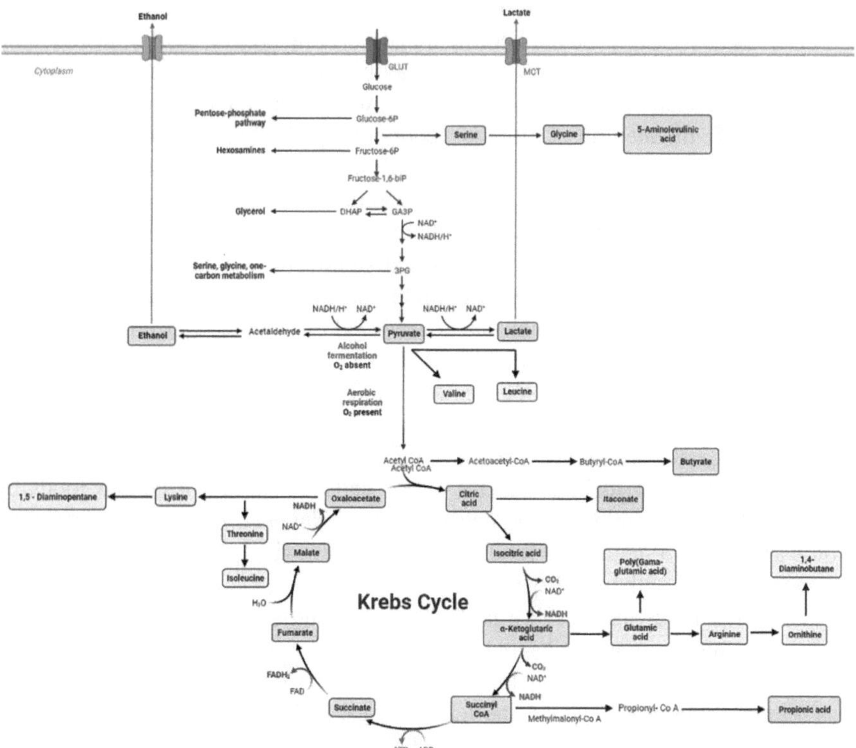

Fig. 12.1 Fermentation and amino synthesis pathway

Fermented foods hold an established place in the diets of humans as they have been used for centuries to preserve food and enhance its flavour. In times there has been an increasing fascination, with the potential health advantages offered by fermented foods along, with their capacity to tackle issues related to food waste and sustainability [26].

Humans have been harnessing the power of fermentation, for centuries to preserve food and enhance its flavours. This natural process involves the use of microorganisms like bacteria and yeasts to break down carbohydrates and other nutrients in our food. Through this process, certain beneficial compounds like vitamins, amino acids and probiotics can be produced. Probiotics, which are microorganisms that offer health benefits when consumed in quantities, have been widely studied for their positive effects, on human well-being. There are advantages associated with probiotics including digestive health boosted immune system and lower chances of specific illnesses. Probiotics have proven to be beneficial, for ailments, like diarrhoea, irritable bowel syndrome and inflammatory bowel disease [27].

On the other hand, some microorganisms can have detrimental effects on human health. Spoilers and pathogens are microorganisms that can cause food spoilage or foodborne illnesses. Spoilers are microorganisms that can cause food to spoil, leading to changes in taste, texture and appearance. These microorganisms can also produce toxins as a by-product. Food spoilage can happen in types of food including meat, dairy products, fruits and vegetables. Pathogens, on the other hand, are microorganisms that can cause foodborne illnesses, which can range from mild to severe and even life-threatening. Common foodborne pathogens include *Salmonella, Escherichia coli* and *Listeria* [28]. It is worth mentioning that not all microorganisms pose a threat to well-being. In reality, numerous microorganisms are actually beneficial to our health and those residing in the gut microbiome. The gut microbiome is a community of microorganisms that significantly influence human health and the development of diseases. It is important to maintain a healthy balance of microorganisms in the gut, which can be achieved through the consumption of prebiotic and probiotic-rich fermented foods.

Foods that have fermentation have had an impact, on the diets and cultural practices of humans for countless centuries. Archaeological evidence suggests that fermented foods have been consumed since the Neolithic period, around 10,000 BCE [29]. Fermentation also improves the safety and palatability of certain foods by reducing the risk of spoilage and enhancing their flavours. Historically, fermentation was used as a method of food preservation, allowing people to store food for extended periods without refrigeration [30].

Fermented foods have been a part of various cultures throughout history, with each region and community having its own unique fermented foods and traditions. For instance, kimchi, a traditional Korean side dish made from fermented vegetables, has been a staple in Korean cuisine for centuries. In Japan, miso, a fermented soybean paste, has been consumed for over 1000 years. Sauerkraut, a dish made by fermenting cabbage has been enjoyed as a food, in Europe, for centuries. In addition to their culinary and practical significance, fermented foods have also held cultural and religious significance in many societies. For example, in some cultures, fermented foods are used in religious ceremonies and rituals. In the Jewish tradition, fermented foods such as sauerkraut and pickles have been used during Passover as a symbol of remembrance of the Jewish people's past enslavement in Egypt. In some African communities, fermented foods are considered sacred and are believed to have spiritual and healing properties. Similarly, in some Asian cultures, fermented foods are believed to have medicinal properties and are used as part of traditional medicine [31].

Today, fermented foods continue to be popular around the world, and there is a growing interest in their potential health benefits. The consumption of foods that have fermentation has been linked to better gut health improved function and a lower likelihood of developing certain diseases. Additionally, fermented foods can help tackle the issues surrounding food waste and sustainability since they can be created using imperfect produce and can prolong the freshness of foods [32].

2 Health Benefits of Fermented Foods

Throughout history, people from different cultures have been enjoying fermented foods for centuries recognizing the health advantages they offer. As depicted in Fig. 12.2, some of the benefits associated with fermented foods include:

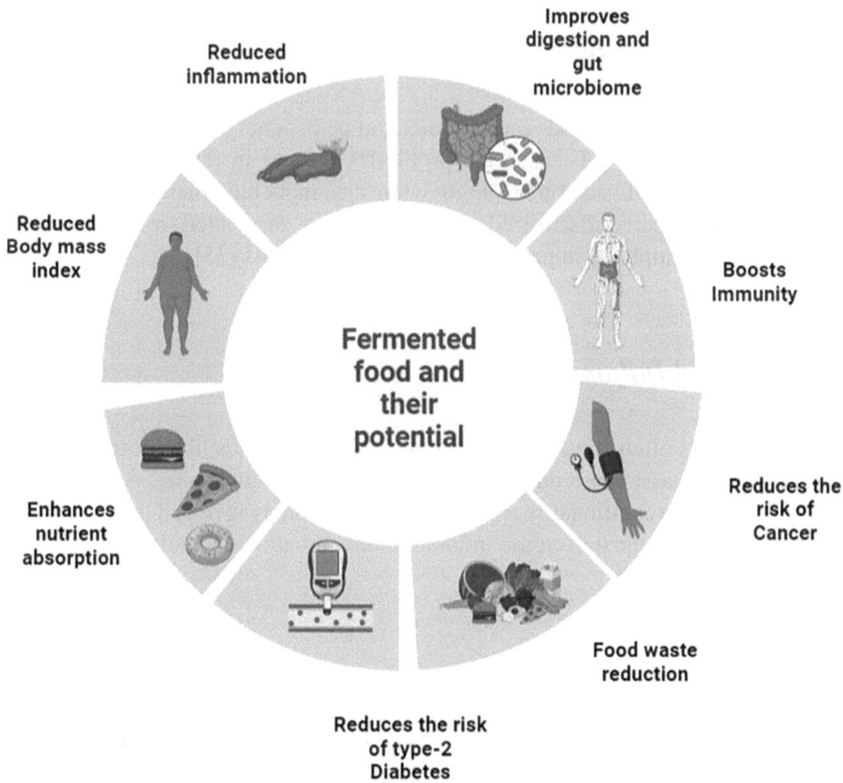

Fig. 12.2 Fermented foods and their potential

2.1 Improved Digestion

Foods that have gone through fermentation contain a number of bacteria that can contribute to a healthy balance, in the gut microbiome and enhance digestion. A comprehensive analysis of studies discovered that probiotics, including the ones found in fermented foods, can have an impact on digestive health, by alleviating symptoms associated with diarrhoea, constipation and inflammatory bowel disease [33].

2.2 Enhanced Nutrient Absorption

The process of fermentation has the ability to break down carbohydrates and proteins found in food making it easier for our bodies to absorb and use nutrients. For instance, fermentation has been proven to enhance the availability of iron and zinc in plant-based foods [34].

2.3 Boosted Immunity

Incorporating fermented foods into our diet can be beneficial for strengthening our immune system. This is because the probiotics present in fermented foods help promote the growth of bacteria in our gut while also reducing inflammation. A study conducted by experts discovered that consuming fermented milk containing Lactobacillus acidophilus improved function among adults [35].

2.4 Reduced Inflammation

Fermented foods often contain anti-inflammatory compounds like short-chain fatty acids, which can assist in reducing inflammation within our bodies. Another study published in the International Journal of Dairy Technology revealed that consuming fermented kimchi helped decrease inflammation among obese adults [36].

2.5 Improved Mental Health

There are indications that consuming fermented foods may positively impact mental health as well. International Journal of Environmental Research and Public Health stated that consuming fermented milk enhanced mood and alleviated symptoms of anxiety and depression, among healthy individuals [37].

Table 12.2 summarizes the target food types, study characteristics and outcomes of various studies investigating the health benefits of fermented foods and probiotics. The research consists of types of studies such, as randomized controlled trials, systematic reviews, analyses and prospective cohort studies. These studies primarily examine how these interventions impact the enhancement of health.

In general, adding fermented foods to your diet can provide health advantages. However, it is crucial to seek advice, from a healthcare expert before making modifications, to your eating habits.

Table 12.2 Studies on the health effects of fermented foods and probiotics

Target	Food type	Study characteristics	Results/outcome	References
Gut microbiota	Probiotic yogurt	Double blind, randomized, placebo-controlled trial, with healthy adults, $N = 60$ for a period of 6 weeks	Probiotic yogurt improved gut microbiota composition compared to placebo	[38]
Cardiovascular disease	Probiotics/fermented milk products	Systematic review and meta-analysis of 14 RCTs, hypertension/high cholesterol, $N = 702$	Probiotics/fermented milk products showed potential for reducing blood pressure and cholesterol	[39]
Depression/anxiety	Probiotic milk drink	Double blind, randomized, placebo-controlled trial, with healthy adults, $N = 150$ for a period of 8 weeks	Probiotic milk drink showed potential for reducing symptoms of depression and anxiety	[40]
Inflammation	Fermented vegetable juice	Double blind, randomized, placebo-controlled trial, with healthy adults, $N = 32$ for a period of 8 weeks	Fermented vegetable juice did not show significant effects on inflammatory markers	[41]
Allergic rhinitis	Kimchi	Double blind, randomized, placebo-controlled trial, with allergic rhinitis adults, $N = 30$ for a period of 8 weeks	Kimchi did not show significant effects on symptoms of allergic rhinitis	[42]
Immune function	Fermented dairy	Double blind, randomized, placebo-controlled trial, with healthy adults, $N = 60$ for a period of 4 weeks	Fermented dairy did not show significant effects on immune function	[43]

(continued)

Table 12.2 (continued)

Target	Food type	Study characteristics	Results/outcome	References
Nonalcoholic fatty liver disease	Kefir	Double blind, randomized, placebo-controlled trial, with NAFLD adults, $N = 60$ for a period of 8 weeks	Kefir showed potential for improving liver function in NAFLD patients	[44]
Cognitive function and memory	Fermented milk product	Double blind, randomized, placebo-controlled trial, with elderly adults, $N = 52$ for a period of 6 weeks	Fermented milk product showed potential for improving cognitive function and memory	[45]
Colorectal cancer risk	Kimchi	Prospective cohort study, 6 years, adults, $N = 1477$	Kimchi consumption was associated with reduced risk of colorectal cancer	[46]
Obesity	Fermented dairy drink	Double blind, randomized, placebo-controlled trial, with obese adults, $N = 80$ for a period of 12 weeks	Fermented dairy drink did not show significant effects on body weight or body composition	[47]
Type 2 diabetes	Fermented milk drink	Double blind, randomized, placebo-controlled trial, with type 2 diabetes adults, $N = 60$ for a period of 12 weeks	Fermented milk drink showed potential for improving blood glucose control and lipid profile	[48]
Inflammatory bowel disease	Kimchi	Placebo-controlled, double blind, randomized trial; patients with mild-to-moderate ulcerative colitis, $N = 36$	Kimchi showed potential for reducing disease activity	[49]
Irritable bowel syndrome	Fermented milk product	Placebo-controlled, double blind, randomized trial; patients with IBS, $N = 52$	Fermented milk product showed potential for reducing symptom severity and improving quality of life	[50]
Atopic dermatitis	Probiotic drink	Placebo-controlled, double blind, randomized trial; children with AD, $N = 56$	Probiotic drink showed potential for reducing disease severity and improving quality of life	[51]
Liver function	Fermented milk product	Double blind, randomized, placebo-controlled trial; patients with non-alcoholic fatty liver disease, $N = 60$	Fermented milk product showed potential for improving liver function	[52]

(continued)

Table 12.2 (continued)

Target	Food type	Study characteristics	Results/outcome	References
Osteoporosis	Fermented milk product	Placebo-controlled, double blind, randomized trial; postmenopausal women, $N = 62$	Fermented milk product showed potential for improving bone density	[53]
Metabolic syndrome	Kefir	Double-blind, placebo-controlled randomized trial; adults with metabolic syndrome, $N = 75$	Kefir showed potential for improving blood glucose, lipid levels and blood pressure	[54]
Eczema	Probiotic drink	Placebo-controlled, double blind, randomized trial; children with eczema	Probiotic drink showed potential for reducing disease severity and improving quality of life	[55]
Hypertension	Fermented milk product	Double-blind, placebo-controlled randomized trial; adults with prehypertension, $N = 50$	Fermented milk product showed potential for reducing blood pressure	[46]

3 How Beneficial Bacteria Improve Gut Health

The gut microbiome is extremely important, for maintaining health as it has an impact on various bodily functions such as digestion, immunity and metabolism. A functioning gut microbiome consists of a well-balanced community of microorganisms. On the other hand, an imbalance, in the gut microbiome, known as dysbiosis has been associated with health issues including inflammatory bowel disease, obesity and diabetes [56, 57].

3.1 Probiotics Helps in Regulating Irritable Bowel (IBS)

Research has shown that consuming probiotics can improve gut health by modulating the gut microbiome. For example, a randomised controlled trial found that consuming fermented milk containing probiotics led to significant improvements in gut microbiota composition, stool consistency, and other digestive parameters compared to a control group. Similarly, a comprehensive stud discovered that adding probiotics to one's diet resulted in improvements, in symptoms associated with irritable bowel syndrome (IBS) including abdominal pain and bloating. One of the mechanisms through which probiotics enhance gut health is by generating acid, which can effectively reduce the pH levels within the gut and create an environment that is less favourable for harmful bacteria. Furthermore, probiotics have the ability to compete with bacteria for nutrients and adhesion sites within the gut thereby diminishing their ability to establish colonies. Additionally, probiotics exhibit an array of substances, such, as bacteriocins, which are capable of inhibiting the growth of harmful bacteria [58, 59].

Extensive research has been conducted on the impact of probiotics, on the system in relation, to inflammatory bowel disease (IBD). This condition involves long term gut inflammation and studies have demonstrated that specific strains of probiotics can help alleviate IBD symptoms by reducing inflammation and enhancing the function of the gut. For example, Lactobacillus plantarum has been shown to reduce inflammation and colitis in animal models of IBD. Similarly, Bifidobacterium bifidum and Bifidobacterium lactis have been shown to improve gut barrier function and reduce inflammation in patients with ulcerative colitis. In addition, to IBD studies have also indicated that probiotics can be helpful in alleviating symptoms associated with bowel syndrome (IBS) a condition characterized by abdominal pain bloating and changes in bowel habits. Research has demonstrated that taking supplements can lead to improvements in IBS symptoms, such as reduced pain and bloating. While the precise mechanisms behind the effects of probiotics, on IBS are not yet fully understood, it is believed that they involve influencing the gut microbiome and enhancing gut barrier function [60].

Moreover, various research studies have indicated that fermented foods containing bacteria can also have an impact, on the functionality of the gut barrier. The gut barrier is composed of a layer of mucus and tight junctions between cells playing a crucial role in guarding gut health. Its primary function is to prevent the passage of substances like toxins and pathogens from crossing into the bloodstream through the wall. Probiotics have been proven to enhance the effectiveness of the gut barrier by facilitating the expression of junction proteins and mucins, which contribute to maintaining its integrity. Furthermore, certain probiotics have demonstrated their ability to generate compounds, like bacteriocins, which can effectively impede the growth of bacteria within the gut [35].

3.2 Fermented Foods Beneficial for Weight Management

One study investigated the association between fermented food consumption and obesity in a Korean population. The research discovered that individuals who included a number of fermented foods, in their diet tended to have a body weight, body mass index (BMI), and body fat percentage as compared to those who consumed fewer fermented foods [61]. The researchers suggested that the probiotics and bioactive peptides found in fermented foods may help to regulate appetite and energy balance, leading to lower calorie intake and ultimately lower body weight. Another study conducted in a Japanese population found similar results, with higher consumption of fermented soy products being associated with lower body weight and BMI [62]. The researchers indicated that this could be due to the fact that fermented soy products contain an amount of fibre and protein, which can contribute to a sense of satiety and ultimately lead to a decrease, in calorie consumption.

3.3 Fermented Foods Help to Reduce the Risk of Type 2 Diabetes

Type 2 diabetes is a long-term metabolic condition marked by high blood sugar levels, insulin resistance and decreased insulin production. Various elements, like genetics, lifestyle choices and dietary habits can influence the onset of type 2 diabetes. Recent findings indicate that incorporating fermented dairy items like yogurt and kefir, into one's diet might potentially lower the chances of developing type 2 diabetes.

Research has indicated that including fermented dairy products like yogurt and kefir in your diet could potentially lower the likelihood of developing type 2 diabetes. The presence of probiotics and bioactive peptides in these food items might contribute to insulin sensitivity, glucose metabolism and reduced inflammation—all factors in the prevention of type 2 diabetes. A comprehensive review of studies discovered a correlation, between increased consumption of fermented dairy products and a decreased risk of type 2 diabetes. The researchers conducted an analysis and their findings revealed that people who regularly consumed quantities of fermented dairy products had a reduced risk, about 28% lower of developing type 2 diabetes compared to those who consumed minimal amounts. Moreover, research conducted in Japan revealed a correlation, between increased intake of fermented soy products, like yogurt and kefir and a reduced likelihood of developing type 2 diabetes [63].

Ongoing research is still exploring the ways in which fermented foods can help lower the risk of type 2 diabetes. One theory suggests that the probiotics present in these foods could enhance the balance of gut bacteria and decrease inflammation leading to better glucose metabolism and insulin sensitivity. Furthermore, fermented doenjang and chungkookjang containing peptides might possess properties, such, as reducing insulin resistance and enhancing insulin secretion [64].

However, it is worth mentioning that more research is required to gain an understanding of how fermented foods affect the risk of type 2 diabetes. Various factors, including the types and quantities of fermented foods consumed, as well as individual lifestyle and genetic aspects can impact the connection, between fermented foods and disease risk. Nevertheless, incorporating fermented foods into a nutritious diet might offer potential advantages in reducing the likelihood of developing type 2 diabetes.

3.4 Fermented Foods Help in Reducing the Risk of Cancer

Including fermented foods in your diet could potentially lower the chances of developing types of cancer. Multiple studies indicate that consuming fermented foods on a daily basis may reduce the risk of cancer. This is thought to be because fermented foods contain antioxidants and anti-inflammatory compounds, which are beneficial for fighting cancer. Additionally, the presence of probiotics in fermented foods further contributes to their cancer properties.

Moreover, during the fermentation process bioactive compounds, like butyrate can be generated, which has been scientifically proven to possess cancer properties. As an illustration, a study conducted discovered that those who consumed quantities of dietary fibres associated with high butyrate concentrations products had a decreased risk of developing colorectal cancer in comparison to individuals who consumed smaller amounts [65]. Similarly, research conducted on adults indicated that the consumption of kimchi—a fermented cabbage dish—was associated with a lowered risk of gastric cancer [66]. Furthermore, various studies have suggested that the consumption of fermented soy products like miso and tempeh might be linked to a reduced risk of breast cancer. In addition to this, regular consumption of fermented foods has been linked with a likelihood of types of cancer. Multiple studies have hinted at the benefits of consuming fermented foods in reducing the risk of colorectal cancer. For instance, in another study involving women participants, it was found that those who incorporated amounts of dairy products into their diets had a lower susceptibility, to colorectal cancer when compared to those who consumed lesser amounts [67, 68].

It is important to keep in mind that although certain fermented foods are linked to health advantages, not all fermented foods are equal. Some commercially made fermented foods can have amounts of sugar or salt, which can be detrimental, to your health. Therefore, it is crucial to opt for fermented foods that are minimally processed and have added sugars or salt.

3.5 Fermented Foods as a Source of Nutrients

Foods that go through fermentation generally contain nutrients compared to their fermented counterparts. This is because during the fermentation process, helpful bacteria break down molecules into forms that are easier for our bodies to absorb. For example, when soybeans are fermented to make miso, it boosts the availability of protein and minerals, like calcium, iron and zinc. Fermentation can also enhance the body's ability to absorb vitamins and minerals. For instance, fermenting milk to create yogurt increases the absorption of calcium, which is crucial for maintaining bones and teeth. Moreover, fermenting vegetables to produce sauerkraut and kimchi enhances the absorption of vitamin C, a nutrient for an immune system and healthy skin. Additionally, fermenting vegetables to produce sauerkraut and kimchi enhances the absorption of vitamin C, a nutrient, for an immune system and healthy skin [34].

In addition to their sensory attributes, fermented foods are also known to be rich sources of various nutrients. Some examples of different fermented foods and their nutrient content are listed below.

Yogurt

Yogurt is made by fermenting milk with lactic acid bacteria. It is a good source of protein, calcium, vitamin B12 and probiotics. Probiotics are actually living microorganisms that can provide health advantages when consumed in quantities. A 100-g serving of plain, low-fat yogurt provides approximately 10 g of protein, 200 mg of calcium and 0.5 μg of vitamin B12 [69].

Kefir

Kefir is a fermented dairy product that is similar to yogurt but is made by fermenting milk with a combination of yeasts and bacteria. It is also a good source of protein, calcium, vitamin B12, and probiotics. A 100-g serving of kefir provides approximately 3.5 g of protein, 120 mg of calcium and 0.4 μg of vitamin B12 [70].

Kimchi

Kimchi a side dish is created by fermenting cabbage and an assortment of vegetables together, with garlic, ginger and chili pepper. It is a good source of fibre, vitamin C and probiotics. A 100-gram serving of kimchi provides approximately 1.8 g of fibre, 18 mg of vitamin C and 107–108 colony-forming units of lactic acid bacteria [71].

Sauerkraut

Sauerkraut is a fermented cabbage dish that is commonly eaten in German cuisine. It is a good source of fibre, vitamin C and probiotics. A 100-g serving of sauerkraut provides approximately 2 g of fibre, 15 mg of vitamin C and 106–107 colony-forming units of lactic acid bacteria [71].

Miso

Miso a seasoning deeply rooted in Japanese tradition is created through the fermentation process of soybeans, with salt and koji—a type of fungus. This remarkable ingredient boasts protein content, valuable vitamin K and beneficial probiotics. When consumed in a 100-g portion miso offers around 10 g of protein 200 µg of vitamin K, and an impressive range of 109–1010 colony-forming units of acid bacteria [72].

Kombucha

Kombucha is a type of tea that undergoes fermentation and is known for its abundance of probiotics, antioxidants and B vitamins [73].

Tempeh

Tempeh a soybean product provides a deal of protein, fibre and vitamins B2, B3 and B6 [74].

Sourdough Bread

Sourdough bread offers an amount of fibre, protein as well as minerals, like iron and selenium [75].

Pickles

When it comes to fermented pickles, they are packed with probiotics, vitamin K and beneficial antioxidants [74].

Natto

Natto is a dish made from fermented soybeans. It serves as a source of protein, fibre and vitamin K2 [72].

Kvass

Kvass is a traditional Slavic beverage made by fermenting rye bread that is rich in probiotics, B vitamins and minerals such as potassium and magnesium [74].

Katsuobushi

Katsuobushi is an ingredient that comes from fermented and smoked skipjack tuna. It is known for being rich in protein, omega 3 acids and important minerals like calcium and iron [75].

Fermented Vegetables

Fermented vegetables such as sauerkraut, pickled beets and fermented carrots are good sources of probiotics, fibre and vitamins such as vitamin C and vitamin K [74].

Cheese

Certain types of cheese, such as blue cheese, gouda and cheddar, are made by fermenting milk with bacteria or fungi and are good sources of protein, calcium and vitamin B12 [76].

In contrast, non-fermented foods may be more processed and have a lower nutrient content due to loss during processing. For example, commercial dairy products like cheese and butter have lower levels of calcium compared to their fermented counterparts like yogurt and kefir, as well as fewer beneficial bacteria [69].

3.6 Fermented Foods and Food Waste Reduction

Fermented foods are not only beneficial for our health, but they can also play a role in reducing food waste. Food waste is a problem, on a scale, where approximately one-third of all food intended for human consumption is lost or wasted annually. In this essay, we will explore how fermented foods can contribute to reducing food waste, as well as the potential challenges and opportunities associated with this approach. Fermentation is a natural process that has been used for thousands of years to preserve food. The process includes the utilization of bacteria or yeast to decompose sugars and other nutrients present, in food. As a result, lactic acid, alcohol, or other by-products are produced, which aid, in preserving the food [76]. Fermentation has been used to make a wide variety of foods, including bread, cheese, yogurt, beer, wine and pickles, among others.

Fermentation offers an advantage by prolonging the freshness of food thereby reducing wastage caused by spoilage or expiration. For example, fermenting vegetables like cabbage to make sauerkraut or kimchi can extend their shelf life from a few days to several weeks or even months. Similarly, fermenting fruits and vegetables to make jams, chutneys and relishes can help preserve them for longer periods. Another way in which fermented foods can contribute to reducing food waste is by utilizing food scraps and by-products that would otherwise be discarded. For example, whey, a by-product of cheese making, can be fermented to make a nutritious drink called kefir. Similarly, vegetable scraps like carrot tops, onion skins and celery leaves can be fermented to make flavourful and nutritious broths [31]. Fermentation can also help reduce food waste by making use of imperfect or 'ugly' produce that may not meet aesthetic standards for sale in supermarkets. Fermenting these fruits and vegetables can help preserve them and make them more appealing to consumers. This approach has been used by several food companies and startups, such as Imperfect Foods and Misfit Juicery, which specialize in making products from 'ugly' produce that would otherwise go to waste.

Despite the potential benefits of fermented foods for reducing food waste, there are also some challenges and considerations that need to be considered. For example, fermentation requires specific environmental conditions, such as temperature, pH and oxygen levels, that may not be feasible in all settings. In addition, there may be safety concerns associated with fermentation, such as the risk of contamination with harmful bacteria or pathogens. Proper sanitation and hygiene practices are therefore crucial when fermenting food [77].

Another challenge is the need to develop efficient and scalable fermentation processes that can meet the demand for fermented foods. This requires knowledge and expertise in microbiology, food science and engineering, among other fields. It also requires investment in research and development to improve and optimize fermentation techniques and develop new products and applications. There are also opportunities for innovation and collaboration in the fields of fermented foods and food waste reduction. For example, there is growing interest in using fermentation to develop new plant-based meat alternatives that can help reduce the environmental impact of meat production [78]. Fermentation can also be used to develop new functional foods and ingredients that have specific health benefits, such as probiotics, prebiotics and bioactive peptides [79].

In summary, fermented foods have the capacity to make an impact, on reducing food waste and promoting food production. By utilizing fermentation techniques, we can effectively preserve food. Prolonging its shelf life ultimately results in a reduction of food waste. Additionally, fermented foods are rich in nutrients and can provide health benefits, making them a desirable addition to a healthy and sustainable diet. With the growing interest in sustainable food practices, fermented foods are likely to become even more popular in the future. It is important for researchers and food producers to continue exploring the potential benefits of fermented foods and to develop new ways to incorporate them into our diets.

However, there are also potential challenges and limitations to using fermentation as a means of reducing food waste and promoting sustainability. One of the main challenges is ensuring food safety and preventing the growth of harmful bacteria during the fermentation process. To create an environment for the growth of bacteria and to prevent the growth of harmful ones, it is crucial to implement appropriate sanitation and quality control measures. In addition, there are concerns that the increased popularity of fermented foods may lead to overconsumption, particularly of sodium-rich products like kimchi and sauerkraut [71]. It is important to balance the potential health benefits of fermented foods with the risks of excessive consumption.

Fermentation has the potential to make a contribution, in addressing food waste and promoting sustainability within our food system. By using fermentation to convert food scraps into delicious products, we can minimize the effects of wasted food and encourage a more circular and sustainable approach to our food production. Moreover, fermented foods provide health advantages such, as digestion, better absorption of nutrients and enhanced immunity. Nevertheless, it is crucial to acknowledge the obstacles and constraints associated with fermentation. It is imperative to establish and implement measures for food safety and quality control, which will facilitate the proliferation of bacteria while inhibiting the growth of ones. By providing education and support, fermentation can serve as an asset, in fostering a sustainable and wholesome food system.

4 Future Directions for Research on Fermented Foods

Current research on fermented foods has focused on the microbial communities involved in the fermentation process and their potential effects on human health. Many studies have identified the diverse microbial communities present in fermented foods and their role in producing unique flavours, aromas and textures. Moreover, research has shown that fermented foods can also have positive effects on mental health. The gut-brain axis serves as a pathway of communication connecting the gut microbiota with the system. Research has shown that the gut microbiota has the ability to influence brain function and behaviour. Fermented foods, especially those containing probiotics, have been found to improve symptoms of anxiety, depression and stress in some studies [37].

However, there are still many gaps in knowledge when it comes to fermented foods. Further exploration is needed in the field of research regarding the impact of fermentation, on ensuring food safety. Although there are advantages associated with fermented foods concerns regarding their safety persist. The risk of pathogen contamination in fermented foods is a major concern, particularly for foods that are not pasteurized or cooked before consumption. Several studies have reported cases of foodborne illness outbreaks linked to the consumption of fermented foods [80].

Studies have found that including microorganisms in fermented foods can potentially inhibit the growth of bacteria. However, it is important to note that there is still a risk of pathogen contamination. This can be avoided using cultures that are non-pathogenic and ensuring controlled fermentation conditions may help minimize this risk. Nevertheless, further studies are required to grasp the factors that influence the safety of fermented foods and devise strategies, for ensuring their overall safety.

One potential approach to improving the safety of fermented foods is the use of novel preservation technologies. For example, high-pressure processing (HPP) has been shown to be an effective method for reducing pathogen levels in fermented foods like kimchi and sauerkraut [81]. Similarly, pulsed electric field (PEF) technology has been shown to be effective in reducing pathogen levels in fermented dairy products. These technologies may offer a promising approach for improving the safety of fermented foods without compromising their nutritional or sensory qualities.

Fermentation has the potential to address various public health and sustainability challenges. In terms of food waste reduction, fermented foods could serve as a means of preserving perishable foods that would otherwise be discarded. In addition, fermentation has the potential to address malnutrition in developing countries by improving the nutritional content of locally available food sources. For instance, fermented maize porridge has been found to have increased levels of bioavailable iron, which could address iron deficiency (anaemia), a common problem in many African countries [34]. Moreover, fermented foods could offer a natural alternative to synthetic preservatives that are associated with health risks. Fermentation can provide a natural preservation method by creating an environment that is inhospitable to spoilage and pathogenic microorganisms. However, there is still concern regarding the safety of fermented foods and more research is required to gain an understanding of the possible risks linked to their consumption.

To fully grasp the advantages of fermented foods, it is crucial to explore areas through future research. One significant aspect that requires investigation is finding approaches to study the communities involved in fermentation. By utilizing cutting-edge techniques, like metagenomics and high throughput sequencing, we can gain an understanding of the diversity and functionality of these microorganisms [82]. Another crucial area for exploration is optimizing existing fermentation methods. Developing techniques enhances the quality and safety of fermented foods. Furthermore, there is a need for research to comprehend both the health benefits and risks associated with consuming fermented foods.

5 Conclusion

In summary, fermented foods are a culturally significant group of foods that have a lot to offer in terms of health benefits and reducing food waste. The friendly bacteria present in fermented foods can improve our gut health support, our system help with weight management, and potentially lower the risk of diseases. These foods are also packed with nutrients. Can enhance absorption. Additionally, through fermentation, we can extend the shelf life of food and combat global food waste challenges. However, there is still more to explore through research to fully understand the health advantages offered by types of fermented foods and how large-scale fermentation can contribute to reducing global food waste. All fermented foods show great potential for promoting individual well-being as well as global health and sustainability. Continued research and advancements in this field are necessary for progress.

References

1. Banerjee R, Mukherjee G, Patra KC (2005) Microbial transformation of tannin-rich substrate to gallic acid through co-culture method. Bioresour Technol 96(8):949–953. https://doi.org/10.1016/j.biortech.2004.08.004
2. Steinkraus KH (2002) Fermentations in world food processing. Compr Rev Food Sci Food Saf 1(1):23–32
3. Rajagopalan G, Krishnan C (2008) α-Amylase production from catabolite derepressed Bacillus subtilis KCC103 utilizing sugarcane bagasse hydrolysate. Bioresour Technol 99(8):3044–3050
4. Yang SQ, Xiong H, Yang HY, Yan QJ, Jiang ZQ (2015) High-level production of β-1, 3-1, 4-glucanase by Rhizomucor miehei under solid-state fermentation and its potential application in the brewing industry. J Appl Microbiol 118(1):84–91
5. Dhillon GS, Kaur S, Brar SK, Verma M (2012) Potential of apple pomace as a solid substrate for fungal cellulase and hemicellulase bioproduction through solid-state fermentation. Ind Crop Prod 38:6–13
6. Ali MN, Mazhar B, Mazhar A (2018) Improvement in cysteine production by different bacterial strains. J Sci Tech Res 5:1–3
7. Rashad MM, Nooman MU (2009) Production, purification and characterization of extracellular invertase from Saccharomyses cerevisiae NRRL Y-12632 by solid-state fermentation of red carrot residue. Aust J Basic Appl Sci 3(3):1910–1919
8. Park SH, Kim HU, Kim TY, Park JS, Kim SS, Lee SY (2014) Metabolic engineering of Corynebacterium glutamicum for L-arginine production. Nat Commun 5(1):4618
9. Badoei-Dalfard A (2016) L-asparaginase production in the Pseudomonas pseudoalcaligenes strain JHS-71 isolated from Jooshan hot-spring. Mol Biol Res Commun 5(1):1
10. Takumi K, Ziyatdinov MK, Samsonov V, Nonaka G (2017) Fermentative production of cysteine by Pantoea ananatis. Appl Environ Microbiol 83(5):e02502–e02516
11. Kulis-Horn RK, Persicke M, Kalinowski J (2014) Histidine biosynthesis, its regulation and biotechnological application in Corynebacterium glutamicum. Microb Biotechnol 7(1):5–25
12. Feng LY, Xu JZ, Zhang WG (2018) Improved L-leucine production in Corynebacterium glutamicum by optimizing the aminotransferases. Molecules 23(9):2102
13. Nærdal I, Netzer R, Irla M, Krog A, Heggeset TMB, Wendisch VF, Brautaset T (2017) L-lysine production by Bacillus methanolicus: genome-based mutational analysis and L-lysine secretion engineering. J Biotechnol 244:25–33
14. Ali S, Haq I (2010) Production of 3, 4-dihydroxy L-phenylalanine by a newly isolated Aspergillus niger and parameter significance analysis by Plackett-Burman design. BMC Biotechnol 10(1):1–8
15. Liu F, Chen Z, Shao J, Wang C, Zhan C (2017) Effect of fermentation on the peptide content, phenolics and antioxidant activity of defatted wheat germ. Food Biosci 20:141–148
16. Sadeghiyan-Rizi T, Fooladi J, Heravi MM, Sadrai S (2014) Optimization of L-tryptophan biosynthesis from L-serine of processed Iranian beet and cane molasses and indole by induced Escherichia coli ATCC 11303 cells. Jundishapur J Microbiol 7(6):e10589
17. Zhang H, Li Y, Wang C, Wang X (2018) Understanding the high L-valine production in Corynebacterium glutamicum VWB-1 using transcriptomics and proteomics. Sci Rep 8(1):3632
18. Akolkar SK, Sajgure A, Lele DS (2005) Lactase production from Lactobacillus acidophilus. World J Microbiol Biotechnol 21:1119–1122
19. Bolumar T, Sanz Y, Aristoy MC, Toldrá F (2008) Purification and characterisation of proteases A and D from Debaryomyces hansenii. Int J Food Microbiol 124(2):135–141
20. Atalah J, Cáceres-Moreno P, Espina G, Blamey JM (2019) Thermophiles and the applications of their enzymes as new biocatalysts. Bioresour Technol 280:478–488
21. Kaprasob R, Kerdchoechuen O, Laohakunjit N, Somboonpanyakul P (2018) B vitamins and prebiotic fructooligosaccharides of cashew apple fermented with probiotic strains Lactobacillus spp., Leuconostoc mesenteroides and Bifidobacterium longum. Process Biochem 70:9–19

22. Poonam, Pophaly SD, Tomar SK, De S, Singh R (2012) Multifaceted attributes of dairy propionibacteria: a review. World J Microbiol Biotechnol 28:3081–3095
23. Rosenberg J, Ischebeck T, Commichau FM (2017) Vitamin B6 metabolism in microbes and approaches for fermentative production. Biotechnol Adv 35(1):31–40
24. Deptula P, Chamlagain B, Edelmann M, Sangsuwan P, Nyman TA, Savijoki K et al (2017) Food-like growth conditions support production of active vitamin B12 by Propionibacterium freudenreichii 2067 without DMBI, the lower ligand base, or cobalt supplementation. Front Microbiol 8:368
25. O'Connor E, Mølgaard C, Michaelsen KF, Jakobsen J, Lamberg-Allardt CJ, Cashman KD (2007) Serum percentage undercarboxylated osteocalcin, a sensitive measure of vitamin K status, and its relationship to bone health indices in Danish girls. Br J Nutr 97(4):661–666
26. Marco ML, Heeney D, Binda S, Cifelli CJ, Cotter PD, Foligné B, Hutkins R (2017) Health benefits of fermented foods: microbiota and beyond. Curr Opin Biotechnol 44:94–102
27. Beena Divya J, Kulangara Varsha K, Madhavan Nampoothiri K, Ismail B, Pandey A (2012) Probiotic fermented foods for health benefits. Eng Life Sci 12(4):377–390
28. Ryan V, Bhunia AK (2017) Mitigation of foodborne illnesses by probiotics. In: Foodborne pathogens, pp 603–634
29. Katz SE (2012) The art of fermentation: an in-depth exploration of essential concepts and processes from around the world. Chelsea Green Publishing, Chelsea
30. Tamang JP, Shin DH, Jung SJ, Chae SW (2016) Functional properties of microorganisms in fermented foods. Front Microbiol 7:578. https://doi.org/10.3389/fmicb.2016.00578
31. Hesseltine CW, Wang HL (1980) The importance of traditional fermented foods. Bioscience 30(6):402–404
32. Gibson GR, Hutkins R, Sanders ME, Prescott SL, Reimer RA, Salminen SJ (2017) Expert consensus document: the International Scientific Association for Probiotics and Prebiotics (ISAPP) consensus statement on the definition and scope of prebiotics. Nat Rev Gastroenterol Hepatol 14(8):491–502. https://doi.org/10.1038/nrgastro.2017.75
33. Stiemsma LT, Nakamura RE, Nguyen JG, Michels KB (2020) Does consumption of fermented foods modify the human gut microbiota? J Nutr 150(7):1680–1692
34. Samtiya M, Aluko RE, Puniya AK, Dhewa T (2021) Enhancing micronutrients bioavailability through fermentation of plant-based foods: a concise review. Fermentation 7(2):63
35. Kaur H, Kaur G, Ali SA (2022) Dairy-based probiotic-fermented functional foods: an update on their health-promoting properties. Fermentation 8(9):425
36. Santiago-López L, Hernández-Mendoza A, Garcia HS, Mata-Haro V, Vallejo-Cordoba B, González-Córdova AF (2015) The effects of consuming probiotic-fermented milk on the immune system: a review of scientific evidence. Int J Dairy Technol 68(2):153–165
37. Karbownik MS, Mokros Ł, Kowalczyk E (2022) Who benefits from fermented food consumption? A comparative analysis between psychiatrically ill and psychiatrically healthy medical students. Int J Environ Res Public Health 19(7):3861
38. Karamali M, Dadkhah F, Sadrkhanlou M, Jamilian M, Ahmadi S, Asemi Z (2016) Effects of probiotic supplementation on glycaemic control and lipid profiles in gestational diabetes: a randomized, double-blind, placebo-controlled trial. Diabetes Metab 42(4):234–241. https://doi.org/10.1016/j.diabet.2016.04.009
39. Nikfar S, Rahimi R, Rahimi F, Derakhshani S, Abdollahi M (2008) Efficacy of probiotics in irritable bowel syndrome: a meta-analysis of randomized, controlled trials. Dis Colon Rectum 51(12):1775–1780. https://doi.org/10.1007/s10350-008-9335-z
40. Huang R, Wang K, Hu J (2016) Effect of probiotics on depression: a systematic review and meta-analysis of randomized controlled trials. Nutrients 8(8):483. https://doi.org/10.3390/nu8080483
41. Kang EJ, Kim SY, Hwang IH, Ji YJ (2013) The effect of probiotics on prevention of common cold: a meta-analysis of randomized controlled trial studies. Korean J Fam Med 34(1):2–10. https://doi.org/10.4082/kjfm.2013.34.1.2

42. Wang YH, Yao N, Wei KK, Jiang L, Hanif S, Wang ZX, Pei CX (2016) The efficacy and safety of probiotics for prevention of chemoradiotherapy-induced diarrhea in people with abdominal and pelvic cancer: a systematic review and meta-analysis. Eur J Clin Nutr 70(11):1246–1253. https://doi.org/10.1038/ejcn.2016.102
43. Song Y, Garg S, Girotra M, Maddox C, von Rosenvinge EC, Dutta A et al (2013) Microbiota dynamics in patients treated with fecal microbiota transplantation for recurrent Clostridium difficile infection. PLoS One 8(11):e81330. https://doi.org/10.1371/journal.pone.0081330
44. Borges NA, Carmo FL, Stockler-Pinto MB, De Brito JS, Dolenga CJ, Ferreira DC, Mafra D (2018) Probiotic supplementation in chronic kidney disease: a double-blind, randomized, placebo-controlled trial. J Ren Nutr 28(1):28–36. https://doi.org/10.1053/j.jrn.2017.06.010
45. Park SK, Park DI, Choi JS, Kang MS, Park JH, Kim HJ et al (2007) The effect of probiotics on Helicobacter pylori eradication. Hepatogastroenterology 54(79):2032–2036
46. Mohammadi AA, Jazayeri S, Khosravi-Darani K, Solati Z, Mohammadpour N, Asemi Z et al (2016) The effects of probiotics on mental health and hypothalamic–pituitary–adrenal axis: a randomized, double-blind, placebo-controlled trial in petrochemical workers. Nutr Neurosci 19(9):387–395. https://doi.org/10.1179/1476830515Y.0000000023
47. Mohammadi AA, Jazayeri S, Khosravi-Darani K, Solati Z, Mohammadpour N, Asemi Z et al (2015) Effects of probiotics on biomarkers of oxidative stress and inflammatory factors in petrochemical workers: a randomized, double-blind, placebo-controlled trial. Int J Prev Med 6:82. https://doi.org/10.4103/2008-7802.164146
48. Kim HJ, Vazquez Roque MI, Camilleri M, Stephens D, Burton DD, Baxter K et al (2005) A randomized controlled trial of a probiotic combination VSL# 3 and placebo in irritable bowel syndrome with bloating. Neurogastroenterol Motil 17(5):687–696. https://doi.org/10.1111/j.1365-2982.2005.00695.x
49. Kim YG, Moon JT, Lee KM, Chon NR, Park H (2006) The effects of probiotics on symptoms of irritable bowel syndrome. Korean J Gastroenterol 47(6):413–419
50. Kim SO, Ah YM, Yu YM, Choi KH, Shin WG, Lee JY (2014) Effects of probiotics for the treatment of atopic dermatitis: a meta-analysis of randomized controlled trials. Ann Allergy Asthma Immunol 113(2):217–226. https://doi.org/10.1016/j.anai.2014.05.021
51. Javadi L, Ghavami M, Khoshbaten M, Safaiyan A, Barzegari A, Pourghassem Gargari BP (2017) The effect of probiotic and/or prebiotic on liver function tests in patients with nonalcoholic fatty liver disease: a double blind randomized clinical trial. Iran Red Crescent Med J 19(4):e46017. https://doi.org/10.5812/ircmj.46017
52. Zeng L, Yu G, Yang K, Hao W, Chen H (2021) The improving effect and safety of probiotic supplements on patients with osteoporosis and osteopenia: a systematic review and meta-analysis of 10 randomized controlled trials. Evid Based Complement Alternat Med 2021:9924410. https://doi.org/10.1155/2021/9924410
53. Salari A, Ghodrat S, Gheflati A, Jarahi L, Hashemi M, Afshari A (2021) Effect of kefir beverage consumption on glycemic control: a systematic review and meta-analysis of randomized controlled clinical trials. Complement Ther Clin Pract 44:101443. https://doi.org/10.1016/j.ctcp.2021.101443
54. Bélanger SA, Vanasse M, Spahis S, Sylvestre MP, Lippé S, l'Heureux F et al (2009) Omega-3 fatty acid treatment of children with attention-deficit hyperactivity disorder: a randomized, double-blind, placebo-controlled study. Paediatr Child Health 14(2):89–98. https://doi.org/10.1093/pch/14.2.89
55. Rice TW, Wheeler AP, Bernard GR, Vincent JL, Angus DC, Aikawa N (2010) A randomized, double-blind, placebo-controlled trial of TAK-242 for the treatment of severe sepsis. Crit Care Med 38(8):1685–1694. https://doi.org/10.1097/CCM.0b013e3181e7c5c9
56. Bull MJ, Plummer NT (2014) Part 1: the human gut microbiome in health and disease. Integr Med 13(6):17–22
57. Peterson CT, Sharma V, Elmén L, Peterson SN (2015) Immune homeostasis, dysbiosis and therapeutic modulation of the gut microbiota. Clin Exp Immunol 179(3):363–377

58. Fuller R, Gibson GR (1997) Modification of the intestinal microflora using probiotics and prebiotics. Scand J Gastroenterol 32(sup222):28–31
59. Azcarate-Peril MA (2019) Beneficial modulation of the gut microbiome: probiotics and prebiotics. How fermented foods feed a healthy gut microbiota: a nutrition continuum, pp 309–324
60. Dale HF, Rasmussen SH, Asiller ÖÖ, Lied GA (2019) Probiotics in irritable bowel syndrome: an up-to-date systematic review. Nutrients 11(9):2048
61. Jayachandran M, Xu B (2019) An insight into the health benefits of fermented soy products. Food Chem 271:362–371
62. Jalili M, Nazari M, Magkos F (2023) Fermented foods in the management of obesity: mechanisms of action and future challenges. Int J Mol Sci 24(3):2665
63. Awwad SF, Abdalla A, Howarth FC, Stojanovska L, Kamal-Eldin A, Ayyash MM (2022) Invited review: potential effects of short-and long-term intake of fermented dairy products on prevention and control of type 2 diabetes mellitus. J Dairy Sci 105(6):4722–4733
64. Kwon DY, Daily JW III, Kim HJ, Park S (2010) Antidiabetic effects of fermented soybean products on type 2 diabetes. Nutr Res 30(1):1–13
65. Chen J, Zhao KN, Vitetta L (2019) Effects of intestinal microbial–elaborated butyrate on oncogenic signaling pathways. Nutrients 11(5):1026
66. Kim J, Kang M, Lee JS, Inoue M, Sasazuki S, Tsugane S (2011) Fermented and non-fermented soy food consumption and gastric cancer in Japanese and Korean populations: a meta-analysis of observational studies. Cancer Sci 102(1):231–244. https://doi.org/10.1111/j.1349-7006.2010.01770.x
67. Smyth D (2016) High-fibre diet associated with reduced risk of breast disease. Cancer Nurs Pract 15(3):14–14. https://doi.org/10.7748/cnp.15.3.14.s18
68. Kontou N, Psaltopoulou T, Panagiotakos D, Dimopoulos MA, Linos A (2011) The Mediterranean diet in cancer prevention: a review. J Med Food 14(10):1065–1078. https://doi.org/10.1089/jmf.2010.0244
69. Deeth HC, Tamime AY (1981) Yogurt: nutritive and therapeutic aspects. J Food Prot 44(1):78–86. https://doi.org/10.4315/0362-028X-44.1.78
70. Bourrie BC, Willing BP, Cotter PD (2016) The microbiota and health promoting characteristics of the fermented beverage kefir. Front Microbiol 7:647. https://doi.org/10.3389/fmicb.2016.00647
71. Yoon HS, Son YJ, Han JS, Lee JS, Han NS (2005) Comparison of D-and L-lactic acid contents in commercial kimchi and sauerkraut. Food Sci Biotechnol 14(1):64–67
72. Fukushima D (1979) Fermented vegetable (soybean) protein and related foods of Japan and China. J Am Oil Chem Soc 56(3Part2):357–362
73. Jayabalan R, Malbaša RV, Lončar ES, Vitas JS, Sathishkumar M (2014) A review on kombucha tea—Microbiology, composition, fermentation, beneficial effects, toxicity, and tea fungus. Comp Rev Food Sci Food Safety 13(4):538–550. https://doi.org/10.1111/1541-4337.12073
74. Szutowska J (2020) Functional properties of lactic acid bacteria in fermented fruit and vegetable juices: a systematic literature review. Eur Food Res Technol 246(3):357–372. https://doi.org/10.1007/s00217-019-03425-7
75. Sakandar HA, Hussain R, Kubow S, Sadiq FA, Huang W, Imran M (2019) Sourdough bread: a contemporary cereal fermented product. J Food Process Preserv 43(3):e13883. https://doi.org/10.1111/jfpp.13883
76. Renner E (1993) Nutritional aspects of cheese. In: Cheese: chemistry, physics and microbiology, General aspects, vol 1. Springer US, Boston, pp 557–579
77. Holzapfel WH (2002) Appropriate starter culture technologies for small-scale fermentation in developing countries. Int J Food Microbiol 75(3):197–212
78. Boukid F, Hassoun A, Zouari A, Tülbek MÇ, Mefleh M, Aït-Kaddour A, Castellari M (2023) Fermentation for designing innovative plant-based meat and dairy alternatives. Foods 12(5):1005. https://doi.org/10.3390/foods12051005
79. Hayes M, Ross RP, Fitzgerald GF, Stanton C (2007) Putting microbes to work: dairy fermentation, cell factories and bioactive peptides. Part I: overview. Biotechnol J 2(4):426–434

80. Sivamaruthi BS, Kesika P, Chaiyasut C (2018) Toxins in fermented foods: prevalence and preventions—a mini review. Toxins (Basel) 11(1):4
81. Sohn KH, Lee HJ (1998) Effects of high-pressure treatment on the quality and storage of kimchi. Int J Food Sci Technol 33(4):359–365. https://doi.org/10.1046/j.1365-2621.1998.00138.x
82. Karlsson FH, Tremaroli V, Nookaew I, Bergström G, Behre CJ, Fagerberg B, Bäckhed F (2013) Gut metagenome in European women with normal, impaired and diabetic glucose control. Nature 498(7452):99–103. https://doi.org/10.1038/nature12198

Chapter 13
Fermentation Technology and Food Enzymes

Yallappa M., Gangadhar B. Megeri, Basavaraj S. Hungund, and Jayachandra S. Yaradoddi

1 Introduction

Using microbes' growth and metabolic processes, fermentation process helps to preserve and transform food ingredients. Perishable product has a longer shelf life during food fermentation because the metabolites synthesized by the organisms involved in fermentation can prevent the growth of pathogenic and rotting microbes. Lactic acid bacteria, for example, can able to produce carbon dioxide, lactic acid, acetic acid, bacteriocins, antimicrobial peptides hydrogen peroxide and ethanol during lactic acid fermentation [1]. These metabolites could inhibit the growth and survival of spoilage microbes as well as pathogenic organisms.

Besides this, by preserving foods, the fermentation process generally slow down to produces distinct flavour, texture, aroma and nutritional effects. Fermentation was therefore initially developed by ancient civilizations by means of preserving fresh agro-based products, but this phenomenon now advanced beyond preservation to become a tool for enhancing the organoleptic characteristics and palatability of foods. The main purpose in utilization of fermentation technology which is

Yallappa M.
Department of Agriculture Microbiology, College of Agriculture Bheemarayanagudi, UAS, Raichur, Karnataka, India

G. B. Megeri
Environmental Science Division, Department of Civil Engineering, Basaveshwar Engineering College, Bagalkote, Karnataka, India

B. S. Hungund
Department of Biotechnology, KLE Technological University, Hubballi, Karnataka, India

Jayachandra S. Y. (✉)
Department of Biotechnology, Basaveshwar Engineering College, Bagalkote, Karnataka, India

J. S. Yaradoddi et al. (eds.), *Frontiers in Food Biotechnology*,
https://doi.org/10.1007/978-981-97-3261-6_13

anticipated to yield finished product with respect to its distinctive organoleptic, structure and texture, characteristics.

Additionally, food items' nutritional profiles are improved and antinutritional elements and pollutants are eliminated through fermentation. For example, fermentation of soybean brings about the hydrolysis process of complex soy proteins into more digestible, amino acids bioavailable peptides and in addition which reduces the effects of antinutritional factors like trypsin inhibitors and phytic acid. Fermentation of soybean also produces products like natto, tempeh and soy sauce.

2 Food Fermentation Processes

2.1 Methods of Fermenting Food

Alkaline, fungal and lactic acid fermentation are the three main categories into which traditional food fermentation procedures can be divided. Products largely fermented by lactic acid bacteria, such as cheese, sauerkraut, sausages, yoghurt and kimchi, are few examples of lactic acid fermented products. Many other lactic acid-fermented items, such as kombucha (a fermented sweetened tea from China) and kefir (a mildly alcoholic dairy beverage from the Caucasus), are also produced using the yeast species. With the exception of natto, which is made via alkaline fermentation, the majority of popular Asian fermented foods made from soy, such as tempeh and soy sauce, are generated by fungal fermentation [2].

Bioreactors that are submerged or solid-state and may be run in batch, semibatch or continuous modes are used in industrial fermentation operations. The majority of food fermentation procedures, ranging from kimchi and sauerkraut to miso and tempeh, employ batch-operated solid-state fermentation methods in which microbes are grown on the surface of a substrate that is insoluble in water [3]. Fermentation via submersion is employed in the manufacturing of alcoholic drinks, food condiments like vinegar and dairy products like yoghurt.

2.2 Current Trends

Nowadays, there is a push to use fermented foods to treat or prevent a variety of illnesses, from cancer to obesity. Kim et al. (2011) [4], for example, assert that kimchi has antiobesity, antiaging, anticancer and anticonstipation activity, whereas Guzel-Seydim et al. (2011) [5] assert that kefir has antimutagenic and anticarcinogenic properties, lowers cholesterol, boosts the immune system and reduces lactose intolerance symptoms. While there is no scientific evidence to support the majority of the health claims made about fermented foods [6–9], some of these claims are supported by recent findings from studies using human intervention, in vitro and animal models.

For instance, a recent study conducted by a Korean group [4] found that eating kimchi for 2 weeks reduced blood pressure, body mass index, waist–hip ratio, total cholesterol, fasting blood sugar, body fat in obese and overweight people.

According to Riley (2015) [10], one of the top ten dietary trends for 2016 was fermented foods, which carry over from previous years' trends. In response to this expanding trend, food companies are either commercializing traditional fermented foods (like kombucha and kefir, which had respective market values of $130 million and $480 million in North America in 2014) or creating innovative fermented foods that are based on the traditional ones (like Bionade, which are flavoured malt-based beverages fermented using kombucha starter culture and Rythem, which are coconut milk- and fruit juice-based beverages fermented using kefir grains). In response to the rising incidence of lactose and gluten intolerances, allergies to dairy proteins and lifestyle choices like veganism, a number of soy- and cereal-based probiotic products are also available on the market [11, 12]. There are also trends in preservation and transportation of food items using the biodegradable packaging material [13, 14].

Using a microbial biotechnology called fermentation, natural renewable substrates can be transformed into products of additional value such polymers, enzymes, organic acids and alcohols. End-products of fermentation, such lactic acid and ethanol, operate as proton sinks, recycling NADH into NAD+ so the cell can keep generating energy through glycolysis through substrate-level phosphorylation. In order to preserve energy balance, microbes produce a large number of end- or by-products. Targeted genetic engineering approaches applied to well-established industrial microbial strains have today improved production of economically significant fermentation products [15].

The microbial strain and the environmental conditions used determine whether end- or by-products are formed. The intended product should guide the selection and development of a microbial strain for the best possible fermentation process. Recombinant DNA technology and mutation are examples of strain development technologies. Temperature, pH, dissolved oxygen and the makeup of the fermentation medium all have an impact on the growth of microbes and the creation of products. Consequently, it's necessary to optimize the growth conditions and fermentation material. Perceptive the biochemical modification which occurs in fermented food can help farmers in enhancing the production capacity through strain improvement and environment adjustments. Apart from growing conditions, media and strain types, fermentation modes also impact productivity. For maximum productivity, fed-batch, batch and continuous modes of fermentation can be chosen. During fermentation processes, fed-batch and continuous modes can get around the substrate constraint. Cell immobilization is another method for increasing productivity since it raises the concentration of biomass in the bioreactor and, consequently, the concentration of biocatalysts in the reactor.

Following the fermentation process, strategies for recovering the product from the fermentation medium must be assessed and improved; these strategies are frequently limited financially. In fact, one of the most expensive stages of the process is frequently purifying the final product. The biomass itself, extracellular products, or intracellular products are examples of the microbial end products. Recovery techniques include homogenization, extraction (liquid and solid) and filtering. The catalytic properties of enzymes, which are derived from living things, have led to their long-standing application in industry. Enzyme activity should be adjusted for every process, taking into account factors such as pH, substrate, temperature and inhibitors. Since recovering enzymes is challenging, enzyme immobilization could be used to lower process costs. Microorganisms have the ability to create enzymes, or they can be extracted from plant and animal tissues. Yet, because they are more readily available and particular, microbial enzymes are recommended. More than 500 commercial goods are made using enzymes [16]. They are used in a wide variety of products, including detergents and food. The majority of enzymes are used in the food, cosmetics, paper, detergent and pharmaceutical industries to improve the quality of their products. They are also sold commercially. Pectic enzymes, lipase, proteases, amylases and rennet—an enzyme that clots milk—are examples of commercial enzymes.

This chapter provides an overview of the criteria for microbial strain generation and selection, as well as the optimization of growth conditions and media. Fermented food production is explained, along with the strains, circumstances and biochemical changes involved. There is discussion of the benefits and drawbacks of various fermentation processes. Lastly, fundamental recovery techniques and the principles of immobilized fermentation are also outlined.

3 Available Technologies in Fermentation

3.1 *Batch Process of Fermentation*

Batch fermentation is performed in a CSTR (continuous stirred tank reactor) using a starting medium volume, neither medium addition nor removal takes place during the fermentation process [17]. Sterilization of the medium and pH correction with either acid or alkali precede fermentation. Product creation occurs by regulating temperature, pH, agitation and dissolved oxygen following the inoculation of bacteria or yeast, depending on the traits of the cultured microorganism. Microorganisms grow according to the four main phases of the growth curve: the lag phase, the log (exponential) phase, the stationary phase and the death phase. This is because there is no medium addition.

Cell growth is low during the lag phase, which is an adaptation period for the cells to get used to their new surroundings (such as the substrates in the culture medium, the start of metabolic pathways, or the synthesis of specialized enzymes like cellulases and amylases) [18]. The maximal rate of microbial growth occurs in the log (exponential) phrase. When the rate of microbial growth and death is equal, the phase is known as the stationary phase. Cells can still create secondary metabolites during this phase. The death phase, which occurs at the end of batch fermentation, is characterized by a higher rate of death than growth because of factors such an excess of inhibitory primary or secondary metabolites or a lack of nutrients.

3.2 *Fed-Batch Fermentation Process*

In this case, the fermentation broth is either removed semi-constantly or not at all, while single or multiple nutrients are given to a reactor either sporadically or continuously (Fig. 13.1). Two advantages can be achieved by adding medium: (1) microbial growth won't be hampered by a shortage of nutrients, and (2) substrate inhibition will be removed if it's a process limitation. In the fed-batch technique, ethyl alcohol will be eliminated by gradually lowering the quantity of end product in the fermentation process. Ethanol is a metabolic inhibitor for yeast (>10–12% based on the type of strain) and a cause for worry during fermentation. However, correct feeding optimization is necessary to increase productivity and ethanol yield [17].

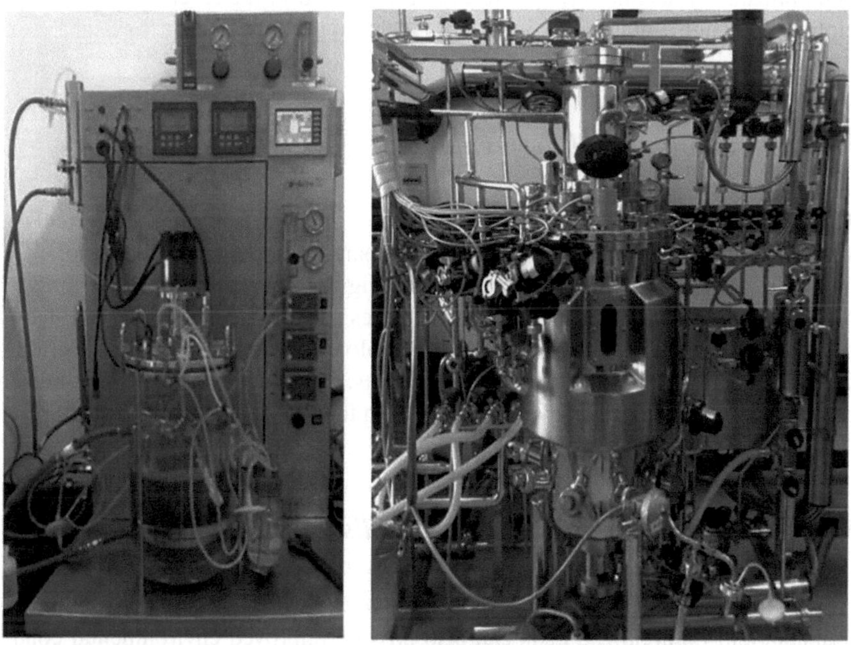

Typical picture of pilot scale fermenter

Typical picture of industrial scale fermenter

Fig. 13.1 The pilot scale fermenter and industrial scale fermenter

In Brazil's ethanol business, the fed-batch cultivation process is most widely used technology [17].

3.3 Continuous Fermentation

Fresh, sterile media are continuously delivered into a reactor during continuous fermentation, sometimes referred to as chemostats or continuous stirred tank reactors (CSTR). The effluent is taken out of the reactor and its volume is kept constant, along with being supplied with new nutrients. Feeding and removal rates do not change. The growth rate of the microbes must match the dilution rates (rate at which cells are removed from the reactor) in order to prevent wash-out, which would require removing all of the cells.

Continuous fermentation has several advantages over batch fermentation, such as lower bioreactor building costs, less maintenance and operating requirements, increased production and improved process control [17]. However, culture stability is a problem for ongoing fermentation. Yield can be lowered by making even slight adjustments to any parameter (such as temperature, feed substrate content, dilution rate). There are a variety of ways to get around the disadvantages of continuous fermentation. Sanchez and Cardona (2008) [17], for example, recommend using immobilized cell technology, where cells are trapped in the reactor by chrysolite, biofilms, calcium alginate, etc. in order to increase production.

3.4 Immobilized Cell Fermentation

The process of stopping cells in a fermentation medium from migrating freely is known as immobilized cell fermentation. Compared to suspended cell reactors, cell immobilization has the advantage of increasing the concentration of biomass or biocatalyst in the reactor, which in turn increases enzyme activity and fermentation byproducts. Immobilization can also enable prolonged fermentation plant operation by reusing cells without incurring recovery costs. Generally speaking, immobilized cell systems are easier to recover products from than suspended cell bioreactors.

3.5 Fermentation of Immobilized Cells

At high dilution rates, immobilization can solve cell wash-out issues in continuous fermentation. Immobilized cells can also provide improved environmental conditions for the growth of microorganisms. Four types of immobilization can be distinguished based on the physical mechanism of interest: mechanical containment behind a barrier, aggregation (cell flocculation), entrapment within a porous matrix and immobilization on solid carrier surfaces [19, 20].

3.6 Downstream Process

Any fermentation process ends with downstream processing, which separates the result from other unwanted ingredients in the fermentation broth and performs additional purification. The biomass itself, extracellular (found in the fermentation broth) or intracellular (enclosed inside the cell) products are all possible outcomes. Due to the complexity of fermentation broth, recovering the desired product necessitates extensive recovery techniques and can cost between 50 and 60% of the entire production cost [21]. The following features must be taken into account for product recovery and typically restrict the techniques of separation. Typically, the products are temperature-sensitive and present in low concentrations within a complicated liquid media.

If the products are intracellular, more difficult purification and separation procedures would be required. Furthermore, the product's chemical or physical characteristics can be strikingly similar to those of the other fermentation broth by-products, which would make downstream processing very challenging. Generally speaking, downstream processing is carried out in four primary processes. The removal of insolubles, which involves separating solid particles like biomass from the liquid media, is typically the initial step. The second phase involves breaking down the biomass's cell wall and removing the intracellular substances. The third stage can be thought of as the removal of soluble from the liquid phase, particularly for items that are in the medium. The product is further purified as the last step.

4 Removal of Insolubles

Filtration or centrifugation is typically used to remove insolubles. Through the use of a filter or solid support that has particles deposited on it, the fluid is forced through filtration to separate the solids from the liquid. This is a really simple procedure creating a diluted suspension of big, hard particles. Small particles and the cells' malleability, however, have the potential to complicate the procedure. Although there are many different kinds of filters, the most widely used ones are rotary vacuum filters and plate and frame pressure filters. Pressure filters with plates and frames are used to extract bacteria and fungi from fermentation broth on a modest scale.

Because rotary hoover filters need less labour, they are the favoured choice for large-scale commercial processes [22]. Although filtration is simple, pretreatments could be necessary for some goods. Pretreatments include heating to denaturize proteins, adding electrolytes to encourage flocculation and coagulation and adding filter aids (earths and perlites) to increase porosity and reduce cakes' compressibility [23].

5 Cell Disruption

However, because of the robust cell walls and high internal osmotic pressure, breaking the cells frequently proves to be challenging [24]. The methods employed must be strong enough to breach the cell wall while remaining delicate enough to avoid causing harm to the final result. Physical techniques include homogenization, ultrasonication and grinding as mechanical disruption techniques. Animal cells or tissues, as well as mycelia organisms, respond well to homogenization. However, for the majority of the suspension cells, ultrasonication and grinding can be used. Conversely, bigger scales are more suited for crushing and homogenization using orifice type [23].

6 Removal of Soluble

After cell disruption, the next stage of product recovery is to separate the extracellular and intracellular products from the broth using a variety of techniques. Only a few techniques will be briefly discussed in this section due to the vast scope of recovery and purification [23]. According to Dutta (2008) [24], extraction is the process of separating the solutes in a liquid solution by coming into contact with the solvent, another insoluble liquid. By choosing the right solvent, the desired product can be removed from the solution and placed into the solvent phase selectively. A good solvent should be inexpensive, non-toxic, have a big density difference between the two phases, and be simple to recover. The liquid's remnant is referred to as the raffinate, whereas the solvent-rich phase is known as the extract [24].

7 Purification

The finished product must be carefully refined, particularly for medications. As a result, several crystallization, chromatographic and ultrafiltration techniques, are used for ultimate purification. Low temperature crystallization reduces heat-sensitive materials' thermal deterioration [18]. Solvent is extracted from a purified wet product through drying. To carry out a drying process, the product's physical characteristics, heat sensitivity and desired final moisture content must be taken into account. The main types of dryers used for purification are tray driers, rotary drum driers, freeze driers, vacuum-, pneumatic conveyor driers and spray drum driers [18].

8 Fermented Foods

Foods have been fermented for thousands of years; grapes and barley were first fermented to produce alcoholic beverages [15]. Foods that have undergone fermentation tend to have better sensory qualities, a longer shelf life and higher nutritional content. This section highlights some of the outcomes of the biochemical and physical changes that microorganisms bring about during the fermentation process.

9 Fermented Dairy Products

A key component in the creation of fermented dairy products is lactic acid bacteria. The lactic acid they make from the fermentation of lactose lowers the pH of milk. The primary protein in milk, casein, has its isoelectric point reached when pH hits this value. The entire amount of positive and negative charges on the amino acids are balanced at the isoelectric point. At that moment, coagulum forms and casein precipitates [25, 26].

This alters the product's composition both chemically and physically. Apart from lactic acid, organic compounds including acetaldehyde, diacetyl, acetic acid, ethanol and exopolysaccharides can also be produced by starting cultures and can impact the product's texture and flavour [27]. As a result, distinct cultures and industrial environments result in unique qualities for particular items. Dairy products that have undergone fermentation include cheese, kefir, yoghurt and cultured buttermilk.

10 Vinegar

According to Adams (1985) [28], vinegar is made by aerobically fermenting different sources of ethanol to create a diluted acetic acid solution. Vinegar should have a minimum acetic acid content of 4% (w/v). According to Hutkins (2006) [27], the vinegar's name—such as red wine vinegar, apple cider vinegar and malt vinegar—indicates the juice source.

11 Beverage Productions

11.1 Wine

Crushing the grape, extracting the juice, removal of the pomace, and inoculation of *Saccharomyces cerevisiae* leading to the grape fermentations to produce wine. While the sugar level and acid in fruit juice could lower's pH (~4.5) but boost the growth of yeasts converting it to ethanol. The ethanol generated inhibits the growth of fungus and bacteria. Several yeast species, such as *Hanseniaspora, Kloeckera, Metschnikowia* and *Candida*, may proliferate during the early stages of fermentation and create trace amounts of glycerol, acids and other possible flavouring ingredients [28]. For the fermentation of white wine, the ideal temperature range is 10–18 °C, and for red wine, it is 20–30 °C shall be maintained.

11.2 Beer

Malted barley ferments when heated with hops to make wort, which is then fermented by yeast to produce beer. The preliminary stages include the malting- milling-mashing-brewing-cooling- fermentation-maturation-filtering, and packaging are followed in the beer-making process. Yeast ferments the carbohydrates via the fermentation process, releasing CO_2 and ethanol. *S. cerevisiae* strains are employed to make 'ale4 or 'top-fermenting yeast' beers. Lager beers are made with *Saccharomyces pastorianus*, sometimes known as 'bottom-fermenting yeast'. Yeasts that ferment at the top produce flocs, or low-density clumps, that trap CO_2 and ascend to the surface. *S. pastorianus* flocculates during the fermentation of lager beer, and the flocs sink to the bottom of the fermentation tank.

11.3 Bread

The yeast *Saccharomyces cerevisiae* is used to make bread. It should possess unique qualities, such as the capacity to yield flavourful bread and the ability to remain viable while being stored. Molasses, different ammonium salts like magnesium sulphate, calcium sulphate and ammonium phosphate, as well as trace minerals like iron and zinc in smaller amounts, are all part of the seed development media. The ideal growth conditions are 30 °C, pH 4.0–5.0, and aerobic. After that, 20% or so of total yeast solids are added to flour in the form of yeast cream. Glucose is used throughout the fermentation process through either anaerobic or aerobic pathways. CO_2 is produced by yeast, giving food an elastic, airy feel. For sour dough rye bread, *Lactobacillus fructivorans*, *Lactobacillus brevis* var. lindneri, and *Lactobacillus farciminis* are employed as starter cultures in addition to yeast. A few flavouring chemicals and acetic acid are also produced by these cultures [29].

11.4 Tempeh

Popular fermented soybean products like tempeh are widespread in Far Eastern nations like Indonesia. Four primary procedures are involved in the manufacturing of tempeh: soaking, boiling, inoculating soybeans with *Rhizopus microsporus* subsp. *oligosporus*, and room-temperature incubation [30]. The process of making tempeh takes roughly 48 h. Endogenous lactic acid bacteria proliferate and lower the pH during the steeping time. As a result, harmful pathogens and spoilage bacteria are inhibited from growing. One kilogram of beans is injected with 107–108 *Rhizopus oligosporus* spores. Apart from *R. oligosporus*, other species utilized in Indonesia include *R. arrhizus*, *R. stolonifer* and *R. oryzae*, [31]. As an alternative, soybeans can be inoculated with 'usar', which is created by applying wild *Rhizopus* spores to the surface of leaves taken from the native Indonesian hibiscus plant.

11.5 Enzyme Applications

In biological systems (microbes, animals and plants), enzymes are proteins with molecular weights ranging from 15,000 to several million Daltons. They catalyse particular biochemical reactions [18, 32, 33]. Numerous proteins that are conjugated and have certain prosthetic groups are enzymes. The term 'apoenzyme' refers to the protein portion of the enzyme, 'haloenzyme' refers to the entire enzyme and 'prosthetic group' refers to the non-protein component (deMan 1996). Dialysis can be used to extract the prosthetic group from the protein, which is frequently just tangentially connected. Coenzymes, such as nicotinamide adenine dinucleotide (NAD), dissociate from the enzyme during or at the end of a catalytic reaction because they bind to the enzyme less firmly than prosthetic groups [34].

Under reaction conditions, the complex is unstable and decomposes fast. It regenerates the enzyme. When the substrate is released from the reaction, it may be in its fully transformed state or it may be highly active. Only the 'active site', or certain areas of the protein molecule, engages in the reaction between an enzyme and its substrate. Active sites are unique collections of amino acid residues created by the sequence and specific folding of the enzyme protein (Fig. 13.2). The backbone and support for the active sites are provided by the portion of the protein molecule that is not within the active areas.

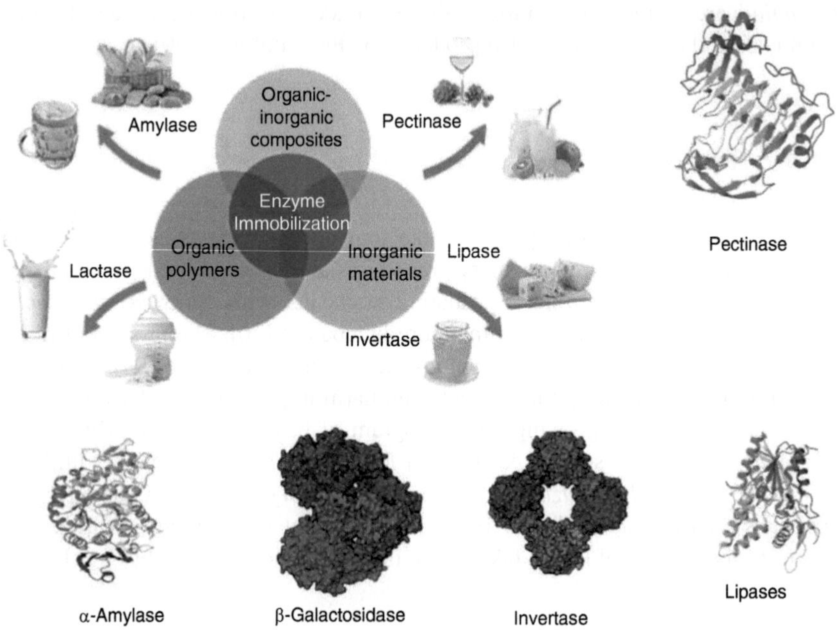

Fig. 13.2 Various applications of food enzymes and their three-dimensional structures

12 Oxidoreductase

The oxidation–reduction processes are catalysed by oxidoreductase enzymes. The enzymes referred to as reductases, oxidases, dehydrogenases, hydroxylases, oxygenases and catalases are included in this group [35]. For instance, lipoxygenase is employed in the baking sector could brings about the glucose oxidase that reduces the Maillard reaction in food system, and catalase the elimination of H_2O_2 in the milk industry [36].

13 Transferases

Enzymes known as transferases are responsible for the transfer of several chemical groups from one substrate to another, including methyl, aldehyde, acetyl, amine, ketone and phosphate residues [37]. Typical examples of this group are kinases and transaminases.

14 Hydrolases

When a water molecule is added, hydrolases cause the hydrolytic breakage of bonds. For instance, pectinases are used to clarify liquids, proteases are used to tenderize meat and amylases are utilized to produce glucose syrups from cornflour.

15 Lyases

By eliminating bonds and leaving rings or double bonds behind, or by adding groups to double bonds, lyases catalyse the non-hydrolytic cleavage of C-C, C-O and C-N bonds as well as other bonds. They leave behind double bonds after removing groups from their substrates. Berk et al. (1976) [33] list aldolases, carboxylases, hydratases and phenylalanine ammonia lyase as examples. Pectate lyases may be used to enhance the surface qualities of natural fibres in the pharmaceutical, culinary and textile industries [38].

16 Isomerases

When a water molecule is added, hydrolases cause the hydrolytic breakage of bonds. For instance, pectinases are used to clarify liquids, proteases are used to tenderize meat and amylases are utilized to produce glucose syrups from cornflour. The substrates of isomerases are changed from one isomeric form to another. Members of this group include racemases and epimerases. For instance, high-fructose corn syrups are made using glucose isomerase. The cost of enzyme preparations is high. Even though the enzyme regenerates during the process and can be used repeatedly, it is typically not economically possible to separate it from the reaction mixture.

Enzymes that have been immobilized can be utilized in batch operations and then extracted for additional use using centrifugation or filtration. Utilizing the similar kind of enzymes preparation repeatedly within a batch operation is made easier by applications of this technology. In specifically constructed fomenters, mainly constant nourished/feed stirred tanks, fluidized bed, packed-bed reactors and others, immobilized enzymes can also be employed in continuous processes. In general, enzymes those are immobilized are more thermostable over the soluble enzymes. The food sector has made use of immobilized enzymes. As an illustration, an immobilized lactase system was created through the adsorption of the enzyme on low-lactose milk is produced by entrapment in gels or fibres, covalent bonding to organic polymers, microencapsulation in nylon or cellulose microcapsules and stainless steel and porous glass beads.

17 Enzymes Used in the Industries

The food business has been using microbial enzymes from centuries. These enzymes could also be used in the leather industry; for example, dung was used to prepare hides [16]. Enzyme technology was originally applied in the food business in the 1930s to clarify fruit juice [39]. Since then, because of their advantages—specificity, moderate conditions, environmental friendliness and reduced waste generation—enzymes have found applications in the detergent, food, cosmetic, leather, pharmaceutical and textile industries [40]. Several important industrial enzymes will be discussed in this section.

17.1 Lipases

When acting on carboxylic ester bonds, lipases do not need a co-factor. Lipase's natural substrates are long-chain fatty acids [41]. Because lipases naturally hydrolyse triglycerides into monoglycerides, fatty acids, glycerol and diglycerides, they are of importance to the industrial sector. Moreover, the industrial community is interested in lipases because of their regioselectivity, substrate specificity and enantio selectivity [42]. Lipase's specific activity, active site, calcium binding, specificity, surfactant compatibility and stability can all be improved through protein engineering [43]. Lipases are low molecular weight proteins measures about 19–60 kDa. Fatty acids form into a glycerol backbone, the chain length, and other factors that greatly influence the properties of the enzyme.

17.2 Amylases

Amylases are enzymes that have the ability to hydrolyse the α-1,4-glucosidic bonds in starch [44]. Though they can be derived from the animals, plants and microbial sources, among these enzymes amylases are more frequently encountered in industrial settings. Two important amylases are glucoamylase and α-Amylase [45]. These enzymes' ability to break down starch has been used by industry [31]. Higher plants are the primary producers of starch, which is insoluble in water [46]. It is made up of two parts: amylose and amyl pectin. Amylose is primarily a linear polysaccharide composed of certain α-1,6-branching points and α-1,4 connected D-glucose residues. The structure of amylopectin resembles a massively branching tree. The percentage of branches is a crucial component of the substrate since enzymes can hydrolyse various substrates with various specificities.

17.3 Pectin Digestive Enzymes

The typical nature of the pectic digestive enzymes is to hydrolyse pectins, which are intricate and lengthy molecules especially the structural polysaccharides of plant cells, play important role in safegaurding the integrity of the cell wall [43]. Pectic enzymes are often utilized in commerce and make up 25% of the global market for food enzymes [33]. The most prevalent microbiological source of these enzymes is *Aspergillus* species; however, other producers include *Penicillum* spp., *Sclerotinia libertiana, Coniothyrium diplodiella* and *Rhizopus* spp. [44]. Pectic compounds are complex polysaccharides with a high molecular weight (30,000–300,000 Da) that are negatively charged. They consist of galacturonic acid residues connected by α-1, 4-linkages at their backbone [47]. According to Bigelis (1993) [40], methyl groups or galactans, arabinogalactans, arabinans and rhamnogalacturonans esterify the monomers of galacturonic acid.

17.4 Proteases

Enzymes known as proteases break down proteins by catalysing the breakage of peptide bonds inside proteins. According to Adler-Nissen (1993) [46], proteins are generally made up of linear polymers of amino acids with the general formula R-CHNH2-COOH. The peptide bonds are created when the nitrogen atoms of one amino group covalently connect to the carbon atoms of the previous carboxyl group, releasing water in the process. Proteases are divided into endopeptidases and exopeptidases based on the catalytic activity of the enzymes. The enzymes are subcategorized by the EC system as exopeptidases (3.4.11–19) and endopeptidases (3.4.21–24). Exopeptides hydrolyse the release of a single amino acid at a time from either the N or C terminus, whereas endopeptidases act on the polypeptide chain at certain vulnerable peptide bonds [45].

Metalloproteases, aspartic proteases, cysteine proteases and serine proteases are examples of endopeptidases. While metalloproteases consists of essential metal element, typically zinc (Zn), at the active sites, cysteine, aspartic acid and serine, where as the proteases contain cysteine, and aspartic acid serine, respectively [46].

In contrast, exopeptidases are classified into three main categories: omega peptidases, carboxypeptidases and aminopeptidases. Aminopeptidases target the polypeptide chain's N terminus to produce a single amino acid residue, either a di- or tripeptide. On the other hand, carboxypeptidases function at the polypeptide's C terminus and release a single amino acid or dipeptide [47]; the screening of these extracellular enzymes was revealed by Jayachandra et al. (2012) [48].

18 Conclusion

The food sector uses fermentation extensively to create goods with additional value. The choice of microbe or strain and the adjustment of growth parameters for the chosen microorganism is critical to the fermentation's success. Food quality and flavour have also been improved through the use of fermentation. For a fermented product that is acceptable to consumers, fermentation optimization should also be carefully studied. However, because of their catalytic activity, enzymes have their own uses in the industry. To achieve high fermentation yields, novel microbial strain isolation and strain development are still required. Furthermore, strains that are capable of producing heterologous products can be developed with the use of genetic and metabolic engineering.

Products that aren't now available for industrial usage may therefore become reasonably priced for such application. Immobilization reactors should be evaluated for each product to determine how immobilization affects yield because some microbes do not make exopolysaccharide, which causes issues with attachment to supports. Studying the strains and production characteristics of traditional fermented foods is important, in addition to yield. The fermented food's flavour and texture are enhanced by the natural flora. As a result, microbes found in traditional foods are ought to be identified and their roles investigated. Thus, it is possible to preserve the variety of traditional fermented dishes.

References

1. Di Cagno R et al (2013) Exploitation of vegetables and fruits through lactic acid fermentation. Food Microbiol 33(1):1–10
2. Dirar HA (1992) Sudan's fermented food heritage. In: Applications of biotechnology to traditional fermented foods. National Academy Press, Washington, DC, pp 27–35
3. Leona P, Petra P, Tomá B (2013) Advanced fermentation processes. In: Engineering aspects of food biotechnology. CRC Press, Boca Raton, pp 89–110
4. Kim EK et al (2011) Fermented kimchi reduces body weight and improves metabolic parameters in overweight and obese patients. Nutr Res 31(6):436–443
5. Guzel-Seydim ZB et al (2011) Review: functional properties of kefir. Crit Rev Food Sci Nutr 51(3):261–268
6. Leroy F, De Vuyst L (2014) Fermented food in the context of a healthy diet: how to produce novel functional foods? Curr Opin Clin Nutr Metab Care 17(6):574–581
7. Marsh AJ et al (2014) Fermented beverages with health-promoting potential: past and future perspectives. Trends Food Sci Technol 38(2):113–124
8. Yaradoddi JS, Sulochana MB, Kontro Merja H, Parameshwar AB, Dayanand A (2020) The occurrence of potential and novel isolates of *Oceanobacillus* sp. JAS12 and *Salinicoccus* sp. JS20 recovered from west coast of Arabian Sea, India. Res J Biotech 15(9):133–140
9. Yaradoddi JS, Sulochana MB (2020) Screening and characterization of bioactive compounds produced by the moderate halophile Halobacillus sp. JS6. Res J Biotech 15(12):131–136
10. Riley C (2015) Top 10 food trends to know in 2016. www.wholefoodsmarket.com
11. Gupta S, Abu-Ghannam N (2012) Probiotic fermentation of plant based products: possibilities and opportunities. Crit Rev Food Sci Nutr 52(2):183–199

12. Hugenholtz J (2013) Traditional biotechnology for new foods and beverages. Curr Opin Biotechnol 24(2):155–159
13. Shellikeri A, Kaulgud V, Yaradoddi J, Ganachari S, Banapurmath N, Shettar A (2018) Development of neem based bioplastic for food packaging application. IOP Conf Ser: Mater Sci Eng 376:012052. https://doi.org/10.1088/1757899X/376/1/012052
14. Yaradoddi J, Shoba H, Banapurmath N, Hunashyal A, Sulochana M, Shettar A, Ganachari S (2018) Alternative and renewable bio-based and biodegradable plastics. In: Martínez LMT et al (eds) Handbook of ecomaterials. Springer International Publishing AG. https://doi.org/10.1007/978-3-319-68255-6_150
15. Campbell-Platt G (1994) Fermented foods—a world perspective. Food Res Int 27:253–257
16. Johannes TW, Zhao H (2006) Directed evolution of enzymes and biosynthetic pathways. Curr Opin Microbiol 9:261–267
17. Sanchez OJ, Cardona C (2008) Trends in biotechnological production of fuel ethanol from different feedstocks. Bioresour Technol 99(13):5270–5295
18. Shuler ML, Kargi F (2008) Bioprocess engineering, 2nd edn. Prentice-Hall, New York
19. Kourkoutas Y, Bekatorou A, Banat I et al (2004) Immobilization technologies and support materials suitable in alcohol beverages production: a review. Food Microbiol 21:377–397
20. Verbelen PJ, Schutter D, Delvaux F, Vestrepen K (2006) Immobilized yeast cell systems for continuous fermentation applications. Biotechnol Lett 28:1515–1525
21. Cliffe K (1988) Downstream processing. In: Scragg A (ed) Biotechnology for engineers. Ellis Horwood, Chichester, pp 302–321
22. Perry RH, Chilton C (1973) Chemical engineers' handbook, 5th edn. McGraw-Hill, New York
23. Belter PA, Cussler EL, Hu WS (1988) Bioseperations downstream processing for biotechnology. Wiley-Interscience, New York
24. Dutta R (2008) Downstream processing. In: Fundamentals of biochemical engineering. Ane Books, New Delhi
25. Üçüncü M (2005) Süt vemamülleri teknolojisi. Meta Basım, İzmir
26. Underkofler LA, Barton R, Rennert R (1958) Production of microbial enzymes and their applications. Appl Microbiol 6:212
27. Hutkins RW (2006) Microbiology and technology of fermented foods. Blackwell Publishing, Ames
28. Adams RD (1985) Vinegar. In: Wood B (ed) Microbiology of fermented foods. Elsevier, Philadelphia, pp 1–49
29. Battcock M, Azam-Ali S (1998) Fermented fruits and vegetables. A global perspective. Food and Agriculture Organization of the United Nations, Rome
30. Sugihara TF (1985) Microbiology of breadmaking. In: Wood B (ed) Microbiology of fermented foods, vol 1. Elsevier, New York, pp 249–261
31. Astuti M, Meliala A, Dalais F, Lwahlqvist M (2000) Tempe, a nutritious and healthy food from Indonesia. Asia Pac J Clin Nutr 9(4):322–332
32. Aurand LW, Woods AE, Wells R (1987) Enzymes. In: Food composition and analysis. Van Nostrand Reinhold, New York, pp 283–346
33. Berk Z (1976) Enzymes. In: Braverman's introduction to the biochemistry of foods. Elsevier, Amsterdam, pp 27–40
34. Eskin NAM (1990) Biochemistry of foods. Academic, San Diego, pp 467–527
35. Adler-Nissen J (1987) Newer uses of microbial enzymes in food processing. TIBTECH 5:170–174
36. Hasan F, Shah AA, Hameed A (2006) Industrial applications of microbial lipases. Enzym Microb Technol 39:235–251
37. Svendsen A (2000) Lipase protein engineering. Biochim Biophys Acta 1543:223–238
38. Han Z, Zhao Q, Wang H, Kang Y (2011) Characteristics of the secretory expression of pectate lyase A from Aspergillus nidulans in Escherichia coli. Afr J Microbiol Res 5(15):2155–2159
39. Pandey A, Poonam N, Soccol C et al (2000) Advances in microbial amylases. Biotechnol Appl Biochem 31:135–152

40. Bigelis R (1993) Carbohydrases. In: Nagodawithana T, Reed G (eds) Enzymes in food processing. Academic, San Diego, pp 121–147
41. Ghosh PK, Saxena R, Gupta R et al (1996) Microbial lipases: production and applications. Sci Prog 79:119–157
42. Aiyer P (2005) Amylases and their applications. Afr J Biotechnol 4:1525–1529
43. Jayani RS, Saxena S, Gupta R (2005) Microbial pectinolytic enzymes: a review. Process Biochem 40:2931–2944
44. Vihinen M, Mantsiila P (1989) Microbial and amylolytic enzymes. Crit Rev Biochem Mol Biol 24:329–418
45. Kashyap DR, Vohra PK, Chopra S, Tewari R (2001) Applications of pectinases in the commercial sector: a review. Bioresour Technol 77:215–227
46. Adler-Nissen J (1993) Proteases. In: Tucker G, Woods L (eds) Enzymes in food processing. Chapman and Hall, New York, pp 191–218
47. Rao MB, Tanksale A, Ghatge M, Deshpande V (1998) Molecular and biotechnological aspects of microbial proteases. Microbiol Mol Biol Rev 62(3):597–635
48. Jayachandra SY, Parameshwar AB, Mohan Reddy K, Sulochana MB (2012) Characterization of extracellular hydrolytic enzymes producing extremely halophilic bacterium *Virgibacillus* sp. JS5. World J Sci Technol 2(2):23–26

Chapter 14
Food Industry By-Products and Waste Management

Bharati S. Meti, Spoorthi R. Kulkarni, Shilpa K. Jigajinni, and Basavaraj Nainegali

1 Introduction: Various By-Products from the Food Processing Industry

In a world experiencing a growing population, the responsible and efficient utilization of raw materials has become increasingly vital. As the demand for food continues to rise, it generates greater amounts of waste and by-products. Consequently, this prompts us to consider how to tackle this pressing issue. There is a need to valorize these agricultural and food by-product wastes into food products as they serve as a cheap source of dietary fiber, protein, and bioactive compounds such as phenolic compounds, antioxidants, minerals, and vitamins.

The composition and functional attributes of these by-products exhibit substantial variation, and as a result, they exert a substantial influence on the properties of fortified products. Considerable research has focused on methods to enhance the value of food waste, encompassing strategies such as integrating them into diverse food items, separating them into distinct components, and extracting valuable compounds from the waste. Food industries worldwide generate a significant volume of food waste or by-products annually. Approximately 38% of these food wastes occur during the food processing stage. These food wastes originate from various sources, including animal-derived processing, such as by-products from bred animals like carcasses, hides, hoofs, heads, feathers, manure, offal, viscera, bones, fat, and meat trimmings, as well as blood. Seafood processing also contributes to food waste with items like skin, bones, oils, and blood. Additionally, dairy processing industries

B. S. Meti (✉) · S. R. Kulkarni · S. K. Jigajinni · B. Nainegali
Department of Biotechnology, Basaveshwar Engineering College, Bagalkote, Karnataka, India

J. S. Yaradoddi et al. (eds.), *Frontiers in Food Biotechnology*,
https://doi.org/10.1007/978-981-97-3261-6_14

Fig. 14.1 Use of industrial food by-products as bioactive compounds and functional foods

generate food waste in the form of whey, curd, and milk sludge resulting from the separation process. Vegetable-derived processing food waste comprises peelings, stems, seeds, shells, bran, and residual materials remaining after the extraction of oil, starch, juice, and sugars (Fig. 14.1). The development of functional foods involves the inclusion of specific compounds that offer health benefits. In today's world, the primary causes of mortality include cardiovascular diseases, diabetes, and cancer. These diseases could potentially be prevented by incorporating functional foods into one's diet. Many fruits and plants contain bioactive compounds, particularly phenolics and flavonoids, which can counter excessive free radical formation and lower the risk of heart attacks and cancer. Furthermore, increasing the consumption of dietary fiber's in one's diet can contribute to reducing plasma and low-density lipoprotein (LDL)-cholesterol levels, as well as decreasing the risk of diabetes and gastrointestinal disorders. Functional foods can be found in a variety of sources, including fruits, vegetables, marine products, meats, and dairy items [1]. Among all food categories, fruits, vegetables, roots, and tubers exhibit the highest rates of wastage. The integration process should involve both extracting valuable compounds and manufacturing other components using the residues generated from those extraction procedures. Furthermore, it's possible to treat the waste and by-products entirely, making them suitable as functional ingredients without further processing. To ensure the efficiency of the entire integrated system and prevent the oversaturation of a single market, it's essential to identify diverse market applications for these products. Food by-products produced in the course of manufacturing and processing have attracted considerable attention because of their potential use in technical, medical, and pharmaceutical contexts, in addition to their positive ecological and environmental effects [2–4]. During the fruit processing, a substantial quantity of by-product, known as pomace, is generated. There is a desire to find additional applications for pomace in food products to address sustainability concerns. However, its utilization is often restricted due to its unique physicochemical and techno-functional properties. The dietary fiber found in these by-products often possesses notable attributes, such as a strong ability to bind water and swell, as well as a high capacity for binding oil, all of which significantly impact the properties of the resulting food product [5].

2 By-Products of Cereals

Cereal grains have been a fundamental part of the human diet for centuries. The processing of these grains is a crucial aspect of food production, involving various intricate methods. Common cereal processing techniques encompass dry milling for grains like wheat and rye, pearling for rice, oats, and barley, wet milling for corn and wheat, and malting for barley, corn, and wheat (Fig. 14.2). Throughout the cereal processing journey, various by-products with distinct physical properties and chemical compositions are generated. As cereals are a vital source of carbohydrates, proteins, lipids, as well as essential vitamins such as those from the B-complex and vitamin E, as well as inorganic and trace elements, finding innovative ways to reuse and add value to these by-products is a significant challenge in advancing the sustainable development of the agricultural and food industry [6].

Cereal industry by-products, like bran and brewers' spent grain, are often not fully utilized and end up being used for animal feed, and fuel, or discarded as waste. Fermentation is an exciting biotechnological approach that holds promise in enhancing the nutritional and functional attributes of these by-products. It can enhance their nutrient absorption, digestibility, and even result in the production of bioactive compounds that may have health benefits. While incorporating milling by-products into food processing presents certain technological challenges, integrating the fiber and protein-rich components of cereal by-products into food formulations can enrich our diets with valuable substances. It's essential to develop technologies that enable the utilization of the entire by-product while eliminating undesirable

Fig. 14.2 Utilization of cereals by-products to a functional food

characteristics and improving its nutritional quality. This is crucial for creating a more sustainable and nutritious food production chain [7].

1. **Wheat Bran:** The outer layers of wheat, which consist of the outer and inner pericarp, seed coat, and nucellar epidermis, are commonly known as bran. In the conventional wheat roller milling process, the majority of the endosperm is separated and further processed into wheat flour. Consequently, what remains is bran, which includes the aleurone layer and remnants of the endosperm, and this constitutes a by-product of the milling process. There are various types of bran, including coarse bran or regular bran, coarse weatings or fine bran, fine weatings or middlings or shorts, and low-grade flour, often referred to as "red dog." These different types are distinguished based on their particle size and the amount of endosperm they contain. Wheat bran, a by-product of wheat milling, can undergo fermentation to improve its nutritional and functional characteristics. This fermentation process can enhance phytase activity, which, in turn, reduces the levels of phytic acid, a known antinutrient. Research has demonstrated that through fermentation, coupled with the application of enzymes, it is possible to effectively bioprocess wheat bran, reducing phytic acid content. Fermented wheat bran has found application in the production of bread, where it helps balance essential amino acids and boost the protein content of wheat-based products [7].
2. **Wheat Germ:** Wheat germ, extracted as a by-product during the milling of wheat, is a nutritional powerhouse. It serves as the main reservoir of vitamin E within the wheat kernel and is abundant in B-group vitamins, proteins, dietary fiber, and essential minerals. Wheat germ proteins contain a majority of the essential amino acids in concentrations exceeding those found in the standard egg protein pattern. One of the most notable and encouraging aspects of fermented wheat germ is its ability to exhibit cytotoxic or cell-killing activity against cancer cell lines. This promising attribute has led to the development of a commercially available wheat germ extract called Avemar. Avemar is produced by fermenting a water-soluble extract of wheat germ using Saccharomyces cerevisiae (a type of yeast), followed by a process of concentration and drying [8]. These characteristics primarily arise from the presence of two quinones, namely, 2-methoxy benzoquinone and 2,6-dimethoxybenzoquinone, which naturally occur in wheat germ but are initially in glycosylated and non-physiologically active forms. To unlock their potential, specific strains of lactic acid bacteria with elevated β-glucosidase activity were deliberately employed in the fermentation of wheat germ. These strains have the capacity to release the two quinones from their less active forms during the fermentation process [9]. Wheat germ is infrequently employed in food processing primarily because it has a limited shelf life. This short shelf life can be attributed to the high levels of unsaturated fatty acids, as well as the presence of hydrolytic and oxidative enzymes within the germ.

3. **Rye Bran:** Rye, a significant provider of dietary fiber in European Nordic nations, is frequently employed as whole grain flour in the production of cereal-based items. However, it's worth noting that rye bran, a by-product of traditional milling processes, can also serve as an ingredient to enhance the nutritional content of food products. The bran component contains an abundance of various bioactive compounds, such as phenolic acids, phytosterols, and tocopherols. Among these, alkylresorcinols and steryl ferulates have garnered attention for their potential to prevent cancer and provide antioxidant benefits. It's important to note that the levels of these bioactive compounds can be influenced by factors such as fermentation conditions and the type of bran used. Rye bran has demonstrated its effectiveness as a substrate for the production of polyunsaturated fatty acids and β-carotene through solid-state fermentation by fungi. This process involved the use of four selected strains of Mucor spp. known for their ability to synthesize γ-linolenic acid (GLA) and β-carotene. When compared to other grains like oat flakes, barley groats, and wheat bran, rye bran that underwent fermentation with Mucor circinelloides CCF-2617 exhibited the highest concentration of both compounds. Furthermore, the addition of brewers' spent grain further enhanced the content of these valuable compounds in the rye bran. The rye bran is particularly well-suited for in situ dextran production, achieving concentrations of 2–3% on dry matter [10, 11].
4. **Rice Germ and Bran:** Asian countries play a central role in rice production, contributing significantly to the daily energy intake of their local populations, accounting for about 50% of their dietary energy supply. Rice bran is rich in various components, including proteins, crude fiber, lipids, carbohydrates, oryzanols, phenolic compounds, polyunsaturated fatty acids, and monounsaturated fatty acids. The removal of germ and bran from commercial white rice is primarily due to the oils they contain being prone to rapid rancidity, which diminishes the rice's shelf life. The protein content in rice bran offers significant health benefits and holds great promise in the field of food processing. Researchers have recently been concentrating on harnessing the phytochemicals present in rice bran. Rice bran contains phytochemicals that have exhibited diverse health advantages. Among these phytochemicals, anthocyanin stands out as a potent antioxidant found in the rice bran of colored rice varieties. Additionally, anthocyanin monomers, dimers, and polymers such as apigenin, cyanidin rutinoside, cyanidin glucoside, epicatechins, eriodictyol, luteolin, isorhamnetins, hesperetin, hermneutics, peonidin glucoside, and tricin are also present in rice bran. The presence of plant sterols in rice bran could offer nutritional support in managing various health conditions, including alopecia, hypercholesterolemia (elevated cholesterol levels), diabetes (with a patent for this application), hepatitis C, herpes, allergies, persistent fatigue, tuberculosis, and breast cancer (specifically related to hair loss) [12].

2.1 Trends and Perspectives

The future bio-economy is focused on sustainable utilization of agricultural by-products and mitigating their environmental consequences. Cereal by-products are currently under investigation for extracting functional compounds across different sectors. Nevertheless, this may result in a rise in by-product production. To tackle this issue, our primary goal should be maximizing the efficient utilization of these by-products, particularly those suitable for human consumption, by reintroducing them into the food supply chain.

3 By-Products of Legumes

Legumes, also referred to as pulses, are members of the Fabaceae family. They are rich in protein content and possess a favorable amino acid profile, often with sufficient levels of lysine, while sulfur-containing amino acids and tryptophan are typically the amino acids that may be limited within this plant family. For centuries, legumes have been recognized for their excellent nutritional qualities. Regrettably, during the processing of legumes, between 5% and 25% of the production goes to waste, resulting in by-products that remain valuable sources of beneficial compounds, notably proteins, which can still find application in food and animal feed formulations. It is crucial to select the appropriate extraction method carefully to safeguard the nutritional integrity of these proteins, as extreme conditions of pH and/or temperature could potentially harm them. Legumes have certain anti-nutritional components like trypsin inhibitors, lectins, phytic acid, and tannins. These substances are typically deactivated through cooking [13]. However, when producing protein extracts for products not undergoing sufficient heat treatment, these factors must be taken into account. Legume by-products are gaining attention for their potential to recover high-quality proteins suitable for livestock feed and to develop functional food ingredients with improved properties like digestibility and bioactivity (Fig. 14.3).

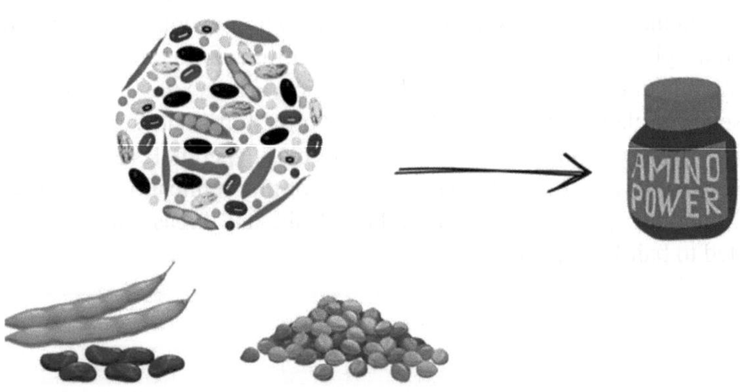

Fig. 14.3 Legumes and utilization of legume by-products

Fig. 14.4 Utilization of oilseed by-products

4 By-Products of Oilseeds

Oilseed by-products, including husks, hulls, cakes, and seeds, can be effectively repurposed rather than discarded, reducing environmental pollution. These by-products are versatile, finding applications in the food and pharmaceutical industries due to their rich content of biologically active ingredients. They can also be used to produce biofuels and biochar through chemical and thermochemical processes. For instance, the lignocellulosic components of olive pulp, a by-product of olive oil extraction, have been explored for medical uses like antimicrobial hydrogel production (Fig. 14.4).

Additionally, oilseed by-products can play roles in sustainable oil production, serve as fertilizers and soil enhancers, and be used as raw materials for soap and biodegradable surfactants [14, 15]. In essence, leveraging oilseed by-products offers a sustainable approach to waste reduction while generating value-added products. Utilizing by-products from soybean, sunflower, and peanut processing has yielded comparable carcass and meat attributes. Consequently, these by-products can be recommended as potential protein sources for lambs in feedlot systems without negatively affecting carcass quality or the fatty acid profile. Diets incorporating these by-products have demonstrated adequate fatty acid deposition, with a higher presence of essential fatty acids linked to improved meat composition and quality [16].

5 By-Products of Dairy

Dairy by-products/wastes have the potential to be transformed into valuable substances such as bioplastics, exopolysaccharides, biofuels, organic acids, bioactive peptides, single-cell proteins, and biosurfactants. The valorization of dairy waste through microbial treatment has been proposed as a sustainable and green method to generate biofertilizers, biofuels, power, and other biobased products. Various techniques have been developed to process dairy by-products, leading to economic and environmental advantages. These by-products have the potential to be utilized in the dairy industry for the production of a variety of dairy products (Fig. 14.5).

Skim milk is a secondary product obtained during cream production, known for its substantial non-fat solids and nutritional richness. In dairy facilities, it's commonly used for standardizing primary dairy product manufacturing or dehydrated into spray-dried form for preservation. Skim milk is not classified as a by-product when used in these ways or consumed as a liquid. It's considered a by-product only when it's either not economically viable or when it's used to derive secondary products like casein, co-precipitates, protein hydrolysates, and similar items.

Buttermilk is a by-product created during the butter-making process. Sweet cream buttermilk is similar to skim milk in terms of its overall chemical composition and is often blended with a significant portion of skim milk for subsequent spray drying or as an ingredient in dairy product manufacturing within dairy facilities. In contrast, Desi buttermilk has been a traditional and significant beverage in India, offering both nutritional and therapeutic benefits. It is also commonly used in the preparation of various dishes like kadhi, dhokla, and idli. Furthermore, some federations and private plants sell flavored and salted buttermilk in 200 mL pouches [17].

Whey is a thin, highly perishable, greenish-yellow liquid, and it represents the most significant by-product in the dairy industry, originating from processes involving cheese, casein, chhana, paneer, chakka, and co-precipitates. Whey contains approximately half of the total solids found in milk and serves as a valuable source of essential nutrients such as lactose, whey proteins, minerals, and vitamins. Although whey proteins are present in relatively small quantities, they possess a high protein efficiency ratio (3.6), biological value (104), and net protein utilization (95), making them one of the most nutritionally valuable proteins, second only to egg protein [17].

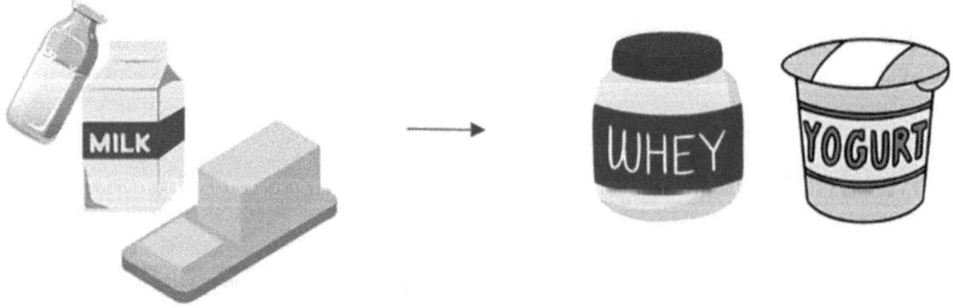

Fig. 14.5 Utilization of dairy by-products

6 By-Products of Fruit and Vegetable Processing

The fruit and vegetable processing industries generate substantial waste and by-products, including peels, seeds, pomace, and unused flesh, all of which contain valuable compounds (Fig. 14.6). These by-products are rich sources of antioxidants, dietary fiber, enzymes, pectin, organic acids, essential oils, and food additives. Leveraging these by-products offers cost-effective and natural access to these compounds. Furthermore, these by-products can serve as the foundation for natural food additives, colorants, proteins, flavorings, antimicrobials, and antioxidants.

They also provide a source of natural pigments, like anthocyanins, betalains, carotenoids, and chlorophylls, with potential applications in the food, pharmaceutical, and cosmeceutical sectors [18]. Through green extraction and processing techniques, the valorization of fruit and vegetable by-products can contribute to the creation of sustainable and environmentally friendly products. Approximately one-third of vegetables can go to waste during preparation, and interestingly; some parts of vegetables are intentionally discarded due to their undesirable taste or texture. This includes vegetable components like hulls, bagasse, and seeds, which are commonly disposed of during production. Additionally, for specific vegetables like broccoli, cauliflower, and pumpkin, the stems and leaves are typically not consumed and are discarded. The seed coat, which serves as a protective outer layer for seeds or beans, is another vegetable by-product. In most cases, seed coats are not eaten because of their texture and taste, with the exception of certain thin seed coats, such as those found on peanuts [19].

Fruit peels or rinds are typically discarded during consumption or manufacturing and form one of the common fruit by-products. These peels serve as the outer protective layer for fruits against environmental factors. While some fruits have thick and rugged skins like pomegranates, others, like mangoes, have thinner peels. Most fruit peels are not consumed due to their coarse texture and bitter taste. On the other hand, certain fruit by-products, like seeds, are often eaten along with the fruit itself, as is the case with watermelon and kiwi. However, larger seeds are usually discarded after consuming the flesh. Fruit seeds are a valuable source of nutrients and have been utilized in the development of functional foods. Fruit pomace represents the solid residue left behind after extracting juice from fruits and may include seeds, skin, pulp, and stems, depending on the fruit type. For instance, apple pomace constitutes approximately 25% of apples, and the waste generated from the apple juice and cider industry can yield substantial amounts of pomace [20].

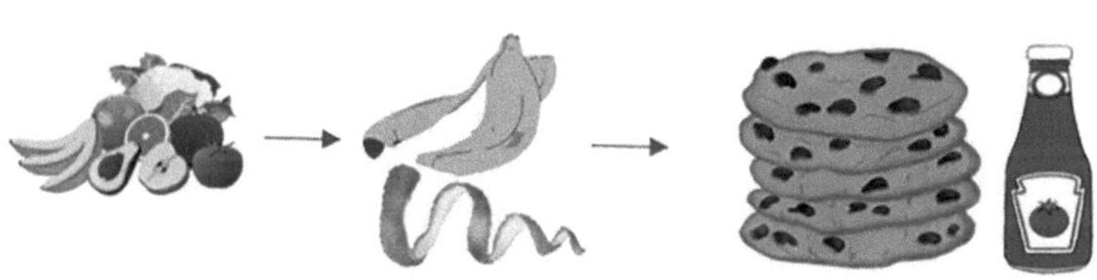

Fig. 14.6 Utilization of fruits and vegetables by-products

Fruit and vegetable by-products are being used to boost the nutritional value of various foods like bread, cookies, noodles, and dairy products. While they enhance nutrition, they can affect the texture and appearance of these foods. Still, with careful incorporation, they offer a balance between health benefits and taste. Additionally, these by-products can be used as natural colorants and fat replacers in dairy products, and they've even improved the fiber content of traditional dishes and jams.

7 By-Products of Meat and Fish Processing Industry

Significant quantities of protein-rich by-products are generated during the processing of meat and fish. These by-products contain valuable proteins and nutrients that can be repurposed for higher-value products, including human food, pet food, and animal feed. Various techniques have been developed to create appealing food products from fish meat by-products, offering fresh and delicious options (Fig. 14.7).

Fish co-products are intended for human consumption, such as the liver, while fish by-products include parts with valuable components that aren't suitable for human consumption. In the meat processing industry, both by-products and co-products can be efficiently managed for eco-friendly disposal. They are valuable for functions like gelation, emulsification, and the production of bioactive peptides with health benefits. The processing sector generates a substantial volume of by-products, which can be upgraded to yield food, feed, or valuable biomolecules. Many regions worldwide recover various fish parts for human consumption, and there is a growing interest in utilizing underutilized fish species in these products [21].

Gelatine is relatively rich in lysine and methionine, two amino acids that are lacking in cereal proteins. Gelatine has widespread applications in the food industry and is also used in the formulation of certain industrial products. It can be obtained from the skin and bones of fish and meat. Fish insulin, unlike human insulin, is more stable because it is not broken down by the protein-splitting enzymes in the pancreas. It is used to treat diabetes mellitus in humans by regulating blood sugar levels [21]. Fish albumin is a product that shares similar physical and chemical properties with egg albumin. It can be derived from protein-rich remnants of fish scrap or fish waste. Two types of fish albumin are manufactured: technical grade and food and

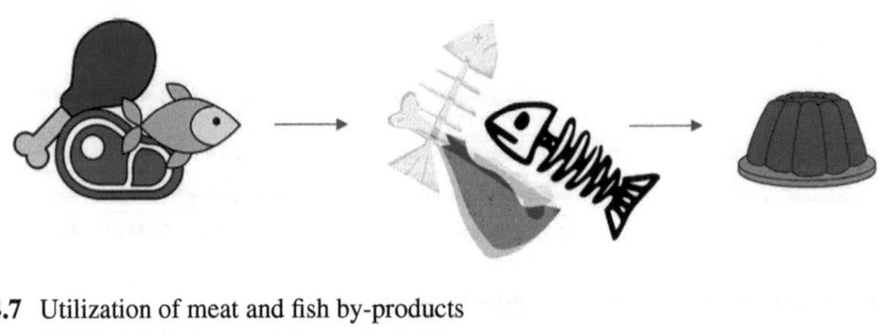

Fig. 14.7 Utilization of meat and fish by-products

pharmaceutical grade. Fish albumin is extensively employed in the food and pharmaceutical industries as an agent for whipping, suspending, or stabilizing. In the food sector, food-grade albumin is used as an additive in various products such as ice cream, soup powder, puddings, confectionery, bakery items, mayonnaise, and custard powder, among others. Edible by-products from meat are rich in essential nutrients and have applications in medicine due to their unique content of amino acids, hormones, minerals, vitamins, and fatty acids. These by-products, like blood and various organs such as lung, kidney, brains, spleen, and tripe, often have higher moisture levels compared to meat. Additionally, certain organ meats like liver and kidney contain more carbohydrates than other meat components [22].

8 Uses of By-Products of Agro-Based Industries in Various Sectors

By-products from agro-based industries have diverse applications across various sectors. They can serve as a nutrient source for insect rearing, promoting insects as a safe protein supply and facilitating nutrient bioconversion [23]. Agro-industrial fruit by-products can be incorporated into baked goods to enhance their fiber content, bioactive properties, and antioxidant capacity, while preserving good sensory quality. Solid-state fermentation is an effective method to convert agro-waste into valuable bio products. By-products from horticultural goods find utility as animal feed, additives for nutraceuticals and functional food, and organic residues for waste valorization. Effective management of agro-industrial waste is vital for maximizing benefits and minimizing environmental impact. The efficient conversion of waste into value-added products and safe disposal methods are crucial for sustainable production [24, 25].

9 Laws and Regulations for Waste Management in Food Processing Industries

Food processing industries are subject to a range of waste management regulations with the goal of preventing or minimizing waste production. Food waste, encompassing both solid and liquid waste, poses a significant challenge in this sector. The composition of waste generated by food processing facilities is varied and depends on the product and production methods. Recycling and repurposing these residues offer opportunities to transform by-products into useful resources instead of disposing of them in the environment. Furthermore, there are various formal and informal mechanisms to address food waste, including laws, social norms, markets, and infrastructure. Initiatives such as composting regulations, changes in date labeling, food waste reduction targets, tax incentives for food donations, and food waste apps exemplify the multifaceted and comprehensive approach to addressing food waste. In summary, there is a pressing need for sustainable waste management practices in the food industry to curtail waste production and its ecological impact [26].

10 Food Industry Wastes Treatment Methods

The food industry generates substantial waste from various sources, including fruit and vegetable industries, food processing facilities, kitchens, restaurants, and cafeterias. This waste can be harnessed in multiple ways to promote sustainability and reduce environmental impact. For instance, waste from fruits and vegetables contains valuable compounds like bioactive substances with antioxidant, antimicrobial, and antibrowning properties, which can enhance food quality [27]. Moreover, food waste can serve as raw material for bioenergy production using methods like anaerobic fermentation and transesterification, addressing energy challenges and contributing to a circular economy [28]. Additionally, food waste from agro-food production, particularly plant food waste, is rich in dietary fibers capable of binding toxic heavy metals, making them suitable for adsorption and mitigating health risks. In essence, utilizing food industry waste plays a vital role in achieving sustainability and a zero-waste global approach, with various valorization methods and biorefinery strategies under exploration for this purpose.

11 Waste Treatment Methods for Cereals

Cereal processing waste can be treated using various methods, with physico-chemical and biological approaches being the most common. Physico-chemical treatments are primarily used for pretreating cereal processing waste to reduce sludge formation during subsequent biological treatments. Biological treatments are favored for their effectiveness in removing pollutants and cost-efficiency compared to physico-chemical methods [29].

These treatments also offer the opportunity to convert cereal waste into valuable products like biohydrogen, bioethanol, biogas, biocoal, industrial enzymes, biomass, biofertilizer, proteins, organic acids, and polysaccharides. Another waste treatment method involves breaking down the waste using a decomposition tank, saturated steam, and continuous stirring. For removing heavy metals from cereals, techniques such as low-pressure vacuum washing, high-pressure washing, acid-alcohol solutions, settling separation, and recycling residues containing heavy metals are employed [30, 31].

Converting wheat waste into biomass or biogas is seen as a promising solution to address energy issues and reduce environmental pollution. Waste treatment techniques like composting, pyrolysis, gasification, and combustion are utilized for corn and rice waste, with potential applications including the production of fertilizers, biomass, and biogas/biofuel.

12 Waste Treatment Methods for Fruits and Vegetables

Waste treatment methods for fruits and vegetables include various approaches. One method involves using a refrigerating and freezing device to facilitate the removal of peels from vegetables and fruits, simplifying the process for users and avoiding nutrient loss. Another method involves processing and recovering vegetable and fruit waste by smashing it, dewatering and filtering it, and then drying and collecting the residues while using the juice for various purposes [32]. In the fruit and vegetable processing industry, waste and wastewater treatment practices are implemented, including biological wastewater treatment plants and valorization of waste by-products for feed or energy production. Additionally, a method for quickly returning greenhouse waste fruits, vegetables, and straw to fields involves smashing the plants, adding sterilized farm manure, performing rotary cultivation, and using a decomposition agent for high-temperature airtight fermentation and decomposition.

13 Waste Treatment Methods for Meat and Fish Industry

Waste treatment methods for the meat and fish industry encompass a range of techniques, offering solutions for both solid waste and wastewater management.

13.1 Preliminary Treatment

The initial step involves the physical removal of solids from wastewater effluent. This is achieved using various methods like screens, air flotation, and physicochemical treatments. These techniques help separate solid particles from the liquid effluent, facilitating further treatment.

13.2 Secondary Treatment

Biological treatment systems are employed in the secondary treatment phase. Microorganisms are used to break down and consume the organic matter present in the effluent, converting it into new microbial cells and less harmful substances. This process aids in reducing the pollutant load in the wastewater [33].

13.3 Anaerobic Digestion

Anaerobic digestion is a method that rapidly reduces organic materials through a two-stage fermentation process. One of the valuable by-products of this process is biogas, which can be used as an energy source. Anaerobic digestion is particularly effective in breaking down complex organic compounds.

13.4 Aerobic Treatment

Aerobic treatments utilize oxygen to facilitate the decomposition of organic matter. Methods like the activated sludge process, trickling filters, lagoons, evaporation, and irrigation are employed to further treat wastewater. These processes encourage the growth of beneficial aerobic microorganisms that consume remaining organic materials [34].

13.5 Combined Anaerobic and Aerobic Treatment

A combination of anaerobic and aerobic treatment methods is often considered the most suitable approach for treating waste in meat industries. This combination maximizes the efficiency of organic matter removal and reduces the environmental impact of effluent discharge. Once treatment is completed, the treated effluent can be safely discharged into the environment, provided that all contaminants have been effectively removed. The remaining sludge can be managed through various means, including landfill disposal, composting to produce organic fertilizers, or soil conditioning. Moreover, there are emerging opportunities to utilize animal-based waste streams as valuable resources through microbial and enzymatic processes. These innovative approaches can potentially transform waste into products with economic and environmental benefits, such as bio-based materials and energy [35].

14 Waste Treatment Methods for Dairy Processing

Dairy processing waste can be effectively treated through a range of methods. Physico-chemical treatment focuses on altering milk fat and protein colloids but can be costly with limited Chemical Oxygen Demand (COD) removal. Biological treatments employ microorganisms for organic matter removal, albeit with potential sludge formation. Combining aerobic and anaerobic processes aids in meeting effluent discharge limits [36]. Biotechnological processes are innovative, yielding valuable by-products like whey-derived products, biofuels, and enzymes.

Baromembrane processes concentrate secondary dairy materials, while ABR + UASB combined processes ensure stable treatment. An alkaline pH adjustment, reverse osmosis, and urea separation process offer an efficient means of dairy wastewater purification, demonstrating diverse approaches to address dairy waste treatment and resource recovery in the industry [37].

15 Waste Treatment Methods for the Brewery Industry

The brewery industry employs a variety of advanced waste treatment methods to manage wastewater and solid residues effectively, prioritizing both environmental sustainability and cost efficiency. Membrane filtration techniques, including microfiltration, ultrafiltration, nanofiltration, and reverse osmosis (RO), are widely adopted for treating brewery effluents due to their compact design, eco-friendliness, automation capabilities, and reduced chemical usage. In addressing wastewater with high suspended solid concentrations, aerobic/anaerobic membrane bioreactors are employed, integrating the advantages of aerobic/anaerobic processes with membrane filtration to enhance treatment efficiency. A novel approach involves hydrothermal carbonization (HTC), which transforms brewery residues into hydrochars suitable for agricultural use. These HTC materials are nutrient-enriched and serve as renewable sources of nitrogen and essential fertilizer microelements [38]. Additionally, a cyanobacterial-bacterial consortium demonstrates promise for bioremediation of brewery wastewater, exhibiting high pollutant removal efficiency and the potential for biomass production for bioethanol. These diverse methods underscore the brewery industry's commitment to sustainable waste management and resource recovery [39].

16 Industrial Wastewater Treatment

Preliminary experiments play a crucial role in the field of waste treatment as they serve as a foundational step in identifying the most suitable methods and parameters for dealing with different types of waste materials. These experiments are essential for optimizing the waste treatment process, ultimately leading to cost reduction and more efficient resource utilization. For example, one waste liquid treatment method involves injecting hydrogen peroxide water into chemical cleaning waste liquid, which initiates an oxidation process. Following this, the pH of the waste liquid is adjusted, and a flocculant is introduced to facilitate the sedimentation of heavy metal components, effectively separating them from the liquid. In another approach, waste reduction is achieved through a method that entails cropping materials to specific lengths and employing lap joint techniques and attachments. This results in the creation of qualified materials, reducing overall waste generation in the process.

Efforts to improve waste treatment equipment have also been made, with a focus on enhancing stirring performance and reducing processing time. This is particularly valuable in the treatment of organic waste, where efficient drying and composting are essential. Furthermore, waste treatment systems have been designed to operate efficiently, with a specific emphasis on energy conservation during initial activation. These systems make use of supercharging means to enhance their overall performance. Lastly, a comprehensive waste treatment system has been developed to address complex and hazardous waste streams. This system incorporates various stages, including pretreatment, plasma sintering melting treatment, and tail gas treatment, to ensure the safe and effective management of these challenging waste materials. These innovative approaches highlight the ongoing efforts to advance waste treatment processes, making them more efficient and environmentally friendly across a range of applications and industries [40].

17 Final Treatment: Zero-Discharge and Zero-Emission System

The provided details a comprehensive system designed for the final treatment of wastewater, emphasizing a zero-discharge and zero-emission approach. This system incorporates various treatment components and tanks to enable real-time recycling and complete elimination of wastewater and waste liquid. It is versatile in its capability to address a wide range of flue gas constituents and harmful substances, offering notable advantages such as reliability, energy efficiency, and cost-effectiveness. The system's components include a grid decontamination machine, fiber recovery system, sedimentation tanks, acid-base wastewater treatment devices, heavy metal wastewater treatment devices, and regeneration liquid treatment devices. To further enhance its performance, the system utilizes integrated chemical softening water purifiers, ultra-filtration membrane filters, and RO membrane filters, ensuring nearly zero wastewater emissions and highly efficient wastewater recycling.

References

1. Helkar PB, Sahoo A, Patil N (2016) Review: food industry by-products used as a functional food ingredients. Int J Waste Resour 6:1–6. https://doi.org/10.4172/2252-5211.1000248
2. Socaci SA, Rugină DO, Diaconeasa ZM, Pop OL, Fărcaş AC, Păucean A, Pintea A (2017) Chapter 1 antioxidant compounds recovered from food wastes in book functional food—improve health through adequate food. InTechOpen, pp 1–21. https://doi.org/10.5772/intechopen.69124
3. Difonzo G, de Gennaro G, Pasqualone A, Caponio F (2022) Potential use of plant-based by-products and waste to improve the quality of gluten-free foods. J Sci Food Agric 102(6):2199–2211. https://doi.org/10.1002/jsfa.11702

4. Galali Y, Omar ZA, Sajadi SM (2020) Biologically active components in by-products of food processing. Food Sci Nutr 8(7):3004–3022. https://doi.org/10.1002/fsn3.1665
5. Struck S, Rohm H (2020) Chapter 1—Fruit processing by-products as food ingredients. In: Galanakis CM (ed) Valorization of fruit processing by-products. Academic Press, pp 1–16. ISBN: 9780128171066. https://doi.org/10.1016/B978-0-12-817106-6.00001-0
6. Papageorgiou M, Skendi A (2018) 1—Introduction to cereal processing and by-products. In: Galanakis CM (ed) Woodhead Publishing series in food science, technology and nutrition,sustainable recovery and reutilization of cereal processing by-products. Woodhead Publishing, pp 1–25. ISBN: 9780081021620. https://doi.org/10.1016/B978-0-08-102162-0.00001-0
7. Verni M, Rizzello CG, Coda R (2019) Fermentation biotechnology applied to cereal industry by-products: nutritional and functional insights. Front Nutr 2019(6):42. https://doi.org/10.3389/fnut.2019.00042
8. Boros LG, Nichelatti M, Shoenfeld Y (2005) Fermented wheat germ extract (Avemar) in the treatment of cancer and autoimmune diseases. Ann N Y Acad Sci 1051:529–542. https://doi.org/10.1196/annals.1361.097
9. Rizzello CG, Mueller T, Coda R, Reipsch F, Nionelli L, Curiel JA, Gobbetti M (2013) Synthesis of 2-methoxy benzoquinone and 2,6-dimethoxybenzoquinone by selected lactic acid bacteria during sourdough fermentation of wheat germ. Microb Cell Factories 12(1):105. https://doi.org/10.1186/1475-2859-12-105
10. Certík M, Adamechová Z, Guothová L (2013) Simultaneous enrichment of cereals with polyunsaturated fatty acids and pigments by fungal solid-state fermentations. J Biotechnol 168(2):130–134. https://doi.org/10.1016/j.jbiotec.2013.03.016
11. Kajala I, Mäkelä J, Coda R, Shukla S, Shi Q, Maina NH (2016) Rye bran as fermentation matrix boosts in situ dextran production by Weissella confuse compared to wheat bran. Appl Microbiol Biotechnol 100(8):3499–3510. https://doi.org/10.1007/s00253-015-7189-6
12. Patel S (2012) Cereal bran: the next super food with significant antioxidant and anticancer potential. Med J Nutr Metab 5(2):91–104. https://doi.org/10.1007/s12349-012-0091-1
13. Prandi B, Zurlini C, Maria CI, Cutroneo S, Di Massimo M, Bondi M, Brutti A, Sforza S, Tedeschi T (2021) Targeting the nutritional value of proteins from legumes by-products through mild extraction technologies. Front Nutr 8:695793. https://doi.org/10.3389/fnut.2021.695793
14. Thangaraju S, Pulivarthi MK, Natarajan V (2020) Waste from oil-seed industry: a sustainable approach. In: Thakur M, Modi VK, Khedkar R, Singh K (eds) Sustainable food waste management. Springer, Singapore, pp 177–190. https://doi.org/10.1007/978-981-15-8967-6_10
15. Sut D, Chutia RS, Bordoloi N, Narzari R, Kataki R (2016) Complete utilization of non-edible oil seeds of Cascabela thevetia through a cascade of approaches for biofuel and by-products. Bioresour Technol 2016(213):111–120. https://doi.org/10.1016/j.biortech.2016.02.066
16. Santos VC, Ezequiel JMB, da Silva Morgado E, de Sousa Júnior SC (2013) Carcass and meat traits of lambs fed by-products from the processing of oil seeds. Acta Sci Anim Sci 35(4):387–394. https://doi.org/10.4025/ACTASCIANIMSCI.V35I4.20403
17. Rafiq SM, Rafiq SI (2019) Milk by-products utilization. In: Ibrahim SA, Zimmerman T, Gyawali R (eds) Current issues and challenges in the dairy industry. InTechOpen. https://doi.org/10.5772/intechopen.85533
18. Dubey P (2021) An overview of fruit by-products valorization: a step towards sustainable utilization. Indian J Pure Appl Biosci 9(1):46–55. https://doi.org/10.18782/2582-2845.8565
19. Sharma M, Usmani Z, Gupta VK, Bhat R (2021) Valorization of fruits and vegetable wastes and by-products to produce natural pigments. Crit Rev Biotechnol 41(4):535–563. https://doi.org/10.1080/07388551.2021.1873240
20. Lau KQ, Sabran MR, Shafie SR (2021) Utilization of vegetable and fruit by-products as functional ingredient and food. Front Nutr 8:661693. https://doi.org/10.3389/fnut.2021.661693
21. Aditi M, Varsha L (2020) Fish processing: product and by-product, processing and marketing. Int J Fish Aquat Stud 8(1):74–77

22. Ayathilakan K, Sultana K, Radhakrishna K, Bawa AS (2012) Utilization of byproducts and waste materials from meat, poultry and fish processing industries: a review. J Food Sci Technol 49(3):278–293. https://doi.org/10.1007/s13197-011-0290-7
23. Bordiean A, Krzyżaniak M, Stolarski MJ (2022) Bioconversion potential of agro-industrial byproducts by Tenebrio molitor—long-term results. Insects 13(9):810–810. https://doi.org/10.3390/insects13090810
24. Patel A, Temgire S, Borah A (2021) Agro-industrial waste as source of bioactive compounds and their utilization: a review. Pharma Innov J 10(5):192–196. https://doi.org/10.22271/TPI.2021.V10.I5C.6197
25. Singh R, Das R, Sangwan S, Rohatgi B, Khanam R, Peera SK, Das S, Lyngdoh YA, Langyan S, Shukla A, Shrivastava M, Misra S (2021) Utilisation of agro-industrial waste for sustainable green production: a review. Environ Sustain 4:619–636. https://doi.org/10.1007/s42398-021-00200-x
26. Poovazhahi A, Thakur M (2020) Environmental standards & regulations for waste management in food industries. In: Thakur M, Modi VK, Khedkar R, Singh K (eds) Sustainable food waste management. Springer, Singapore, pp 21–39. https://doi.org/10.1007/978-981-15-8967-6_2
27. Aqilah NMN, Rovina K, Felicia WXL, Vonnie JM (2023) A review on the potential bioactive components in fruits and vegetable wastes as value-added products in the food industry. Molecules 28(6):2631. https://doi.org/10.3390/molecules28062631
28. Rajkumar R, Kurinjimalar C (2022) Chapter 7—Food wastes/residues: valuable source of energy in circular economy. In: Sahay S (ed) Handbook of biofuels. Academic Press, pp 147–163. ISBN: 9780128228104. https://doi.org/10.1016/B978-0-12-822810-4.00007-5
29. Hassan G, Shabbir MA, Ahmad F, Pasha I, Aslam N, Ahmad T, Rehman A, Manzoor MF, Inam-Ur-Raheem M, Aadil RM (2021) Cereal processing waste, an environmental impact and value addition perspectives: a comprehensive treatise. Food Chem 363:130352. https://doi.org/10.1016/j.foodchem.2021.130352
30. Arvanitoyannis IS, Tserkezou P (2008) Wheat, barley and oat waste: a comparative and critical presentation of methods and potential uses of treated waste. Int J Food Sci Technol 43(4):694–725. https://doi.org/10.1111/j.1365-2621.2006.01510.x
31. Arvanitoyannis I, Tserkezou P (2008) Corn and rice waste: a comparative and critical presentation of methods and current and potential uses of treated waste. Int J Food Sci Technol 43(6):958–988. https://doi.org/10.1111/j.1365-2621.2007.01545.x
32. Zhu Y, Luan Y, Zhao Y, Liu J, Duan Z, Ruan R (2023) Current technologies and uses for fruit and vegetable wastes in a sustainable system: a review. Food Secur 12(10):1949. https://doi.org/10.3390/foods12101949
33. Irshad A, Sureshkumar S, Raghunath BV, Rajarajan G, Mahesh Kumar G (2016) Treatment of waste water from meat industry. In: Prashanthi M, Sundaram R (eds) Integrated waste management in India. Environmental science and engineering. Springer Cham, pp 251–263. https://doi.org/10.1007/978-3-319-27228-3_23
34. Arvanitoyannis I, Ladas D (2008) Meat waste treatment methods and potential uses. Int J Food Sci Technol 43(3):543–559. https://doi.org/10.1111/J.1365-2621.2006.01492.X
35. Kosseva MR (2020) Assessment and recuperation of commodities. In: Kosseva MR, Webb C (eds) Food industry wastes, 2nd edn. Academic Press, pp 67–85. ISBN: 9780128171219
36. Al-Tayawi AN, Sisay EJ, Beszédes S, Kertész S (2023) Wastewater treatment in the dairy industry from classical treatment to promising technologies: an overview. PRO 11(7):2133. https://doi.org/10.3390/pr11072133
37. Ahmad T, Aadil RM, Ahmed H, Rahman UU, Soares BCV, de Souza SLQ, Pimentel TC, Scudino H, Guimarães JT, Esmerino EA, Freitas MQ, Almada RB, Vendramel SMR, Silva MC, Cruz AG (2019) Treatment and utilization of dairy industrial waste: a review. Trends Food Sci Technol 88:361–372. https://doi.org/10.1016/J.TIFS.2019.04.003
38. Meshksar M, Roostaee T, Rahimpour MR (2020) Chapter 10—Membrane technology for brewery wastewater treatment. In: Basile A, Comite A (eds) Current trends and future developments on (bio-) membranes. Elsevier, pp 289–303

39. Skrzypczak D, Mironiuk M, Witek-Krowiak A et al (2023) Innovative fertilizers and soil amendments based on hydrochar from brewery waste. Clean Techn Environ Policy 26:1–17. https://doi.org/10.1007/s10098-023-02557-5
40. Gehrmann H-J, Hiebel M, Simon F-G (2017) Methods for the evaluation of waste treatment processes. Hindawi J Eng 3:1–13. https://doi.org/10.1155/2017/3567865

Chapter 15
Food Additives and Preservatives

Harshitha T., Akshay H. Dasalkar, Parinitha A., and Sudheer Kumar Yannam

1 Introduction

Food biotechnology rapidly evolves and transforms how we produce, process, and consume food. At the heart of this transformation lies a profound exploration of food additives and preservatives, which are essential components that underline the intricate relationship between innovation, safety, and culinary excellence. This chapter examines how biotechnology can satisfy consumers' demands for high-quality and safe food using biotech-based additives and preservatives. A food additive affects the characteristics of food. They are used for producing, manufacturing, packing, processing, preparing, treating, packaging, transporting, or holding food. Food additives have various functions, including improving texture, color, flavor, shelf life, and nutritional content. The combination of traditional wisdom and modern biotechnology expands possibilities. Preservation has been a long-standing practice, but recent advancements in biotechnology have sparked a revival. The need for food security, sustainability, and waste reduction has led to the exploration of preservatives that maintain food integrity while meeting modern demands. This exploration covers the evolution of food preservation from ancient times to present-day laboratories. Food engineers play a vital role in creating a trustworthy and appealing culinary environment. This chapter explores the investigation of food additives and preservatives. It includes the developments in biopreservation and personalized additives and delves into the interplay of science, tradition, and progress that shapes our culinary experiences.

Harshitha T. · A. H. Dasalkar · Parinitha A. · S. K. Yannam (✉)
Department of Traditional Foods and Applied Nutrition, CSIR-Central Food Technological Research Institute, Mysuru, India

Academy of Scientific and Innovative Research (AcSIR), Ghaziabad, India
e-mail: syannam@cftri.res.in

J. S. Yaradoddi et al. (eds.), *Frontiers in Food Biotechnology*,
https://doi.org/10.1007/978-981-97-3261-6_15

2 Fundamentals of Food Additives

Globalization has led to greater availability and diversity in food consumption. Consumers enjoy food for its aroma, flavor, taste, texture, and other sensory attributes. Food additives enhance food's appeal to consumers and have become increasingly prominent in food production and consumption. As processed foods become increasingly common, food additives are also on the rise. Consumers want natural and healthy additives, such as fat replacers, sweeteners, bulking agents, stabilizers, and natural flavors. Blends of additives will be the industry's future, with synergistic effects making them even more effective while using less. The industry's driving forces are health, nutrition, naturalness, quality, safety, convenience, and price. People prefer natural ingredients that cut calories, fat, cholesterol, and sodium in their food. Fortifying food with amino acids, trace minerals, and vitamins is vital. Natural additives such as gelatin, pectin, and vitamins are preferred over chemicals such as potassium benzoate and butylated hydroxyanisole. To stay ahead of the competition, food additive suppliers and manufacturers create their blends in secret to keep their edge. The food additives industry faces more regulation, research, legal expenses, and lengthy approval processes for new products that have to be approved by different regulators. The platform providing one stop solution shall be an added advantage for the newer developments in the food additive sector.

Nevertheless, convenience, health, and flavor are what consumers want, and additives help with that. Specialized additives make a preferable choice for consumers, as these are better food and last longer on shelves. Low-fat, low-calorie, and vitamin-and mineral-fortified options bring in more consumers for food processors. Expect more additives in processed foods, especially in regions where people eat ethnic foods and pursue vegetarianism.

2.1 Definition and Classification of Food Additives

A food additive is a substance intentionally added to improve the quality and safety of food. It is not typically consumed or used as an ingredient and can affect the food's characteristics throughout various stages of production and distribution. It is important to note that contaminants and nutritional substances are not food additives.

Classification

Classification Based on Their Function Includes [1]

- ***Preservatives***, e.g., antioxidants, pH control agents, thickeners, stabilizers, sugars, salt, essential oils
- ***Technological enhancers/process improvers***, e.g., anticaking agents, emulsifiers, enzymes, pH control agents, leavening agents, processing aids, and sugars
- ***Sensory enhancers/appeal modifiers***, e.g., anticaking agents, colors, antioxidants, emulsifiers, flavors, enzymes, humectants, sweeteners, leavening agents, thickeners and stabilizers
- ***Nutrition enhancers***, e.g., antioxidants, emulsifiers, fat substitutes, sweeteners, thickeners, stabilizers, vitamins and minerals

Classification Based on Determined Function [2]

Direct food additives are a group of substances that serve determined functions in food. These include inorganic chemicals such as sulfites, phosphates, and calcium chloride; synthetic organic chemicals such as dyes, aroma chemicals, benzoates, and vitamin A; extractants and derivatives of natural sources such as essential oils, pectin, and vitamin E; and fermentation-derived products including citric acid, enzymes, and xanthan gum.

Indirect food additives are a group of substances that do not have a stated function in food. These include components of adhesives such as calcium ethyl acetoacetate 1,4-butanediol modified with adipic acid; components of coatings including acrylate ester copolymer coatings and polyvinyl fluoride resins; components of paper and paperboard such as slimicides, sodium nitrate/urea complex, and alkyl ketone dimers; basic components of single and repeated-use food contact surfaces such as cellophane, ethylene acrylic acid copolymers, isobutylene copolymers, and nylon resins; components of articles intended for repeated use such as ultrafiltration membranes and textiles and textile fibers; compounds controlling the growth of microorganisms such as sanitizing solutions; antioxidants and stabilizers such as octyl tin stabilizers in vinyl chloride plastics; and certain adjuvants and production aids including animal glue, hydrogenated castor oil, synthetic fatty alcohols, and petrolatum.

2.2 *Significance of Food Additives and Preservatives*

Food additives play a crucial role in enabling the production and global distribution of various products, despite consumer concerns. They also help to maintain quality and safety by enhancing "designed" products rather than using organic modifications. The use of additives can preserve both the quality and appearance of food.

Table 15.1 Importance of food additives and preservatives

• Reduces the risk of foodborne illness by inhibiting the growth of pathogenic microbes, **e.g.**, antimicrobials
• Inhibits growth of microbes causing spoilage of food and extends shelf life, **e.g.**, benzoates, sulfites
• Facilitates transportation and distribution of foods in various regions and helps to address food scarcity **e.g.**, preservatives
• Contributes to food quality by maintaining texture, appearance, color and flavor throughout storage and transportation, **e.g.**, humactants, stabilizers, colorants, flavorings
• Improves sensory appeal and overall acceptability of the product, **e.g.**, colorants, flavorants, taste improver
• Retain the nutritional content by reducing the rate of biochemical reactions, **e.g.**, antioxidants
• Enhance the nutritive value of foods and promote consumer health[a], **e.g.**, vitamins, minerals and amino acids
• Enables new product developments with convenience, unique texture, flavor, and nutritional profile to cater to consumer demands, **e.g.**, preservatives, stabilizers, thickners, flavorants, nutrients[a]

[a]CODEX definition of food additive excludes food fortifying agents

Table 15.1 details the importance of food additives in food processing and preservation. Food additives can be a safe and effective solution to combat world food scarcity as long as their safety and food-grade status are proven. However, excessive use of food additives is a topic for criticizing processed food products.

Production of Food Additives

Food additive manufacturing protocols involve various methods and technologies to create substances that enhance food products' flavor, texture, appearance, and shelf life. These additives can be natural or synthetic, and their production methods differ depending on the type of additive. The process begins with sourcing raw materials, either natural or synthetic, followed by extraction, chemical synthesis, or fermentation, depending on the nature of the additive. Regardless of the source, additives require purification and refinement to meet strict quality standards. Quality control measures are implemented throughout production to ensure that the additives meet safety and quality standards. Once approved, the additives are packaged and distributed to food processors and manufacturers. Adequate documentation and labeling are essential for tracking additive use in food products and informing consumers of their presence. Ongoing research and development efforts are vital to improve existing additives, develop new ones, and explore innovative technologies in food additive production. Food safety and consumer health assurance demand a highly regulated process requiring stringent quality control, safety assessments, and adherence to legal requirements for producing food additives in various countries and regions.

The process of manufacturing food additives necessitates the implementation of stringent sanitary conditions and the attainment of high levels of purity. The production of chemical additives requires substantial capital investment and the establishment of specialized plant units that meet food-grade quality standards. Chemical methods come with both advantages and disadvantages. However, the disadvantages of using toxic and costly catalysts, high processing temperatures and pressures, high energy consumption, the generation of toxic intermediates, and difficulties in producing stereospecific chemicals make them less desirable for manufacturing food additives. Microbial production has emerged as a viable and economically feasible alternative for manufacturing a broad spectrum of value-added ingredients.

2.3 Adverse Effects of Food Additives

The use of chemical additives in food production has a long history. The American Academy of Pediatrics reported the potential risks associated with food additives, especially for children [3]. Due to their smaller size, children's "dose" of a given chemical may be higher than that of adults, and their behavior can increase their likelihood of ingesting harmful chemicals. Additionally, children's developing bodies are more at risk of harm and have more time for damage to occur. Specific food additives of concern include bisphenols, phthalates, perfluoroalkyl chemicals, perchlorate, artificial food colors, and nitrates/nitrites. Bisphenols have been shown to interfere with puberty and fertility, increase body fat, and cause problems with the immune and nervous systems. Phthalates can increase the risk of obesity and cardiovascular disease and interfere with male genital development. Perfluoroalkyl chemicals can cause low birth weight babies, immune system and thyroid problems, and fertility issues. Perchlorate in some dry food packaging and drinking water can disrupt thyroid function and early brain development. Artificial food colors may increase symptoms in children with attention-deficit/hyperactivity disorder, and nitrates/nitrites can affect thyroid function and increase cancer risk and are commonly found in processed meats. However, the recent prevalence of additives in processed foods has raised concerns over their long-term safety. Despite this, it is widely acknowledged in the scientific community that the incorporation of additives plays a vital role in the production and marketing of food, and their prudent application contributes significantly to food stability, safety, flavor, and appearance. Although utilizing food additives is crucial for the palatability and attractiveness of numerous food items, their consumption in the human diet is negligible. Furthermore, ensuring food additive safety through extensive research and the differentiation between natural and synthetic additives has often become trivial. CODEX Alimentarius restricts the use of certain additives due to their harmful effects. Examples of these additives are shown in Fig. 15.1.

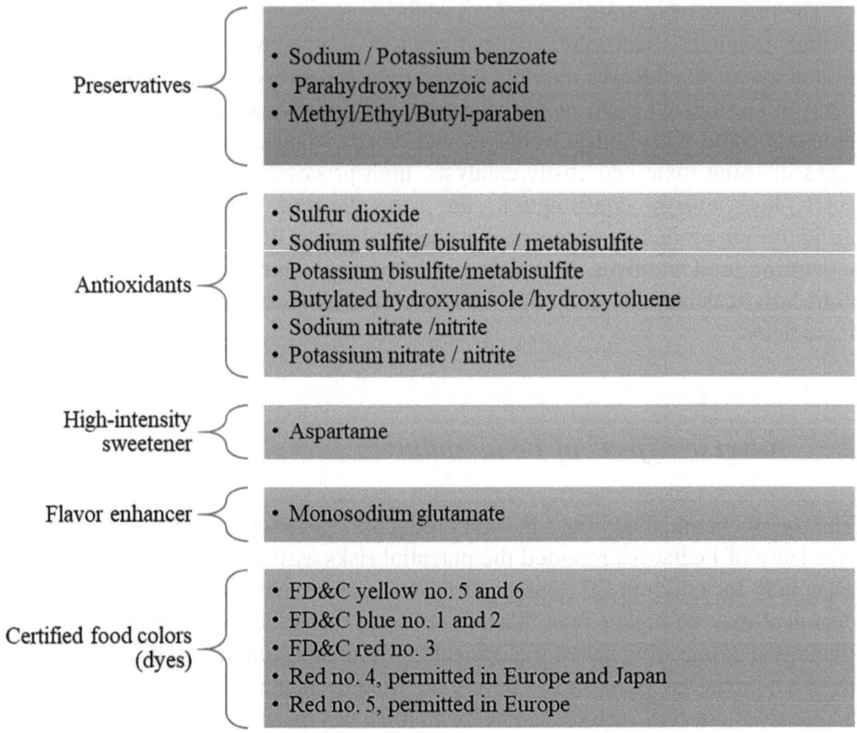

Fig. 15.1 Food additives suspected of causing adverse reactions [2]

Many food additives, such as monosodium glutamate (MSG), synthetic and natural vitamins, and citric acid, are chemically identical despite their sources. Regulations are in place to monitor the safety of food additives, including the Joint Experts Committee on Food Additives set up by the Food and Agriculture Organization and World Health Organization. The Joint Experts Committee on Food Additives considers the safety of additives and sets their specifications and limits. These limits take the form of an acceptable daily intake (ADI). However, most consumers are in a dilemma in using processed food due to incomplete information on the safety of using two or more additives. This indicates the need for toxicology and safety studies on the effect of two or more additives.

The risks posed to the consumer by the food supply are rated in decreasing order of severity as follows:

1. Microbiological hazards include food poisoning from bacteria or bacterial toxins, salmonellosis, and botulism.
2. Nutritional hazards involve excessive sodium and saturated fat consumption.
3. Environmental pollutants such as mercury in fish and lead from car exhaust.
4. Natural toxicants include mushroom poisoning, solanines in potatoes and other solanaceous plants, and shellfish toxins.

5. Pesticide residues protect plants and fruit against insects but may pose toxic effects on humans upon oral, nasal, or skin contact.
6. Food additives that have documented cases of poisoning, although rare due to noncompliance with federal regulations.

While the perils to human well-being from food additives are at the bottom of the list amidst food hazards, it is not without the possibility of perilous risks. This promotes research and industrial interest in safer natural sources in additive production.

3 How Biotechnology Influences the Creation and Utilization of Additives

The food and pharmaceutical industries are turning toward biological sources for additive extraction and production due to the recent scrutiny of the use of chemical additives. Although this approach may come with increased costs, it is a more environmentally friendly and sustainable way to produce additives. Natural food additives can be produced through microbial biosynthesis/biotransformation as a safer alternative to chemically synthesized substances, meeting the demands of consumers for natural and safe products. Biotechnology tools and processes, such as metabolic engineering, can be used to produce high-quality bioproducts, enzymes, biofuels, and healthcare products. Biotechnological advancements in the microbial production of food additives can enhance productivity and safety and improve quality. The list of food additives produced through biotechnological routes is shown in Table 15.2.

Table 15.2 List of food additives processed from microorganisms [4]

Food additive	Microorganism(s) involved
Bio pigments	
Yellow, orange, red	*Monascus pilosus*, M. *purpureus*, M. *ruber*
Astaxanthin	*Xanthophyllomyces dendrorhous*
A pink red	*Penicillium oxalicum*
β-Carotene	*Blakeslee trispora, Mucor circinelloides, Phycomyces blakesleeanus, Eremothecium ashbyii*
Monascorubramine	*Monascus purpureus*
Lycopene	*Flavobacterium sporotrichioides, Bradyrhizobium* spp.
Zeaxanthin	*Flavobacterium* spp.
Isorenieratene	*Brevibacterium aurartiacum*
Bioflavors	
Monosodium glutamate	*Micrococcus glutamicus*
Glutamic acid	*Corynebacterium glutamicum*

(continued)

Table 15.2 (continued)

Food additive	Microorganism(s) involved
Terpenes	*Ceratocystis* spp., *Ascoidea hylecoeti, Cronartium fusiforme, Phellinus* spp., *Pleurotus euosmus*
Esters	*Saccharomyces* spp., *Hansenula* spp., *Candida utilis*
Lactones	*Sporobolomyces odorus, Trichoderma viride*
Pyrazines	*Bacillus subtilis, Corynebacterium glutamicum*
Vanillin	*Aspergillus niger*
Organic acids	
Lactic acid	*Lactobacillus* spp.
Acetic acid, gluconic acid	*Acetobacter* spp.
Citric acid	*Aspergillus* spp.
Benzoic acid	*Streptomyces maritimus*
Propionic acid	*Propionibacterium freudenreichii*
Gums	
Xanthan	*Xanthomonas campestris*
Pullulan	*Aureobasidium pullulans*
Dextran	*Leuconostoc mesenteroides, L. dextranicum, Pseudomonas* spp., *Azotobacter* spp.
Alginates	*Sphingomonas paucimobilis, Pseudomonas elodea*
Gellan gum	*Sphingomonas paucimobilis, Sphingomonas elodea*
Biopreservatives	
Nisin	*Lactococcus lactis*
Lacticin 3147	*Lactococcus lactis*
Pediocin	*Pediococcus acidilactici*
Biosurfactants	
Rhamnolipids	*Pseudomonas* spp.
Sophorolipids	*Torulopsis* spp.
Trehaloses	*Mycobacterium* spp., *Rhodococcus* spp.
Phospholipids	*Arthrobacter* spp., *Corynebacterium* spp.
Fatty acids	*Candida* spp., *Pseudomonas* spp., *Micrococcus* spp., *Candida* spp., *Penicillium* spp., *Aspergillus* spp.
Lipopeptides	*Bacillus* spp., *Streptomyces* spp., *Corynebacterium* spp., *Mycobacterium* spp.

3.1 Role of Biotechnology in the Development and Application of Additives

Food biotechnology uses gene splicing in plants, animals, and microorganisms to enhance food quality. It also utilizes packaging technology, microbial fermentation processes, enzymes, and three-dimensional (3D) printing. Microbial production of additives is feasible owing to a better understanding of the metabolic pathways of relevant organisms. Metabolic engineering techniques embody preconceived approaches for a particular end-product and may not be suitable for others. A majority of microbial-derived products originate from metabolites. Hence, manipulating metabolic pathways is pivotal for synthesizing a desired product and a comprehensive array of products derived from the same metabolite. Biotechnological tools can efficiently produce native and novel products using various cultures, including LAB strains, *Escherichia coli*, *Bacillus subtilis*, and *Corynebacterium glutamicum*.

Biotechnological techniques can modify metabolic and biosynthetic pathways, resulting in the production of various compounds, such as aroma compounds, pigments, flavor enhancers, texturizers, amino acids, organic acids, vitamins, carbohydrates, low-calorie sugars, and bacteriocins, more efficiently. Engineered microbes have the potential to produce sweeteners such as stevia-derived molecules, xylitol, and erythritol. Furthermore, yeast can be modified to achieve more exotic and hoppy flavors in beer. Notably, microbes contribute to the production of aroma compounds such as rose (2-Phenylethanol), orange/lemon (limonene), mint (menthol), and peach (gamma-decalactone). Additionally, microbes produce colored molecules that are useful as pigments for food and beverages. Examples of such pigments include orange (beta-carotene, canthaxanthin), red (lycopene, astaxanthin, prodigiosin), yellow (riboflavin), blue (phycocyanin), purple (violacein), and black (melanin) colorants.

3.2 Exploring Innovative Approaches

Microorganisms convert energy sources into food substances through fermentation. Refined food substances act as functional elements or processing aids in other food preparations. The process involves recovery and purification steps to remove unwanted substances. After refining or purifying, the food substance may undergo supplementary finishing stages, such as concentration, stabilization, dehydration, or addition of other materials, based on its need. The extracted compounds were screened for purity and safety. Modern fermentation facilities use methods similar to those used for thousands of years in food production, but the modern processes are well monitored and controlled accordingly. For example, the production of wine involves the involvement of microorganisms, specifically yeast, in converting sugars in grape juice to ethanol. It also entails implementing recovery and purification steps and finishing procedures, such as the aging of the wine in oak barrels before bottling.

In the generation of food products, multiple strains of microorganisms utilize genetic modification only if they possess the necessary traits to produce the intended substance on a large scale. Conventional food processes, such as the creation of bread, wine, beer, and yogurt, typically employ unmodified microorganisms, occasionally using some food enzymes. The modification of genetics is essential in achieving a desired substance, optimizing efficiency, and reducing the cost of manufacturing an endogenous enzyme or other food substances. Some methods to modify microorganisms [5] for food production are as follows:

Nontargeted Mutagenesis

Genetic modification of microorganisms using nontargeted methodologies that rely on classical mutagenesis and selection is one of the oldest approaches. These methodologies entailed the introduction of random alterations to DNA, necessitating laborious screening endeavors to pinpoint characteristics. However, with the advent of modern biotechnology, targeted genetic modifications are predictable and less uncertain in the final genetic makeup of modified microbes.

This approach is limited to modifying traits with poorly understood genetic bases. Molecular biology is likely to replace this method with precise genetic modification methods. However, random mutagenesis followed by repeated testing has been the basis for developing several safe, robust, and well-characterized microbial production platforms for expressing new traits through genetic engineering.

Genetic Engineering/Bioengineering/Genome Editing

Scientists use in vitro nucleic acid techniques to create genetically modified organisms (GMOs) for food production. They insert genes into a selected, safe microorganism via recombinant DNA or related techniques to impart enhanced or new functionality. The process includes the elements that control the expression of functional genes, and the choice of DNA sequences can be random in a microorganism's genome but often inserted into specific loci, with genome sequencing used to confirm the insertion site. Genetically engineered microorganisms can source DNA from the same organism or a different organism, which may be closely or distantly related. Bioengineered organisms can produce functional compounds (food additives) commercially through genome editing, resulting in an organism that is not found in nature or produced by traditional breeding. Synthetic DNA sequences can be inserted into microorganisms to create genetically engineered microorganisms that produce modified enzyme proteins for testing in target applications.

The process of genetic engineering involves the insertion or deletion of specific DNA sequences within the genome of a microorganism. This manipulation can lead to newer or enhanced functionalities. With the use of synthetic DNA, it is now possible to efficiently create genetically engineered microorganisms that have different enzyme proteins. It is essential to consider safety considerations during this process, as there is a risk of inadvertently introducing sequences from potentially harmful pathogens.

Protein Engineering

Protein engineering is a commonly employed technique to optimize enzyme functionality and enhance catalytic activity. Modified enzymes function optimally under unfavorable temperature, pH, and salt concentrations. For example, the engineering of baking amylases can lead to their increased tolerance to high oven temperatures, enabling them to achieve the same level of reactions with lower initial enzyme concentrations before inactivation occurs.

In practical terms, protein engineering typically involves generating multiple variants of a wild-type enzyme through genetic engineering of the production organism (particularly when expression levels need to be enhanced) or gene editing (to evaluate the impact of specific amino acid changes in an endogenous protein). These variants are then evaluated regarding their application and performance to produce multiple "hits" that exhibit improved characteristics. These "hits" are sometimes combined in successive generations to maximize the improvement achieved. A final variant enzyme protein is selected, which may differ from the wild-type sequence in one or multiple amino acids.

Gene and Genome Editing

CRISPR represents a groundbreaking technological advance employed in precision gene editing. This technique utilizes specialized proteins to excise and rejoin DNA and is available in three distinct forms: zinc fingers, TALENs, and CRISPR–Cas. CRISPR–Cas is the most widely utilized and effective, distinguished by its elegant design and straightforward cell delivery. Its applications encompass treating genetic disorders, cultivating crops with enhanced climate resilience, and the production of designer materials, foods, and drugs. CRISPR–Cas represents a complex of enzymes and genetic guides that work together to locate and modify DNA. Scientists generate guide RNA sequences recognizing specific DNA codes in any living cell. These RNA sequences excise one or more base pairs from the DNA code and introduce replacement nucleic acids at the precise location where a target sequence was interrupted. It aims to repair DNA, modify the sequence, or introduce a new beneficial gene rather than merely knock out endogenous gene function.

3.3 Biosynthesis of Natural Flavors, Sweeteners, and Colorants

Cyclodextrins for Odor/Flavor Removal [6]

Complex sugars called cyclodextrins (CDs) can remove odor and flavor compounds from food products. However, their high cost limits their use in the food industry. CDs are cyclic oligosaccharides made by starch degradation via cyclodextrin glycosyltransferase (CGTase) and are released into extracellular environments by microorganisms. They can form inclusion complexes with various compounds, solubilizing valuable (e.g., bioactive) compounds, increasing stability/bioavailability, or reducing undesirable tastes/flavors.

Production Approaches and Application of CDs

Wild-type Bacillus strains or genetically engineered microorganisms are involved in producing cyclodextrins (CDs). Recent advancements in the field include the development of mutant *E. coli* strains to improve CGTase production and the utilization of engineered CGTases from Bacillus spp. for selective production of larger CDs. Furthermore, immobilization of cells or enzymes has proven to enhance CD production. However, in cyclical processes, the immobilization of CDs is less practical than immobilizing microorganisms or CGTases. Recovery of immobilized CDs entails using organic solvents unsuitable for food preparation.

Immobilized CDs in food packaging with CD-containing films can lower cholesterol levels in milk. Before or during CD application, flavor/odor-active compounds must be unbound from proteins and dissolved in water. CDs may not exhibit higher affinities to most of these compounds than proteins, resulting in significant amounts remaining bound post-CD application. An alternate approach, such as LAB fermentation or acid treatment, may be combined with CD treatment to increase the amounts of unbound compounds. After treatment, the protein in the product is precipitated and removed by adjusting the pH before proceeding to the next stage.

3.4 Case Studies

Legumes [7]

Legumes are protein-rich foods popular in Asia, the Middle East, South America, and the West. However, their beany or grassy aromas, bitter or astringent flavors, and flatulence-causing compounds still limit their acceptance. Removing unpleasant flavors and aromas from legumes is challenging due to various chemical compounds contributing to undesirable tastes and smells. The oxidation of lipids, either enzymatic or nonenzymatic, leading to the breakdown of unsaturated fatty acids, can cause the formation of low-molecular-weight alcohols, ketones, and aldehydes; these compounds significantly contribute to the unique "beany" aroma. Moreover, pyrazine derivatives, which arise from the degradation of proteins and amino acids, are also typically present in legumes. It is worth noting that some of these derivatives possess unfavorable "pea-like" or "vegetable-like" aromas/flavors. In contrast, others have "nutty" or "chocolate-like" aromas/flavors that can serve to mask unpleasant tastes and smells. Furthermore, it is essential to acknowledge that odor/flavor-active compounds and their precursors may bind to legume proteins, polysaccharides, and lipids, leading to complications in eliminating undesirable flavors/aromas.

Various approaches have been studied to reduce/mask undesirable flavors in legume-based food products, including physical, chemical, and biotechnological approaches. The most commonly used method is heating treatment to inactivate enzymes that cause flavor degradation, but it may only be effective for some compounds. Other nonthermal physical treatments and chemical methods could be more successful. Biotechnological methods, including germination, enzymatic treatment, fermentation, and genetic engineering, have gained interest due to their environmentally friendly nature. In recent times, scholarly investigations have emphasized creating enzymatic and fermentation-based techniques to direct protein-bound odor or flavor-active compounds and their antecedents.

Naturally occurring and processed organic compounds have a discernible impact on the odor and flavor of legumes and legume products. Some of these compounds, such as aldehydes, ketones, and pyrazines, are deemed less desirable, while others, such as alcohols and specific pyrazines, have desirable flavors. Low-molecular-weight organic acids can also contribute to desirable aromas. The principal cause of objectionable aromas and flavors in legumes and their production is the degradation of unsaturated fatty acids. Linoleic acid (18:2), linolenic acid (18:3), and oleic acid (18:1) are the most prominent precursors of these compounds.

Nevertheless, these fatty acids are also vital components of the human diet. Due to hydrophobic interactions and other intermolecular forces, most odor- and flavor-active compounds and their precursors tend to bind to legume proteins. Oleosins, a protein fraction found in legumes, exhibit a high binding affinity to lipids and long-chain fatty acids, which are generally odor- and flavor-neutral. Undesirable odors and flavors may arise with storage time due to the degradation of lipids and long-chain fatty acids. Aldehydes and ketones bind to vicilin and legumin, the most common legume globulins.

Fermentation can enhance the nutritional, functional, and sensory characteristics of legume-based products. Compared to other treatment methods, fermentation-based approaches are more cost-effective, as they require minimal or no usage of chemicals and involve only moderate heating. Microbial growth in fermentation occurs at 30–37 °C, which is significantly lower than the 60–80 °C required for lipoxygenase and hydroperoxide lyase deactivation or 190 °C for decreasing alcoholic compound contents. The spontaneous fermentation process has the advantage of lower processing costs, as it eliminates the need for pure microbial strains. Fermentation-based methods also eliminate the requirement for enzyme extraction and purification, reducing associated expenses. During fermentation, microbially mediated protein hydrolysis occurs, improving digestibility without denaturing the protein. Depending on the strains and conditions used, fermentation can eliminate undesirable compounds while producing desirable ones. The use of local single-species or multispecies microbial starter cultures or controlled spontaneous fermentation processes can lead to the creation of unique legume-derived products. Recent studies have better understood the crucial metabolic pathways and enzymatic activities involved in microbially mediated flavor- and odor-active compound transformations.

One successful example of legume fermentation is the production of flavor-rich soy sauce. Despite soy sauce having almost 300 aroma/flavor-active compounds, including alcohols, aldehydes, esters, and organic acids, it typically lacks aroma/flavor. Soy sauce's liquid nature and complex, prolonged fermentation process allow for the effective elimination of any aroma/flavor compounds. Soy sauce comprises minimal constituents that can maintain unappealing volatile compounds, thus rendering it a fitting medium for manufacturing a sauce that lacks flavor and aroma. By applying a comparable technique to soy sauce preparation, it is plausible to generate a corresponding product with a flavor and aroma by fermenting lupine seeds [8].

Various microorganisms, such as lactic acid bacteria, molds, and yeasts, have historically been utilized to ferment legumes, including soybeans, fava beans, and peas. Nevertheless, selecting an appropriate starter culture for legume fermentation is formidable. Most commercially available strains were from milk products, and their influence on legume flours/protein isolates differs from that of milk constituents. Microorganisms may struggle to metabolize carbon and energy sources in legume flours/protein isolates as they would in their natural habitats. Consequently, they may face competition for carbon and energy sources from the autochthonous legume microflora.

Meat Substitute

The case study explores how the Impossible Foods company has significantly impacted the growth of the plant-based meat sector, both within their home country and on a global scale, by leveraging innovative biotechnological methods [9]. Impossible foods have a mission to revolutionize the food industry by creating alternatives to meat that closely mimic its taste, texture, and nutritional composition. Using biotechnology, they produce heme—a flavor molecule found in meat—through genetically modified yeast and replicate meat-like texture and taste using plant-based ingredients. This biotechnological breakthrough has driven the popularity of their flagship product, the Impossible Burger, among consumers and led to partnerships with well-known restaurants, driving demand for environmentally conscious dietary choices domestically. Moreover, Impossible Foods' successful expansion into international markets underscores its ability to navigate regulatory complexities and collaborate with global distributors, expanding awareness and acceptance of plant-based options. The case study accentuates how biotechnological advancements are propelling the growth of the plant-based meat industry, shaping dietary preferences.

Lab-grown meat, or cultured meat, has revolutionized the food industry by cultivating animal cells in controlled environments to create sustainable and ethical meat products. This innovation addresses resource-intensive traditional meat production, reducing environmental footprints and greenhouse gas emissions. Lab-grown meat caters to ethical consumer demands, eliminates animal suffering, and offers customizable nutritional profiles [10]. It holds promise for global food security and disruptive market change. Despite challenges, its increasing acceptance highlights a shift toward environmentally conscious and humane food choices. As technology advances, lab-grown meat's transformative impact on sustainability, ethics, and industry dynamics continues to reshape the trajectory of food production and consumption in the future.

4 Preservation Through Biotechnology

Perishable food materials need to be processed into more useful and shelf-stable products in this regard, and various technologies are being used. The processed food must be physically, chemically, and microbiologically safe. Preserving food is an integral part of the food industry, as it ensures safety and quality during storage. Preservation methods can be broadly divided into two categories: physical and chemical. Physical methods involve thermal processing, such as sterilization, pasteurization, and cooking, and nonthermal methods, such as irradiation, pressure processing, ultraviolet radiation, and cold plasma. Chemical methods are employed as a means of enhancing the safety of food by introducing synthetic or natural substances. Although each technique presents its own challenges [11], selecting the most suitable and effective preservation techniques is crucial to ensure food safety, stability, and consumer acceptability.

Thermal processing of food can result in the loss of essential nutrients and changes in taste and texture. Additionally, thermal processes consume significant amounts of energy [12]. Smoking can introduce polycyclic aromatic hydrocarbons (PAHs) into fish, a known carcinogenic substance that impacts the product's taste and quality. Refrigeration technology, while effective, requires significant energy consumption and may involve refrigerants that are harmful to the environment. Traditional drying techniques can alter the nutritional and sensory composition of food. While ionizing radiation can be helpful in certain circumstances, it is not recommended for conductive materials and may negatively impact vitamins at higher doses. High-pressure processing affects food properties and can decrease process efficiency due to microbial stress responses [13]. Adopting processing technologies such as radiofrequency, ohmic heating, spray drying, freeze drying, irradiation, and ozone treatments involves a significant upfront investment, especially for newer and more advanced technologies. Furthermore, the approaches need consideration regarding environmental safety.

Certain chemical preservatives pose a health risk and may lead to fetal abnormalities, neurotoxicity, asthma, and cancer. Excessive consumption of monosodium glutamate (MSG) can trigger allergies, migraine headaches, and general weakness. Nitrite-containing foods can lead to digestive irritation, elevated cholesterol levels, and type-1 diabetes. Propyl gallate and tertiary butyl hydroquinone can increase cancer risk. As a result, it is crucial to regulate or abstain from consuming these preservatives. Compounds possessing preservative properties or those that compete with microbes have the potential to replace conventional chemical preservatives in various products. Several factors, including low yield, the use of solvents, and high processing costs, impede the extraction of natural substances for preservation. These challenges encourage researchers to address the issue and ensure that the process is economically viable and sustainable. Overcoming obstacles to preserving natural resources requires a collaborative approach in developing and implementing innovative strategies. Advancements in biotechnology have made it possible to produce an extensive range of products from microbial sources at a lower cost despite the challenges. As a result, the food industry must maintain a positive outlook and strive to employ the most effective methods for preserving food with confidence.

4.1 Biopreservation Techniques to Control Spoilage and Pathogens

Fermentation

Fermentation of food by microorganisms extends the shelf life and functional properties of food. Lactic acid bacteria are widely used in fermentation worldwide. *Lactobacillus, Lactococcus, Streptococcus, Leuconostoc, Pediococcus*, and *Enterococcus* are examples of lactic acid bacteria genera. Lactic acid bacterial fermentation has previously been acknowledged for its ability to improve human health and gut microflora. Due to their safety and antimicrobial and antifungal properties, lactic acid bacteria are becoming increasingly popular as biopreservatives. They compete for resources and space, and they create antimicrobial compounds such as lactic acid, ethanol, and bacteriocins.

Bacteriocins

Bacteriocins are small, ribosomally synthesized antimicrobial peptides that are immune to other bacteria and have bactericidal or bacteriostatic effects on them [14]. Their ability to limit potential spoilage organisms and pathogens can be quite helpful [15]. Growing interest in bacteriocins as substitutes for antibiotics and chemical food preservatives has resulted in the development of novel methods to detect bacteriocin producers.

Acetic Acid Bacteria

Lactic acid bacteria (LAB) and propionic acid bacteria (PAB) are effective against spoilage fungi in bakery products. The study identified a mixed culture of *Propionibacterium freudenreichii* D6 and *Lactiplantibacillus plantarum* L9 as the most effective in inhibiting the growth of *Aspergillus niger* [16]. Acetic acid is an active agent for the inactivation of various fungi.

Antimicrobial Peptides

Antimicrobial peptides (AMPs) are naturally occurring proteins found in animals, plants, insects, and bacteria. They are critical in protecting the host against infections. Chemical, enzymatic, or recombinant methods can generate AMPs, which provide an alternative to chemical preservatives [17]. Presently, nisin is a single antimicrobial peptide commonly used in food preservation.

Essential Oils

The use of natural agents with antimicrobial activity, such as essential oils (EOs), as preservatives in the food industry is interesting. EOs derived from plants have been found to be effective against foodborne pathogens and spoilage bacteria in various food products, such as ready-to-eat foods, milk, apple juice, cheese and meat-based products [18]. EOs from plants such as *Cinnamomum verum, Cymbopogon martini, Cymbopogon citratus, and Cymbopogon flexuosus* exhibit inhibitory activity against *E. coli* strains and repress genes responsible for biofilm formation. EOs from peppermint, coriander, and anise have been shown to inhibit the biofilms of *S. aureus* and *E. coli* [19].

4.2 Bacteriophages

Bacteriophages, also known as phages, are emerging as a category of food biopreservatives. They are viruses that can specifically target and destroy bacteria, offering a potential alternative to antimicrobials. Endolysins are antibacterial enzymes derived from phages that have shown broad-spectrum activity against various bacterial species, including food-related bacteria. Bacteriophages and enzybiotics are feasible strategies for maintaining the long shelf life of food products. These techniques involve the use of protective bacterial cultures, antimicrobial products, antimicrobial compounds derived from plants, and bacteriophages or their bacterial enzymes called phage lysins or enzybiotics. The use of phages and their lytic enzymes in food preservation has been studied, with the *Lactobacillus helveticus* bacteriophage 0303 showing a broad spectrum of activity against different bacterial species. Phages have a unique host range and can specifically control strains of pathogenic bacteria, making them an alternative to antimicrobials in the fight against bacteria. Phages have evolved strategies that defeat antiviral defenses in bacteria and then effectively increase, making it easier to discover new phages that can infect bacteria instead of discovering new antibiotics [20].

5 Natural Antioxidants

Lipid oxidation is a crucial concern within the food industry, given its negative impact on food items' nutritional content, flavor, texture, appearance, and shelf life. Even during the lipid extraction and refining procedures, heat application while cooking, frying, and prolonged storage can further exacerbate lipid oxidation products compared with those found in biological lipids. These oxidation products (hydroperoxides and fatty acid breakdown products) can trigger various biochemical reactions in the human body. Free radicals, which are highly reactive molecules, can damage cells and cause aging, cardiovascular diseases, and cancer [21]. The imbalance of ROS and antioxidants causes oxidative stress, leading to senescence and diseases. While the human body has antioxidant systems to prevent damage from oxidation products, unhealthy eating habits and exposure to environmental stressors may hinder their effectiveness [22]. The adverse effects of trans-fatty acid oxidation products can increase the risk of various physiological threats and illnesses, such as cardiovascular issues, immune dysfunction, cancer, oxidative stress, and premature aging.

Maintaining stable lipids is crucial in the food industry because it ensures food safety and promotes good health. Three primary oxidation pathways—autooxidation, thermal oxidation, and enzymatic oxidation—can compromise lipid stability. High levels of unsaturated fatty acids make lipids particularly vulnerable to oxidation. Scientific studies show that trace ingredients such as vitamin E, carotenoids, and polyphenols can effectively prevent oxidation, while environmental factors and metal ions can accelerate it. Encapsulation and light protection methods are also employed to enhance lipid stability. Antioxidants are widely used in food manufacturing to prevent spoilage, highlighting their practical significance in maintaining lipid stability.

5.1 Role of Antioxidants in Preventing Oxidation and Maintaining Food Quality

Antioxidants prevent lipid oxidation by removing harmful radicals and inhibiting oxidation initiators. Their effectiveness lies in their ability to neutralize unpaired electrons, reducing reactive oxygen species and lipid peroxidation while enhancing redox mechanisms. By scavenging free radicals, antioxidants can reduce cell damage, low-density lipoprotein, platelet aggregation, thromboxane B2, and prostaglandin D2 [11]. Moreover, antioxidants have the added benefit of slowing aging and preventing diseases such as atherosclerosis, cardiovascular disease, diabetes, and cancer. Ultimately, using antioxidants to prevent lipid oxidation is a valuable technique with significant benefits for human health and well-being, achieved by neutralizing free radicals and reducing oxidative stress.

To preserve their products, the food industry has traditionally utilized synthetic antioxidants such as butylated hydroxyanisole, butylated hydroxytoluene, tert-butylhydroquinone, and propyl gallate [23]. However, these antioxidants can potentially cause cancer and are therefore restricted. Synthetic antioxidants used as food additives may contain harmful and carcinogenic components. Therefore, relying on safer and more potent natural antioxidants is crucial to prevent any potential health hazards. As a result, there has been an increasing use of natural antioxidants derived from plants, animals, and microorganisms in nutrition, cosmetics, and pharmaceuticals. Bioactive compounds from natural products such as fruits, vegetables, spices, grains, oils, seeds, algae, fish/shellfish, marine bacteria, and microorganisms have shown promising therapeutic potential. These properties make them potential raw materials in various fields, including pharmaceuticals, dietary supplements, biofertilizers, and crop pest control. These natural products also have varying levels of antioxidant activity.

5.2 Bioresources for Antioxidant Production

Microbial diversity provides a vast pool of novel chemicals, including antioxidants. Actinomycetes and fungi are the primary producers of microbial secondary metabolites. Microbial metabolites are potential ingredients to develop life-saving drugs for bacterial and fungal infections, cancer, transplant rejection, and high cholesterol. Actinomycetes produce a variety of bioactive molecules with antioxidant potential, such as benthocyanins, carazostatin, and carbazole compounds [24]. Phenazine derivatives produced by *streptomycetes* also possess efficient antioxidant activity. Streptomyces exfoliates and *Streptomyces* spp. produce antioxidant isoflavonoids and terpenoids, respectively. *Streptomyces* spp. produce carbazole compounds, identified as a significant class of antioxidants. Nitrogen-containing metabolites such as carbazole and phenazine heterocycles also demonstrate antioxidant activity. *Streptomyces* spp. and other actinomycetes isolated from diverse habitats possess antioxidant activity—marine actinobacteria, such as *Streptomyces* sp. and *Streptomyces* sp. VITTK3, also exhibit antioxidant activity [25]. Phenolic compounds isolated from *Streptomyces* sp. demonstrate DPPH scavenging activity as potent free radical scavengers.

Eubacteria are a group of tiny, one-celled microorganisms that lack a nucleus and can create various extracellular substances. They can be found close to all living organisms. Probiotic foods contain bacteria that can reside in the gastrointestinal tract and create exopolysaccharides (EPS) that exhibit antioxidant activity [26]. Microbial sources have provided a variety of compounds with potent antioxidant properties. Additionally, the EPS produced by *Lactobacillus paracasei* and *Lactobacillus plantarum* have remarkable antioxidant properties and have demonstrated in vitro immunomodulation [27, 28]. These findings suggest that microbial sources have significant potential for developing novel antioxidant therapies and immunomodulatory agents. Furthermore, fermented milk that contains *Lactobacillus casei*, *Lactobacillus acidophilus*, and *Lactococcus lactis* has the potential for antioxidant activity and cholesterol assimilation [29]. Lactic acid bacteria in the gastrointestinal tract, such as *Bifidobacterium longum* and *Lactobacillus acidophilus*, exhibit potent antioxidant activity by preventing linoleic acid peroxidation. Lactobacillus strains obtained from fermented foods in Nigeria demonstrate antioxidant activity and can inhibit lipid peroxidation. EPS produced by *Lactobacillus fermentum* and *Lactobacillus plantarum* contain different sugars and functional groups that are reactive with free radicals. Probiotic bacteria such as *Bifidobacterium* spp., *Leuconostoc* spp., and *Lactobacillus* spp. exhibit radical scavenging activity and inhibition of lipid peroxidation. Fermented milk and whey portions display enhanced antioxidant activity due to the production of EPS and other metabolites during fermentation. *Lactobacillus casei* can reduce oxidative stress and liver damage induced by carbon tetrachloride. Resorstatin produced by Pseudomonas and xanthine inhibits lipoperoxidation and oxidation of linoleic acid. Bacteria associated with marine brown algae and endophytic bacteria from plants produce bioactive molecules with antioxidant properties [30]. These antioxidants have potential applications in the food and health industries.

Cyanobacteria, known as blue–green algae, are photosynthetic bacteria that produce carotenoids. These carotenoids possess potent antioxidant and anti-inflammatory properties [31]. Furthermore, blue–green algae contain phycocyanin, a pigment with hepatoprotective effects that can inhibit inflammation. Astaxanthin, another carotenoid pigment found in blue–green algae, serves various metabolic functions and is particularly effective at neutralizing singlet oxygen and free radicals [32]. Some cyanobacteria produce phenolic compounds that exhibit robust antioxidant activity [33]. Lichens, symbiotic associations between fungi and algae, produce unique secondary metabolites that can be potential sources of natural antioxidants [34, 35].

Fungi play a dual role in their impact on humans. While they can lead to food spoilage, they also have the potential to produce a range of beneficial substances, such as organic acids, pigments, enzymes, vitamins, antibiotics, and alcohol [36]. Edible mushrooms are an example of functional food with medicinal value and are rich in antioxidants, phenolics, and compounds with potential therapeutic applications [37]. Filamentous fungi are also a valuable source of metabolites, including potent antioxidants that outperform synthetic counterparts and phytochemicals. The *Pestalotiopsis microspora* fungus, for instance, produces compounds that demonstrate both antifungal and antioxidant activity [38].

5.3 Fermentation as a Biotechnological Tool to Enhance Antioxidant Production

Fermentation is an age-old practice used for centuries to preserve food and create beneficial bioactive compounds. Biotechnology has been proven to enhance the productivity of metabolites during fermentation. Two primary fermentation methods exist: SmF (Submerged fermentation) and SSF (solid-state fermentation). SSF requires less water and energy, making it beneficial for solid waste management and enzyme production, while SmF is superior in process control and handling. However, establishing an SSF system requires meticulous microorganism selection and parameter control. Moreover, fermentation elevates the antioxidant activity of food products by augmenting the concentration of phenolic compounds and flavonoids. Lactic acid bacteria contribute to food fermentation and improve functional properties.

Fermenting pineapple waste can boost its phenolic content and antioxidant properties. A solid-state fermentation (SSF) process that utilizes soy flour and microorganisms can also enhance the antioxidant features of pineapple and pomegranate waste. Additionally, after fermentation, certain fungi can increase the total phenolic content and antioxidant activity of grape pomace. The wine industry can also utilize various waste materials to ferment and retain phenolic compounds. The SSF of black grape, apple, and pitahaya residues increases the total phenolic content and antioxidant activity. These fermented waste products can be valuable sources of natural antioxidants for the food industry. The extraction of antioxidant compounds from food waste is an ongoing endeavor with numerous advantages and potential applications. Various food waste sources, including cheese whey, citrus peels, olive mill waste, and tomato waste, have been optimized using unique proprietary techniques and product-specific applications. Lycopene has been extracted from tomato peels using enzymes that breakdown cell walls, demonstrating the potential for completely recovering natural antioxidants. Phenolic compounds can be obtained through solid-state fermentation and used as edible coatings and films to prolong the shelf life of foods. Some examples in Table 15.3 are food additives produced through fermentation. They can also fulfil consumers' desires for natural additives. More extensive surveys and research, including broad surveys and analyses of specific products and components, are necessary to understand consumer acceptance of additives derived from food waste and by-products.

Preserving food and preventing contamination is crucial, but conventional chemical fungicides carry potential health and environmental hazards. Natural alternatives such as essential oils and microorganisms can act as preservatives. Moreover, the fermentation of food by-products to obtain antifungal compounds is a sustainable and waste-reducing solution. Antifungal compounds can take on different forms, such as organic and fatty acids, peptides, volatiles, and enzymes. Phenolic compounds in rice bran fermented with *R.oryzae* have antifungal properties, impeding growth by nearly 40%. It is effective against various fungi and can extend bread's shelf life by 3 days. Fermented apple by-products in breadmaking can also delay mold contamination. Recent research suggests that fermented whey, a

Table 15.3 By-product fermentation to yield antifungal and antioxidant activities

Microorganism	Substrate	Fermentation type	Role	Reference
Aspergillus niger	Grapefruit by-products	SSF	Antioxidant compounds	[39]
Bacillus subtilis	Peanut meal, whey proteins	SmF	Antioxidant peptides	[40, 41]
Bacillus spp.	Rice starch extraction by-product	SmF	Antioxidant peptides	[42]
Monascus purpureus	Porcine liver proteins	SmF	Antioxidant peptides	[43]
Rhizopus oryzae	Grape waste, beer waste, olive oil waste	SSF	Antioxidant compounds	[44]
Rhizopus oligosporus	Apricot pomace, chokeberry pomace, cranberry pomace	SSF	Enhanced antioxidant activity	[45–47]
Blakeslea trispora (+) MTCC 884	Orange, carrot, and papaya peels	SSF	Antioxidant activity	[48]
Lactobacillus isolates: L. *casei* L-26, L. *fermentum* 56, *L. paracasei* 106, and *L. plantarum* 53	Guava	SmF	Antioxidant activity	[49]
P. acidilactici	Shrimp waste	SmF	Antioxidant compounds	[50]
L. plantarum	Whey, sea bass by-products	SmF	Antifungal compounds	[51]
B. cereus	Shrimp and crab shell	SmF	Antioxidant and antifungal compound	[52, 53]

by-product of the dairy industry, has the potential to generate antifungal agents that can effectively inhibit fungal growth in pita and sliced bread, thereby reducing the risk of contamination.

Animal by-products can potentially possess antifungal properties. Research has shown that sea bass and Indian major carp by-products fermented using *Lactobacillus plantarum, E. faecium*, and *P. acidilactici* exhibit antifungal activity against various fungi. Similarly, shrimp and crab shell powder fermented with *Bacillus cereus* has also demonstrated antifungal properties against specific types of fungi.

6 Regulatory Considerations and Safety Assessment

According to OECD-FAO Agricultural Outlook 2023–2032 [54], 79% of global crop production growth, 15% cropland expansion, and 6% higher cropping intensity are due to continued progress in plant breeding and the transition toward more intensive production systems. Food production and consumption methods are swiftly evolving owing to biotechnological progress. However, ensuring the consumption of safe and high-quality food remains of paramount importance. Each year, worldwide, approximately one out of every 10 individuals become sick due to ingesting contaminated food [55].

The fifth World Food Safety Day in 2023 theme emphasizes the significant responsibility of food standards in preserving lives. It highlights the pressing necessity to address the growing global fatalities linked to food safety concerns. This can be achieved by implementing and strictly enforcing food standards rooted in scientific risk assessments. These standards should encompass a broad spectrum of hazards, covering biological, chemical, and physical risks, to guarantee the security of our food source.

6.1 Regulatory Frameworks Governing Food Additives and Preservation

Most governments and institutions adhere to and implement food standards that stem from scientific evaluation of physical, chemical, and biological hazard risks. Individual government organizations, regional entities, or international standard-setting bodies may formulate these norms. An international body dedicated to setting quality and safety standards for food is the Codex Alimentarius Commission, also known as the Codex. Codex unites representatives from 188 member countries and one-member organization, the European Union, to ensure food safety. Codex functions with the mission of safeguarding consumer well-being and promoting equitable food trade practices. These standards, employed by governments and the food industries, steer domestic food safety laws and uphold optimal practices.

Other key international regulatory bodies include the FDA (U.S. Food and Drug Administration) in the United States, EFSA (European Food Safety Authority) in the European Union, ANVISA (Agência Nacional de Vigilância Sanitária) in Brazil, FSANZ (Food Standards Australia New Zealand) covering both New York and Australia, MFDS (Ministry of Food and Drug Safety) in South Korea, FSSAI (Food Safety and Standards Authority of India) in India and CAC (Central Asian Council) comprising Central Asian countries. These regulatory bodies are dedicated to formulating guidelines for food additives and conducting safety assessments using scientific evaluations.

The Joint FAO/WHO Expert Committee on Food Additives (JECFA) constitutes an autonomous panel of scientific experts, offering risk evaluations and recommendations to the FAO, WHO, and Codex. It offers guidance on food additives, contaminants, and veterinary drug residues, shaping Codex Committee standards. Given the need for reliable global risk assessments, JECFA is a crucial resource for countries lacking expertise and funds, offering independent expert advice aids in setting international health and trade standards. Its role aligns with the WTO's SPS agreement, making Codex-established standards crucial for fair trade. JECFA's guidance is instrumental in shaping national food safety programs and legislation.

The "Codex Ad Hoc Intergovernmental Task Force on Foods Derived from Biotechnology" was created to formulate global standards, guidelines, or similar guidance concerning biotechnologically produced foods and food additives. The decision to establish this task force was made during the 23rd Session of the Joint FAO/WHO Codex Alimentarius Commission, which took place in Rome from June 28 to July 3, 1999.

6.2 Safety Evaluation and Risk Assessment of Biotechnologically Derived Additives

Codex CAC/GL 45-2003 offers a suggested method for assessing the safety of foods originating from recombinant-DNA plants. Similarly, CAC/GL 46-2003 and CAC/GL 68-2008 were used for recombinant microorganisms and animals, respectively. Safety evaluation is a part of the risk assessment framework (Table 15.4). The principles of risk analysis established by the Codex are designed to be used for individual chemical substances such as food additives and pesticide residues, as well as specific chemical or microbial contaminants that have identifiable hazards and risks. These principles are not meant to be applied to entire food products. CAC/GL 44-2003 enumerates the principles of risk analysis that pertain to the safety and nutritional qualities of food products obtained through advanced biotechnology.

Traditionally, new food plant varieties undergo limited safety evaluation before marketing, except for foods such as infant formulas. Common crops undergo an assessment by breeders and lack rigorous safety testing, such as food additives or pesticide residues. The safety assessment methods provided by Codex standards for recombinant plants (CAC/GL 45-2003), microorganisms (CAC/GL 46-2003), and animals (CAC/GL 68-2008) include a comparison of genetically modified species with their conventional counterparts for composition, nutrients and potential allergens. Toxicological testing, allergenicity assessment, and stability assessment are mandatory assessments to be followed. They are further investigated for unintended effects and environmental impact.

Table 15.4 Risk assessment framework according to CAC/GL 44-2003

Safety assessment: Check for any new safety concerns or hazards in the modified food compared to the regular one
Comparative analysis: Compare the modified food with the regular one to find any differences in composition, nutrients, and potential risks
Risk characterization: If new concerns are found, understand how risky they are for human health
Intended and unintended effects: Look at both what's planned and any unexpected effects in the modified food
New hazards identification: Identify any new risks introduced by the modification
Changes in nutrients: Spot changes in key nutrients that could impact health.
Structured approach: Assess each case individually, with good-quality data and methods.
Science-based data: Gather info from various sources and evaluate using reliable methods
Multidisciplinary data: Use different scientific info considering various factors
Testing procedures: Make sure the tests are scientifically sound and comparable
Proportional risk management: The safety measures for biotech foods should match the risks identified
Equivalent protection: Different measures can provide the same level of safety
Uncertainties consideration: Manage uncertainties identified during risk assessment
Risk management measures: Might include labeling, approval conditions, and monitoring after market entry
Post-market monitoring: In some cases, ongoing monitoring is needed after the food is in the market

The European Union has recently implemented the concept of the qualified presumption of safety (QPS), a safety evaluation framework for microorganisms utilized in European food and feed contexts [56]. QPS operates at the species level and complements the European Novel Food System, which assesses safety at the strain level, facilitating information exchange between the two systems. Furthermore, the United Nations FDA has also approved microbial-derived food additives as GRAS (generally recognized as safe) status listed in 21 CFR Parts 131, 133, 136, and 137 for human consumption. The strain-level evaluation system is a novel food system compared to existing global regulations on microbial-derived food products [57]. Moreover, the safety assessment of food enzymes used in industrial applications adheres to the guidelines outlined in "Environmental Health Criteria 240" during evaluation by JECFA.

6.3 *Importance of Labeling and Consumer Transparency*

Labeling and consumer transparency hold immense significance in biotechnologically derived foods. These products, often containing genetically modified ingredients, necessitate clear and accurate labeling to empower consumers with informed choices. Such transparency is essential for several reasons. First, it enables individuals to make decisions aligned with their preferences, dietary needs, and ethical beliefs. Second, it aids those with allergies or sensitivities in avoiding potential health risks by clearly identifying allergenic components. Moreover, transparent labeling supports traceability, allowing for effective recall procedures and accountability in case of any issues. Beyond individual concerns, it contributes to regulatory compliance and builds consumer trust by showcasing the food industry's commitment to openness and safety. The codex CAC/GL 76-2011 provides a complete compilation of codex standards addressing the labeling of foods stemming from advanced biotechnology.

Labeling of GEM-Produced Food Substances as "GMO"

Regulatory agencies establish criteria for evaluating the safety of GEM-produced food substances. They also determine whether these substances or products containing them necessitate disclosure as "GMOs." It may seem daunting to interpret the regulatory requirements for labeling GEM-produced food substances, particularly if one is only acquainted with how regulations apply to agricultural products or commodities derived from them. Nonetheless, under most regulatory frameworks, GEM-produced food substances are not classified as "GMOs" for GMO labeling, contingent upon one or more of the following conditions:

- Food substances "produced with" a GEM are exempt from "GMO" labeling.
- Food substances without detectable DNA are exempt from "GMO" labeling.
- Food products created through specific genetic modifications are exempt from "GMO" labeling if they can be obtained through traditional breeding or occur naturally.
- Foods produced with GEM-derived processing aids/incidental additives are exempt from "GMO" labeling. It may necessitate no detectable rDNA in the final food in some jurisdictions.

7 Future Trends and Emerging Technologies

As mentioned in a report jointly issued by the Food and Agriculture Organization (FAO) and the World Health Organization (WHO), biotechnological innovations are propelling the progress of novel food additives and preservation methods, revolutionizing the way we enhance food safety, quality, and sustainability [46]. Bioengineered enzymes and precision fermentation are reshaping the food industry by delivering tailored, sustainable, and premium quality food products that meet the demands of modern consumers who prioritize both health and sustainability. Precision fermentation, as highlighted by a study published in "Nature Food," allows the targeted production of specific compounds, enabling personalized nutrition and efficient preservation techniques [47]. In recent years, the primary focus of many startup ventures in precision fermentation has centered on creating substitute proteins. However, there are also emerging companies exploring recombinant lipids and oligosaccharides. This is driven by the substantial growth opportunities in the market for animal-free protein alternatives, including meat, dairy, and egg substitutes [58].

Personalized nutrition, also known as precision nutrition, represents an emerging field of lifestyle intervention. It involves tailoring dietary recommendations at an individual level by incorporating a diverse set of data sources. These include demographic details, lifestyle-related information such as dietary habits and physical activity, phenotype-related data such as measurements of body characteristics, and standard clinical biomarkers for assessing disease risk [59]. Additionally, gene-related information, including genetic testing of single nucleotide polymorphisms and omics-based data, such as the gut microbiome composition, are integrated into the approach.

Nutrigenetics, metabolomics, and metagenomics are actively employed within precision nutrition. These technologies collectively enable a comprehensive exploration of how dietary and lifestyle factors contribute to assessing the risk factors associated with conditions such as obesity and type 2 diabetes. Notably, specific genes and genetic variations have been identified as critical factors influencing how individuals metabolically respond to particular diets [60].

Among its various applications, metabolomics examines how individuals metabolize identical diets differently in both health and disease states, enabling the customization of dietary interventions based on metabolic patterns [61]. This is particularly useful in improving the accuracy of dietary assessments by pinpointing food-derived biomarkers as intervention targets.

Metagenomics plays a crucial role in precision nutrition, as it facilitates a comprehensive examination of the interplay between diet and the microbiome [62]. This analysis helps identify distinct metabotypes that characterize metabolic risks, enabling tailored dietary interventions to improve overall health.

Precision fermentation offers the potential to enhance traditional food fermentations by improving efficiency and sustainability in producing valuable food ingredients. However, challenges such as limited genetic knowledge, low yields, purification issues, and environmental concerns persist. CRISPR–Cas9 holds promise for metabolic engineering and creating novel flavors in fermented foods, but regulatory approval remains a significant hurdle due to varying GMO-related laws worldwide [63]. Future adoption of precision methods, such as high-throughput screening and modeling, may lead to healthier and tastier products with increased production efficiency. Consumer acceptance may evolve over time due to factors such as food security, public trust in science, and media coverage.

8 Ethical and Societal Implications

8.1 *Potential Ethical Concerns Related to Biotechnological Additives and Preservatives*

The development and use of biotechnological additives and preservatives in the food industry raise several ethical concerns, including:

- **Safety and Health:** Ensuring the safety of biotechnological additives and preservatives is a primary ethical concern. Consumers have the right to know that the substances added to their food are safe for consumption. Evidence of potential long-term health effects or the absence of proper testing raises ethical questions about additives in the food supply.
- **Labeling and Transparency:** It is essential to pay attention to the importance of transparency in food labeling. The lack of transparency can lead to ethical concerns, particularly regarding biotech additives. These additives may have adverse effects on individuals with allergies or dietary restrictions, making it imperative to ensure that the labeling is accurate and informative. It is vital to address these concerns to guarantee the safety and well-being of consumers.
- **Environmental Impact:** The production of biotechnological additives and preservatives may have environmental consequences. Large-scale cultivation of genetically modified crops for food additives could lead to deforestation, soil degradation, or excessive water use, raising ethical concerns about environmental sustainability. It is crucial to prioritize sustainable production methods to minimize these impacts.
- **Biodiversity:** Biotechnological production processes involving GMOs can have unintended consequences for biodiversity, including the spread of modified genes to wild relatives and the potential for crossbreeding with wild relatives, which can lead to genetic contamination. When dealing with GMOs, it is crucial to prioritize ethical considerations to protect natural ecosystems and reduce potential risks.
- **Consumer Choice:** Ethical debates can arise when consumers have limited food choices due to the prevalence of biotechnological additives and preservatives. It can lead to economic challenges for traditional and artisanal food producers who avoid such additives and raise questions about respecting diverse dietary preferences.
- **Dependency on Technology:** There is concern that the widespread use of biotechnological additives may lead to dependence on technology in food production. Traditional food preservation and flavor enhancement methods may decline if biotechnological additives dominate the market. It could marginalize traditional food production methods and local food systems, which have cultural and ethical significance. This could lead to a reliance on limited technologies and the erosion of traditional food cultures.

- **Socioeconomic Impact:** The development and control of biotechnological additives and preservatives are often in the hands of large corporations. Large biotechnology corporations hold patents on certain genetically modified organisms in food production. This may limit access to these technologies for small-scale farmers or indigenous communities, exacerbating socioeconomic disparities. Ethical issues arise when this technology widens the economic gap between large agribusinesses and small-scale farmers or displaces traditional food producers from the market.
- **Access and Equity:** Accessing safe and nutritious food is a core ethical preoccupation. Biotechnological supplements incur more significant costs than their conventional counterparts, which could escalate inequity, particularly among the underprivileged. If such supplements result in augmented food prices, they may unfairly impact vulnerable populations grappling with food insecurity.
- **Long-Term Effects:** There are ethical considerations that arise regarding the enduring effects that biotechnological additives may have on both human health and the environment. Predicting the consequences of widespread adoption over extended periods may take time. Although a new biotechnological additive has gained widespread use, its long-term health and environmental effects still need to be fully understood. Potential risks that become apparent only after extensive use may raise ethical concerns.

Achieving ethical standards in food production requires a thoughtful approach that balances innovation and responsible stewardship. It includes conducting rigorous safety testing, providing transparent labeling, adopting sustainable production practices, and considering societal impacts. Engaging a diverse group of stakeholders in the decision-making process is imperative to ensure the best outcomes.

8.2 Social Perception and Favor of Additives from Biotechnological Facilities

Social perception and favor of genetically modified additives vary widely among individuals, communities, and regions. Various factors, including scientific literacy, cultural beliefs, economic considerations, and trust in regulatory agencies, influence this complex issue. Here are some key points to understand the social perception and acceptance of genetically modified additives:

Key consideration	Positive perception	Negative perception
Knowledge and awareness	Some consumers may have a positive perception of GM additives in regions with a strong understanding of biotechnology and its applications. They may view these additives as improving food quality, reducing the need for chemical additives, and enhancing crop yields	In areas where knowledge about biotechnology is limited, there may be more significant skepticism and fear associated with GM additives. The lack of awareness and understanding can lead to distrust and negative perceptions
Safety concerns	Consumers may be more willing to accept GM additives that have undergone safety assessments and have a history of safe consumption and regulation	Concerns about the long-term health effects of GM additives can lead to negative perceptions. Some individuals worry about allergenicity, age dependent issues, physiological status, unintended consequences, or unforeseen risks associated with consuming GM ingredients
Labeling and transparency	Providing clear and transparent labeling of GM additive products can positively influence consumer perception and acceptance. Numerous consumers desire the option to exercise their discretion in making informed determinations regarding their dietary preferences	If GM additives are not clearly labeled or labeled as inadequate or misleading, it can erode trust and lead to negative perceptions
Environmental and sustainability concerns	Some patrons believe that the integration of GM technology into crops has the potential to mitigate environmental impact through the amplification of crop yield. It can also reduce the need for chemical pesticides and herbicides	Concerns about the unintended ecological consequences of GM crops, such as the development of resistant pests, can lead to negative perceptions among those who prioritize environmental sustainability
Cultural and ethical beliefs	In cultures where food security is a significant concern, GM additives act as a means to increase food production and alleviate hunger	Cultural and ethical beliefs related to naturalness and traditional food production methods can lead to resistance against GM additives, particularly in cultures with solid agricultural traditions
Media and advocacy groups	Positive or negative media coverage and advocacy efforts can significantly influence public perception. Proponents and opponents of GM additives often use media campaigns to shape public opinion	Sensationalized or biased media coverage can contribute to negative perceptions and create consumer confusion
Regulation and policy	Strong regulatory frameworks and government oversight can reassure consumers about the safety of GM additives	Concerns about regulatory capture or inadequate government oversight can undermine trust in GM additives

9 Conclusion

The employment of biotechnology in preserving food possesses great potential in generating natural, sustainable, and safe additives and preservatives. This method utilizes customized microbial strains, precise fermentation, and gene editing to decrease allergens and pathogens, ensuring food safety and establishing regulatory protocols that ensure the safety of biotechnological processes. It is crucial to undertake further research and invest in exploring novel methods and applications. Adherence to safety and innovation is crucial for attaining optimal outcomes. Thus, it is crucial to perform extensive investigation and boost the commercialization of food preservatives and additives to advance the caliber of processed foods, curtail waste, and strengthen food safety. Such undertakings will expedite the overarching aim of guaranteeing the feasible and secure utilization of biotechnology for conserving food.

References

1. Varghese SM, Parisi S, Singla RK, Begum ASA (2022) Food additives. In: Trends in food chemistry, nutrition and technology in Indian sub-continent. Springer International Publishing, Cham, pp 25–29. https://doi.org/10.1007/978-3-031-06304-6_4
2. Hui YH, Sherkat F (2005) Handbook of food science, technology, and engineering—4 volume set. CRC Press, p 3618
3. Trasande L, Shaffer RM, Sathyanarayana S (2018) Food additives and child health. [Online]. http://publications.aap.org/pediatrics/article-pdf/142/2/e20181408/1064574/peds_20181408.pdf
4. Shetty, K., Paliyath, G., Pometto, A., Levin, R.E. (Eds.). (2006). Functional Foods and Biotechnology (1st ed.). CRC Press. https://doi.org/10.1201/9781420007725
5. Hanlon P, Sewalt V (2020) GEMs: genetically engineered microorganisms and the regulatory oversight of their uses in modern food production. Crit Rev Food Sci Nutr 61(6):959–970. https://doi.org/10.1080/10408398.2020.1749026
6. Tao X, Su L, Wang L, Chen X, Wu J (2020) Improved production of cyclodextrin glycosyltransferase from *Bacillus stearothermophilus* NO2 in *Escherichia coli* via directed evolution. Appl Microbiol Biotechnol 104(1):173–185. https://doi.org/10.1007/S00253-019-10249-8
7. Kryachko Y, Tanaka T, Nickerson MT, Korber DR (2023) Novel biotechnological approaches to improving aromas and flavors of legume-derived food products. Cereal Chem 100(3):556–566. https://doi.org/10.1002/CCHE.10641
8. Lülf RH, Vogel RF, Ehrmann MA (2021) Microbiota dynamics and volatile compounds in lupine based Moromi fermented at different salt concentrations. Int J Food Microbiol 354:109316. https://doi.org/10.1016/J.IJFOODMICRO.2021.109316
9. Rogers E (2023) Impossible foods: growing the plant-based meat sector at home and abroad. J Glob Bus Commun 14(1). https://doi.org/10.56020/001c.71493
10. Bhat ZF, Morton JD, Bekhit AE-DA, Kumar S, Bhat HF (2021) Cultured meat: challenges in the path of production and 3D food printing as an option to develop cultured meat-based products. In Alternative proteins. CRC Press, pp. 271–295. doi: https://doi.org/10.1201/9780429299834-9

11. Rahaman MM et al (2023) Natural antioxidants from some fruits, seeds, foods, natural products, and associated health benefits: an update. Food Sci Nutr 11(4):1657–1670. https://doi.org/10.1002/fsn3.3217
12. Kubo MTK et al (2021) Thermal processing of food: challenges, innovations and opportunities. A position paper. Food Rev Int 39(6):3344–3369. https://doi.org/10.1080/87559129.2021.2012789
13. Balasubramaniam VM, Ting EY, Stewart CM, Robbins JA (2004) Recommended laboratory practices for conducting high-pressure microbial inactivation experiments. Innov Food Sci Emerg Technol 5(3):299–306. https://doi.org/10.1016/j.ifset.2004.04.001
14. Hegarty JW, Guinane CM, Ross RP, Hill C, Cotter PD (2016) Bacteriocin production: a relatively unharnessed probiotic trait? F1000Res 5;2587. [version 1; referees: 2 approved]. doi: https://doi.org/10.12688/f1000research.9615.1
15. Gálvez A, López RL, Abriouel H, Valdivia E, Ben Omar N (2008) Application of bacteriocins in the control of foodborne pathogenic and spoilage bacteria. Crit Rev Biotechnol 28(2):125–152. https://doi.org/10.1080/07388550802107202
16. Ran Q et al (2022) A mixed culture of Propionibacterium freudenreichii and Lactiplantibacillus plantarum as antifungal biopreservatives in bakery product. Food Biosci 47:101,456. https://doi.org/10.1016/J.FBIO.2021.101456/REFERENCES
17. Rai M, Pandit R, Gaikwad S, Kövics G (2016) Antimicrobial peptides as natural biopreservative to enhance the shelf-life of food. J Food Sci Technol 53(9):3381–3394. https://doi.org/10.1007/s13197-016-2318-5
18. Premanath R, James JP, Karunasagar I, Vaňková E, Scholtz V (2022) Tropical plant products as biopreservatives and their application in food safety. Food Control 141:109185. https://doi.org/10.1016/j.foodcont.2022.109185
19. Khorshidian N, Yousefi M, Khanniri E, Mortazavian AM (2018) Potential application of essential oils as antimicrobial preservatives in cheese. Innov Food Sci Emerg Technol 45:62–72. https://doi.org/10.1016/j.ifset.2017.09.020
20. Ramos-Vivas J et al (2021) Phages and enzybiotics in food biopreservation. Molecules 26(17):5138. https://doi.org/10.3390/MOLECULES26175138
21. Chandra P, Sharma RK, Arora DS (2020) Antioxidant compounds from microbial sources: a review. Food Res Int 129:108849. https://doi.org/10.1016/j.foodres.2019.108849
22. Valentão P, Fernandes E, Carvalho F, Andrade PB, Seabra RM, de Lourdes Bastos M (2002) Antioxidant activity of Hypericum androsaemum infusion: scavenging activity against superoxide radical, hydroxyl radical and hypochlorous acid. Biol Pharm Bull 25(10):1320–1323. https://doi.org/10.1248/bpb.25.1320
23. Tresina PS, Paulpriya K, Mohan VR, Jeeva S (2017) Effect of gamma irradiation on the nutritional and antinutritional qualities of Vigna aconitifolia (Jacq.) Marechal: an underutilized food legume. Biocatal Agric Biotechnol 10:30–37. https://doi.org/10.1016/j.bcab.2017.02.002
24. Kato S, Kawasaki T, Urata T, Mochizuki J (1993) In vitro and ex vivo free radical scavenging activities of carazostatin, carbazomycin B and their derivatives. J Antibiot (Tokyo) 46(12):1859–1865. https://doi.org/10.7164/antibiotics.46.1859
25. Thenmozhi M Characterization of antioxidant activity of Streptomyces species VITTK3 isolated from Puducherry Coast, India [Online]. http://www.megasoftware.net
26. Saadat YR, Khosroushahi AY, Movassaghpour AA, Talebi M, Gargari BP (2020) Modulatory role of exopolysaccharides of Kluyveromyces marxianus and Pichia kudriavzevii as probiotic yeasts from dairy products in human colon cancer cells. J Funct Foods 64:103675. https://doi.org/10.1016/j.jff.2019.103675
27. Jain S, Yadav H, Sinha PR (2009) Antioxidant and cholesterol assimilation activities of selected lactobacilli and lactococci cultures. J Dairy Res 76(4):385–391. https://doi.org/10.1017/S0022029909990094
28. Wang K et al (2019) Physicochemical characteristics and in vitro and in vivo antioxidant activity of a cell-bound exopolysaccharide produced by Lactobacillus fermentum S1. Int J Biol Macromol 139:252–261. https://doi.org/10.1016/j.ijbiomac.2019.07.200

29. Aguilar-Toalá JE et al (2019) Modulatory effect of the intracellular content of Lactobacillus casei CRL 431 against the aflatoxin B1-induced oxidative stress in rats. Probiotics Antimicrob Proteins 11(2):470–477. https://doi.org/10.1007/s12602-018-9433-8
30. Horta A et al (2014) Antioxidant and antimicrobial potential of the Bifurcaria bifurcata epiphytic bacteria. Mar Drugs 12(3):1676–1689. https://doi.org/10.3390/md12031676
31. Dose J et al (2016) Free radical scavenging and cellular antioxidant properties of Astaxanthin. Int J Mol Sci 17(1):103. https://doi.org/10.3390/ijms17010103
32. Naguib YMA (2000) Antioxidant activities of Astaxanthin and related carotenoids. J Agric Food Chem 48(4):1150–1154. https://doi.org/10.1021/jf991106k
33. Blagojević D, Babić O, Rašeta M, Šibul F, Janjušević L, Simeunović J (2018) Antioxidant activity and phenolic profile in filamentous cyanobacteria: the impact of nitrogen. J Appl Phycol 30(4):2337–2346. https://doi.org/10.1007/s10811-018-1476-4
34. Ranković B, Kosanić M (2015) Lichens as a potential source of bioactive secondary metabolites. In Lichen secondary metabolites. Springer International Publishing, pp. 1–26. doi: https://doi.org/10.1007/978-3-319-13374-4_1
35. Tomović J, Kosanić M, Ristić S, Ranković B, Stanojković T, Manojlović N (2018) Chemical composition and bioactive properties of the lichen, *Pleurosticta acetabulum*. Trop J Pharm Res 16(12):2977. https://doi.org/10.4314/tjpr.v16i12.23
36. Dey TB, Chakraborty S, Jain KK, Sharma A, Kuhad RC (2016) Antioxidant phenolics and their microbial production by submerged and solid-state fermentation process: a review. Trends Food Sci Technol 53:60–74. https://doi.org/10.1016/j.tifs.2016.04.007
37. Tapal A, Tiku PK (2019) Nutritional and nutraceutical improvement by enzymatic modification of food proteins. In Enzymes in food biotechnology. Elsevier, pp 471–481. doi: https://doi.org/10.1016/B978-0-12-813280-7.00027-X
38. Harper JK et al (2003) Pestacin: a 1,3-dihydro isobenzofuran from Pestalotiopsis microspora possessing antioxidant and antimycotic activities. Tetrahedron 59(14):2471–2476. https://doi.org/10.1016/S0040-4020(03)00255-2
39. Larios-Cruz R et al (2019) Valorization of grapefruit by-products as solid support for solid-state fermentation to produce antioxidant bioactive extracts. Waste Biomass Valorization 10(4):763–769. https://doi.org/10.1007/s12649-017-0156-y
40. Zhang Y et al (2014) Isolation and identification of an antioxidant peptide prepared from fermented peanut meal using *Bacillus subtilis* fermentation. Int J Food Prop 17(6):1237–1253. https://doi.org/10.1080/10942912.2012.675605
41. Pérez YA, Urista CM, Cerda AM, Sánchez JÁ, Rodríguez FR (2019) Antihypertensive and antioxidant properties from whey protein hydrolysates produced by encapsulated *Bacillus subtilis* cells. Int J Pept Res Ther 25(2):681–689. https://doi.org/10.1007/s10989-018-9714-9
42. Piu LD et al (2014) Exploitation of starch industry liquid byproduct to produce bioactive peptides from rice hydrolyzed proteins. Food Chem 155:199–206. https://doi.org/10.1016/j.foodchem.2014.01.055
43. Yu HC, Hsu JL, Chang CI, Tan FJ (2017) Antioxidant properties of porcine liver proteins hydrolyzed using Monascus purpureus. Food Sci Biotechnol 26(5):1217–1225. https://doi.org/10.1007/s10068-017-0166-3
44. Leite P, Silva C, Salgado JM, Belo I (2019) Simultaneous production of lignocellulolytic enzymes and extraction of antioxidant compounds by solid-state fermentation of agro-industrial wastes. Ind Crop Prod 137:315–322. https://doi.org/10.1016/j.indcrop.2019.04.044
45. Dulf FV, Vodnar DC, Dulf EH, Pintea A (2017) Phenolic compounds, flavonoids, lipids and antioxidant potential of apricot (Prunus armeniaca L.) pomace fermented by two filamentous fungal strains in solid-state system. Chem Cent J 11(1):92. https://doi.org/10.1186/s13065-017-0323-z
46. Dulf FV, Vodnar DC, Dulf E-H, Diaconeasa Z, Socaciu C (2018) Liberation and recovery of phenolic antioxidants and lipids in chokeberry (Aronia melanocarpa) pomace by solid-state bioprocessing using *Aspergillus niger* and Rhizopus oligosporus strains. LWT 87:241–249. https://doi.org/10.1016/j.lwt.2017.08.084

47. Vattem DA, Shetty K (2002) Solid-state production of phenolic antioxidants from cranberry pomace by Rhizopus oligosporus. Food Biotechnol 16(3):189–210. https://doi.org/10.1081/FBT-120016667
48. Kaur P, Ghoshal G, Jain A (2019) Bioutilization of fruits and vegetables waste to produce β-carotene in solid-state fermentation: characterization and antioxidant activity. Process Biochem 76:155–164. https://doi.org/10.1016/j.procbio.2018.10.007
49. de Oliveira SD et al (2020) Improvement in physicochemical characteristics, bioactive compounds and antioxidant activity of acerola (Malpighia emarginata D.C.) and guava (Psidium guajava L.) fruit byproducts fermented with potentially probiotic lactobacilli. LWT 134:110200. https://doi.org/10.1016/j.lwt.2020.110200
50. Sachindra NM, Bhaskar N (2008) In vitro antioxidant activity of liquor from fermented shrimp biowaste. Bioresour Technol 99(18):9013–9016. https://doi.org/10.1016/j.biortech.2008.04.036
51. Martí-Quijal FJ et al (2020) Isolation, identification and investigation of fermentative bacteria from sea bass (Dicentrarchus labrax): evaluation of antifungal activity of fermented fish meat and by-products broths. Food Secur 9(5). https://doi.org/10.3390/FOODS9050576
52. Ismail SA (2019) Microbial valorization of shrimp byproducts via the production of thermostable chitosanase and antioxidant chitooligosaccharides. Biocatal Agric Biotechnol 20:101269. https://doi.org/10.1016/j.bcab.2019.101269
53. Choksawangkarn W, Phiphattananukoon S, Jaresitthikunchai J, Roytrakul S (2018) Antioxidative peptides from fish sauce byproduct: isolation and characterization. Agric Nat Resour 52(5):460–466. https://doi.org/10.1016/j.anres.2018.11.001
54. OECD-FAO Agricultural Outlook 2023–2032. doi: https://doi.org/10.1787/08801ab7-en
55. A guide to World Food Safety Day 2023: food standards save lives. https://www.who.int/publications/i/item/WHO-HEP-NFS-AFS-2023.6. Accessed 13 Sep 2023
56. Koutsoumanis K et al (2023) Update of the list of qualified presumption of safety (QPS) recommended microbiological agents intentionally added to food or feed as notified to EFSA 17: suitability of taxonomic units notified to EFSA until September 2022. EFSA J 21(1):e07746. https://doi.org/10.2903/j.efsa.2023.7746
57. Heo S et al (2022) Safety assessment systems for microbial starters derived from fermented foods. J Microbiol Biotechnol 32(10):1219–1225. https://doi.org/10.4014/jmb.2207.07047
58. Augustin MA, Hartley CJ, Maloney G, Tyndall S (2023) Innovation in precision fermentation for food ingredients. Crit Rev Food Sci Nutr:1–21. https://doi.org/10.1080/10408398.2023.2166014
59. Ordovas JM, Ferguson LR, Tai ES, Mathers JC (2018) Personalized nutrition and health. BMJ 361:bmj.k2173. https://doi.org/10.1136/bmj.k2173
60. Tremblay BL, Cormier H, Rudkowska I, Lemieux S, Couture P, Vohl M-C (2015) Association between polymorphisms in phospholipase A2 genes and the plasma triglyceride response to an n-3 PUFA supplementation: a clinical trial. Lipids Health Dis 14(1):12. https://doi.org/10.1186/s12944-015-0009-2
61. Tebani A, Bekri S (2019) Paving the way to precision nutrition through metabolomics. Front Nutr 6. https://doi.org/10.3389/fnut.2019.00041
62. Brouwer-Brolsma EM et al (2017) Combining traditional dietary assessment methods with novel metabolomics techniques: present efforts by the Food Biomarker Alliance. Proc Nutr Soc 76(4):619–627. https://doi.org/10.1017/S0029665117003949
63. Chai KF, Ng KR, Samarasiri M, Chen WN (2022) Precision fermentation to advance fungal food fermentations. Curr Opin Food Sci 47:100,881. https://doi.org/10.1016/j.cofs.2022.100881

Part IV
Recent Development in Genetically Engineered Foods

Chapter 16
Prospectus of Genetically Modified Organisms in Food

Shivaleela V. B. and Avitha K. M.

1 Introduction

Biotechnological processes have been used for millennia to make human consumables like cheese, yogurt, beer, wine, and bread. Ancient peoples used microorganisms like yeast and bacteria without even realizing they existed. We now know that there are many different strains of bacteria and yeast, some of which are used in commercial fermentation processes after being chosen based on certain qualities to enhance the quality of the final product or production procedures [1]. The genetic origins of the distinctive characteristics of the majority of the microorganisms used in fermentation processes are still mostly unknown. Genetic engineering is becoming a necessary tool for biotechnology as well as many other contemporary biological and medical disciplines. Through the use of genetic engineering, modern biotechnology has made it possible to transfer certain genes from one organism to another, as well as between unrelated species. Genetically modified organisms (GMOs) are the resultant organisms, which can be plants, animals, or microbes whose genetic material has been changed in a way that does not happen normally through natural recombination and mating.

Ten thousand years ago, human ancestors first domesticated and consumed plants. During this time, they engaged in "selective breeding" and "artificial selection," two Darwinian concepts that broadly refer to selecting parent organisms with desirable traits (such as harder stems) and breeding them to propagate those traits. [2]. It has also been documented that similar methods can be used to produce modern, highly desirable-for-human eating varieties of apples, broccoli, and bananas

Shivaleela V. B.
Pomla Natuals LLP, BEC STEP, Bagalkot, Karnataka, India

Avitha K. M. (✉)
P.G Department of Botany, BVVS Akkamahadevi Women's College,
Bagalkote, Karnataka, India

J. S. Yaradoddi et al. (eds.), *Frontiers in Food Biotechnology*,
https://doi.org/10.1007/978-981-97-3261-6_16

that differ from their ancestral plant forms. The discovery that genetic material could be transferred between various animals in 1946 marked the beginning of the advancements that led to modern genetic manipulation. The discovery of the DNA double helical structure and the formulation of the fundamental dogma—that DNA is transcriptionally converted to RNA and then translated into proteins—by Watson and Crick in 1954 came next. As such, a set of pioneering studies conducted in 1973 by Boyer and Cohen entailed "cutting and pasting" DNA between various species by means of restriction endonucleases and DNA ligase—a.k.a. "molecular scissors and glue" [3] created the world's first genetically modified organism with success. The first genetically modified plants (GM) in agriculture were petunias and tobacco, which were successfully developed in 1983 by three separate research groups. China was the first nation to commercialize genetically modified tobacco for viral resistance in 1990. The Flavr Savr tomato, grown in California, was the first genetically modified plant (GM) to be approved for human consumption by the Food and Drug Administration (FDA) in 1994. Through the use of antisense technology, this tomato was genetically altered to impede the development of the polygalacturonase enzyme, which delayed ripening and made it resistant to rot. Since then, in 1995 and 1996, a number of transgenic crops were approved for widespread human production. Corn/maize, cotton, and potatoes (*Bacillus thuringiensis* (Bt) gene modification, Ciba-Geigy and Monsanto), canola (Calgene: improved oil output), and cotton (Calgene: bromoxynil resistance) were among the first plants approved by the Food and Drug Administration (FDA) and Monsanto's Roundup Ready soybeans (glyphosate resistance). The genetically modified crop pipeline has now spread to include other fruits, vegetables, and cereals like lettuce, strawberries, eggplant, sugarcane, rice, wheat, carrots, and so on. Its intended applications include boosting the bioproduction of vaccines, providing nutrients for animal feed, and conferring salinity and drought resistance traits to plants that grow in unfavorable environments and climates. The economy and the environment have profited from genetically modified crops since they were first commercialized. Over a rather limited region, the worldwide food crop yield grew by more than 370 million tons between 1996 and 2013. Moreover, genetically modified crops have been shown to lessen their effects on the environment and ecology, which has increased the diversity of species (Fig. 16.1). Thus, it should come as no surprise that the majority of environmentalists worldwide, growers, and agricultural scientists have praised genetically modified crops.

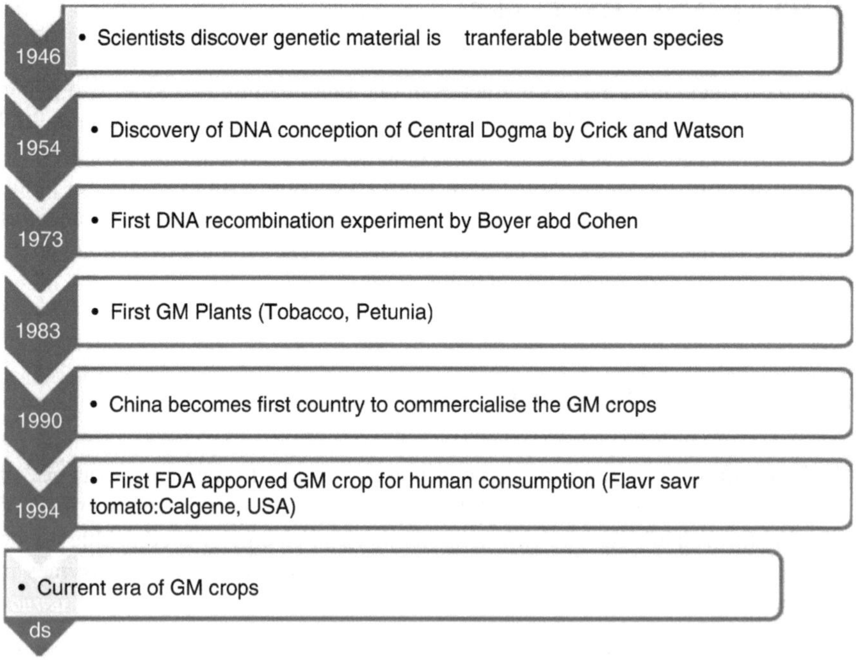

Fig. 16.1 A timeline of development events up to the current G.M. crop era

2 Why Plants Are Genetically Modified?

It is feasible to exert tighter control over the genes by genetically altering the plant in a lab. The favorable genes and the unfavorable genes can be distinguished using GM.

Additionally, GM allows for the crossing of species that would not naturally breed with one another. Some reasons for genetic modification include:

- Increased yields.
- Disease resistance.
- Insect damage resistance.
- Herbicide tolerance and better nutritional value.
- Improved climate survival through increased resistance to drought, flood, or cold conditions.
- Allowing for the utilization of formerly unsuitable territory, longer shelf life, improved health, increased agricultural yields, less farming expenses, and increased farm profit and environmental outcomes.
- Foods with an additional gene sequence.
- Food containing a gene deletion.

3 Why to Make GM Foods?

In the past, integrating the desired genes in one plant required crossing two plants of the same species or closely related species, which was a difficult task that required more time and care. Growing crops with higher yields or better quality, resistance to pests or diseases, or tolerance is useful from an economic and agricultural point. Desirable genes may offer plants a way to resist these issues. The development of transgenic technology makes it possible to harvest beneficial genes from a variety of living sources and combine them quite simply.

4 What Allows for Genetic Modification?

All living things have the same genetic makeup thanks to the same sequence of amino acids that make up DNA. Molecular cloning and genetic engineering and recombinant technologies are two methods that can change or modify this amino acid sequence from one organism to another. There are two essential enzymes in this process:

1. Restriction enzyme is employed to "CUT" DNA segments from a single genome.
2. To "PASTE" them into another genome, DNA. Ligases are utilized.

DNA transfers from one organism to another organism that possesses the required attribute and identifies the gene sequence that codes for it. Insert the gene sequence into the genome of the plant cell. Allow the genetically engineered cell to grow into a plant and the Plant to propagate than, a vector can carry DNA. The vector can be a gene gun, viruses and bacteria also can be utilized to transfer gene.

5 Method of Transformation

The entire procedure consists of the following three fundamental phases, which should enable stable DNA integration into the host genome without structural modification:

A specific amount of copies of the transforming DNA must be incorporated, and the new phenotype must remain stable for numerous generations. Additionally, the inserted gene must also be character-regulated.

Transformation techniques:

1. *Agrobacterium tumefaciens.*
2. Electroporation.
3. Polyethylene Glycol.
4. Fiber made with silicon carbide.

5. Gene Gun.
6. Microinjection.

1. *Agrobacterium tumefaciens*: *Agrobacterium tumefaciens* mediates plant transformation. *Agrobacterium tumefaciens* and *Agrobacterium rhizo* genes are virulent strains, the large Ti plasmid. These bacteria can transmit T-DNA, resulting in the development of the Crown of Gall Disease.

   The preliminary outcomes of the T-DNA transfer procedure to plant cells show three crucial characteristics: The incorporation of T-DNA leads to tumor development. It is possible to transfer any DNA sequence into plant cells, where the transcription of T-DNA genes occurs solely.
2. Electroporation: In this method, the material is exposed to high-voltage electric pulses while being incubated in a buffer solution containing DNA which migrates through plasma membrane pores created by high voltage, then integrates into the genome. This method Mainly Used for: - Rice, Wheat and Maize.
3. Polyethylene glycol: The plant protoplast with bare DNA can be altered by polyethylene glycol treatment in the presence of the divalent cation Ca^{2+}. Both contribute to the instability of the plant's plasma membrane and facilitate DNA penetration. The DNA is integrated into the genome once it has reached the nucleus.
4. Fiber made with silicon carbide: In this method, plant material is introduced into a buffer containing silicon carbide fibers and DNA. Then it fibers penetrate the cell wall and plasma membrane allowing the DNA to gain access to enter inside the cell.
5. Gene gun: A particle gun is used to shoot small bits of metals coated with the gene into the plant.
6. Microinjection: The method entails injecting growing embryos with DNA material delivered as a solution using a very fine needle.

6 Genetic Modification Applications

The first reports of genetic alteration in a lab date back to the early 1970s. Since then, genetic engineering has been the subject of several research projects and applications in the fields of industrial, agriculture, medicine, the environment, and food production. The crops listed below are resistant to viruses, insects, and herbicides.

6.1 In the Field of Medicine

- Insulin to treat diabetic mellitus
- Production of human growth hormone
- Production of blood clotting factors VIII and IX

- The treatment of cystic fibrosis
- Research into human and animal diseases

Other fields include environmental cleanup of soil spills, treatment of contaminated land and water, and production of useful compounds like enzymes and plants that produce industrial chemicals from renewable sources.

7 Crops Being Developed

Sweet potato for animal feed that has higher concentrations of protein and the amino acid lysine. Golden rice that has been Further developed to boost levels vitamin A, bioavailability of iron, zincand vitamin E, to improve the quality of protein. Transgenic rice was created by a California-based business to improve the oral rehydration therapy for diarrhea.

The following phases are included in the process of plant transformation by particle bombardment:

1. Take out the protoplasts from the leaf tissue.
2. Introduce DNA-coated particles into protoplasts using particle cannon.
3. Grow back into complete plants.
4. Acclimate the transgenic plants in a greenhouse. A review of the mechanism of production and labeling issues regarding genetically modified food.

8 Common Foods with Genetic Modification

1. Soybean, corn and sugar-beet: resistant to glyphosate by inserting herbicide resistant gene.
2. Cottonseed oil: by introducing the gene for the pest-resistant B.T. crystal protein.
3. Tomato: eliminating the polygalacturonase gene, which is the agent that causes fruits to soften after harvest.
4. Potatoes: By inhibiting the GBSS (granule-bound starch synthase) gene, which generates amylose, amylopectin-rich types can be created.
5. Rapeseed (canola): introducing a new gene increased oleic acid concentration.
6. Rice: Added daffodil gene for high vitamin A content.

9 Golden Rice

Ingo Potrykus invented it. Golden rice is a type of rice created by means of genetic engineering to produce beta-carotene (pro-vitamin A) through biosynthesis precursors in the endosperm, the edible section of the rice grain. There are above

120 million undernourished children worldwide. Golden rice may lessen the one- to two-million annual fatalities brought on by nutritional deficiencies.

In the part of rice that people eat the endosperm, the rice plant can naturally produce beta-carotene, which is a carotenoid pigment that occurred in the leaves and involved in the photosynthesis. Since photosynthesis does not take place in the endosperm, the plant does not typically develop the pigment. However, the plant does not normally produce the pigment in the endosperm since photosynthesis does not occur in the endosperm. Golden rice was created by incorporating rice with two beta carotene biosynthesis genes i.e Psy (Phytoene synthase) & Lyc (lycopene cyclase). The psy and lyc genes were transformed into the rice nuclear genome and placed under the control of an endosperm specific promoter, so that they are only expressed in the endosperm.

Lycopene is the end result of the planned process; the rice would have been red if the plant had gathered lycopene. According to a recent study, rice gets its unique yellow hue from an endogenous enzyme that transforms lycopene in the endosperm into beta-carotene.

In 2005, a new variety called golden rice-2 was announced which produces up to 23 times more beta carotene than the original variety of golden rice.

10 Iron-Fortified Rice

Numerous strategies have been employed to improve iron accumulation and modify iron metabolism because cereal grains are low in a number of critical mineral elements, including iron. Since ferritin is a universal iron store protein found in all living things, ferritin genes have been put into wheat and rice plants to increase their iron content.

11 Tolerant of Cold

By incorporating the antifreeze gene from a cold water fish into tomato plants, scientists have created a type that is resistant to frost. The hard water flounder, a fish that can withstand extremely low temperatures, is the source of the antifreeze genes.

A gene in the flounder allows in order to generate chemical antifreeze. This is separated from the antifreeze DNA and attached to a DNA fragment known as a plasmid. This hybrid DNA, which is a combination of DNA from two different sources, is known as recombinant DNA. The recombinant DNA, including the antifreeze gene, is placed in a bacterium. The bacterium is allowed to reproduce many times producing lots of copies of the recombinant DNA. The bacteria produce a lot of copies of the recombinant DNA as it is allowed to multiply again; as a result, bacteria are present in the cells of the tomato plant. The antifreeze gene in the bacteria's plasmids integrates into the DNA of tomato plant cells. The growth medium is used to cultivate tomato cells and enhance plant growth. In opposition to the hp. gene, the OG-c (crimson) gene and the hp. (high pigment) gene, which expresses beta-carotene, these tomato seedlings have now been planted. If high pigment gene is expressed, beta carotene content increases by 40% and the fruits get a red colour. If OG-c gene is also expressed, beta-carotene content is decreased by 40 and the fruit gets a crimson color. The fruit becomes scarlet and loses 40% of its beta-carotene content if the OG-c gene is also activated.

12 Tomato Is High in Vitamins

The Agrobacterium normally infects plants and causes a variety of diseases; replacing that gene with one that is favorable results in a new genetic structure with beneficial features.

Carrots' vibrant orange hue results from beta-carotene, an amino acid that is needed to synthesize vitamin A. Therefore, adding the color gene to the tomato also improves both its look and the level of vitamin A in it to the appropriate level.

13 NASA's Protein-Fortified Sweet Potatoes

By adding a synthetic protein storage gene, researchers at Tuskegee University led by Dr. Marceline Egnin have produced a sweet potato that is enhanced with protein. NASA is looking for nutritionally rich crops provide sufficient protein and amino acids to astronauts with their complete nutritional requirements. Sweet potato is a good source of energy because it has high level of minerals and carbohydrates.

14 Purple Potatoes

The gene for granule bound starch synthase (GBSS) (the key enzyme for the synthesis of amylose) was switched off by inserting antisense copy of the GBSS gene.

15 Modified Potato Protein

The Potato is a non-cereal food cropbalanced in the amount amino acids like lysine, tyrosine, methionine and cysteine. Potatoes have been modified using the tuber-specific protein amaranth seed albumin (AmA1). In actuality, its amino acid content exceeds averages for other organisms. The WHO has recommended genetically modified food as a source of a nutrient-dense protein because of this protein's non-allergenicity in its pure form. It has been observed that transgenic tubers of the potato with the AmA1 gene had 35–45% more total protein and 2.5–4 times more lysine, tyrosine, methionine, and cysteine.

16 B.T. Soybean

The soybean looper and the velvet bean caterpillar are the two insects that transgenic soybeans resistant to insects are designed to withstand. These pests can significantly reduce yield because they feed on the soybean plant's leaves. Populations of velvet bean caterpillars can quickly reach harmful proportions. Dimilin is frequently used as a prophylactic measure by farmers in regions where the velvet bean caterpillar is a serious issue. Scientists have introduced the *Bacillus thuringiensis* gene into soybeans to sustain the crop's productivity. This gene produces an insecticidal protein. A common soil bacterium called *Bacillus thuringiensis* is the insect-resistant gene source.

17 Soybean Tolerant to Roundup

A genetically altered soybean with received an immunity to the herbicide Roundup is known as a Roundup Ready soybean. The only significant issue in soybean growing was weeds because the inherent trypsin inhibitors in soybeans fight against pests, making them revolutionary. The herbicide's glyphosate would disrupt the soybean plant's ESPSP gene, which is essential for maintaining the "biosynthesis of aromatic metabolites." A plasmid which was transferred to the soybean cells through the cauliflower mosaic virus was soon developed to provide immunity to glyphosate-containing herbicides, and, after this process was perfected, the Roundup Ready soybean was ready, first debuted the US market in 1996.

18 Genetically Modified Corn

Corn is purposefully genetically modified (GM) to have features advantageous to agriculture. Herbicide resistance and insect pest resistance are two traits engineered into maize; the latter is attained by incorporating a gene that produces the toxins produced by Bacillus thuringiensis (Bt).

Transgenic maize was grown commercially in 11 countries, including the United States (where 85% of the maize crop was genetically modified), Brazil (36% GM), Argentina (83% GM), South Africa (57% GM), Canada (84% GM), the Philippines (19% GM) and Spain (20% GM) along with 11 nations, there are glyphosate-resistant corn cultivars, such as Liberty and Roundup. Pioneer Hi-Bred has marketed corn hybrids with tolerance to imidazoline herbicides under the trademark

"Clearfield" -though in these hybrids, the herbicide-tolerance trait was bred without the use of genetic engineering.

As a result, imidazoline-tolerant maize is exempt from the regulatory framework that applies to the acceptance, application, exchange, and ingestion of transgenic crops. In the United States, GM corn that resists herbicides is planted.

19 Canola

In 1978, "Canadian oil, low acid" became the moniker "canola." Because chlorophyll was present, rapeseed oil had an unpleasant greenish tint and peculiar flavor. Additionally, it has a significant amount of erucic acid. Erucic acid consumption in big doses has the potential to harm the heart, according to animal experiments. Because it contains a lot of the bitter compounds known as glucosinolates, rapeseed feed grain did not appeal to animals very much and they refused to consume it. The most resilient Canola type of rapeseed to disease and drought to date is thought to be a cultivar created in 1998.

Canola oil: About 43% of a canola seed oil the remainder is hemp meal which is well-tolerated by animals and is now used to produce high-quality animal feed. Canola oil may be produced from 22.68 kg of rapeseed. Many foods have canola oil as a crucial ingredient. Markets all around the world are seeing increased demand for it because of its reputation as a healthy oil.

Numerous health professional associations like the American Dietetic Association and American Heart Association acknowledge that canola oil has a high monounsaturated fat content, a low saturated fat content (less than 7%), and a favorable omega-3 fatty acid profile and a low content of trans fat (which has well-established heart benefits). Although wild rapeseed oil contains significant amounts of a known toxin, a erucic acid, the cultivar used to produce commercial food-grade canola oil was bred to contain less than 2% erucic acid, levels that are not believed to cause harm in humans and no ill health effects has been associated with consumption by humans of the genetically modified oil.

20 Papaya

The virus that causes papaya ring spot, drastically reducing fruit output, threatens papaya cultivation. The College of Hawaii papaya resistant to the ring spot virus illness was produced. This was accomplished by inserting certain viral genes into the papaya genome that encode capsid proteins. Papaya plant exhibits what might be called an "immune response" in reaction to these viral capsid proteins. In Hawaii in 1999 the first resistant papaya varieties were grown commercially. These genetically modified papayas are approved for consumption both in US and in Canada.

21 Genetically Modified Food Labeling

On the fruits-the Price Look-Up (PLU) code found can be used to determine if the fruit was cultivated organically, using chemical fertilizers, fungicides, or herbicides, or was genetically engineered. For fruits cultivated, the PLU code on the sticker consists of four numbers with chemicals.

Organically grown fruit has a five-numeral PLU prefaced by the number 9. Genetically engineered fruit has a fivenumeral PLU prefaced by the number 8. traditionally cultivated banana, for instance, would be: 4011.

The code for an organic banana is 94011. A banana that has been genetically modified would be 84011.

22 Federal Guidelines

Except for meat and poultry products, the United States Department of Agriculture (USDA) the FDA primarily regulate food additives. The FDA, USDA, and Environmental Protection Agency collaborate on issues related to food safety, such as setting tolerance limits for pesticide residues in foods.

The FDA published the recommendations to assist businesses in determining whether they require FDA approval for a genetically modified food product. The FDA decided because specific safety issues, like if a tomato's peanut protein gene was added, would only necessitate a special review of a genetically modified food product.

23 Current Labeling Regulations

The GM/GE food labeling policy is made public by the US Food & Drug Administration.

In order to lower the quantity of saturated fat in soybeans, biotechnology was used to manufacture the high oleic acid soybean oil and produce the cornmeal used in this product. Genetically modified ingredients are not used by us or produced using biotechnology.

24 Risks Associated with the Use of Genetic Engineering

(a) The technology involves playing with live things and interacting with other living things; therefore, the hazards are naturally higher and less predictable than in trials involving chemicals. The GMO cannot be removed from the environment after its release due to the unpredictable nature of living things to manage because genetically modified organisms are living things.

GMOs can spread to new environments, colonize those locations, and reproduce if they become diluted. As long as the environment is conducive to new activities, such as synthesizing metabolic products, enzymes, and toxins, GMOs continue to function metabolically. Living things can't be recalled after they've been established.

(b) Genetically modified food dangers.

Although improving human life quality is genetic engineering's main objective, it is important to discuss any potential negative effects this technology may have on individuals. Genetic material can enter the human body through food, bacteria, viruses, vaccines, and medications. Usually, a marker gene is added to foods derived from the desired attribute together with the gene responsible for GMOs.

The authors of these experiments assert that if the marker genes were successfully transferred, the organism will exhibit a novel resistance to particular antibiotics. People who eat food that has these genes may have problems, particularly if they are unaware that the food contains the genes. When a person consumes a product, their gene for antibiotic resistance may cause any antibiotics they take to become ineffective or less effective.

Consuming food that has genes resistant to antibiotics on a frequent basis can also cause people to become resistant to antibiotics. Food that has undergone genetic alteration may be less nutritious than food that occurs naturally.

For many people, genetically modified organisms can cause allergies, in addition to any issues that may arise from marker genes and declining nutritional quality. People may consume things detrimental to them if they are not entirely aware of the nature of the food they are eating. Even if a person is familiar that they should avoid a certain drug, they might be unaware that introducing A similar chemical was induced to be expressed in the product by a new gene. Folks wouldn't anticipate the introduction of meat genes into tomatoes, for instance. Unaware that the tomato is the same chemical causing the allergic reaction, a person allergic to meat may also be allergic to tomatoes. This is why foods containing G.M.O.s must have clear labels.

Religious objections to ingesting G.M.O.s are another possibility. Some of the potential issues that consuming such meals for religious people who follow dietary regulations claim Chaudry and Regenstein. Genetic engineering is acceptable under Jewish law (Halacha) if it improves the quality or quantity of the global food supply. However, in the Muslim world, there is no need for food crops to be genetically modified because God made everything perfectly and man has no right to interfere with anything that God has created with His creative wisdom. Contrarily, there is no broad agreement among Christians over whether G.M. technology, G.M. research, or G.M. food use are acceptable.

25 Biodiversity Risks

Genetically modified plants could have a negative impact on biodiversity in the ecosystem. Birds, insects, and other animals that depend on specific crops for survival may not be able to consume the genetically modified crops as a result of the inserted gene or mutation. They can be poisoned by the new qualities or allergic to them. These animals would, therefore, need to discover other food sources or risk going hungry. The entire food chain as well as the predator–prey interactions would be affected by this.

Indigenous fauna and flora may be displaced due to the introduction of a changed creature into the environment. If the new strain outperforms the parent strain, it can saturate the habitat or drive out the natural populations.

26 Conclusion

The topic of genetically modified food is still relatively new in a few countries, and its acceptance has been constrained primarily by unfounded myths that downplay significant benefits. India is the second most popular country in the world, making it difficult to feed the sizable number of hungry and malnourished people. The advancement of biotechnology is the best option for building a more resilient country with healthier population. In order to create high-yield, nutrient-rich grain and generate income, genetically modified crop propagation seems to be a viable option because it requires less reliance on pesticides and herbicides and does no harm to the environment or agriculture. This technology is still quite young, despite all of its benefits. Genetic modification, organic farming, and crop rotation are among the most recent innovations in field farming that have been made available to farmers through village-level awareness programs. This has caused a rise in output per acre with wholesome crops.

Compared to traditional approaches, these technologies have a great deal of promise to meet the world's food needs in the future.

References

1. Hui YH, Khachatourians GG (1995) Food biotechnology: microorganisms, Series: Food Science and Technology. V.C.H, New York, NY
2. Doebley J, Gaut B, Smith B (2006) The Molecular genetics of crop domestication. Cell. 127(7):1309–1321. https://doi.org/10.1016/j.cell.2006.12.006
3. Rangel G (2015) From Corgis to Corn: A brief look at the long history of G.M.O. Technology—Science in the New Harvard University: The Graduate School of Arts and Sciences. http://sitn.hms.harvard.edu/flash/2015/from-corgis-to-corn-abrief-look-at-the-long-history

Further Reading

Butler D (1996) EU urges national action to police imports of genetically altered maize. Nature 384:502–503

Genetically Modified Food (2014) A review on mechanism of production and labeling concern. Article in Advances in Plants and Agriculture Research. https://www.researchgate.net/publication/280323826

Gerald MD, Claire AJ, Stephen RP (2008) Glyphosate resistant crops: adoption, use and future considerations. Pest Manag Sci 64(4):326–331

Pitzschke A, Hirt H (2010) New insights into an old story: *Agrobacterium*-induced tumour formation in plants by plant transformation. EMBO J 29(6):1021–1032

Sheweta B, Deepak M, Singh KB (2010) Genetically modified food: benefits, safety aspects and concerns. As J Food Ag-Ind 3(6):548–560

Singh A, Kumar V, Poonam, Gupta HR (2014) Review of the mechanism of production and labeling issues for genetically modified food. Adv Plants Agric Res 1(4):00020. https://doi.org/10.15406/apar.2014.01.00020

Ufaz S, Galili G (2008) Improving the content of essential amino acids in crop plants: goals and opportunities. Plant Physiol 147(3):954–961

Chapter 17
Genetically Engineered Foods and Health Benefits

Sarvamangala S. Cholin, Chaitra C. Kulkarni, and Chandan B. M.

1 Introduction

Genetically modified (GM) foods, commonly referred to as genetically engineered (GE) foods or bioengineered foods, are those developed in the laboratory by altering the DNA makeup of crop plants or any organism by altering a base/s by, adding, or deleting a specific fragment of DNA using GE techniques rather than conventional crossbreeding [1]. Genetic engineering allows the selection of a single gene to integrate with the genome of interest, avoiding other genes with undesired properties. Furthermore, it accelerates the development of a product or variation with desirable qualities. Many genetically modified organisms (GMOs) incorporate genes from unrelated species that would be difficult or impossible to develop through traditional selective breeding approaches.

GE is the most rapidly developing and most discussed technology in the history of agriculture. It has been 40 years since Herbert Boyer and Stanley Cohen created human insulin from bacteria, which was approved for commercial manufacture by the US Food and Drug Administration (FDA) to treat diabetes in 1982. In 1983, the first genetically engineered plants, tobacco and petunia, with antibiotic resistance, were developed in China [2]; later in 1992, they became the first to commercialize virus-resistant transgenic tobacco [3]. Since then, several GM foods have been developed for nutritional value, and insect, pest, and herbicide tolerance and have been approved in different countries. According to FDA policy, GM plants must meet the same safety standards and procedures as conventionally bred plants. Before

S. S. Cholin (✉) · Chandan B. M.
Plant Molecular Biology Lab (DBT-BIOCARe), Centre for Biotechnology Research, College of Horticulture, University of Horticultural Sciences, Bagalkot, Karnataka, India

C. C. Kulkarni
Kitturu Rani Channamma College of Horticulture, Arabhavi, University of Horticultural Sciences, Bagalkot, Karnataka, India

J. S. Yaradoddi et al. (eds.), *Frontiers in Food Biotechnology*,
https://doi.org/10.1007/978-981-97-3261-6_17

being released to the market, GE foods are subjected to extensive regulations to examine the risk factors associated with adverse effects on human or animal health and environmental safety.

2 History and Current Status of GM Crops

The twentieth-century DNA discoveries and genetic engineering advancements greatly influenced transgenic technology [4]. In 1988, the first genetically altered microbial enzymes for food production was approved [5]. Several enzymes are involved together in the digestion and fermentation of foods. GE technology could produce recombinant enzymes from genetically altered microorganisms, including lipase and chymosin for cheese manufacturing and alpha-amylase in the beer industry for flavor enhancement [6]. An enzyme mixture called Rennet used to coagulate milk into cheese was accessible solely from the calf's stomach earlier and hence, was very expensive. Other microbial sources that produced rennet had an unpleasant taste [7]. GE technology makes it possible to precisely isolate rennet-producing genes from animals which can then be cloned into bacteria, fungi, or yeasts to produce chymosin-a crucial rennet enzyme. Several organisms such as *Escherichia coli,* Kluyveromyces lactis, yeast (*Komagataella (Pichia*), and fungi (*Aspergillus niger)* have been cloned to produce recombinant chymosin [8, 9]. In the 1990s, only a few nations used recombinant rennet [10].

The first GM crop, Flavr Savr tomato, was altered to be more rot-resistant and had delayed ripening. Calgene released the Flavr Savr, the first commercial GM food certified for human consumption by the FDA in 1994 [11, 12]. This marked the beginning of commercially selling GM food in the USA [13]. Agricultural cash crops, such as cotton, soybeans, maize/corn, and canola, have received the most attention through genetic modifications [12]. A zebrafish called as GloFish, inserted with a luminous gene to glow in the dark under ultraviolet light, was the first GM animal commercialized [14]. In 2015, AquAdvantage salmon was the first GE animal authorized for culinary use [15]. It was possible to grow fish year-round by altering a growth hormone-regulating gene from a Pacific Chinook salmon and a promoter from an ocean pout instead of restricting to spring and summer [16].

Genes that are encoded for important therapeutic compounds, increased nutrition level, or resistance to pesticides and herbicides were explored often from various organisms and transferred to the target species, giving them better survivability and increased productivity [17–20]. GM crop production has increased over 100-fold in the last 25 years [21, 22]. More than 10 million farmers currently cultivate around 190 million hectares of GM crops, almost similar to the whole surface area of Mexico [22, 23]. International Services for the Acquisition of Agri-Biotech Applications (ISAAA) maintains the records of the global status of commercialized transgenic crops. Soybean, maize, cotton, and canola are the four major cultivated GM crops. Soybean ranks in the top position with a high adoption rate of 48% of the global GM crops (91.9 million hectares) followed by maize with 32% of the global

GM crop area (60.9 Mha). Cotton was planted in 25.7 Mha, covering 79% of the global cotton area, and canola occupied 10.1 Mha, 27% of total canola production worldwide. Other GM crops such as alfalfa, eggplant, papaya, sugar beets, safflower, potatoes, squash, sugarcane, apple, and pineapple covered 1.8 Mha across different countries [23].

Currently, around 71 countries with approximately 32 plants have adopted GM crops. Only 29 countries have grown 190.4 million ha. of GM crops benefitting around 1.95 million people across six continents [24]. The USA, Brazil, Argentina, Canada, and India planted about 91% world's GM crop area. As per the GM crops global market report of 2023, it is expected that the GM crops market size will grow from $21.08 billion in 2022 to $22.29 billion in 2023 at a compound annual growth rate (CAGR) of 5.8% and by 2027 it is expected to reach $28.03 billion at 5.9% CAGR. Recent developments in GE technology make genome editing more promising for changing the desired gene and accelerating crop breeding [24].

A wide variety of techniques have been used to date to enhance the use of genetic engineering. Electroporation, particle bombardment technique, Agrobacterium-mediated gene transfer, antisense RNA, RNA interference (RNAi), Virus-Induced Gene Silencing (VIGS), and Clustered Regularly Interspaced Short Palindromic Repeats-CRISPR associated protein 9 (CRISPR-Cas9) are all powerful tools in the field of genetic engineering and gene regulation. These technologies allow researchers to manipulate gene expression and study gene functions.

CRISPR-Cas9 (Clustered Regularly Interspaced Short Palindromic Repeats-CRISPR associated protein 9) technology has revolutionized genetic engineering and significantly improved food crops. The first step in applying CRISPR-Cas9 in food crops is to identify the target gene or genes needing modification to improve the desired trait. This could include genes involved in yield, quality, disease resistance, or other agronomically important traits. Once the target gene is identified, a guide RNA (gRNA) is designed precisely to target that gene. The gRNA consists of a CRISPR RNA (crRNA) sequence complementary to the target gene sequence and a trans-activating CRISPR RNA (tracrRNA) that acts as a scaffold for the Cas9 protein. The gRNA and Cas9 protein are delivered into the cells of the target crop plant [25]. This can be achieved through various methods such as Agrobacterium-mediated transformation, biolistic, or viral vectors. Inside the plant cells, the Cas9 protein and gRNA form a complex. Further, the gRNA guides the Cas9 protein to the target gene sequence in the plant genome. The Cas9 protein creates a double-strand break (DSB) at the targeted gene sequence. The plants DNA repair mechanisms, viz., homology-directed repair (HDR) or non-homologous end joining (NHEJ), then come into play. The Non-Homologous End Joining (NHEJ) pathway is the predominant repair mechanism in plants. It rejoins the broken DNA ends without relying on a repair template, often resulting in small insertions or deletions (indels) at the site of the break. These indels can disrupt the target gene's function, leading to gene knockout or inactivation [26]. Homology-directed repair (HDR) can be utilized in food crops by providing a repair template and the CRISPR-Cas9 system. The repair template contains the desired genetic modification, such as a specific DNA sequence insertion or replacement. The plant's repair machinery uses the provided repair template to guide the repair process, resulting in precise changes in the target gene sequence. The edited plant cells or whole plants are evaluated for the desired genetic modifications and resulting phenotypic changes. This may involve analyzing the expression of the target gene, assessing the improved traits, such as disease resistance or increased yield, and conducting field trials to evaluate the performance of the edited crops [27].

3 Genetically Modified Food Crops for Nutritional Benefit

3.1 GE for Enhanced Vitamins to Fight Against Vitamin A Deficiency (VAD)

VAD is the most devastating cause of preventable blindness in children worldwide and is associated with an increased risk of disease and death from severe infections. It is estimated that 190 million pre-school-age children and 19 million pregnant women suffer from nutritional blindness caused by VAD [28–30]. Retinol, another name for vitamin A, is a micronutrient crucial for vision, growth and development [31], and the immune system [32]. Mammals cannot thrive, reproduce, or fight off disease without vitamin A [33, 34]. Pregnant women are at increased risk of miscarriage and mortality due to vitamin A insufficiency [35, 36].

β-carotene is an essential precursor of vitamin A. Incorporation of the gene/s responsible for enhanced β-carotene in staple foods like rice may be the greater solution to combat VAD in children and other age groups. Professors Ingo Potrykus and Peter Beyer conceived the idea in the late 1980s at the International Rice Research Institute (IRRI), Philippines, and became the first to get a license for golden rice. In the early 90 s Rockefeller Foundation funded the project on the development of golden rice through genetic engineering approach to express the genes encoding phytoene synthase and carotene desaturase, from daffodil and the bacterium *Erwinia uredovora,* respectively, the key enzymes of carotenoid biosynthetic pathway to express in the endosperm of rice. Golden yellow grains result from the direct product of β-carotene was made possible in 1999. The first golden rice trial was conducted in the United States in 2004, and further studies on the increase in vitamin A content in adult volunteers suggested that GE golden rice could effectively convert the β-carotene into vitamin A. Finally, in July 2021, the Philippines received a biosafety permit for the commercial propagation of golden rice in farmer's fields. It is the first GE rice with enhanced nutritional benefits globally and the first in Asia to get biosafety permission for its commercial cultivation. The genes have been transferred to other rice cultivars through marker-assisted backcross breeding.

Golden rice (GR2) utilizes the phytoene synthase gene from maize instead of the daffodil of GR1. Maize or corn (*Zea mays* L.) is another important, most consumed cereal crop worldwide. A triple vitamin biofortified maize that shows higher expression of β-carotene, ascorbate, and folate was developed through metabolic engineering and is another promising GM food crop. This transgenic maize developed for Africa contains a 169-fold increase in β-carotene, a sixfold higher amount of ascorbate, and double the folate content than conventionally bred maize cultivars. In another attempt, biofortified maize comprising of β-cryptoxanthin resulted in more efficient bioconversion of vitamin A than β-carotene when tested in Mongolian gerbils.

Genetic engineering approach has been widely used to increase Provitamin A content in other staple foods like tomato (*Solanum lycopersicum*) [37], maize (*Zea mays*) [38, 39], wheat [40], canola (*Brassica napus*) [41], potato (*Solanum tuberosum*) [42], flaxseed (*Linum usitatissimum*) [43], cassava (*Manihot esculenta*) [42], and Sorghum [44].

Children aged 5–7 years who consume biofortified maize that accumulates provitamin A have been demonstrated to have significantly higher total body reserves of vitamin A [45]. Additionally, it has been demonstrated that consumption of biofortified maize could enhance visual function in kids who lacked vitamin A [46].

4 GE for L-Ascorbic Acid (Vitamin C)

Humans cannot synthesize or store vitamin C [47]. Thus, they must get their daily requirements of 75–90 mg from food sources [48]. High concentrations of vitamin C improve quality of life and have reportedly been shown to aid in treating cancer, arteriosclerosis, and several other cardiovascular conditions [49]. Vitamin C can help to treat Huntington's and Parkinson's diseases [50].

Several distinct ascorbate production routes in plants have been identified, and manipulation of these pathways to boost vitamin C and improve the nutritional content of horticulture crops has been proposed as a prominent topic of future research [51]. Over-expression of GDP-L-Gal phosphorylase (GMP) enzyme has been shown to increase ascorbate levels in strawberry and tomato fruit by two- to six-fold and potato tubers by up to threefold [52] and minimal fold in acerola fruits [53]. Carrots grown in high CO2 environments accumulate vitamin C due to variations in the expression levels of 12 biosynthetic genes [54]. Celery petioles also showed increased vitamin C with higher expression of ascorbate genes when grown under elevated CO_2 environments [55]. This shows the necessity to biofortify crops for higher photosynthetic efficiency. In transgenic tomatoes, it was discovered that by inhibiting the mitochondrial malate dehydrogenase activity photosynthesis was improved in addition to CO_2 assimilation, and chloroplast electron transport rates, leading to increased dry matter of 11 and 19% and ascorbate by 5.7 times in fruit [56].

4.1 GE for Tocopherols (Vitamin E)

Vitamin E insufficiency in humans is uncommon and typically results from a metabolic condition rather than a deficit in the diet [57]. Immune system dysfunction, retinopathy, myopathies, ataxia (progressive neurological illness), and peripheral neuropathy are all caused by vitamin E deficiencies [58]. To boost the amount of total vitamin E antioxidants (tocotrienols + tocopherols), cDNAs encoding homogentisic acid geranylgeranyl transferase (HGGT) were over-expressed in maize [59] and Sorghum [44].

4.2 GE for Enhanced Minerals

Micronutrient malnutrition is another critical challenge among school children and pregnant women in developing nations in Asia and Africa [60]. Advancements in GE technology have made it possible to develop commercial crop varieties enriched with minerals, fatty acids, enhanced essential micronutrients, and amino acids [61], like Fe-dense rice [62], wheat [63], and sorghum (*Sorghum bicolor*) [64], and aided in improved nutritional status of human population [65]. Due to anti-nutritional elements like phytic acid, several micronutrients, such as Fe and Zn, are susceptible to poor absorption [66]. Genetic manipulation addressed sorghum's lysine, vitamin A, iron, and zinc bioavailability [67].

Edible oils: GM soybeans were developed with improved oils for processing [68]. *Camelina sativa* was modified to accumulate high levels of oils similar to fish oils [69].

The GM food crops commercially approved for various nutritional traits are presented in Table 17.1.

Table 17.1 GM crops approved for commercial traits

Crops	GM traits	Targeted gene	Method	Developer
Maize (*Zea maize* L.)	Phytase production, mannose metabolism	**phy02:** Phytase synthesis	Agrobacterium tumefaciens-mediated plant transformation	**Agrivida, Inc.no**
Canola (*Brassica napus*)	Modified oil/ fatty acid	**te**: Increased level of triacylglycerides with esterified lauric acid (12:0)	*Agrobacterium tumefaciens*-mediated plant transformation	Monsanto Company
	Phytase production	**phyA:** Enhances the breakdown of plant phytates which further binds phosphorus and makes it readily available to monogastric animals		BASF

(continued)

Table 17.1 (continued)

Crops	GM traits	Targeted gene	Method	Developer
Tobacco (*Nicotiana tabacum L.*)	Nicotine reduction	NtQPT1 (antisense): Inhibiting transcription of the QPTase gene, and reduces the production of nicotinic acid	Agrobacterium tumefaciens-mediated plant transformation	Vector Tobacco Inc. (USA)
Safflower (*Carthamus tinctorius L.*)	Modified oil/ fatty acid	**fatB, fad2.2**: Production of FATB enzymes or acyl-acyl carrier protein thioesterases were silenced by RNAi	*Agrobacterium tumefaciens*-mediated plant transformation	Go Resources Pty Ltd
Rice (*Oryza sativa L.*)	Mannose metabolism, enhanced provitamin A content **Golden rice**	**crt1**: Catalyzes the conversion of 15-cis-phytoene to all-trans-lycopene **psy1**: Converts geranylgeranyl diphosphate into phytoene, and acts upstream of CRTISO in the carotenoid biosynthetic pathway	*Agrobacterium tumefaciens*-mediated plant transformation	International Rice Research Institute (IRRI), Philippines
Cotton (*Gossypium hirsutum* L.)	Low gossypol	**dCS: dsRNA** inhibiting the expression of the d-cadinene synthase gene that encodes d-cadinene synthase, a key enzyme involved in gossypol biosynthesis, via RNAi pathway	*Agrobacterium tumefaciens*-mediated plant transformation	Texas A&M AgriLife Research University
Melon (*Cucumis melo*)	Delayed ripening/ senescence	**SAM-k:** Causes delayed ripening by reducing the S-adenosylmethionine (SAM), a substrate for ethylene production	*Agrobacterium tumefaciens*-mediated plant transformation	Agritope Inc. (USA)
Pineapple (*Ananas comosus*)	Delayed ripening/ senescence, modified fruit color	**Psy:** Increases lycopene and beta-carotene content **b-Lyc, e-Lyc:** Increases lycopene accumulation using RNAi technology **Acc:** Catalyzes penultimate step in ethylene biosynthesis	*Agrobacterium tumefaciens*-mediated plant transformation	Del Monte Fresh Produce Company

(continued)

Table 17.1 (continued)

Crops	GM traits	Targeted gene	Method	Developer
Apple (*Malus × Domestica)*	Non-browning	**PGAS PPO suppression gene**: Small interfering RNAs (siRNAs) suppress PPO and produce non-browning apple	*Agrobacterium tumefaciens*-mediated plant transformation	Okanagan Specialty Fruits Incorporated
Tomato (*Lycopersicon esculentum* L.)	Delayed fruit softening FLAVR SAVR™	**Pg (sense or antisense):** Inhibits production of polygalacturonate enzyme that breaks down pectin molecules in the cell wall, resulting in delayed ripening of fruit	*Agrobacterium tumefaciens*-mediated plant transformation	Monsanto Company
Potato (*Solanum tuberosum L.)*	Lowered free asparagine, lowered reducing sugars, reduced steroidal glycoalkaloids	**asn1:** Developed to generate dsRNA to down-regulate Asn1 transcripts to reduce asparagine formation	*Agrobacterium tumefaciens*-mediated plant transformation	J.R. Simplot Co.
		PhL, R1, Vlnv: Targeting to regulate PhL, R1 and Vlnv transcripts to reduce reducing sugars		
		Sterol side chain reductase 2: Inhibit steroidal glycoalkaloids synthesis		
Eucalyptus *(Eucalyptus* sp.)	Altered growth and yield	**cel1:** Promotes a faster growth	*Agrobacterium tumefaciens*-mediated plant transformation	FuturaGene Group
Coffee	Decaffeinating	CaMXMT1 (theobromine synthase) and CaDXMT1 (caffeine synthase) reduction inhibit caffeine synthesis	RNAi	Harvard University [70]
Brassica	Reduced erucic acid	BnFAD2 and BnFAE1: Alter fatty acid profile in edited plants	RNAi	[71]

4.3 Insecticide and Herbicide-Free Food

The problem of soil pollution owing to increased usage of pesticides, herbicides, and insecticides may also be alleviated by the use of recombinant microorganisms [72, 73]. Hence, efforts are being made to produce bacterial and viral insecticides to limit the usage of chemical pesticides [74, 75]. GM bacteria in which toxic genes from *Bacillus thuringiensis* have been cloned are used as biopesticides [68, 76, 77].

4.4 Pest Resistance

Bacillus thuringiensis (*Bt*) is a naturally occurring soil bacterium that produces proteins targeting specific insects [78]. *Bt* crops are plants that are genetically engineered to contain Cry (crystal) proteins code for delta endotoxins [79] and are resistant to several harmful pests of Lepidoptera, Diptera, Coleoptera, Hymenoptera, and Nematoda [80, 81]. The Belgian company "Plant Genetic Systems" was the first to produce a *Bt* plant (tobacco) in the laboratory in 1985, but it was not commercially successful. However, in 1995 the USA's Environmental Protection Agency (EPA) approved the commercial production and distribution of *Bt* plants, corn, cotton, potato, and tobacco [82]. Corn and cotton are the most commonly grown Bt crops [82]. Four insect-resistant *Bt* brinjal (eggplant) varieties were approved for commercial seed production and marketing in Bangladesh in 2014 [83].

Some crops, such as cotton, corn, and soybeans, have been genetically modified to express insecticide-producing *Bt* genes [84]. *Bt* corn has been created to manage pests like European corn borer, corn earworm, and southwestern corn borer [85, 86]. At the same time, *Bt* cotton has been engineered to combat cotton pests, including cotton bollworm, tobacco budworm, and rose-pink bollworm [87]. The use of *Bt* crops has drastically reduced the traditional synthetic insecticides usage [88–90].

4.5 Virus Resistance

Rainbow papaya is one of thè successful examples of GMO intended to be virus-resistant [91]. Ringspot virus threatened the Hawaiian papaya industry and the livelihoods of Hawaiian farmers, scientists produced Rainbow papaya, which is Ringspot virus resistant [92]. The potato plant was genetically modified to withstand the potato virus X and potato virus Y [93] and the Potato Leaf Roll Virus [93]. Bean (*Phaseolus vulgaris)* was modified to Bean Golden Mosaic Virus (BGMV). Plum (*Prunus domestica*) was modified to Plum pox virus (PPV) and antibiotic resistance. Squash (*Cucurbita pepo*) was modified to cucumber mosaic virus (CMV), and Watermelon Mosaic Potyvirus 2 (WMV2), sweet pepper (*Capsicum annuum*), Zucchini Yellow Mosaic Potyvirus (ZYMV), and tomato (*Lycopersicon esculentum*) were also modified to Cucumber Mosaic virus (CMV). Genetic modifications of these crops provided higher adaptability to withstand viral infections. RNAi technology has been applied in bananas to improve resistance against the banana bunchy top virus or banana streak virus by silencing genes involved in viral replication or movement [94]. CRISPR-Cas9 in soybeans successfully edited genes associated with resistance to viruses like Soybean mosaic virus (SMV) and soybean vein necrosis-associated virus (SVNaV), as well as fungal pathogens like *Phakopsora pachyrhizi*, which causes soybean rust [95].

In rice, antisense RNA has enhanced disease resistance by silencing genes involved in susceptibility to specific pathogens, such as bacterial blight or fungal diseases [96]. Disease resistance apples were developed using antisense RNA technology against pathogens like fire blight or apple scab and modified fruit characteristics [97]. RNAi technology has been employed in potatoes to enhance resistance against pathogens like late blight (*Phytophthora infestans*) by silencing genes related to pathogen recognition or replication [98]. RNAi has been applied in apples to improve traits such as fruit color, texture, or resistance against diseases like fire blight (*Erwinia amylovora*) and apple scab (*Venturia inaequalis*) by targeting relevant genes [99]. RNAi technology has been used in soybeans to enhance resistance against pests such as soybean cyst nematode by silencing genes involved in the pest's life cycle or host interaction [100].

5 Herbicide Resistance

Herbicides that destroy weeds without hampering crops can be used on certain herbicide-tolerant GM crops. Glyphosate (commercially offered as Roundup® (Bayer, USA) is the most frequently used chemical herbicide, and its manufacturer (Monsanto Corporation, currently a division of Bayer) claims that it is a safe, inexpensive, effective, and ecological friend [101, 102]. The "Roundup-resistant" GMO crop lines are available in soybean, tobacco, corn, soybean, tomato, petunia, potato, alfalfa, canola, sorghum, Indian mustard, sugar beet, and cotton [102]. These transgenic crops are made to withstand the herbicide glyphosate, a component of the weed killer Roundup, and are the best example of GM herbicide-resistant crops [103]. Thus, these herbicide-tolerant plants have made it possible for farmers to transition from conventional herbicides to glyphosate. But we still need to exercise caution because the World Health Organization recently declared that glyphosate is a possible carcinogen [104, 105]. Glyphosate is one of the most harmful herbicides and is linked to cancer in rats and mice, and experiments have shown to damage DNA in human cells [106].

According to the NIH, pesticide's health effects are poorly understood [87, 107]. Probable effects include cancer and damage to the endocrine, reproductive, and nervous systems [87]. Acute effects of pesticides cause skin and throat irritation and respiratory issues, and nausea, dizziness, and diarrhea are common [108]. Preliminary research indicates that chronic, low-dose pesticide exposure raises the likelihood of cognitive impairments and diseases such as Alzheimer's and Parkinson's later in life [108]. A study involving 50 pesticides and almost 30,000 licensed pesticide applicators connected exposure to seven pesticides containing chlorinated chemicals (including two herbicides, two organophosphate insecticides, and two organochlorines) to an elevated risk of diabetes [109, 110]. Pesticide exposure has also been linked to increased infertility in women and developmental issues in children [111]. As a result, the creation of pesticide and herbicide-resistant crops benefits human health.

5.1 Pharming

The term "Pharming" (combined from the words "farming" and "pharmaceutical") refers to the manufacturing of pharmaceutical compounds from transgenic organisms [102]. Pharmaceuticals can be produced using various modes of expression, including transgenic plants, cell cultures of plants, bacteria, yeast, microalgae, and animals [112]. Various types of biopharmaceuticals produced in recombinant *E. coli* are presented in Table 17.2.

"Molecular pharming" refers to using plants as bioreactors to produce therapeutically and commercially significant compounds in plants, such as proteins, carbohydrates, and lipids. Plastid genome engineering has been established to create recombinant proteins in potato, lettuce, tomato, cabbage, and several other food plants [113, 114]. The chloroplast expression system has been identified as a remarkable promise for oral vaccination [115]. Large quantities of antigens for oral treatment in humans have been shown to express, particularly in lettuce chloroplasts [116, 117].

Table 17.2 An abundance of *E. coli*-made recombinant products

Therapeutic products	Application
Insulin	Treatment of diabetes
Hemoglobin	Used as a blood substitute
Human protein C	Used as an anticoagulant
Alpha-1antitrypsin (AAT)	Treatment of AAT deficiency
Vaccines	Used as antigens
Growth hormones	Treatment of deficiencies
Factor VIII blood clotting factor	Used as a blood clotting factor
Factor IX blood clotting factor	Used as a blood clotting factor
Fibrinogen blood clotting factor	Used as a blood clotting factor
Lactoferrin	Used as an infant formula additive
Tumor necrosis factor	Treatment of tumor cells
Interleukin-2(IL-2)	Treatment of cancer, immune deficiency, and HIV infection
Taxol	Treatment for ovarian cancer
Interferon	Treatment for cancer; viral infections

6 Edible Vaccines

An edible vaccine is a food that produces proteins, vitamins, or other nutrients that act as a vaccination against a specific disease. These foods are typically plants [118]. Edible vaccines stimulate the mucosal and humoral immune systems [119]. GE crops that contain antigens for specific diseases are called edible vaccines [120]. Due to their low cost of production, minimal or no adverse side effects, bio-friendly features, and no storage issues, edible vaccines have numerous advantages over conventional vaccinations. The manufacture of hepatitis B and heat-labile toxin B in potato plants was initiated concurrently with the production of the edible vaccine in tobacco [121].

According to recent data, only 40 acres of land are needed annually for the entire population of China to produce hepatitis B edible vaccines, whereas 200 acres are needed to produce edible vaccinations for all infants worldwide [122].

Banana: Banana and potato plants contain antigen HBsAg, used to treat hepatitis B. The highest HBsAg expression in leaves elicits a systematic reaction in the human gut [123]. The principal component of the commercial hepatitis B virus (HBV) subunit vaccine prepared in yeast was S-HBsAg, which is highly immunogenic. Plant species particularly, edible plants, are being explored for S=HBsAg synthesis. This could potentially enable oral delivery of HBV-derived antigens to enhance the immune response, and bananas are particularly promising in this regard [123].

Rice: RNAi technology improved botulinum vaccine antigen (BoHc) expression in rice seeds. BoHc is involved in the nuclear induction of specific immunoglobulin A (IgA) and IgC antibodies. It showed good mucosal immunogenicity with protective immunity [124]. Newcastle disease virus (NDV) occurs in birds; transgenic rice induces the production of neutralizing antibodies in chickens and can be used as a vaccine [125].

Alfalfa (Lucerne): Alfalfa, a plant commonly used in veterinary medicine, is utilized to make edible vaccinations. Transgenic lucerne with swine fever virus glycoprotein E2 was created in 2005. Echinococcus granulosa Eeg95-EgA31 expressing lucerne plants were created [126]. Alfa-Alfa transgenic plant having eBRV4 antigen is effective against Rotaviral diarrhea [127].

Carrots: Carrots are nutritious and delicious and can also be used to make edible vaccines. Vaccines against *Helicobacter pylori,* HIV, and *E. coli* show promise in transgenic carrots. Consumption of this antigen-containing carrot vaccine benefits people with weakened immune systems [128–130]. Two kinds of transgenic carrot plants produce Mycobacterium TB proteins (ESAT6 and CFP10). Oral immunization of mice generates both cell-mediated and humoral immunity. These proteins are suitable as a potential dietary vaccination against tuberculosis [131].

Tobacco: The VP1 protein, which protects against chicken infectious anemia, is expressed in transgenic tobacco. Tobacco can create a polypeptide that is associated with hepatitis B and also develop a coccidiosis vaccine [132]. Transgenic tobacco with P24 and Nef antigens stimulates IgG antibodies to treat AIDS [133]. Antigen HPV16L1 present in transgenic tobacco is active in producing specific IgG and IgA antibodies and involves induction of cell-mediated immune response against cervical cancer [134]. PA antigen of transgenic tobacco induces specific IgG and IgA antibodies against Anthrax in animals [135]. The nucleus and chloroplast of tobacco have been modified to express several antigens, including those for chikungunya, dengue, Ebola, influenza, and Zika [125]. The relevant genes were introduced into *Nicotiana benthamiana* by Agro bacterium-mediated transformation, generating influenza VLPs with a structure resembling that of natural virions [136]. Likewise, researchers generated transgenic tobacco to produce the S protein of severe-acute-respiratory-syndrome-related coronavirus 2 (SARS-CoV-2), to obtain SARS-CoV-2 vaccines in plant systems [125].

6.1 Bioremediation

Genetically modified organisms are used to clean up oil spills, which pose a major threat to the environment. New strains of *Pseudomonas* have been developed that can degrade various hydrocarbons at the site of oil spills, reducing the usage of toxic chemical dispersants. Some microorganisms involved in the degradation of hydrocarbons are *pseudomonas, corynebacteria,* and some *yeasts* [137–139]. The list of different transgenic bacteria designed specifically for bioremediation is listed below (Table 17.3).

Table 17.3 Toxic waste degradation by transgenic bacteria

Bacterium	Bioremediation	References
Pseudomonas putida	p-Nitrophenol (PNP) occurs as contaminants of industrial effluents	[140]
Recombinant E. coli with surface MTT5	Adsorb Cd from the contaminated wastewater	[141]
Alcaligenes sp.	Ecological stressors like oxidative stress, pollutants, and salinity may be considered for bioremediation of oil-polluted saline areas	[142]
Acinetobacter sp.	Bioremediation of petroleum-contaminated soil	[143]
Pseudomonas cepacia	Biosurfactants proved to be effective for metallic corrosion inhibition	[144]

7 GM Food's Contribution to Food Safety, Sustainability, and Climate Change

After the commercial release of GM crops, about 17 million farmers from developing countries could adopt them and improve their socio-economic conditions. As per **ISAAA 2019** [23], the agriculture sector has seen enormous changes from 1996 to 2018. GM crops have not only benefited the economy but also had a substantial impact on the quality and quantity of food, climate resilience, and sustainable progress as follows.

- Crop productivity was enhanced up to 822 million tons.
- A safer environment was ensured by preventing the release of harmful pesticides of up to 776 million kg into the environment.
- Up to 231 million hectares of land could be protected and preserved biodiversity.
- Poverty elimination was possible and the economic status of 16–17 million families of small farmers was improved and this benefit was extended to >65 million people (2018).
- Drastic reduction in CO_2 emissions by up to 23 billion kg, which equals the removal of 15.3 million automobiles from the road for a year (2018).

8 Conclusion

The potential benefits of GM food crops for improving human health are enormous for sustainable nutritional security. The advances in next-generation sequencing technology, huge information databases, and computational tools for higher prediction accuracy have enabled researchers to discover potential genes. The improved procedures to incorporate the gene of interest, like CRISPR technology, have empowered our knowledge of the genes with their precise function and gene product interaction when inserted into the crop or animal system. However, procedures and policies must be assessed and implemented to evaluate the produced products' efficacy, food safety, risk factors, and consumer benefits. GMOs must be considered case-by-case, given the potential benefits, and risks in addition to social, ethical, and regulatory considerations. The development and use of GMOs would be more rewarding when there is a collaborative effort of scientists, policymakers, and stakeholders from different sectors for the benefit of mankind.

References

1. Mathur R (2018) Genetic engineering and biosafety in the use of genetically modified foods. IJASRM 2018(I):76–82
2. De Block M, Herrera-Estrella L (1983) Chimeric genes as dominant selectable markers in plant cells. Protoplasts 1983: Poster Proceedings, pp 130–131. https://doi.org/10.1007/978-3-0348-6556-2_60
3. James C (1997) Global status of transgenic crops in 1997. ISAAA Briefs 5:1–31
4. Jackson DA, Symons RH, Berg P (1972) Biochemical method for inserting new genetic information into DNA of simian virus 40: circular SV40 DNA molecules containing lambda phage genes and the galactose operon of Escherichia coli. Proc Natl Acad Sci 69(10):2904–2909. https://doi.org/10.1073/pnas.69.10.2904
5. Kaya IH (2020) Genetically modified organisms and regulations concerning biotechnological products. Cambridge Scholars Publishing
6. Singh J, Panesar PS (2023) Industrial enzymes: basic information, assay, and applications. In: Basic biotechniques for bioprocess and bioentrepreneurship. Academic Press, pp 295–309. https://doi.org/10.1016/b978-0-12-816109-8.00020-9
7. Dekker P (2019) Dairy enzymes. Industrial Enzyme Applications, pp 143–166. https://doi.org/10.1002/9783527813780.ch2_3
8. Balabova DV, Rudometov AP, Belenkaya SV, Belov AN, Koval AD, Bondar AA, Bakulina AY, Rukhlova EA, Elchaninov VV, Shcherbakov DN (2022a) Biochemical and technological properties of moose (Alces alces) recombinant chymosin. Vavilov J Genet Breeding 26(3):240. https://doi.org/10.18699/vjgb-22-31
9. Balabova DV, Belenkaya SV, Volosnikova EA, Hermes T, Chirkova VY, Sharlaeva EA, Shcherbakov DN, Belov AN, Koval AD, Elchaninov VV (2022b) Can recombinant tree shrew (Tupaia belangeri chinensis) chymosin coagulate cow (Bos taurus) milk. Appl Biochem Microbiol 58(6):761–770. https://doi.org/10.1134/s0003683822060023
10. Gebeyehu MN (2023) Recent advances and application of biotechnology in the dairy processing industry: a review. Intensive Animal Farming-A Cost-Effective Tactic. https://doi.org/10.5772/intechopen.105859
11. Bruening G, Lyons J (2000) The case of the FLAVR SAVR tomato. Calif Agric 54(4):6–7. https://doi.org/10.3733/ca.v054n04p6
12. Zhang C, Wohlhueter R, Zhang H (2016) Genetically modified foods: a critical review of their promise and problems. Food Sci Human Wellness 5(3):116–123. https://doi.org/10.1016/j.fshw.2016.04.002
13. James C (2011) Global status of commercialized biotech/GM crops, 43rd edn. International Service for the Acquisition of Agri-biotech Applications (ISAAA), Ithaca, NY
14. Vàzquez-Salat N, Salter B, Smets G, Houdebine LM (2012) The current state of GMO governance: are we ready for GM animals? Biotechnol Adv 30(6):1336–1343. https://doi.org/10.1016/j.biotechadv.2012.02.006
15. FDA (2019) Aquabounty Cleared to Sell Salmon in USA for Commercial Purposes
16. Dunham RA, Su B (2020) Genetically engineered fish: potential impacts on aquaculture, biodiversity, and the environment. GMOs: Implications for Biodiversity Conservation and Ecological Processes, pp 241–275. https://doi.org/10.1007/978-3-030-53183-6_11
17. Hefferon KL (2015) Nutritionally enhanced food crops; progress and perspectives. Int J Mol Sci 16(2):3895–3914. https://doi.org/10.3390/ijms16023895
18. Dizon F, Costa S, Rock C, Harris A, Husk C, Mei J (2016) Genetically modified (GM) foods and ethical eating. J Food Sci 81(2):R287–R291. https://doi.org/10.1111/1750-3841.13191
19. Huang K (2017) Safety assessment of genetically modified foods. Springer, Singapore. https://doi.org/10.1007/978-981-10-3488-6
20. Mahgoub SE, Nollet LM (2019) Testing and analysis of GMO-containing foods and feed. CRC Press. https://doi.org/10.1201/9781315178592

21. Pellegrino E, Bedini S, Nuti M, Ercoli L (2018) Impact of genetically modified maize on agronomic, environmental and toxicological traits. AFMA Matrix 27(2):18–23. https://doi.org/10.1038/s41598-018-21284-2
22. Turnbull C, Lillemo M, Hvoslef-Eide TA (2021) Global regulation of genetically modified crops amid the gene edited crop boom–a review. Front Plant Sci 12:630396. https://doi.org/10.3389/fpls.2021.630396
23. ISAAA (2019) Global status of commercialized biotech/GM crops in 2019: biotech crops drive socio-economic development and sustainable environment in the new frontier. Executive Summary ISAAA Brief 55. http://www.isaaa.org/
24. Verma V, Negi S, Kumar P, Srivastava DK (2022) Global status of genetically modified crops. In: Agricultural biotechnology: latest research and trends. Springer Nature, Singapore, pp 305–322. https://doi.org/10.1007/978-981-16-2339-4_13
25. Jiang F, Doudna JA (2017) CRISPR–Cas9 structures and mechanisms. Annu Rev Biophys 46:505–529. https://doi.org/10.1146/annurev-biophys-062215-010822
26. Kosicki M, Allen F, Steward F, Tomberg K, Pan Y, Bradley A (2022) Cas9-induced large deletions and small indels are controlled in a convergent fashion. Nat Commun 13(1):3422
27. Cui Y, Li R, Li G, Zhang F, Zhu T, Zhang Q, Ali J, Li Z, Xu S (2020) Hybrid breeding of rice via genomic selection. Plant Biotechnol J 18(1):57–67. https://doi.org/10.1111/pbi.13170
28. Dutta D, Nayak A, Dutta D (2023) Reconnoitering the usage of agro-industrial waste in carotenoid production for food fortification: a sustainable approach to tackle vitamin A deficiency. Food Bioprocess Technol 16(3):467–491. https://doi.org/10.1007/s11947-022-02888-z
29. Ahenda P, Washburn DJ, Colwell B, Lee S, Gwarzo I, Kellstedt DK, Maddock JE (2023) Shifts in social determinants of vitamin A supplementation among children under five in Kenya, 2003–2014. Matern Child Health J 27:1–9. https://doi.org/10.1007/s10995-023-03663-y
30. Zhao T, Liu S, Zhang R, Zhao Z, Yu H, Pu L, Wang L, Han L (2022) Global burden of vitamin A deficiency in 204 countries and territories from 1990–2019. Nutrients 14(5):950. https://doi.org/10.3390/nu14050950
31. Smirnov VM, Wilmet B, Nassisi M, Condroyer C, Antonio A, Andrieu C, Devisme C, Sancho S, Sahel JA, Zeitz C, Audo I (2022) Large benefit from simple things: high-dose vitamin a improves RBP4-related retinal dystrophy. Int J Mol Sci 23(12):6590. https://doi.org/10.3390/ijms23126590
32. Huang Z, Liu Y, Qi G, Brand D, Zheng SG (2018) Role of vitamin A in the immune system. J Clin Med 7(9):258. https://doi.org/10.3390/jcm7090258
33. Tanumihardjo SA (2011) Vitamin A: biomarkers of nutrition for development. Am J Clin Nutr 94(2):658S–665S. https://doi.org/10.3945/ajcn.110.005777
34. Pasquariello R, Anipchenko P, Pennarossa G, Crociati M, Zerani M, Brevini TA, Gandolfi F, Maranesi M (2022) Carotenoids in female and male reproduction. Phytochemistry 204:113459. https://doi.org/10.1016/j.phytochem.2022.113459
35. WHO (2019) World Health Organisation: Micronutrient Deficiencies. http://www.who.int/nutrition/topics/vad/en/. Accessed 19 Nov 2019
36. WHO. World Health Organization: Global Prevalence of Vitamin a Deficiency in Populations at Risk 1995–2005
37. Fraser PD, Enfissi EM, Bramley PM (2009) Genetic engineering of carotenoid formation in tomato fruit and the potential application of systems and synthetic biology approaches. Arch Biochem Biophys 483(2):196–204. https://doi.org/10.1016/j.abb.2008.10.009
38. Strobbe S, De Lepeleire J, Van Der Straeten D (2018) From in planta function to vitamin-rich food crops: the ACE of biofortification. Front Plant Sci 9:1862. https://doi.org/10.3389/fpls.2018.01862
39. Nie Y, Wang H, Zhang G, Ding H, Han B, Liu L, Shi J, Du J, Li X, Li X, Zhao Y (2023) The maize plastid terminal oxidase (PTOX) gene controls the carotenoid content of kernels. bioRxiv:2023–2003. https://doi.org/10.1101/2023.03.13.532334

40. Socol CT, Mierliţă D, Maerescu CM, Bangar SP, Rusu AV (2023) Genetic manipulation of colored cereals for improved nutritional quality. In: Functionality and application of colored cereals. Academic Press, pp 217–240. https://doi.org/10.1016/b978-0-323-99733-1.00012-1
41. Tian Q, Li B, Feng Y, Zhao W, Huang J, Chao H (2022) Application of CRISPR/Cas9 in rapeseed for gene function research and genetic improvement. Agronomy 12(4):824
42. Beyene G, Solomon FR, Chauhan RD, Gaitán-Solis E, Narayanan N, Gehan J, Siritunga D, Stevens RL, Jifon J, Van Eck J, Linsler E (2018) Provitamin A biofortification of cassava enhances shelf life but reduces dry matter content of storage roots due to altered carbon partitioning into starch. Plant Biotechnol J 16(6):1186–1200. https://doi.org/10.1111/pbi.12862
43. Fujisawa M, Misawa N (2010) Enrichment of carotenoids in flaxseed by introducing a bacterial phytoene synthase gene. Plant Secondary Metabolism Engineering: Methods and Applications, pp 201–211. https://doi.org/10.1007/978-1-60761-723-5_14
44. Che P, Zhao ZY, Glassman K, Dolde D, Hu TX, Jones TJ, Gruis DF, Obukosia S, Wambugu F, Albertsen MC (2016) Elevated vitamin E content improves all-trans β-carotene accumulation and stability in biofortified sorghum. Proc Natl Acad Sci 113(39):11040–11045. https://doi.org/10.1073/pnas.1605689113
45. Gannon B, Kaliwile C, Arscott SA, Schmaelzle S, Chileshe J, Kalungwana N, Mosonda M, Pixley K, Masi C, Tanumihardjo SA (2014) Biofortified orange maize is as efficacious as a vitamin A supplement in Zambian children even in the presence of high liver reserves of vitamin A: a community-based, randomized placebo-controlled trial. Am J Clin Nutr 100(6):1541–1550. https://doi.org/10.3945/ajcn.114.087379
46. Palmer AC, Healy K, Barffour MA, Siamusantu W, Chileshe J, Schulze KJ, West KP Jr, Labrique AB (2016) Provitamin A carotenoid–biofortified maize consumption increases pupillary responsiveness among Zambian children in a randomized controlled trial. J Nutr 146(12):2551–2558. https://doi.org/10.3945/jn.116.239202
47. Dresen E, Lee ZY, Hill A, Notz Q, Patel JJ, Stoppe C (2023) History of scurvy and use of vitamin C in critical illness: a narrative review. Nutr Clin Pract 38(1):46–54. https://doi.org/10.1002/ncp.10914
48. Bedhiafi T, Idoudi S, Fernandes Q, Al-Zaidan L, Uddin S, Dermime S, Billa N, Merhi M (2023) Nano-vitamin C: a promising candidate for therapeutic applications. Biomed Pharmacother 158:114093. https://doi.org/10.1016/j.biopha.2022.114093
49. da Silva Dias JC (2014) Nutritional and health benefits of carrots and their seed extracts. Food Nutr Sci 5(22):2147. https://doi.org/10.4236/fns.2014.522227
50. Harrison FE, May JM (2009) Vitamin C function in the brain: vital role of the ascorbate transporter SVCT2. Free Radic Biol Med 46(6):719–730. https://doi.org/10.1016/j.freeradbiomed.2008.12.018
51. Troesch B, Hoeft B, McBurney M, Eggersdorfer M, Weber P (2012) Dietary surveys indicate vitamin intakes below recommendations are common in representative Western countries. Br J Nutr 108(4):692–698. https://doi.org/10.1017/s0007114512001808
52. Bulley S, Wright M, Rommens C, Yan H, Rassam M, Lin-Wang K, Andre C, Brewster D, Karunairetnam S, Allan AC, Laing WA (2012) Enhancing ascorbate in fruits and tubers through over-expression of the l-galactose pathway gene GDP-l-galactose phosphorylase. Plant Biotechnol J 10(4):390–397. https://doi.org/10.1111/j.1467-7652.2011.00668.x
53. Suekawa M, Fujikawa Y, Inoue A, Kondo T, Uchida E, Koizumi T, Esaka M (2019) High levels of expression of multiple enzymes in the Smirnoff-Wheeler pathway are important for high accumulation of ascorbic acid in acerola fruits. Biosci Biotechnol Biochem 83(9):1713–1716. https://doi.org/10.1080/09168451.2019.1608808
54. Wu XJ, Sun S, Xing GM, Wang GL, Wang F, Xu ZS, Tian YS, Hou XL, Xiong AS (2017) Elevated carbon dioxide altered morphological and anatomical characteristics, ascorbic acid accumulation, and related gene expression during taproot development in carrots. Front Plant Sci 7:2026. https://doi.org/10.3389/fpls.2016.02026
55. Liu JX, Feng K, Wang GL, Wu XJ, Duan AQ, Yin L, Shen D, Xu ZS, Xiong AS (2020) Effect of elevated CO 2 on ascorbate accumulation and the expression levels of genes involved in

ascorbate metabolism in celery. J Plant Growth Regul 39:1046–1060. https://doi.org/10.1007/s00344-019-10045-8
56. Nunes-Nesi A, Carrari F, Lytovchenko A, Smith AM, Ehlers Loureiro M, Ratcliffe RG, Sweetlove LJ, Fernie AR (2005) Enhanced photosynthetic performance and growth as a consequence of decreasing mitochondrial malate dehydrogenase activity in transgenic tomato plants. Plant Physiol 137(2):611–622. https://doi.org/10.1104/pp.104.055566
57. Simkin AJ (2019) Genetic engineering for global food security: photosynthesis and biofortification. Plan Theory 8(12):586. https://doi.org/10.3390/plants8120586
58. Cimmino F, Catapano A, Petrella L, Villano I, Tudisco R, Cavaliere G (2023) Role of milk micronutrients in human health. Front Biosci 28(2):41. https://doi.org/10.31083/j.fbl2802041
59. Cahoon EB, Coughlan SJ, Cahoon RE, Butler KH, EI Du Pont de Nemours and Co. (2006) Compositions and methods for altering tocotrienol content. U.S. Patent 7154029.
60. Kiran A, Wakeel A, Mahmood K, Mubaraka R, Hafsa, Haefele SM (2022) Biofortification of staple crops to alleviate human malnutrition: contributions and potential in developing countries. Agronomy 12(2):452. https://doi.org/10.3390/agronomy12020452
61. Garg M, Sharma N, Sharma S, Kapoor P, Kumar A, Chunduri V, Arora P (2018) Biofortified crops generated by breeding, agronomy, and transgenic approaches are improving lives of millions of people around the world. Front Nutr 5:12. https://doi.org/10.3389/fnut.2018.00012
62. Matres JM, Arcillas E, Cueto-Reaño MF, Sallan-Gonzales R, Trijatmiko KR, Slamet-Loedin I (2021) Biofortification of rice grains for increased iron content. Rice improvement, 471. https://doi.org/10.1007/978-3-030-66530-2_14
63. Ludwig Y, Slamet-Loedin IH (2019) Genetic biofortification to enrich rice and wheat grain iron: from genes to product. Front Plant Sci 10:833. https://doi.org/10.3389/fpls.2019.00833
64. Elkonin L, Italyanskaya J, Panin V (2018) Genetic modification of sorghum for improved nutritional value: state of the problem and current approaches. J Investig Genom 5:39–48. https://doi.org/10.15406/jig.2018.05.00076
65. Pérez-Massot E, Banakar R, Gómez-Galera S, Zorrilla-López U, Sanahuja G, Arjó G, Miralpeix B, Vamvaka E, Farré G, Rivera SM, Dashevskaya S (2013) The contribution of transgenic plants to better health through improved nutrition: opportunities and constraints. Genes Nutr 8:29–41. https://doi.org/10.1007/s12263-012-0315-5
66. Kumar A, Lal MK, Kar SS, Nayak L, Ngangkham U, Samantaray S, Sharma SG (2017) Bioavailability of iron and zinc as affected by phytic acid content in rice grain. J Food Biochem 41(6):e12413. https://doi.org/10.1111/jfbc.12413
67. Zhao ZY, Che P, Glassman K, Albertsen M (2019) Nutritionally enhanced sorghum for the arid and semiarid tropical areas of Africa. Methods Mol Biol 1931:197–207. https://doi.org/10.1007/978-1-4939-9039-9_14
68. Li Q, Cai S, Xu D, Lin J, Shen G, Feng J (2023) Comparative analysis of nutritional composition between GM and non-GM soybeans and soybean oils by NMR and GC-FID techniques. Food Anal Methods 16(3):478–490. https://doi.org/10.1007/s12161-022-02435-8
69. Huang D, Gao L, McAdams J, Zhao F, Lu H, Wu Y, Martin J, Sherif SM, Subramanian J, Duan H, Liu W (2023) Engineered Cleistogamy in Camelina sativa for bioconfinement. Hortic Res 10(2):uhac280. https://doi.org/10.1093/hr/uhac280
70. Ogita S, Uefuji H, Yamaguchi Y, Koizumi N, Sano H (2003) Producing decaffeinated coffee plants. Nature 423(6942):823–823. https://doi.org/10.1038/423823a
71. Shi J, Ni X, Huang J, Fu Y, Wang T, Yu H, Zhang Y (2022) CRISPR/Cas9-mediated gene editing of BnFAD2 and BnFAE1 modifies fatty acid profiles in brassica napus. Genes 13(10):1681. https://doi.org/10.3390/genes13101681
72. Bokade P, Gaur VK, Tripathi V, Bobate S, Manickam N, Bajaj A (2023) Bacterial remediation of pesticide polluted soils: exploring the feasibility of site restoration. J Hazard Mater 441:129906. https://doi.org/10.1016/j.jhazmat.2022.129906
73. Veerapagu M, Jeya KR, Sankaranarayanan A (2023) Role of plant growth-promoting microorganisms in phytoremediation efficiency. Plant-Microbe Interaction-Recent Advances in

Molecular and Biochemical Approaches, pp 45–61.https://doi.org/10.1016/b978-0-323-91875-6.00020-7
74. Bravo A, Soberón M (2023) Can microbial-based insecticides replace chemical pesticides in agricultural production? Microb Biotechnol 16:2011–2014. https://doi.org/10.1111/1751-7915.14316
75. Xu J, Zheng L, Tan M, Wu H, Yan S, Jiang D (2023) The susceptibility of Hyphantriacunea larvae to microbial pesticides Bacillus thuringiensis and Mamestra brassicae nuclear polyhedrosis virus under Cd stress. Pestic Biochem Physiol 191:105383. https://doi.org/10.1016/j.pestbp.2023.105383
76. Fatima N, Rehman A, Bukhari DA (2022) Biotoxicity assessment of cloned cry 11 protein gene from Bacillus thuringiensis 9NF. Saudi J Biol Sci 29(11):103463. https://doi.org/10.1016/j.sjbs.2022.103463
77. Baranek J, Jakubowska M, Gabała E (2023) Insecticidal activity of Bacillus thuringiensis towards Agrotisexclamationis larvae–A widespread and underestimated pest of the Palearctic zone. PLoS One 18(3):e0283077. https://doi.org/10.1371/journal.pone.0283077
78. Guerrero M, G.G. (2023) Sporulation, structure assembly, and germination in the soil bacterium bacillus thuringiensis: survival and success in the environment and the insect host. Microbiol Res 14(2):466–491. https://doi.org/10.3390/microbiolres14020035
79. Azizoglu U, Salehi Jouzani G, Sansinenea E, Sanchis-Borja V (2023) Biotechnological advances in Bacillus thuringiensis and its toxins: recent updates. Rev Environ Sci Biotechnol 22:1–30. https://doi.org/10.1007/s11157-023-09652-5
80. Abbas MST (2018) Genetically engineered (modified) crops (Bacillus thuringiensis crops) and the world controversy on their safety. Egypt J Biol Pest Control 28:1–12. https://doi.org/10.1186/s41938-018-0051-2
81. Ma G, Chen Y, Liu X, Gao Y, Deavila JM, Zhu MJ, Du M (2022) Vitamin a supplementation during pregnancy in shaping child growth outcomes: a meta-analysis. Crit Rev Food Sci Nutr 63:12240–12255. https://doi.org/10.1080/10408398.2022.2099810
82. Yang F, Wang Z, Kerns DL (2022) Resistance of Spodoptera frugiperda to Cry1, Cry2, and Vip3Aa proteins in Bt corn and cotton in the Americas: implications for the rest of the world. J Econ Entomol 115(6):1752–1760. https://doi.org/10.1093/jee/toac099
83. Shelton AM, Hossain MJ, Paranjape V, Prodhan MZ, Azad AK, Majumder R, Sarwer SH, Hossain MA (2019) Bt brinjal in Bangladesh: the first genetically engineered food crop in a developing country. Cold Spring Harb Perspect Biol 11(10):a034678. https://doi.org/10.1101/cshperspect.a034678
84. Raman R (2017) The impact of genetically modified (GM) crops in modern agriculture: a review. GM Crops Food 8(4):195–208. https://doi.org/10.1080/21645698.2017.1413522
85. Hutchison WD, Hunt TE, Hein GL, Steffey KL, Pilcher CD, Rice ME (2011) Genetically engineered Bt corn and range expansion of the western bean cutworm (Lepidoptera: Noctuidae) in the United States: a response to Greenpeace Germany. J Integr Pest Manag 2(3):B1–B8. https://doi.org/10.1603/ipm11016
86. Ruiu L, Lentini A (2022) Sustainable silage maize integrated protection against the European corn borer Ostrinia nubilalis and the corn earworm *Helicoverpa armigera* employing the farm irrigation system. Agronomy 12(2):362. https://doi.org/10.3390/agronomy12020362
87. Hsaio J (2015) GMOs and pesticides: helpful or harmful. Science in the News
88. Jehangir N, Ali S (2023) The insecticidal efficacy and performance of Bt cotton under variable abiotic stresses–a review on recent findings. Plant Stress 8:100151. https://doi.org/10.1016/j.stress.2023.100151
89. Yang X, Zhao S, Liu B, Gao Y, Hu C, Li W, Yang Y, Li G, Wang L, Yang X, Yuan H (2023) Bt maize can provide non-chemical pest control and enhance food safety in China. Plant Biotechnol J 21(2):391–404. https://doi.org/10.1111/pbi.13960
90. Reavey CE, Walker AS, Joyce SP, Broom L, Willse A, Ercit K, Poletto M, Barnes ZH, Marubbi T, Troczka BJ, Treanor D (2022) Self-limiting fall armyworm: a new approach in

development for sustainable crop protection and resistance management. BMC Biotechnol 22(1):1–16. https://doi.org/10.1186/s12896-022-00735-9
91. Gonsalves C, Lee DR, Gonsalves D (2004) Transgenic virus-resistant papaya: the Hawaiian 'Rainbow'was rapidly adopted by farmers and is of major importance in Hawaii today. APSnet Features. https://doi.org/10.1094/apsnetfeature-2004-0804
92. Gonsalves CV, Gonsalves D (2014) The Hawaii papaya story. In: Handbook on agriculture, biotechnology and development. Edward Elgar Publishing, pp 642–660. https://doi.org/10.4337/9780857938350.00047
93. Liu J, Yue J, Wang H, Xie L, Zhao Y, Zhao M, Zhou H (2023) Strategies for engineering virus resistance in potato. Plan Theory 12(9):1736. https://doi.org/10.3390/plants12091736
94. Collinge DB, Sarrocco S (2022) Transgenic approaches for plant disease control: status and prospects 2021. Plant Pathol 71(1):207–225. https://doi.org/10.1111/ppa.13443
95. Nagy ED, Stevens JL, Yu N, Hubmeier CS, LaFaver N, Gillespie M, Gardunia B, Cheng Q, Johnson S, Vaughn AL, Vega-Sanchez ME (2021) Novel disease resistance gene paralogs created by CRISPR/Cas9 in soy. Plant Cell Rep 40:1047–1058. https://doi.org/10.1007/s00299-021-02678-5
96. Wang M, Dean RA (2022) Host induced gene silencing of magnaportheoryzae by targeting pathogenicity and development genes to control rice blast disease. Front Plant Sci 13:959641. https://doi.org/10.3389/fpls.2022.959641
97. Patel RR, Sundin GW, Yang CH, Wang J, Huntley RB, Yuan X, Zeng Q (2017) Exploration of using antisense peptide nucleic acid (PNA)-cell penetrating peptide (CPP) as a novel bactericide against fire blight pathogen Erwinia amylovora. Front Microbiol 8:687. https://doi.org/10.3389/fmicb.2017.00687
98. Kieu NP, Lenman M, Wang ES, Petersen BL, Andreasson E (2021) Mutations introduced in susceptibility genes through CRISPR/Cas9 genome editing confer increased late blight resistance in potatoes. Sci Rep 11(1):4487. https://doi.org/10.1038/s41598-021-83972-w
99. Rocafort M, Arshed S, Hudson D, Sidhu JS, Bowen JK, Plummer KM, Bradshaw RE, Johnson RD, Johnson LJ, Mesarich CH (2022) CRISPR-Cas9 gene editing and rapid detection of gene-edited mutants using high-resolution melting in the apple scab fungus, Venturia inaequalis. Fungal Biol 126(1):35–46. https://doi.org/10.1016/j.funbio.2021.10.001
100. Alkharouf NW, Klink VP, Matthews BF (2007) Identification of Heterodera glycines (soybean cyst nematode [SCN]) cDNA sequences with high identity to those of Caenorhabditis elegans having lethal mutant or RNAi phenotypes. Exp Parasitol 115(3):247–258. https://doi.org/10.1016/j.exppara.2006.09.009
101. Funke T, Han H, Healy-Fried ML, Fischer M, Schönbrunn E (2006) Molecular basis for the herbicide resistance of roundup ready crops. Proc Natl Acad Sci 103(35):13010–13015. https://doi.org/10.1073/pnas.0603638103
102. Narayanan Z, Glick BR (2023) Biotechnologically engineered plants. Biology 12(4):601. https://doi.org/10.3390/biology12040601
103. Azeem A, Ul-Allah S, Sher A, Ijaz M, Sattar A, Ahmad W, Qayyum A (2023) Biosafety and biosecurity in genetically modified crops. In: Sustainable agriculture in the era of the OMICs revolution. Springer International Publishing, Cham, pp 501–510. https://doi.org/10.1007/978-3-031-15568-0_23
104. International Agency for Research on Cancer (IARC) (2015) IARC monographs volume 12: evaluation of 5 organophosphate insecticides and herbicides. World Health Organization
105. Tarazona JV, Court-Marques D, Tiramani M, Reich H, Pfeil R, Istace F, Crivellente F (2017) Glyphosate toxicity and carcinogenicity: a review of the scientific basis of the European Union assessment and its differences with IARC. Arch Toxicol 91:2723–2743. https://doi.org/10.1007/s00204-017-1962-5
106. Cressey D (2015) Widely used herbicide linked to cancer. Nature 24:1–3. https://doi.org/10.1038/nature.2015.17181
107. Varmus H, Klausner R, Zerhouni E, Acharya T, Daar AS, Singer PA (2003) Grand challenges in global health. Science 302(5644):398–399. https://doi.org/10.1126/science.1091769

108. Kamel F (2013) Paths from pesticides to Parkinson's. Science 341(6147):722–723. https://doi.org/10.1126/science.1243619
109. NIEH (2008) Long-term pesticide exposure may increase risk of diabetes. National Institute of Environmental Health Sciences. http://www.niehs.nih.gov/news/newsroom/releases/2008/june04/
110. Mukah FE, Chinedu-Ndukwe PA, Imarhiagbe O, Nwaubani DA (2023) Agrochemical use and emerging human and animal diseases. In: One health implications of agrochemicals and their sustainable alternatives. Springer Nature, Singapore, pp 53–76. https://doi.org/10.1007/978-981-99-3439-3_2
111. Chiu YH, Williams PL, Gillman MW, Gaskins AJ, Mínguez-Alarcón L, Souter I, Toth TL, Ford JB, Hauser R, Chavarro JE, EARTH Study Team (2018) Association between pesticide residue intake from consumption of fruits and vegetables and pregnancy outcomes among women undergoing infertility treatment with assisted reproductive technology. JAMA Intern Med 178(1):17–26. https://doi.org/10.1001/jamainternmed.2017.5038
112. Barbosa MJ, Janssen M, Südfeld C, D'Adamo S, Wijffels RH (2023) Hypes, hopes, and the way forward for microalgal biotechnology. Trends Biotechnol 41:452. https://doi.org/10.1016/j.tibtech.2022.12.017
113. Zhou YX, Lee MYT, Ng JMH, Chye ML, Yip WK, Zee SY, Lam E (2006) A truncated hepatitis E virus ORF2 protein expressed in tobacco plastids is immunogenic in mice. World J Gastroenterol: WJG 12(2):306. https://doi.org/10.3748/wjg.v12.i2.306
114. Cardi T, Lenzi P, Maliga P (2010) Chloroplasts as expression platforms for plant-produced vaccines. Expert Rev Vaccines 9(8):893–911. https://doi.org/10.1586/erv.10.78
115. Davoodi-Semiromi A, Schreiber M, Nalapalli S, Verma D, Singh ND, Banks RK, Chakrabarti D, Daniell H (2010) Chloroplast-derived vaccine antigens confer dual immunity against cholera and malaria by oral or injectable delivery. Plant Biotechnol J 8(2):223–242. https://doi.org/10.1111/j.1467-7652.2009.00479.x
116. Lakshmi PS, Verma D, Yang X, Lloyd B, Daniell H (2013) Low cost tuberculosis vaccine antigens in capsules: expression in chloroplasts, bio-encapsulation, stability and functional evaluation in vitro. PLoS One 8(1):e54708. https://doi.org/10.1371/journal.pone.0054708
117. van Eerde A, Gottschamel J, Bock R, Hansen KEA, Munang'andu HM, Daniell H, Liu Clarke J (2019) Production of tetravalent dengue virus envelope protein domain III based antigens in lettuce chloroplasts and immunologic analysis for future oral vaccine development. Plant Biotechnol J 17(7):1408–1417. https://doi.org/10.1111/pbi.13065
118. Concha C, Cañas R, Macuer J, Torres MJ, Herrada AA, Jamett F, Ibáñez C (2017) Disease prevention: an opportunity to expand edible plant-based vaccines? Vaccine 5(2):14
119. Mishra N, Gupta PN, Khatri K, Goyal AK, Vyas SP (2008) Edible vaccines: a new approach to oral immunization
120. Abeysundara AT, Aponso M, De Silva GO (2017) A review on edible vaccines: a novel approach to oral immunization as a replacement of conventional vaccines. Int J Food Sci Nutr 2(4):19–22
121. Penney CA, Thomas DR, Deen SS, Walmsley AM (2011) Plant-made vaccines in support of the millennium development goals. Plant Cell Rep 30:789–798. https://doi.org/10.1007/s00299-010-0995-5
122. Hudu SA, Shinkafi SH, Umar S (2016) An overview of recombinant vaccine technology, adjuvants and vaccine delivery methods. Int J Pharm Pharm Sci 8(11):19–24. https://doi.org/10.22159/ijpps.2016v8i11.14311
123. Rukavtsova EB, Rudenko NV, Puchko EN, Zakharchenko NS, Buryanov YI (2015) Study of the immunogenicity of hepatitis B surface antigen synthesized in transgenic potato plants with increased biosafety. J Biotechnol 203:84–88. https://doi.org/10.1016/j.jbiotec.2015.03.019
124. Yuki Y, Mejima M, Kurokawa S, Hiroiwa T, Kong IG, Kuroda M, Takahashi Y, Nochi T, Tokuhara D, Kohda T (2012) RNAi suppression of rice endogenous storage proteins enhances the production of rice-based Botulinum neutrotoxin type A vaccine. Vaccine 30:4160–4166. https://doi.org/10.1016/j.vaccine.2012.04.064

125. Su H, Van Eerde A, Rimstad E, Bock R, Branza-Nichita N, Yakovlev IA, Clarke JL (2023) Plant-made vaccines against viral diseases in humans and farm animals. Front Plant Sci 14:1170815. https://doi.org/10.3389/fpls.2023.1170815
126. Kurup VM, Thomas J (2020) Edible vaccines: promises and challenges. Mol Biotechnol 62:79–90. https://doi.org/10.1007/s12033-019-00222-1
127. Wigdorovitz A, Mozgovoj M, Santos MJD, Parreno V, Gomez C, Perez-Filgueira DM, Trono KG, Ríos RD, Franzone PM, Fernandez F, Carrillo C (2004) Protective lactogenic immunity conferred by an edible peptide vaccine to bovine rotavirus produced in transgenic plants. J Gen Virol 85(7):1825–1832. https://doi.org/10.1099/vir.0.19659-0
128. Karasev AV, Foulke S, Wellens C, Rich A, Shon KJ, Zwierzynski I, Hone D, Koprowski H, Reitz M (2005) Plant-based HIV-1 vaccine candidate: tat protein produced in spinach. Vaccine 23(15):1875–1880. https://doi.org/10.1016/j.vaccine.2004.11.021
129. Zhang H, Liu M, Li Y, Zhao Y, He H, Yang G, Zheng C (2010) Oral immunogenicity and protective efficacy in mice of a carrot-derived vaccine candidate expressing UreB subunit against Helicobacter pylori. Protein Expr Purif 69(2):127–131. https://doi.org/10.1016/j.pep.2009.07.016
130. Ye Y, Li W (2010) Immunoprotection of transgenic alfalfa (Medicago sativa) containing Eg95-EgA31 fusion gene of Echinococcus granulosus against egprotoscoleces. J Trop Med (Guangzhou) 10(3):235–237
131. Uvarova EA, Belavin PA, Permyakova NV, Zagorskaya AA, Nosareva OV, Kakimzhanova AA, Deineko EV (2013) Biomed Res Int 2013:1–8. https://doi.org/10.1155/2013/316304
132. Mathew M, Thomas J (2022) Tobacco-based vaccines, hopes, and concerns: a systematic review. Mol Biotechnol 65:1023–1051. https://doi.org/10.1007/s12033-022-00627-5
133. Gonzalez-Rabade N, McGowan EG, Zhou F, McCabe MS, Bock R, Dix PJ, Gray JC, Ma JK-C (2011) Plant. Biotechnol J 9:629–638. https://doi.org/10.1111/j.1467-7652.2011.00609.x
134. Hongli L, Xukui L, Ting L, Wensheng L, Lusheng S, Jin Z (2013) Transgenic tobacco expressed HPV16-L1 and LT-B combined immunization induces strong mucosal and systemic immune responses in mice. Hum Vaccin Immunother 9(1):83–89. https://doi.org/10.4161/hv.22292
135. Gorantala J, Grover S, Rahi A, Chaudhary P, Rajwanshi R, Sarin NB, Bhatnagar R (2014) Generation of protective immune response against anthrax by oral immunization with protective antigen plant-based vaccine. J Biotechnol 176:1–10. https://doi.org/10.1016/j.jbiotec.2014.01.033
136. Sepotokele KM, O'Kennedy MM, Wandrag DB, Abolnik C (2023) Optimization of infectious bronchitis virus-like particle expression in Nicotiana benthamiana as potential poultry vaccines. PLoS One 18(7):e0288970. https://doi.org/10.1371/journal.pone.0288970
137. Saha R, Nag S (2022) Microbial remediation of petroleum hydrocarbons in liquid wastes. In: Development in wastewater treatment research and processes. Elsevier, pp 117–129. https://doi.org/10.1016/b978-0-323-85839-7.00006-2
138. Amran RH, Jamal MT, Pugazhendi A, Al-Harbi M, Ghandourah M, Al-Otaibi A, Haque MF (2022) Biodegradation and bioremediation of petroleum hydrocarbons in marine ecosystems by microorganisms: a review. Nat Environ Poll Technol 21(3). https://doi.org/10.46488/nept.2022.v21i03.019
139. Gouthami K, Mallikarjunaswamy AMM, Bhargava RN, Ferreira LFR, Rahdar A, Saratale GD, Bankole PO, Mulla SI (2023) Microbial biodegradation and biotransformation of petroleum hydrocarbons: progress, prospects, and challenges. Genomics Approach to Bioremediation: Principles, Tools, and Emerging Technologies, pp 229–247. https://doi.org/10.1002/9781119852131.ch13
140. Zhou Z, Liu Y, Zanaroli G, Wang Z, Xu P, Tang H (2019) Enhancing bioremediation potential of Pseudomonas putida by developing its acid stress tolerance with glutamate decarboxylase dependent system and global regulator of extreme radiation resistance. Front Microbiol 10:2033. https://doi.org/10.3389/fmicb.2019.02033

141. Lu CW, Ho HC, Yao CL, Tseng TY, Kao CM, Chen SC (2023) Bioremediation potential of cadmium by recombinant Escherichia coli surface expressing metallothionein MTT5 from Tetrahymena thermophila. Chemosphere 310:136850. https://doi.org/10.1016/j.chemosphere.2022.136850
142. Durán RE, Méndez V, Rodríguez-Castro L, Barra-Sanhueza B, Salvà-Serra F, Moore ER, Castro-Nallar E, Seeger M (2019) Genomic and physiological traits of the marine bacterium *Alcaligenes aquatilis* QD168 isolated from Quintero Bay, Central Chile, reveal a robust adaptive response to environmental stressors. Front Microbiol 10:528. https://doi.org/10.3389/fmicb.2019.00528
143. Liu X, He L, Zhang X, Kong D, Chen Z, Lin J, Wang C (2022) Bioremediation of petroleum-contaminated saline soil by Acinetobacter baumannii and Talaromyces sp. and functional potential analysis using metagenomic sequencing. Environ Pollut 311:119970. https://doi.org/10.1016/j.envpol.2022.119970
144. da Silva Faccioli YE, da Silva GO, Sarubbo LA (2022) Application of a biosurfactant from Pseudomonas cepacia CCT 6659 in bioremediation and metallic corrosion inhibition processes. J Biotechnol 351:109–121. https://doi.org/10.1016/j.jbiotec.2022.04.009

141. Lu CW, Ho HC, Yao CL, Tseng TY, Kao CM, Chen SC (2021) Bioremediation potential of cadmium by recombinant Escherichia coli surface expressing metallothionein MTT5 [illegible] Chemosphere [illegible]
142. [illegible]
143. [illegible]
144. [illegible]

Chapter 18
Biosafety Regulations in GM Foods

Veeranna S. Hombalimath and Shivalingsarj V. Desai

1 Introduction

Genetic engineering and contemporary biotechnology-related recombinant DNA (rDNA) technologies have become a very effective mechanism with a wide range of potential applications in the healthcare and agriculture sectors. These techniques have produced cutting-edge plant varieties known as transgenic plants in agriculture. The main goals behind the development of such plants include raising production, decreasing reliance on agricultural chemicals, changing the natural features of crops, and improving the nutrient content of food and animal feed. However, when genetically modified (GM) plants are developed, and the ensuing food crops become widely available, concerns about their safety and possible adverse effects on human health have been raised. To modify an organism's genetic makeup, genes must be transferred from one species to another. This genetic code comprises the same fundamental building components and is encoded similarly in various organisms' DNA. It is feasible to introduce a certain DNA sequence or gene that is responsible for a certain feature into the cells of a different organism. The resulting plant is characterized as having been genetically altered after the gene has been integrated into the genome of the receiving plants. Subsequent plant generations inherit the novel features that the inserted gene has determined. Given that closely related species may not naturally possess the necessary features, genetically engineered (GE) plants are developed by combining useful genes from several sources to improve plant kinds. There are numerous distinct steps involved in creating a genetically modified plant:

V. S. Hombalimath (✉) · S. V. Desai
Department of Biotechnology, KLE, Technological University, Hubballi, Karnataka, India
e-mail: hombalimath@kletech.ac.in; desaisv@kletech.ac.in

J. S. Yaradoddi et al. (eds.), *Frontiers in Food Biotechnology*,
https://doi.org/10.1007/978-981-97-3261-6_18

Gene Identification: Finding the genes responsible for the desired attribute in an organism—which could be a plant, an animal, or a microorganism—is the first stage in the gene identification process. Once discovered, these genes are duplicated and extracted utilizing molecular biology methods.

Gene Modification for Insertion: After being isolated and cloned, the target gene is modified with additional parts to create what is known as a gene construct. After that, the modified gene construct is ready to be inserted into the host plant. The addition of promoter and termination sequences, which act as markers for the start and finish of the genome sequence, could be among the alterations. A marker gene may also be included to make it easier to recognize genetically modified cells and tissues during experimental procedures.

Transformation: Transformation is the process by which an altered gene of interest is inserted into the host plant. Utilizing techniques like the gene gun or the agrobacterium approach, this transfer is accomplished. The integration of the inserted gene construct into the host plant's genetic structure, which results in inheritable alterations, is the fundamental component of transformation. Selection plant genetic material is transferred to a growing medium with unique features after the gene insertion stage. This selective environment typically contains antibiotics or pesticides.

Selection: Depending on the particular sort of selectable marker used, a certain medium is selected. Only plants with successful integration of the selectable marker gene may flourish in this unique environment. These plants' ability to survive is a sign that they have inherited the desired transgenic characteristic. Then, in lab settings and greenhouses, complete plants are produced using tissue culture methods and are given a thorough examination. This assessment includes both the confirmation of the introduced genes' steady inheritance and the evaluation of how well those genes function. Additionally, any unintentional effects on elements like plant growth and yield quality are carefully analyzed throughout this procedure.

Field Trials and Safety Assessment: The next stage of this process comprises conducting trials that last for several years and take place in various locations. Both controlled greenhouse habitats and true outdoor settings are used for these investigations. Their main goal is to thoroughly analyze the performance and effects of the transgenic creature. The potential environmental effects of the introduction of the transgenic species are also carefully taken into account. Furthermore, during this stage, a thorough analysis of food safety-related factors is done. The significance of protecting the security of our food supply has been underlined by the increased awareness surrounding this issue. To put it another way, it is crucial to ensure that the food produced by these organisms is safe for human consumption and meets its intended uses. These unexpected changes may have an impact on the nutritional value and health effects of the food items that customers depend on. Adopting an organized and methodical methodology known as risk analysis is crucial if we want to protect the human health component of foods made from genetically modified plants. Three essential elements of risk analysis are risk assessment, risk management, and risk communication. This analytical approach provides a structured basis for decision-making and allows for a methodical assessment of safety issues relating to the effects of (GM) foods on human health. It also lays the framework for revising safety standards as soon as new information becomes available. Through legislative actions, the government establishes rules that control the creation, import, use, research, and introduction of genetically modified organisms (GMOs) and the goods resulting from them.

RDAC	Recombinant DNA Advisory Committee
RCGM	Review Committee on Genetic Manipulation
GEAC	Genetic Engineering Approval Committee
IBSC	Institutional Biosafety Committees
SBCC	State Biotechnology Coordination Committees
DLC	District Level Committees
MoHFW	The Ministry of Health and Family Welfare

To ensure the safety assessment of foods derived from genetically engineered (GE) plants, comprehensive guidance is essential, especially in light of their possible influence on human health. This significance is reinforced by the fact that various GE food crops are currently undergoing field trials in India, as well as by the rise in international trade in foods made from GE crops that have been given the go-ahead for production in other countries. The Indian Council of Medical Research (ICMR), which serves as the scientific and technical advisory body to the Ministry of Health and Family Welfare (MoHFW), has developed guidelines in response to this requirement. By taking into account the worldwide Guideline for Conduct, these rules aim to develop methods for the safety assessment of foods derived from GE plants. For the safety evaluation of genetically modified (GM) foods, an alternative strategy has been adopted due to the difficulties in applying conventional toxicological testing and risk assessment methodologies to food as a whole. This approach acknowledges that the primary goal of the estimate is to determine whether the GM product is as safe as comparable conventional foods rather than to establish absolute safety. A given food item's nutritional needs, potential risks, and other safety-related factors are all identified under the framework for safety evaluation. If such issues are found, the procedure includes gathering and analyzing data relevant to their type and severity. Each case's assessment is carried out in a methodical, comprehensive manner. An organized series of questions serves as the basis for a methodical, step-by-step approach to the safety evaluation of foods derived from genetically modified (GE) plants. The identity of the food, its source, composition, the effects of processing or cooking, the transformation procedure used to produce the genetically engineered product, and properties of the recombinant DNA, such as insertion stability and potential gene transfer, are all taken into consideration during this evaluation. Due to the variety of foods that are now accessible, evaluating the prospective intake and dietary impact of introducing genetically modified (GM) food demands varied degrees of information. These recommendations provide applicants with guidance by outlining the typical data needed to evaluate the safety of foods derived from GE plants. But, not all requirements will always be applicable, and explanations and interpretations may change as new knowledge and experience are learned. Developers are in charge of providing scientific data, which should come from a variety of sources, including scientific literature, technical data, and input from independent scientists, regulatory agencies, international organizations, and other stakeholders. This data should not solely be derived from experiments. Data evaluation should follow risk assessment procedures that are based on science.

Experiments that show the safety of foods produced by GE plants should follow strict scientific guidelines and, when appropriate, Good Laboratory practices. Access to primary data may be requested by regulatory agencies, which must be produced using reliable scientific processes and analyzed using acceptable statistical techniques, as necessary. All applied analytical techniques should have their sensitivity described, and references to these techniques must be easily accessible. Before submitting their research, applicants are urged to speak with the relevant regulatory bodies to understand the requirements for submitting the main whole food product made from a genetically modified plant.

Application: The information in this paper applies to all foods, food items, and ingredients made from genetically modified (GM) plant sources. Its goal is to offer direction on regulatory issues to both applicants and reviewers. However, as additional data requirements can arise during the safety assessment process, it does not aim to completely identify all the data that may be necessary.

2 Definitions

Antinutrient: A material (like oxalate and phytate, which prevent the body from absorbing calcium) that prevents the body from properly utilizing one or more nutrients. Traditional equivalent associated non-genetically modified plant varieties, their parts, and/or finished goods have a track record of proven safety based on widespread usage as food. The organism from which genetic material is collected and transferred to the recipient organism is referred to as the donor organism. Food that has been genetically altered includes both the food itself and the ingredients that have been genetically altered through the use of contemporary biotechnology. It also includes food and food ingredients made from genetically modified animals or plants derived from contemporary biotechnology but not containing such materials.

Genetically Engineered Plant (GE Plant): A plant whose genetic makeup has been transformed using in vitro nucleic acid methods, such as recombinant DNA and direct nucleic acid injection into cells or organelles. It is also known as a transgenic plant, recombinant DNA, or genetically modified (GM) plant. A biological, chemical, or physical agent in a food or a food condition that has the potential to have a negative impact on health if consumed is a hazard. The use of in vitro nucleic acid techniques, such as recombinant DNA and direct nucleic acid injection into plant or crop cells or organelles, is known as modern biotechnology. It also covers approaches not used in conventional plant breeding and crop selection, cell fusion across taxonomic families, and overcoming physiological or recombinant obstacles.

Risk: Refers to the likelihood that a food hazard will have a negative impact on consumers'.

Health and the Seriousness of That Impact: Risk analysis is a procedure that includes risk management, risk communication, and risk assessment in connection to any food product. Risk assessment is a procedure with a scientific foundation that includes hazard identification, hazard characterization, exposure evaluation, and risk evaluation.

Transgenic Plant: A plant whose genome has been altered to incorporate a transgene. **Transformation**: The particular DNA recombination that took place as a result of the integration of transgenes for genetic modification in one plant cell, which was then employed to produce fully transgenic plants.

The document's scope includes all whole foods, food items, and ingredients made from genetically modified (GM) plants. Its goal is to provide regulatory context guidance for both applicants and reviewers. It does not, however, attempt to fully define every piece of information that might be required throughout the safety assessment because more information might be needed as the assessment process progresses. According to Section 6.2.1, the information and data presented should be related to the plant parts used as a food source.

Definitions: A chemical known as an antinutrient prevents the body from utilizing one or more nutrients (for instance, oxalate and phytate prevent calcium absorption). The non-genetically modified plant variety, its components, and/or goods with a track record of demonstrated safety based on long-standing use as food. The organism from which genetic material is obtained and transferred to the recipient organism is known as the donor organism.

Foods and ingredients that have been genetically altered through the use of contemporary biotechnology are referred to as genetically modified foods. It also includes foods and food ingredients made from living things or plants that were grown using contemporary biotechnology, even if such things or plants do not include genetically engineered material. The plant that has been genetically altered utilizing in vitro nucleic acid methods, such as recombinant DNA and direct insertion of nucleic acid into cells or organelles, is referred to as a genetically engineered plant (GE plant). It is also known as a transgenic plant, recombinant DNA plant, or genetically modified plant (GM plant).

Hazard: A component of food, such as a biological, chemical, or physical substance, that, when exposed to, has the potential to have a negative impact on health.

3 GM Food Safety Evaluation

Food safety has traditionally been determined based on a food's history of safe ingestion. It's interesting that a lot of the common items in our modern diet haven't gone through thorough toxicological testing. Nevertheless, it is generally agreed that eating these items is safe. Changes in specific features may take place when traditional breeding techniques are utilized to create new plant or animal kinds, either positively or negatively. However, thorough safety analyses are often not performed on these.

Modifications: The safety of genetically modified (GM) foods derived from particular classes of genetically modified organisms (GMOs) is evaluated differently. Every GMO, without exception, is thoroughly examined to make sure they don't endanger the environment or human health. The safety of GM foods cannot be assessed uniformly as a single category because it depends on the type of food and the precise genetic alterations used. For instance, if two varieties of maize are genetically altered to display different traits or if they are altered to attain the same trait using various ways, separate safety evaluations would be required for each kind of GM maize. Similar to this, more safety assessments would be necessary if the GM maize underwent further alterations. GMO-derived foods can be used as ingredients in other food items when they have undergone comprehensive safety evaluations and have been found to be safe. For instance, GM soybean or GM maize oils, flours, and similar derivatives can be utilized to make consumables like bread and snacks. Individual items made from these GM sources do not require independent safety studies in such cases. The Cartagena Protocol on Biosafety, the organization for Economic Co-operation and Development (OECD), and other organizations play important roles in this process. A framework of comparative risk assessment is used to investigate the potential alterations brought about by genetic engineering. The foundational idea behind this strategy is that organisms grown traditionally have a history of safe use and serve as the benchmark. As a result, the conclusion of the safety review procedure is whether the genetically modified organism (GMO) is equally safe as its non-GM conventional counterpart. The examination of the safety of food and feed and the assessment of environmental risk are both included in the safety assessment required for the commercial release of a genetically modified (GE) plant.

4 Safety Assessment of GM Foods

Food safety has traditionally been assessed based on a long history of safe ingestion. Surprisingly, a large number of the items that make up our modern diet have not been subjected to in-depth toxicological testing. Nevertheless, it is generally accepted that eating these foods is harmless. Changes in specific features may take place as a result of conventional breeding methods used to create new plant or animal types, either in a good or bad way. These modifications, however, often do not go through a thorough safety assessment process. Contrarily, foods that have undergone genetic modification (GM) and come from specific GMOs are subject to a different safety assessment. Every GMO, without exception, undergoes a thorough evaluation to make sure they don't endanger the environment or human health. Because safety factors differ based on the type of food and the precise genetic alterations used, GM foods cannot be uniformly assessed as a single category. For instance, if two types of GM maize are modified to have different traits or if they are modified to have the same trait but in different ways, separate safety evaluations would be required for each type of GM maize.

Similar to this, more safety evaluations would be required if the GM maize underwent further alterations. GMO-derived foods can be used as ingredients in other food items when they have undergone thorough safety evaluations and have been found to be safe. For instance, GM soybean or GM maize oils, flours, and similar derivatives may be used in the creation of foods like bread and snacks. In such circumstances, independent safety assessments are not required for the specific goods obtained from these GM sources.

The Cartagena Protocol on Biosafety, the organization for Economic Co-operation and Development (OECD), and the Commission are important players in this process. The potential changes brought about by genetic engineering are assessed using a framework for comparative risk assessment. This comparison method's core concept is that conventionally grown organisms have a proven track record of safe usage and serve as the standard. A determination as to whether the genetically modified organism (GMO) is as safe as its conventional non-GM counterpart is made as a result of the safety review process. The evaluation of the safety of food and feed, as well as the assessment of environmental risk, are all parts of the safety assessment required for the commercial release of a genetically modified (GE) plant. Data from the molecular characterization of the GE plant and the characterization of the transgenic proteins it expresses are added to this evaluation. The transformed organism is carefully examined for potential toxicity, allergenicity, nutritional alterations, and other significant features relevant to the particular genetic modification in issue in order to assess the impact on human health. The evaluation of the newly introduced substance's toxicity and allergenicity factors takes into account its chemical characteristics, functions, concentration in edible plant parts, and likelihood of dietary exposure. Depending on the circumstance, laboratory animals are also used in relevant oral toxicity investigations. Evaluation of allergenicity entails gathering information about amino acid homology between produced proteins and recognized allergens from bioinformatics databases, as well as conducting a thorough analysis of heat stability, pepsin digestibility, and other relevant characteristics. By comparing the concentration of important components in GE plants with those in conventional counterparts cultivated and harvested under identical agro-climatic circumstances, a complete compositional analysis is done to establish nutritional equivalency. Studies involving the feeding of these plants to livestock may also be carried out under certain circumstances. Because different crops have varied biological characteristics, each case-by-case environmental risk assessment for GE plants is done. There is no set method or model to follow because the expertise and experience of the unmodified plant serve as the foundation, emphasizing the necessity for a tailored strategy. Foods that have been genetically modified (GM) are evaluated for safety. The evaluation of food products' safety has traditionally been based on their track record of safe consumption.

Surprisingly, despite not having undergone extensive toxicological testing, many foods in our modern diet are nevertheless regarded as safe to consume. When using traditional breeding techniques to create new plant or animal kinds, certain features may change, either positively or negatively. However, a thorough safety examination is typically skipped over when making these improvements. In contrast, foods that have been genetically modified (GM) or made from specific GMOs go through a special safety review process. Every GMO goes through meticulous testing to make sure there are no dangers to the environment or human health. Due to varied safety concerns based on the type of food and the precise genetic alterations produced, GM foods cannot be universally evaluated as a single category. For instance, distinct safety evaluations would be necessary for each GM maize variety if two varieties of the crop were genetically modified to exhibit different traits or if various techniques were employed to produce the same trait. Similar to this, new safety evaluations are required if changes are made to the GM maize. GMO-derived foods can be utilized as ingredients in other food items once they have successfully passed thorough safety assessments and been deemed safe. For instance, items like bread and snacks can contain flour and derivatives from GM soybeans or maize. Individual items made from these GM sources don't require independent safety reviews in such circumstances. The Cartagena Protocol on Biosafety, the Organisation for Economic Co-operation and Development (OECD), and the Commission all play crucial roles in this process. A methodology for comparing risks is used to examine potential changes brought about by genetic engineering. The key principle of this strategy is to employ organisms that have been traditionally farmed and have a track record of safe use as a benchmark.

5 Risks of Modern Biotechnology Research

The following points highlight the four major risks of modern biotechnology.

1. Health Risk.
2. Environmental Risks.
3. Risks to Biodiversity.
4. Socioeconomic Risks.

Health Risks: The primary area of concern within modern biotechnology revolves around health risks. Specifically, the focus lies on assessing and mitigating the potential dangers posed by allergens in genetically enhanced foods. Innovative biotechnological methods have paved the way for the identification, characterization, and reduction of allergenic risks associated with genetically modified organisms. This issue has garnered global attention, particularly in industrialized nations. A significant proportion of food allergens, affecting approximately 2% of adults and 4–6% of children, can be attributed to eight specific food groups. Genetically improved crops and food products may elevate the risk of allergenic reactions either by increasing the presence of inherent allergens or by introducing new allergenic elements. Evaluating allergen-related risks presents a formidable challenge. The International Life Sciences Institute (ILSI) has developed a decision-making framework that relies on specific criteria. These criteria stipulate that an introduced protein in a food product is deemed safe if it meets the following conditions: (1) No history of common allergenicity. (2) No substantial similarity in amino acid sequences to known allergens. (3) Rapid digestion of the protein. (4) Low levels of protein expression

Environmental Risks: The risks policymakers and regulators need to assess include the potential for the spread of traits from genetically improved plants to the same or related species, the build-up of resistance in insect populations, and the potential threat to biodiversity posed by widespread monoculture of genetically improved crops. (i) A transparent, science-based framework is required, which assesses risks on a case-by-case basis and takes account of all stakeholder views. (ii) Environment-related issues to be considered in this case include the possibilities for gene transfer, weediness, specific trait effects, genetic and phenotypic variability, and expression of pathogenic genes. (iii) Risk management needs to consider the prospects for managing any specific risks identified with a proposed release. (iv) Experience is accumulating in the management of the Bt genes in transgenic cotton varieties in several countries, and this needs to be closely monitored. v. An agricultural sustainability protocol that balances risks and benefits may have value for the approval and use of new crop varieties. The risks lie in the management of the cropping system involving soil, water, and other inputs. There is a need for the establishment of baseline information in the environment where such introductions are to be made.

Environmental Risks: The second facet of concern in the realm of modern biotechnology pertains to environmental risks. Policymakers and regulators are tasked with the responsibility of assessing a spectrum of potential hazards, including the dissemination of genetic traits from genetically enhanced plants to closely related species, the development of resistance within insect populations, and the looming danger to biodiversity posed by the widespread cultivation of genetically improved crops. To effectively address these environmental risks, a transparent and science-based framework is imperative. This framework should meticulously evaluate risks on a case-by-case basis while taking into account the perspectives of all stakeholders involved. Each environmental scenario necessitates a careful consideration of issues such as the potential for gene transfer, the propensity for plants to become invasive (weediness), the specific impacts of modified traits, the variability in genetic and phenotypic characteristics, and the expression of potentially harmful genes. An essential component of risk management involves the examination of strategies for mitigating identified risks associated with the proposed release of genetically modified organisms. Experience in managing genetically modified organisms, exemplified by the management of Bt genes in transgenic cotton varieties across multiple countries, is accumulating. Continuous monitoring of these experiences is crucial.

Socioeconomic Risks: There is a risk that modern science may bypass the needs of poor people. Biotechnology is only one tool in addressing the challenges of food security and poverty. There is a need for biotechnology to be integrated with appropriate policies and other conventional R&D programs. The positive and negative impacts of biotechnology should be monitored over time in terms of whom and what are affected and how they are affected. Monitoring impact will provide guidance for public policymakers in the future. Unless countries have policies in place to ensure that small farmers have access to delivery systems, extension services, productive resources, markets, and infrastructure, there is a risk that the introduction of agricultural biotechnology could lead to increased inequality of income and wealth. In such cases, larger farmers are likely to capture most of the benefits through early adoption of the technology, expanded production, and reduced unit costs. Biotechnology has the potential to reduce input use, reduce risk to biotic and abiotic stress, increase yields, and enhance quality traits, which should enable the development of new crop varieties that are appropriate to poor producers and consumers. Modern biotechnology is not a silver bullet for achieving food security, but when used in conjunction with other agricultural research, it may be a powerful tool in the fight against poverty. It has the potential to help enhance agricultural productivity in developing countries in a way.

6 Some of the Controversial Aspects of Recombinant Products

Bt cotton: Bacillus thuringiensis cotton, involves the utilization of bacterial strains that produce a range of more than 200 distinct Bt toxins. Each of these toxins is specifically harmful to various types of insects, with a notable focus on the larvae of moths, butterflies, beetles, cotton bollworms, and flies. Importantly, Bt toxins do not pose harm to another form of life. To confer this insecticidal property onto cotton plants, the gene responsible for Bt toxin production has been introduced into cotton as a transgene, causing the cotton to naturally produce this insecticide within its tissues. In regions where cotton is commercially cultivated, lepidopteron larvae are often the primary pests. These larvae are effectively controlled by the Bt protein produced by genetically modified cotton when they consume it. This obviates the necessity for substantial applications of broad-spectrum insecticides to combat these lepidopteran pests, particularly in cases where some of them have developed resistance to pyrethroids. Such a practice preserves the balance of natural insect predators within the farm ecosystem and further contributes to pest management without relying on insecticides. However, it's important to note that Bt cotton does not effectively combat various other cotton pests, such as plant bugs, stink bugs, and aphids. Depending on specific circumstances, it may still be necessary to employ insecticides for prevention. In 2006, a research investigation conducted by scientists from Cornell University, in collaboration with the Center for Chinese Agricultural Policy and the Chinese Academy of Science, examined the practice of cultivating Bt cotton in China. The findings revealed that over a span of seven years, secondary pests that were traditionally managed through pesticide application experienced a notable increase in population. Consequently, farmers found themselves in a situation where they needed to apply pesticides at levels comparable to those used in non-Bt cotton cultivation. This additional pesticide requirement led to reduced profits for farmers due to the increased cost associated with the utilization of genetically modified (GM) seeds.

Recombinant Bovine Somatotropin (rBST) is a synthetic growth hormone also known as rBGH or artificial growth hormone. It was developed by four major pharmaceutical companies: Monsanto, America Cyanamid, Eli Lilly, and Upjohn, who submitted their commercial rBST products for approval to the US Food and Drug Administration (FDA). Monsanto was the first company to receive approval and marketed their product as "Posilac." In 2008, Monsanto sold this business to Eli Lilly and Company for $300 million, plus additional considerations. Although rBST was approved for commercial use in several countries, including Mexico, Brazil, India, Russia, and others, it has been banned in Canada, Australia, New Zealand, Japan, Israel, and the European Union since 2000. Argentina also prohibited its use. Multiple organizations, including the FDA, World Health Organization (WHO), and National Institutes of Health (NIH), have independently stated that dairy products and meat from cows treated with BST are safe for human consumption. In the United States, due to public concerns, some manufacturers and retailers chose to market only rBST-free milk. A report from the European Union on the animal welfare effects of rBST indicated that its use often leads to significant suffering for cows, including mastitis, foot disorders, and reproductive problems. rBST is found in the milk of both treated and untreated.

Genetically Modified Fish: Often referred to as GM fish, they belong to the taxonomic clade encompassing Agnatha (jawless fish), Chondrichthyes (cartilaginous fish), and Osteichthyes (bony fish). These organisms have undergone genetic alterations through the application of genetic engineering techniques to modify their DNA. In most instances, the objective is to introduce a novel trait into the fish species that does not naturally occur, a process known as transgenesis. GM fish serve various purposes, including their use in scientific research and as pets. They are also being developed for roles such as serving as environmental pollutant indicators and for their potential contribution to aquaculture for food production. Notably, in 2015, the Aqua Advantage salmon received approval from the US Food and Drug Administration (FDA) for commercial production, sale, and consumption. This marked a significant milestone as it became the first genetically modified animal sanctioned for human consumption. Certain GM fish have been engineered with promoters that stimulate excessive production of "all fish" growth hormone, leading to remarkable growth enhancements in several species, including salmonids, carps, and tilapias. Critics have raised objections to the use of GM fish on various grounds.

These include concerns about their ecological impact, considerations for animal welfare, and questions regarding the safety of consuming GM fish as food. Additionally, there is debate regarding the necessity of GM fish in addressing global food.

Further Reading

Asante-Owusu R (1999) GM technology in the forest sector. WWF International, Gland

Balkrishna P (2001) Agriculture and biodiversity. IUCN, Gland

Balakrisna P (2002) Assessing and managing risks: biotechnology and biosafety (IUCN Asia)

Bauer MW, Gaskell G (eds) (2002) Biotechnology: the making of a global controversy. Cambridge University Press

Bereano P (1999) Report on Danish "Citizen Consensus Conference" on Genetically Engineered Foods, March 12–15 (Luka group). http://www.loka.org/pages/DanishGeneFood.html

Bergelson J, Purrington C, Wichmann G (1998) Promiscuity in transgenic plants. Nature 395:25

BioTIK Expert Group (Danish Ministry of Trade and Industry) (1999) An ethical foundation for genetic engineering choices. http://www.em.dk/publikationer/html/english/biotik /appendix_2.htm

Bonalume R (1999) Brazillian farmers fighting against legal barriers to GM crops. Nature 402:344

Tumma MK. PBRD Asia-Pacific Millet India. Pioneer Hibrid Pvt. Ltd., Hyderabad. Poorna Chandrika KSV. ICAR-Indian Institute of Oilseeds Research, Hyderabad, Telangana. https://www.ndsu.edu/pubweb/~mcclean/plsc431/students99/oberoi.htm

Chapter 19
Challenges in Genetically Modified Foods

Rashmi G. Gadwal

1 Introduction

Africa is the only continent where at least one in three people suffer from hunger and malnutrition and where food output per capita is declining [1]. Africa's existing agricultural production management techniques are inadequate to generate enough yields in sufficient quantities to feed the continent's rapidly growing population [2]. Ethiopia's population was 18, 128,000 in 1950, according to reports, and it increased to 104, 957,000 in 2017. It is predicted to reach 139, 620,000 and 190, 870,000 in 2030 and 2050, respectively [3].

During the next few years of the twenty-first century, feeding over nine billion people will be one of the biggest challenges facing humanity [4]. Consequently, to feed the world's rapid expanding population, agricultural yields must be raised. Assuming that crop yield would account for 80% of this frustration, crop yields need to increase by 35.0% during the next 32 years. This corresponds to or translates as a 0.9% yearly growth rate [5].

Increasing the amount of arable land used, improving irrigation techniques, using chemical inputs like pesticides, fertilizers, and herbicides, along with other agricultural practices, all contributed significantly to the twentieth-century Green Revolution's increase in crop output. It was argued, however, that these opportunities were overshadowed by undesirable traits that had a detrimental impact on natural resources, such as the environment, salinization of irrigated areas, excessive groundwater extraction, the development of pests and diseases resistant to chemicals, the decline in the biodiversity of beneficial plants and animals [6].

A large number of the Green Revolution's benefits have been undermined by growing global population, climate change, and inadequate poverty interference initiatives [7]. Any scientific application that uses biological systems, living

R. G. Gadwal (✉)
Department of Biotechnology, Gulbarga University, Kalaburagi, Karnataka, India

J. S. Yaradoddi et al. (eds.), *Frontiers in Food Biotechnology*,
https://doi.org/10.1007/978-981-97-3261-6_19

organisms, or their derivatives to develop or modify anything specific for use in agriculture is known as agricultural biotechnology [8]. It has the potential to significantly affect agricultural sustainability, quality, and productivity. Applications of biotechnology range from low-tech methods like artificial insemination, fermentation processes, and biofertilizers to high-tech strategies and sophisticated DNA-based methods like GM organisms.

The development of agrobacterium-mediated genetic engineering techniques in the late 1980s suggested the successful transfer of gene material to tomato plants' nuclear genomes [9]. While genetic engineering adoption in Africa was relatively low, some countries such as South Africa, Egypt, Sudan, Kenya, and Burkina Faso have seen noticeable advancements in the use of genetically modified (GM) crops, such as Bt cotton (cotton containing toxic protein from a bacterium called *Bacillus thuringiensis*) and Bt maize (maize containing toxic protein from *Bacillus thuringiensis*), which are genetically modified bacteria that produce different toxins that are harmful to various insects and pathogens [10].

Ethiopia's textile manufacturing sector is rapidly growing throughout the nation, beginning with the creation of industrial parks. Additionally, it is projected that the growth of agricultural industrial parks will raise Ethiopia's need for cotton production and supply [11]. Ethiopia imports cotton, particularly Bt cotton, from a number of foreign suppliers since its productivity as a cotton crop is insufficient to fulfill the growing demand of the textile industries [12]. Numerous companies have already opened offices in Ethiopia in recent years due to government-led outreach, relatively cheap labor costs, low electricity costs, unjustified government support, loans, and a sizable tax exemption [13].

Ethiopia intends to start commercializing Bt cotton in response to the need for cotton, having recently implemented a decree granting research rights in both lab and agricultural settings. Therefore, limited experimental trials are now under progress, and Bt cotton commercialization is anticipated to occur in the upcoming years [14]. The National Agricultural Biotechnology Research Centre is founded, and four centers are home to 68 ongoing investigations. Genetically modified organisms (GMOs) have been the subject of heated and significant discussions since the 1990s. The prospective effects on plant biodiversity, global food chain, etc. are the main topics of discussion in these discussions. It is untrue that agricultural genetic engineering can help with the majority of problems on basic food security by crop improvement and quality.

However, the genetic engineering of crop plants gives rise to concerns about possible risks to the environment and living creatures, where the opportunities and difficulties should be carefully considered on a case-by-case basis [15]. In order to establish a country-based regulating framework, it is requested that present biosafety contexts and functioning be taken into account, together with the implications and duties of the Cartagena procedure on biosafety and the then-Convention on Biological Diversity (CBD) [16]. In most nations, people have differing and sometimes dramatically different perspectives on the possible benefits and risks of agrarian biotechnology (Fig. 19.1) [17].

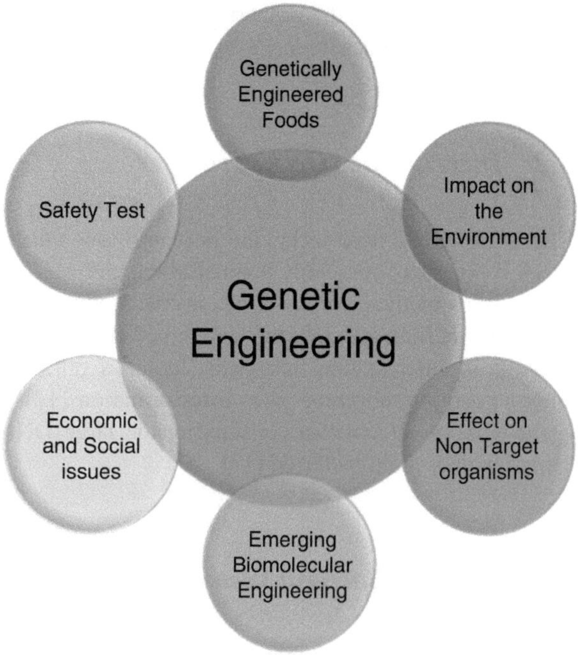

Fig. 19.1 Content discussed concerning to genetic engineering

Research should be done on risk assessment of potential toxicity and allergic effects on human or animal health. Moreover, research must be done before introducing genetically modified crops to ensure that it has no potential to spread plant diseases and has no impact on gene flow within species and to wild families. It was proposed that the advantages of genetically modified food products appear to outweigh the risks and those regulatory bodies and businesses involved in GM food companies should raise public awareness of the technology to increase GM food acceptance globally through transparency, education, and research [18].

In order to guarantee food security for its people, Africa must adopt contemporary agricultural technologies [19]. Nonetheless, it is widely acknowledged that genetically modified crops cannot be sold until a thorough safety analysis has verified their safety [20]. Thus, this review's goal is to examine the potential benefits and difficulties of producing genetically modified crops in Ethiopia while learning from other nations' experiences.

1.1 Generation of GM Crops

For the production of GM foods, researchers introduced the gene(s) with certain traits and transferred into a host cell (plant), thereby regeneration of whose plant using tissue culture tools and techniques. The transferred gene is expressed is inherited to optimize the property of the product. To modify genes of cells, plant tissue culture includes three ways-

1.2 Direct Gene (DNA) Transfer

The most widely used technique in plant tissue culture for delivery of exogenous DNA is the micro-projectile bombardment and was developed in 1980s by Sanford [11]. In this method, foreign DNA is coated with gold or tungsten, and then delivered using helium gas, into the targeted cells or tissues such as meristems. Transfer of gene of interest into host cells, includes other methods like electroporation, microinjection, and chloroplast transformation [12]. However, particle bombardment is more effective in considering the large DNA fragments—even if whole chromosomes are transferred [13].

1.3 Use of a Bacteria Vector

The soil bacterium *Agrobacterium tumefaciens* is able to introduce foreign genes into plant cells. Plants that are infected by the soil bacterium *A. Tumefaciens* develop tumors. The bacteria change the plant's genome, which allows the plant to manufacture modified amino acids as a specialized food source for itself in addition to driving the plant's cells to proliferate. The bacteria harbor a tumor-inducing plasmid known as Ti-plasmid, which contains T DNA and allows the introduction of target genes into the host cell.

1.4 Direct Editing of Genomic DNA

The "CRISPR-Cas9" system was created in 2012. It is a ground-breaking technique for genome editing and offers an additional way to modify genes in different kinds of cells [14, 15]. This method makes working with plants much easier by increasing the efficiency of genetic engineering dramatically [16]. DNA endo-nuclease Cas9 was first discovered in bacteria, where it shields the host bacterium from incoming DNA molecules. A unique "guide RNA" (gRNA), whose sequence is complementary to the invasive sequence, directs the endo- nuclease to the target DNA. Cas9 cleaves the double-stranded DNA by using its two active sites to do so.

The recently discovered DNA double-strand breaks (DSBs) are then repaired by two distinct mechanisms within the cells: the first is called "homologous recombination" (HR), which permits the insertion of donor DNA into the endogenous gene at the break site, and the second is called "non-homologous end joining" (NHEJ), which results in a small deletion or random DNA insertion, leading to a truncated gene or knockout. According to reports, the US Department of Agriculture has ruled that a number of genome-edited crops are ineligible for the present laws. Lastly, the USDA published the tentative proposals to update its criteria for genetically modified crops on November 18, 2015.

1.5 Impacts on the Environment

In order to ensure a high standard of human health and environmental protection, biosafety demands maintaining an equilibrium while simultaneously offering a reliable regulatory mechanism in the food chain [21]. Environmental contamination, accidental gene transfer to wild plants, the potential for invasive weed development, the harm to crop genetic diversity, ethical, cultural, and religious issues, and the fear of unknowable consequences are some other, more general worries. There are gaps in our knowledge regarding the possible consequences of gene flow and the impact of particular features on the performance of wild families or weeds, as well as the possibility of gene flow from engineered crops to sexually compatible wild families and agricultural weeds [22].

In order to monitor and assess newly introduced genetically modified crops in Ethiopia, there needs to be a robust panel of professionals with well-defined norms, regulations, and safety protocols that address the following questions: Has the product been applied as suggested or anticipated? Are the side effects and recognized consequences what were expected? Are there any unanticipated adverse effects from the product [23]. This aids in keeping GM inputs below the 0.9% legal tolerance limit [24]. The biosafety proclamation of Ethiopia, which was stricter in the past, has been updated to encourage the use of technology. Many nations have declined to introduce or produce genetically modified crops due to safety concerns.

For instance, the Scottish Government declared it will employ the new EU laws that forbid the cultivation of genetically modified crops. Looking to biotechnology as a solution is one way to alleviate the world's hunger problem as the fight to secure the food and nutritional security of millions of people rages on, but should this come at the expense of the environment? There is a rising discussion on the possible benefits of contemporary biotechnology.

1.6 Effect on Nontarget Organisms (NTO's)

Animal health may be at risk from genetically modified plants. Other non-target creatures may be negatively impacted by a transgene that has been introduced into a target organism to change its original traits for the advantage of humans [25]. While some of the dangers associated with transgenic crops are well understood, others need further research. According to reports, the usage of genetically modified crops is one of the reasons behind the fast decline in bee populations worldwide

[26]. The Bluetongue Virus (BTV) is causing alarm as it spreads over Northern Europe, providing startling proof of the establishment of a "exotic" vector borne illness in a previously uncharted area. Non-Target Organisms (NTOs):

There is little knowledge about the origin of this virus, which poses a serious and novel risk to animal production [27]. Owing to these concerns, nations like Italy forbid the commercial growing of genetically modified plants or their field testing at their university labs or research facilities [28].

1.7 Social and Economic Issues

People's views are thought to play a major role in determining how biotechnology is used and develops. Most experts predict that public perceptions of GM non-food goods will improve over the next 10–15 years, but that public acceptance of GM food products will remain constant [29]. The change of food quality, probable antibiotic resistance from GM crops, potential toxicity, potential allergy, and carcinogenicity from GM foods are among the concerns raised by GM technology opponents. Coexistence needs to be allowed, and guaranteeing consumers' freedom of choice requires proper labeling [30].

Disputes were investigated in a competitive study of Australian and UK citizens' acceptance of GMO food. Due to user resistance and the impact of EU regulations, the acceptance of transgenic foods at the government and industry levels in the UK has not led to their commercial adoption. In contrast, Australian federal and state governments have gradually accepted genetically modified crops, potentially securing Australia's place in the GM food and farming trajectory [31]. The European community views biotechnology in general and the application of genetic engineering in food production in particular seriously and perceives it as harmful. Disparities have been noted between European and US citizens [32].

Environmentalists and human rights organizations have great concern and blame for Bt cotton and the biotech sector as a whole. Even though biotech crops have shown quantifiable outcomes and are being implemented in India, the nation continues to face strong criticism from rights groups. The Indian Supreme Court has at one point been embroiled in a legal dispute concerning the application of biotechnology efforts involving mustard. African nations like Ethiopia may learn a lot from this about raising awareness and involving concerned individuals and human rights organizations.

Obtaining approved seed is one difficulty in producing genetically modified crops. Because growers are required to buy genetically modified seeds annually, they are dependent on GM seed suppliers and producers who demand higher prices for GM seeds than regular seeds. Furthermore, distributing seeds might be difficult in countries with inadequate infrastructure. Many nations have developed and implemented policies to allay manufacturers' and consumers' concerns about safety [33], although there is a great deal of variation in labeling laws across borders [34]. The quasi-generalized obligation to label products derived from GM crops that are not substantially equal to their conventional counterparts is the single characteristic shared by the nations that have labeling legislation [35].

The two general regulatory approaches for labeling genetically modified food are:

- Mandatory labeling, which requires declaration of characteristics imparted to a food by the use of gene technology (are they healthy and safe and/or process-related), or use of gene technology itself in food production.
- Voluntary labeling, which is driven largely by market forces and lacks legislative requirements to declare the use of GMOs in food production. Guidelines for the voluntary labeling of non-GM products are also in place in some countries where the labeling of GM ingredients is required [36].

It is interesting that comparable advancements are not as noticeable for the genetically modified plants whose nutritional content has been improved, given the promise of plant genetic engineering and the freedom it can bring from reliance on traditional breeding to create diversity. These characteristics, which are frequently referred to as output traits since they modify the harvested product, also typically benefit the consumer rather than meeting the needs of the farmer, as is the case with input features like insect resistance and herbicide tolerance.

Why aren't any genetically modified crops with improved nutrition being produced commercially in fields yet? Maybe this is because nutritional quality is more genetically complex than these single-gene input qualities, and it is also harder to measure and prove effective than herbicide tolerance. For the purposes of this review, we have defined nutritional enhancement as the addition or elevation of a nutrient in a product through genetic modification (GM), as opposed to other techniques of plant breeding or direct supplementation. Nutritional enhancement can be performed in a variety of methods.

The success of the input attributes indicates that the underlying technology functions effectively on several different dimensions. Likewise, it is simple to modify already-existing supporting agricultural infrastructures (with little additional costs) to handle GM crop.

However, many populations still struggle with poor diet and malnutrition, and new issues are becoming increasingly apparent. Millions of people were saved from starvation thanks to the Green Revolution, but metabolic diseases like type 2 diabetes, obesity, and cardiovascular disease—which are caused by consuming excessive amounts of calorie-dense but nutritionally deficient foods—are now a bigger concern [32]. This review examines a number of different factors that have contributed to the delay of these advancements in addition to the development of genetically modified crops that are enhanced nutritionally. We specifically examine the advantages of two well-known instances, omega-3 fish oil crops and Golden Rice, in the face of resistance, raising the question, "If not now, when?" Will genetically modified crops ever be able to provide nutritional enhancement? The reader is directed to Martin and Li [33] recent review for more information on the broader topic of more developing instances of genetically modified nutrition enhancement.

The idea of "Golden Rice," or more accurately, rice endosperm made yellow by artificially accumulating beta-carotene (pro-vitamin A), was reportedly originally proposed in 1984 [33, 34], which puts it at the forefront of plant genetic engineering. Nevertheless, unlike tolerance to herbicides, it took more than 10 years to show even the rudimentary stages of producing a rice grain that could be polished while retaining beta-carotene. Burkhardt et al. demonstrated in 1997 that the daffodil (*Narcissus pseudonarcissus*) phytoene synthase gene, expressed specifically in the endosperm, directed the accumulation of phytoene, the precursor of beta-carotene [35]. Ye et al. also reported in 2000 the first iteration of Golden Rice [36, 38], wherein the endosperm of the grain produced beta-carotene.

With the benefit of hindsight, what was once considered a groundbreaking discovery may now be seen as a "proof-of-principle" study, but even so, this discovery marks the beginning of a journey that could significantly improve millions of people's lives by rescuing them from vitamin A deficiency and the diseases that go along with it, like childhood blindness [37, 38]. Though there were still certain technical obstacles to be resolved (like the slight build-up of beta-carotene in the polished grain), the innovators believed that problems with intellectual property rights, opposition from non-governmental organizations, and the regulatory load placed on any genetically modified crop [39] were what prevented Golden Rice from becoming a commercially successful crop.

Another barrier, in terms of money and experience, was trying to pilot the development of such a GM trait outside of agrobiotechnology companies. Despite this, numerous advancements were made, such as the introduction of the Golden Rice trait into indica rice varieties that are appropriate for growing in the most disadvantaged geographic areas and the creation of a second version (Golden Rice 2 (GR2)) in which the daffodil phytoene synthase gene was substituted with an activity similar to maize [39], increasing the endosperm levels of beta-carotene by at least ten times. Crucially, GR2 was developed in partnership with Syngenta, enabling the initiative to take advantage of this business's knowledge of regulatory approval and intellectual property rights. In 2005, hope was reestablished, with hopes that the crop will be licensed and produced extensively.

1.8 Emerging Bimolecular Engineering

In any other case, biomolecular engineering will be regarded as DNA nanotechnology. This field aims to create molecular assemblies and devices where DNA is used as a manipulation material throughout the entire process. The sequence assemblages enable a distinct, balanced design of the DNA structures in sizes ranging from nanometers to millimeters, and molecular signaling routes transfer the information. Highly distinctive features of DNA base pairing give a controlled mechanism for DNA interactions [40]. Regretfully, there isn't another molecular manipulation technology like this one that can plan the entire biosynthesis of intricate biomolecules from basic precursor molecules and other conventional biomolecular systems [41–44].

The vast array of concepts and benefits associated with nanotechnology are related to correspondingly extensive categories of techniques and materials, as they include unique qualities. One of these units, known as bio-nanotechnology, is concerned with organizing biochemical systems, processes, and biomolecules in novel ways for various applications.

2 Future Direction

Regarding the cultivation and alteration of genetically modified crops, significant advancements have been made in adjusting the nutritional composition and processing properties of food crops. For instance, oilseed rape is engineered to yield large amounts of oils with a variety of properties and a high saturation level of the fatty acids generated, including laurite. Similarly, with the aid of GM crops and a variety of biotechnological tools and techniques, it is now possible to modify, on an experimental size, the balance of sugar and starch in potatoes, which influences the processing quality of potatoes for the manufacturing of snack foods.

The genes encoding biologically active components of food crops, such as natural toxicants like kidney bean lectin and potato glycol-alkaloids, antinutrients like trypsin inhibitors, and allergies like certain nut proteins, are being identified and altered using modern genetic techniques. Over time, the early advancements result in the creation of foods devoid of these unwanted ingredients.

Fruit and vegetable deterioration on an industrial scale is a big issue. For instance, when a plant sustains a lesion, the plant tissue tends to become brown, a condition that can be prevented by using preservatives like sulfite. The polyphenol oxidase enzyme, which is released from distinct sub-cellular compartments and catalyzes the conversion of monophenols to quinones, which oxidize to generate brown polyphenolic pigments, is released by damaged cells. However, in experimental experiments, the polyphenol oxidase gene has been turned off through genetic manipulation, preventing this discoloration and spoiling. The biochemistry of fruit and vegetable ripening and deterioration is being fully uncovered by genetic manipulation and other molecular and biochemical techniques. Numerous innovative approaches in food preservation are also being developed, but without the use of chemical preservatives.

The utilization of crops as renewable sources of valuable resources, such as pharmaceuticals, bioplastics, vaccines, and other industrial materials, is another prospect that is likely to spark a lot of attention. Simultaneously, pharmaceutical compounds are being created in the milk of cattle and sheep by genetic modification, enabling far more economical and efficient drug production. It is possible to produce and distribute vaccinations in tropical areas using food crops like bananas, though a thorough discussion of this is outside the purview of this article.

3 Conclusion

Numerous potential advantages for the environment are presented by genetically modified species. It aids in lessening environmental harm. For example, biodegradable plastics can be made by specific microorganisms that have undergone genetic engineering. A bacteria known as *Ralstonia eutropha* produces a flexible polymer that is used to make plastic by converting glucose and certain other acids.

Genetically modified crops can help commercial agriculture to overcome a number of its existing problems. According to market trends, the cultivation of genetically modified crops is one of the most inventive and quickly expanding business practices, benefiting both growers and customers. To combat unethical research and misinformation, the scientific community and the agriculture sector must, make significant investments in improved science communication and regulation. It is anticipated that GM crops would increase commercial agriculture's profitability and production, paving the way for further advancements [45].

There has been broad innovative progress in the production of GMOs. There are many disputants over GMOs who consider them to be a control of life, and also clashes in regard to the dangers to nature and human well-being. Also, it is important to provide buyers with information that will enable them to comprehend the purpose of present-day innovation in crops and rural creation and particularly to comprehend the noteworthiness of genetic changes. The utilization of GMOs and assessment of the potential dangers explain the effects and consequences of genetic alterations on the earth and on human well-being.

References

1. James C (2008) Global status of commercialized biotech/GM crops. International Service for the Acquisition of Agri-Biotech Applications (ISAAA), Ithaca, NY
2. Blanchfield JR, Lund D, Spiess W. Report on the food security forum. Held in conjunction with the 14th World Congress of Food Science and Technology, Shanghai, China, 19–23 October 2008
3. UN (United Nations, Department of Economic and Social Affairs) (2017) Population Division. World Population Prospects: The 2017 Revision, key findings and advance tables Working paper No.ESA/P/WP/248
4. James C (2014) ISAAA report on global status of biotech/GM crops. International service for the acquisition of Agri-biotech applications
5. Alexandratos N, Bruinsma J (2012) World agriculture towards 2030/2050: the 2012 revision. ESA working paper No. 12–03. FAO, Rome
6. The future of food and agriculture—trends and challenges (2017) Rome www.fao.org/3/a--i6583e.pdf
7. Lisen R, Marja-Liisa TB (2011) Land tenure, climate change mitigation and agriculture mitigation of climate change in agriculture (MICCA) programme June 2011. FAO, Rome
8. Borlaug NE, Ending world hunger. (2000) The promise of biotechnology and the threat of antiscience zealotry. Plant Physiol 124(2):487–490. www.plantphysiol.org
9. Article 2 (2013) Use of terms. Montreal, Canada. https://www.cbd.int/convention/text/default.shtml

10. van Roekel JS, Damm B, Melchers LS, Hoekema A (1993) Factors influencing transformation frequency of tomato (*Lycopersicon esculentum*). Plant Cell Rep 12(11):644–647
11. GM Crops List (2017)
12. Bob K (2016) Ethiopia to commercialize BT cotton in two years Africa Business Communities News. 25-11-2016. 12:57:51
13. Ethiopia—Cotton (2017) Addis Ababa
14. Ethiopia plans to commercialize Bt cotton soon (2016) GAIN Report Number: ET 1637. Addis Ababa
15. Directive EU (March 2015) 2015/412, Article 26b (3). http://eur-lex.europa.eu/legalcontent/EN/ALL/?uri=OJ
16. Consensus document www.icgeb.org/biosafety guidance for a fit-for-purpose regulatory framework for GMOs (2012) International Centre for Genetic Engineering and Biotechnology (ICGEB)
17. Wendt J, Izquierdo J (2003) Management of appropriate agricultural biotechnology for small producers: case study—Ecuador. Electron J Biotechnol 6(1):1–7
18. Uzogara SG (2000) The impact of genetic modification of human foods in the 21st century: a review. Biotechnol Adv 18(3):179–206
19. Technology answer to Africa's food security challenge, say scientists EBC; March 9, 2017
20. Secretariat of the Convention on Biological Diversity (2000) Secretariat of the Convention of Biological Diversity
21. Robert P, Guillermo H (2014) Food safety: state-of-play, current and future challenges. In-depth analysis; IP/A/ENVI/2014-11.PE.536.287
22. Clark SJ, Rothery P, Perry JN (2006) Farm scale evaluations of spring-sown genetically modified herbicide-tolerant crops: a statistical assessment. Proc Biol Sci 273(1583):237–243
23. Devos Y, Demont M, Dillen K, Reheul D, Kaiser M, Sanvido O (2009) Coexistence of genetically modified (GM) and non-GM crops in the European Union: a review. Agron Sustain Dev 29(1):11–30
24. Censored, Center for Environmental Risk Assessment (2007) GM Crop Database: Tomato
25. Holmes B (2010) Altered animals: creatures with bonus features. New Sci 7
26. Jones KE, Patel NG, Levy MA et al (2008) Global trends in emerging infectious diseases. Nature 451(7181):990–993
27. Genetically modified plants and foods Challenges and future issues in Europe Final report Berlin (2009)
28. Dibden J, Gibbs D, Cocklin C (2013) Framing GM crops as a food security solution. J Rural Stud 29:59–70
29. Sinemus K (2008) Communicating science, technical, ethical and social aspects related to GMOs: In 1st Global Conference on GMO Analysis. Villa Erba, Como, Italy 24–27 June 2008
30. Beckmann V, Soregaroli C, Wesselern J (2006) Co-existence rules and regulations in the European Union. Am J Agric Econ 88(5):1193–1199
31. Moghissi AA, Jaeger LM, Shafei D, Bloom LL (2018) Regulatory science requirements of labeling of genetically modified food. Crit Rev Biotechnol 38(3):386–393
32. Gruère GP, Rao SR (2007) A review of international labeling policies of genetically modified food to evaluate India's proposed rule. AgBioforum 10(1):51–64
33. Martin C, Li J (2017) Medicine is not health care, food is health care: plant metabolic engineering, diet and human health. New Phytol 216:699–719
34. Enserink M (2008) Tough lessons from golden rice. Science 320:468–471
35. Burkhardt PK et al (1997) Transgenic rice (*Oryza sativa*) endosperm expressing daffodil (*Narcissus pseudonarcissus*) phytoene synthase accumulates phytoene, a key intermediate of provitamin A biosynthesis. Plant J 11:1071–1078
36. Ye X et al (2000) Engineering the provitamin A (beta-carotene) biosynthetic pathway into (carotenoid-free) rice endosperm. Science 287:303–305
37. Al-Babili S, Beyer P (2005) Golden Rice—five years on the road—five years to go? Trends Plant Sci 10:565–573

38. Dubock A (2014) The present status of Golden Rice. J Huazhong Agric Univ 33:69–84
39. Paine JA et al (2005) Improving the nutritional value of Golden Rice through increased pro-vitamin A content. Nat Biotechnol 23:482–487
40. Yaradoddi J, Kontro M, Ganachari S, Sulochana M, Agsar D, Tapaskar R, Shettar A (2018) DNA nanotechnology. In: Kharisov BI (ed) Handbook of ecomaterials. Springer International Publishing AG. https://doi.org/10.1007/978-3-319-68255-6_191
41. Chen Y-J, Groves B, Muscat RA, Seelig G (2015) DNA nanotechnology from the test tube to the cell. Nat Nanotechnol 10(9):748–760. https://doi.org/10.1038/NNANO.2015.195
42. Yaradoddi J, Kontro M, Ganachari S, Sulochana M, Agsar D (2018) RNA nanotechnology. In: Kharisov BI (ed) Handbook of ecomaterials. Springer International Publishing AG. https://doi.org/10.1007/978-3-319-68255-6_193
43. Yaradoddi J, Kontro M, Ganachari S, Sulochana M, Agsar D, Tapaskar R, Shettar A (2018) Protein nanotechnology. In: Kharisov BI (ed) Handbook of ecomaterials. Springer International Publishing AG. https://doi.org/10.1007/978-3-319-68255-6_192
44. Afonin KA, Lindsay B, Shapiro BA (2015) Engineered RNA nanodesigns for applications in RNA nanotechnology. DNA RNA Nanotechnol 1(1):1–15. https://doi.org/10.2478/rnan-2013-0001
45. Potrykus I (2010) Lessons from the 'Humanitarian Golden Rice' project: regulation prevents development of public good genetically engineered crop products. New Biotechnol 27:466–472

Chapter 20
Nutrigenomics and Nutri-informatics

Jayachandra S. Yaradoddi, Chetan D. M., Bharati S. Meti, Virupakshaiah D. B. M., and S. S. Injaganeri

1 Introduction

1.1 Nutrigenomics

It investigates how our genes and the foods we consume interact [1–3]. It specifically investigates the effects of various meals on individuals with different genetic makeups. With the help of this technology, individuals can be matched with the right foods [4, 5]. To follow the personalized nutrition guidance and create functional meals that optimize health according to individual needs. It combines how we eat with what our ancestors consume and discusses how this affects how our cells and biological systems develop. It is comparable to how the diet will express our anatomical growth along explicit trajectories based on our genetic makeup [6].

J. S. Yaradoddi (✉) · B. S. Meti
Department of Biotechnology, Basaveshwar Engineering College, Bagalkote, Karnataka, India

Chetan D. M.
Department of Biotechnology Engineering, N.M.A.M. Institute of Technology (Deemed to be University), Udupi, Karnataka, India

Virupakshaiah D. B. M.
Department of Microbiology, Davangere University, Davangere, Karnataka, India

S. S. Injaganeri
Environmental Studies Division, Department of Civil Engineering, Basaveshwar Engineering College, Bagalkote, Karnataka, India

J. S. Yaradoddi et al. (eds.), *Frontiers in Food Biotechnology*,
https://doi.org/10.1007/978-981-97-3261-6_20

1.2 Nutrigenetics

Studies on how the person's genetic makeup is affecting, how their body reacts to ingested nutrients. Additionally, persons are thought of as gene variants. For instance, minor genetic makeup variations among individuals led to diverse reactions to specific nutrients. Depending on the foods consumed, unique combinations may trigger the onset of a particular illness condition [6, 7]. Genetics is the study of genes passed down from generation to generation.

Genes produce our bodies' proteins, and it is these proteins that determine individual distinctions [1]. The aforementioned biological distinctions include overt traits, such as blood type and disease susceptibility, and covert characteristics such as eye color and nose shape. Humans have between 20,000 and 25,000 genes [2]. The human genome project was started by scientists in 1990 to identify and sequence those genes [8]. They created the first thorough map of the human genetic composition by the year 2000.

1.3 Relationship Between Nutrients and Gene Expression

The way that diets (nutrigenomics) stimulate the expression of various genes, further the way they follow to change phenotypic and proper cell growth. Nutrients bind to the transcription factor receptors described in pathway B, as shown in path A. Via altering the concentrations of intermediates or substrates, these consequences can be associated with both primary and secondary pathways. Additionally, dieting can disrupt typical cell growth via signal transduction pathways C.

Researchers have discovered a connection between live cells' responses to different surroundings and altered gene expression. Additionally, they have investigated how nutrition can impact cell division and proliferation [8]. The need for transcriptome or genome analysis is growing along with metabolome and proteome research. Consequently, nutrigenomic research on fundamental and advanced nutrition in food science provides fresh perspectives.

DNA methylation is a process that occurs in illnesses such as cancer and significantly impacts how genes are expressed. According to Pufulete and others, folic acid consumption can restore DNA hypomethylation [9]. A significant advancement in cancer treatment and a rapidly developing area of research is the targeting of delayed or suppressed apoptosis. Flavonoids have drawn attention for their capacity to prevent cancer and their high potential as cell-toxic anticancer medicines that support the apoptosis of carcinoma cells. Additionally, research has shown that the flavonoid kaempferol can cause DNA fragmentation, upregulate p53 expression, and start phosphorylation, which disrupts cell proliferative signaling in breast cancer cells [10, 11].

The main bioactive polyphenols in tea are theaflavins [11]. Rheumatoid arthritis, commonly causing inflammatory effects and joint dysfunction, leads to cartilage and bone destruction. In addition, the Mediterranean diet has also been studied as a cardio protective diet, loaded with vegetables, fruits, grains, nuts, omega-3 polyunsaturated fats, and monounsaturated [12].

1.4 Nutrients Serves as Gene Switches

Researchers have examined food intake's effects on disease and widespread gene expression using rat models and cell culture studies [13]. Some illnesses escalating in frequency are associated with a shift in DNA methylation. Methyl donors, including folate, vitamin B12, methionine, betaine, and choline, and foods like garlic, beets, and onions, are transformed into energy during DNA methylation [14]. More hungry and obese mice have the agouti gene, which causes individuals to turn yellow rather than brown. This gene made them more susceptible to conditions like diabetes and cancer.

1.5 Genetic Variations Among Populations

Humans share 99.9% of the same gene sequence. However, common polymorphisms can pinpoint each person's nutritional needs. Typically, phenotypic differences result from 0.1% variations in series. These variances manipulate individuals to behave differently even when eating. Single nucleotide polymorphisms, sometimes SNPs, are the most prevalent polymorphisms [15]. Ninety percent of all variants are caused by SNPs, which change just one nucleotide, and there are three million SNPs that originate in humans.

Different SNPs in a person can cause a group of people to respond differently to the same food [16]. Other than the aforementioned alterations, structural differences in chromosomes and clines (insertions, inversions, duplications, deletions, and copy-number variants) can be spotted infrequently in organisms that can affect the routes used to metabolize nutrients [17].

1.6 Applications

Nutrigenetics and nutrigenomics are entering a new research phase, particularly in nutrition and health. The methods above allow for an extensive assessment of the effects of inappropriate nutrient intakes. Few of these technologies are yet in the growing stage, while some are highly developed, and as a result, the amount of research into health-related problems varies (Fig. 20.1).

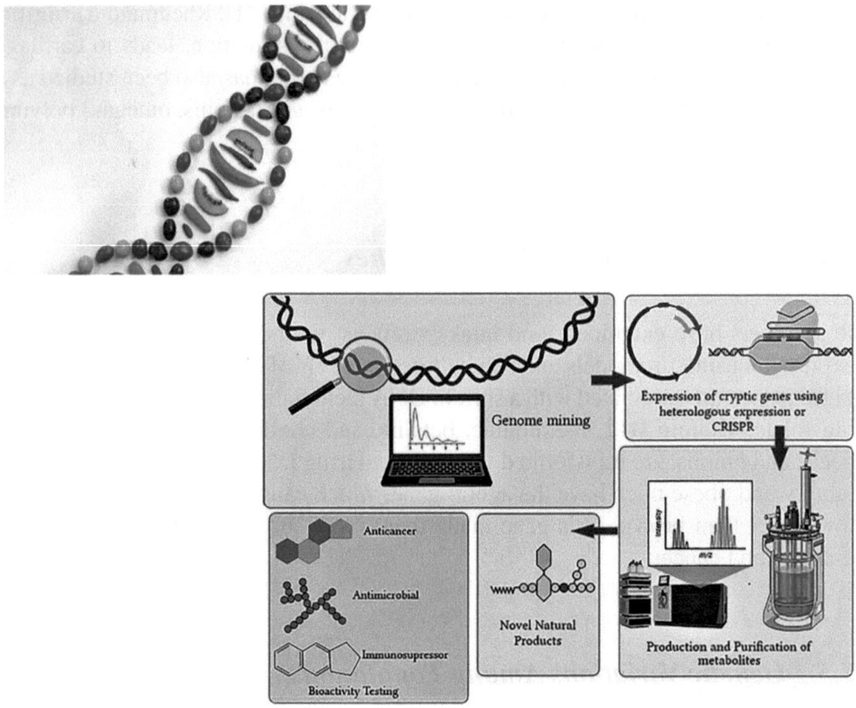

Fig. 20.1 The potentials of nutrigenomics and proteomics; expression of cryptic genes, production; purification of metabolites

Consuming balanced and reasonable nutrition can prevent chronic diseases and numerous kinds of cancer or at least restrict their occurrence [18]. It may be possible to approach important global concerns such as malnutrition with the awareness obtained through looking at diet or gene synergy in many communities. Additionally, genetic diversity can be linked to effects on nutritional uptake, metabolism, consumption, utilization, and excretion, which influence several metabolic processes [19]. Therefore, there is a desire for personalized dietary guidance that will optimize health based on individual needs [20]. By creating individualized nutrition regimens, these technologies and discoveries could also enhance athletes' performance [21].

As a result of nutrigenetics, gluten-free diets for people with celiac disease and pro-biotics for people with lactose intolerance have also been developed in the food industry [22]. Additionally, individuals with galactosemia and PKU are given tyrosine-supplemented diets restricted in phenylalanine and free of galactose, respectively. Therefore, these newly developed nutrigenetic and nutrigenomic technologies are used to create unique food technology products. In manufacturing medicines and nutraceuticals, specialized research investigations are conducted in functional and medical food applications [23].

Additionally, there are ongoing studies of the intricate metabolic pathways affected by several SNPs. Furthermore, by utilizing each person's unique genomic blueprint, genomic medicine aims to create collective decision-making in pharmaceutics and intellectualize drug formulations and biomarker manufacturing in the pharmaceutical companies to help patients and to better the national healthcare organization [24]. Dermagenetics, which tests for specific genetic alterations associated with skin health and produces skin creams or cosmetics, are another application of nutrigenomics that seeks to commercialize cosmetic items for the general public [25].

For instance, scientists have examined the possible effect of the bioactive substance present in the Acmella plant could able to minimize the stimulation of nitric oxide synthase in mRNA and protein expression. This technique may stop the activation of several transcription factors, which make cells more susceptible to Smad downregulation (Smad is a family of structurally related proteins that functions as the main signal transducer in cells). Finally, these modifications result in anti-inflammatory activities that help treat conditions including pancreatitis and dermatitis [25].

In addition, NuGO is a network that links research on the human genome, nutrition, and health that Europe finances. European nutritional science is a network that strategically constructs a virtual "center of superiority" to assemble and advance genetic technology. By using and combining new post genomic technologies, NuGO aimed to reorganize, expanding the European scientific and technological capabilities in nutrigenomics [26].

2 Useful Compounds Associated with Food from the Ocean Environment

The oceans are a vast repository of bioactive substances, mostly unaffected by harmful anthropogenic activity. The food industry can benefit from a wide range of substances, including peptides, enzymes, proteins, polysaccharides, polyunsaturated fatty acids (PUFAs), polysaccharides, phenolics, pigments, and other secondary metabolites from a variety of sources, including bacteria, micro- and macroalgae, crustaceans, sponges, seaweeds, invertebrates and different vertebrates [27–31].

2.1 Advantages and Disadvantages

Both of these technologies attempt to increase the effectiveness of personal dietary intervention techniques that are meant to prevent disease, improve quality of life and support healthy aging. The main benefits of these technologies include enhancing health, preventing diseases by utilizing individual dietary and routine recommendations, and managing chronic conditions. Finding the solution to these new issues is tricky because of the intricacy of food and evolutionary disorders (mainly cancer and diabetes etc.). Available technologies are costly, which will put a lot of pressure on developing countries. In particular, the requirement for sophisticated research investigations restricts the use of these essential technologies in underdeveloped countries.

2.2 Future Challenges and Trends

The description of novel gene-nutrient interactions supports novel diagnostic analyses for diet-related adverse effects. They identify and manage populations by specialized nutrient needs, with the virtue of nutrigenomic advantages to human health for both limited to longest periods. Soon, nutrigenomics shall provide useful platforms to produce biomarkers for societal well-being, early illness predisposition biomarkers, dietary non responders, responders, and bioactive food components. Additionally, developing and developed nations should establish a strong network in contributing technological and scientific expertise in nutrigenomics by implementing and combining new postgenomic technologies to attain the predicted benefits of these bidirectional technologies (Fig. 20.2).

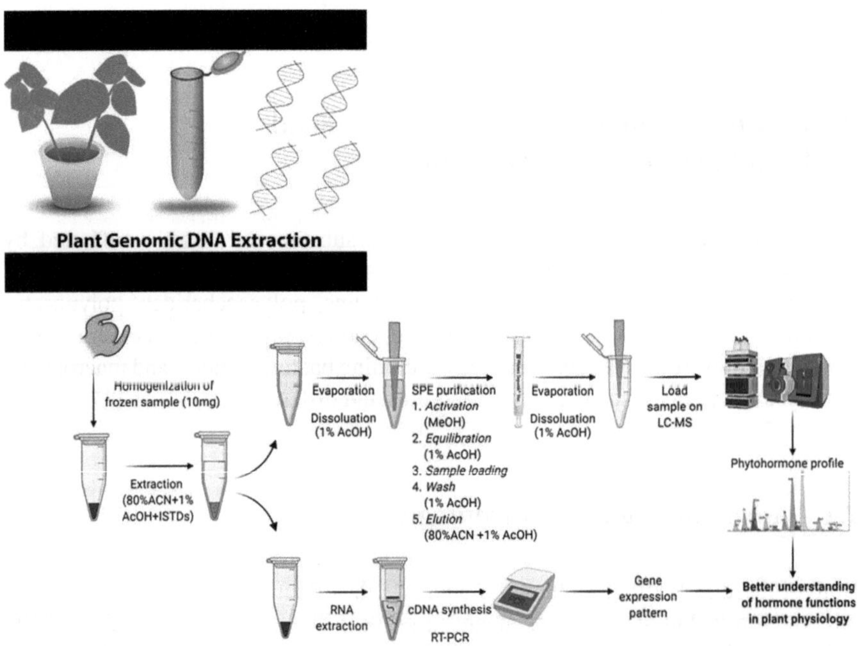

Fig. 20.2 The tools and techniques that are used in DNA extraction, expression of desired gene in plants species

3 Nutri-Informatics

3.1 The Rise of Nutrition Informatics

The term "nutriinformatics" refers to methods that use bioinformatics-based integration of nutrition data report to comprehend interactions among organisms and their nutritional environment [32]. Nutri-informatics strives to integrate and analyze nutrition study data sets computationally. Nutri-informatics is crucial for integrating nutrition with computational biomedical sciences because of curiosity wherein nutrients, food, and nutrition sociology affect health and a contemporary drive toward "big data."

With numerous organizations working on related topics without community development guidelines to assure unity and cohesiveness among nutri-informatics, and biological resources, nutri-informatics lack standardization. Nutrition is underrepresented in many of the resources for nutri-informatics that are accessible. This might be because nutrition is a broad and diverse area, which makes data modeling more challenging. Using computational nutrition data, resources and biomedical ontologies, methods have been started to formalize nutrition study language bound to systematized terminologies over biomedical domains.

Numerous nutrition-associated ontologies are underway; however, they are yet in the early stages of development and need further work from nutritionists and biomedical ontologists. Nutri-informatics research can be continued to struggle with mismatched information if nutrition data production does not standardize terminology, records disclaimer, or wishes for data utility. The community must promote nutrition data production, reuse, and publication standards to support nutri-informatics. The standardization of language and data congruency should be encouraged by academic journals and persons involved in a nutritional investigation and biomedical ontologies groups.

4 Nutrition Research Path for Human Biology

The field of nutrition research covers a wide range of human biology. Although diet and nutrition are important factors in leading a fit life, nutritional science is still very young. The first vitamin isolation in 1926 marked the start of contemporary nutrition study, which has since expanded into a wide field. As per biological standpoint, nutrition is necessary for every living organism. Macronutrients, water, vitamins, and minerals derived via drink and food are important nutrients supporting human life, bodily functions, reproduction, and other species.

Therefore, nutrition studies have majorly concentrated on identifying the necessary nutrients [33], identifying the foods that contain those nutrients [34], identifying the biological processes in which a nutrient may be involved [35], identifying the effects of food processing on nutrient content [36], and determining the paramount nutrient requirements used for persons with particular health situations [37]. Evidence-based nutrition research has informed medical and community health practices, mainly the addition of folic acid to grain commodities, because it is connected to insufficient folate consumption by pregnant women and neural tube abnormalities amongst offspring. For preventing and controlling diseases and improving health, it has consistently been important to define individuals' and populations' nutritional needs [38].

Nutrition is now known to be far more complex than previously assumed due to recent discoveries of nutrient-nutrient interactions [39, 40], drug-food communications [41], molecular mechanisms, and also the role of microbial communities [42]. Beyond biochemical studies, nutrition is unique in how research is applied in practice since food is so individualized and culturally rich. Humans need food to meet their basic needs for energy and nutrition, but biological and cultural factors influence how much food is consumed. A culture or tradition may regulate nutritional intake in various ways, including food selection and preparation, the frequency of meals, when people eat, how they eat [43], religious observance, and individual dietary preferences [44].

Due to location and transit requirements, individuals might have lower access to wholesome and desired food, impacting food selection and consumption [45]. Food and other essentials, such as housing, must often be chosen by people with low financial resources, which can harm their health and safety [46]. Communities with nutritional consumption and class of life were significantly impacted by the sociocultural implications of food in ways that are not fully understood from a merely physiological perspective. The complexity of food and nutrition highlights how essential they are to human everyday existence and biological function while producing a widely diverse concept of perfect nutrition.

Due to the profound complexity of nutrition, it is debatably incomplete to describe health outcomes, including diet, without including sociological data. Nutrition data and research must also be controlled collectively due to the interconnected nature of nutrition, especially in light of the current advances in data creation from numerous nutrition research projects based on -omics.

The thriving purpose of amalgamated informatics methods in the former biomedical area, such as genomics, metabolomics, and transcriptomics, in conjunction employment of statistical and conventional epidemiological techniques, can guide the need to use nutri-informatics methods to interpret nutritional data. Different types of nutrition data are available, such as clinical data, survey data, fundamental science phenomenon data, omics data, and observational data.

4.1 *Advances in Nutri-informatics and its influence in Disease Management*

Despite employing advanced statistics in nutri-informatics at a large-scale examinations of food consumption through surveys may seems to be a fresh trend [47]. Since the beginning of WWEIA in the survey year of 2003–2004, researchers have benefited from having admittance to these nourishment data sets meant for their studies.

Studies were directed toward dairy substitutes' cost and energy consumption for people who do not take dairy products [48]. The occurrence of likely undiagnosed celiac ailment and a possible decline in femur bone mineral density [49] are two examples of the diverse research projects that use data from WWEIA and NHANES. FFQs (Semi quantitative food frequency questionnaires), currently in their third iteration, are also used by the Nurses' Health Study (NHS) to gather longitudinal dietary data from their large cohorts [50]. Since its inception, the NHS has utilized participant clinical outcomes, biological materials, and nutritional data to examine prospective nutrition biomarkers. It was the first study to employ toenail samples to estimate selenium consumption.

Additionally, the NHS has contributed to dietary recommendations by advising people to cut down or eliminate trans fats consumptions from their diets to lower their risk of coronary heart disorder and by emphasizing the link between dietary methods in stopping hypertension, also the preclusion of men's colorectal cancer [51]. Food surveys by NHS and WWEIA made this possible. The landscape of nutri-informatics in food industries has been given in Table 20.1.

Table 20.1 The various segments and their focus of nutri-informatics

Landscape	Information and focus
Food and nutrients	Nutrients, bioactive compounds, sustainability, food products, packaging, food processing and production, food and drug interactions, etc.
Clinical and individual nutrition	Food and drug interactions, nutrition-associated diseases, food-associated metabolites, digestion and metabolism, food preferences, health literacy, public health and nutritional recommendations
Community nutrition	Food preferences, health literacy, nutrition knowledge, resource access, cultural norms, personal nutritional beliefs, location, religious beliefs, etc......
Govt. regulatory bodies	Food safety, food labeling, animal welfare, food processing and production, public health nutritional recommendations

4.2 Current Nutri-informatics Challenges

Several medical systems concentrated on converting to electronic resources for nutritional data records and storage, making stagnant clinical measurements publicly available. Conversely, the use of EHRs for research is constrained by various techniques in gathering dietary data. Among them, a requirement of *standardization* is vital [52]. Furthermore, the availability of nutrition data in EHRs is frequently lacking, making it difficult to assess any potential negative effects of diet on health.

In contrast to larger, additional combinational research, which includes a broad range of data types and outcomes, medical nutri-informatics studies are more frequently concentrated on explicit health conditions, and more specificity about the disease status, and therapeutic biomarkers. A noteworthy example; the application of malnutrition screening tests is crucial for recognizing and managing malnutrition, ask two questions: "Have you recently lost weight without trying? How much weight have you lost, if any? As well as "Have you been eating poorly because your appetite has decreased?" [53].

No nutritional consumption or other information is gathered from the two questions in the patient survey. This allows the clinician in determining malnutrition at a surface level, although it offers little information about particular dietary issues. There have been very few attempts to use omics data, nutrition and clinical survey statistics, and previous varied data types in collaboration with one another in research. The intricate physiological response against the nutrient revelations.

such as potential contaminants, food additives, nutrients, or toxins), incomplete perceptive, instituting the metabolic biomarkers, contradictory documentation language are other significant challenges for personalized nutrition endeavors [53].

Self-report survey methods also raise concerns since individuals may overstate their dietary intake, casting doubt on research conclusions. Compared to estimated consumption determined from 24-h urine salt content, typical 24-h meal recalls in one study underestimated dietary sodium intake [54]. There are also issues with racial discrepancies when using FFQs. Given the racial differences that nutrition surveys do not account for, one analysis found a considerably higher correlation between 24-h dietary recall and FFQ for white women than for black women [55]. This raises questions about how to interpret eating behaviors and formulate nutritional recommendations.

Many different research approaches are accessible, and each has pros and cons that must be considered while conducting nutrition research. While studies focusing on a single dietary element are ideal for assessing nutrition biomarkers, such studies can be difficult to carry out in humans under controlled conditions. They may also result in broad conclusions about the systemic effects of food or nutrient on health. On the other hand, many studies assess food as a whole, limiting the ability to evaluate certain cellular and molecular connections or signaling. Numerous nutrition studies have been undertaken on related theories due to the lack of a "perfect" methodological design, resulting in a variety of data and data types that are insufficient for larger-scale analysis.

4.3 A Call for Improved Nutrition Representation and Standards

Integrating nutrition terminology and linkages into ontology's and knowledge resources is crucial to advancing our understanding of how nutrition and food affect health, human behavior, culture, and more. A focus should be placed on increasing the representation of nutrition in areas already being modeled in ontology's, such as foods, and those still being investigated in the available resources. This category can include food processing, nutrition behavioral counselling, nutrition personal and cultural views, and nutrition biomarkers.

Additionally, not all nutrition materials are created with compatibility in mind, which further restricts interpretation and terminology agreement among resources. The nutrition and ontology groups will need to put in a lot of constant effort to address these issues with nutrition representation and compatibility across resources.

5 Conclusion and Future Scope

Our diet can be customized based on our genetic makeup thanks to the science of nutritional genomics. As this area advances, we will soon eat meals that appeal to our DNA. The future of dietary recommendations could be greatly improved by nutritional genomics. Thanks to nutrigenetics, personalized nutrition advice based on a person's genetic makeup will be possible. This technique has been used over the past 20 years to treat several monogenic disorders.

A new discipline that studies the relationships between dietary processes and genetic variations among people and communities has emerged due to the development of high-throughput sequencing technologies over the past several decades. The fundamental issue of how food and its effects relate to gene expression and whether research into these connections may advance the general public's health and the well-being of people and groups identifiable by their unique genomes is addressed by the burgeoning subject of nutrigenomics. Finding the genotypes or genetic blueprints involved in illness prevention through nutrition and understanding the mechanisms underlying these links are challenges for research.

This area of research necessitates extensive DNA sampling and the gathering of comprehensive phenotypic data, including medical and environmental factors that may interact in various ways. The development of nutrigenomics depends on hiring people from multiple communities, establishing DNA databases, and improving quick access to medical records. This crucial infrastructure raises several ethical and societal issues that must be resolved to ensure the appropriate translation of research into clinically useful knowledge. These include concerns about DNA biobanking's confidentiality, anonymity, and privacy; the clinical accuracy and usefulness of direct-to-consumer products; and the broad implications of genetics on human identity.

Nutrition is essential for humans and nonhuman animals because it influences behavior and culture and how diseases, genes, and phenotypes manifest. Modern nutrition research may be a "younger" branch of biology, but it is by no means unimportant. Strong community standards are needed to promote FAIR practices, transparency, and the greatest possible knowledge gains from research. Nutritional data could be incorporated into the larger framework of biological knowledge bases, allowing interchange and reuse using standardized language and biomedical ontologies.

With the creation of fresh ontology's and working groups with a focus on nutrition, this integration process has already started. Nutri-informatics research will grow to its complete potential, if supporting innovation from the nutrition information go beyond a limited studies, providing comprehensions outside the area of nutrition by constant education. Some exhaustive effort from the researchers of nutrition also from ontology developers to amalgamate nutrition investigation to biomedical ontologies.

References

1. Fenech M, El-Sohemy A, Cahill L et al (2011) Nutrigenetics and nutrigenomics: viewpoints on the current status and applications in nutrition research and practice. J Nutr Nutr 4(2):69–89
2. Kaput J (2005) Decoding the pyramid: a systems-biological approach to nutrigenomics. Ann N Y Acad Sci 1055(1):64–79
3. Breitbart RE, Andreadis A, Nadal-Ginard B (1987) Alternative splicing: a ubiquitous mechanism for the generation of multiple protein isoforms from single genes. Annu Rev Biochem 56(1):467–495
4. International Human Genome Sequencing Consortium (2004) Finishing the euchromatic sequence of the human genome. Nature 431(7011):931–945
5. Murgia C, Adamski MM (2017) Translation of nutritional genomics into nutrition practice: the next step. Nutrients 9(4):366
6. Pavlidis C, Patrinos GP, Katsila T (2015) Nutrigenomics: a controversy. Appl Transl Genom 4:50–53
7. Roberts MA, Mutch DM, German JB (2001) Genomics: food and nutrition. Curr Opin Biotechnol 12(5):516–522
8. Mutch DM, Wahli W, Williamson G (2005) Nutrigenomics and nutrigenetics: the emerging faces of nutrition. FASEB J 19(12):1602–1616
9. Pufulete M (2005) Effect of folic acid supplementation on genomic DNA methylation in patients with colorectal adenoma. Gut 54(5):648–653
10. Li J, Zheng J (2018) Theaflavins prevent cartilage degeneration via AKT/ FOXO3 signaling in-vitro. Mol Med Rep 19:821–830
11. Zhang L, Kajiwara H, Kuboyama N et al (2011) Reduction of CXCR4 expression in rheumatoid arthritis rat joints by low level diode laser irradiation. Laser Ther 20(1):53–58
12. Engler MB (2009) Nutrigenomics in cardiovascular disease: implications for the future. Prog Cardiovasc Nurs 24(4):190–195
13. Uthpala TGG, Navaratne SB (2020) Acmella oleracea plant; identification, applications and use as an emerging food source—review. Food Rev Int 7:1–16
14. Shorter KR, Felder MR, Vrana PB (2015) Consequences of dietary methyl donor supplements: is more always better? Prog Biophys Mol Biol 118(1–2):14–20

15. Syvänen AC (2001) Accessing genetic variation: genotyping single nucleotide polymorphisms. Nat Rev Genet 2(12):930–942
16. Butler JM (2012) Single nucleotide polymorphisms and applications. In: Advanced topics in forensic DNA typing. Elsevier, pp 347–369
17. Maurya NK (2020) Nutrigenomics in functional foods and personalized nutrition. GEDRAG Organ Rev 33(02):670–686
18. Heber D (2006) Nutritional oncology, 2nd edn. Elsevier- Academic Press, Amsterdam, p 822
19. Guest NS, Horne J, Vanderhout SM et al (2019) Sport nutrigenomics: personalized nutrition for athletic performance. Front Nutr 6:8
20. Szopa M, Wybrańska I, Malczewska-Malec M et al (2005) Nutrigenomics-about genetics and nutrition. Przegl Lek 62(4):245–252
21. Thomas DT, Erdman KA, Burke LM (2016) Nutrition and athletic performance. Med Sci Sports Exerc 48(3):543–568
22. Chibbar D (2019) The gut microbiota in celiac disease and probiotics. Nutrients 11(10):2375
23. Nazzaro F, Fratianni F, Coppola R (2012) Microtechnology and nanotechnology in food science. In: Boye JI, Arcand Y (eds) Green technologies in food production and processing. Springer, pp 471–494
24. Pavlidis C, Lanara Z, Balasopoulou A et al (2015) Meta-analysis of genes in commercially available nutrigenomic tests denotes lack of association with dietary intake and nutrient-related pathologies. OMICS J Integr Biol 19(9):512–520
25. Subbiah MTR (2010) Application of nutrigenomics in skin health: nutraceutical or cosmeceutical? J Clin Aesthet Dermatol 3(11):44–46
26. Bakondi E, Singh SB, Hajnády Z et al (2019) Spilanthol inhibits inflammatory transcription factors and iNOS expression in macrophages and exerts anti-inflammatory effects in dermatitis and pancreatitis. Int J Mol Sci 20(17):4308
27. Dewapriya P, Kim S-K (2014) Marine microorganisms: an emerging avenue in modern nutraceuticals and functional foods. Food Res Int 56:115–125. https://doi.org/10.1016/j.foodres.2013.12.022
28. Yaradoddi JS, Sulochana MB, Kontro Merja H, Parameshwar AB, Dayanand A (2020) The occurrence of potential and novel isolates of *Oceanobacillus* sp. JAS12 and *Salinicoccus* sp. JS20 recovered from West Coast of Arabian Sea, India. Res J Biotech 15(9):133–140
29. Yaradoddi JS, Sulochana MB (2020) Screening and characterization of bioactive compounds produced by the moderate halophile *Halobacillus* sp. JS6. Res J Biotech 15:12
30. Jayachandra SY, Parameshwar AB, Mohan Reddy K, Sulochana MB (2012) Characterization of extracellular hydrolytic enzymes producing extremely halophilic bacterium *Virgibacillus* sp. JS5. World J Sci Technol 2(2):23–26
31. Sulochana MB, Jayachandra SY, Anil Kumar S, Dayanand A (2014) Siderophore as a potential plant growth-promoting agent produced by *Pseudomonas aeruginosa* JAS-25. Appl Biochem Biotechnol 174(1):297–308
32. Döring F, Rimbach G (2014) Nutri-informatics: a new kid on the block? Genes Nutr 9:1–3
33. Jew S, Antoine J-M, Bourlioux P et al (2015) Nutrient essentiality revisited. J Funct Foods 14:203–209
34. Schmid A, Walther B (2013) Natural vitamin D content in animal products. Adv Nutr 4:453–462
35. Chung HR (2014) Iodine and thyroid function. Ann Pediatr Endocrinol Metab 19:8–12
36. Reddy MB, Love M (1999) The impact of food processing on the nutritional quality of vitamins and minerals. In: Jackson LS, Knize MG, Morgan JN (eds) Impact of processing on food safety. Springer US, Boston, MA, pp 99–106
37. Bush CL, Blumberg JB, El-Sohemy A et al (2020) Toward the definition of personalized nutrition: a proposal by the American nutrition association. J Am Coll Nutr 39:5–15
38. Tosti V, Bertozzi B, Fontana L (2018) Health benefits of the Mediterranean diet: metabolic and molecular mechanisms. J Gerontol Ser A 73:318–326

39. Rosanoff A, Dai Q, Shapses SA (2016) Essential nutrient interactions: does low or suboptimal magnesium status interact with vitamin D and/or calcium status? Adv Nutr 7:25–43
40. Combet E, Gray SR (2019) Nutrient–nutrient interactions: competition, bioavailability, mechanism and function in health and diseases. Proc Nutr Soc 78:1–3
41. Dahan A, Altman H (2004) Food–drug interaction: grapefruit juice augments drug bioavailability—mechanism, extent and relevance. Eur J Clin Nutr 58:1–9
42. Xu K-Y, Xia G-H, Lu J-Q et al (2017) Impaired renal function and dysbiosis of gut microbiota contribute to increased trimethylamine-N-oxide in chronic kidney disease patients. Sci Rep 7:1445
43. Fieldhouse P (2013) Food and nutrition: customs and culture. Springer US, New York, NY
44. Boatemaa S, Badasu DM, De-graft Aikins A (2018) Food beliefs and practices in urban poor communities in Accra: implications for health interventions. BMC Public Health 18:434
45. Coveney J, O'Dwyer LA (2009) Effects of mobility and location on food access. Health Place 15:45–55
46. Knowles M, Rabinowich J, Ettinger de Cuba S et al (2016) "Do you wanna breathe or eat?": parent perspectives on child health consequences of food insecurity, trade-offs, and toxic stress. Matern Child Health J 20:25–32
47. Group FSR, Health USDO, Control CFD, Statistics NCFH (2015) What we eat in America (WWEIA)
48. Cifelli C, Auestad N, Fulgoni V III (2020) Replacing the nutrients in dairy foods with non-dairy foods will increase cost, energy intake and require large amounts of food: National Health and nutrition examination survey 2011–2014. Public Health Nutr 27:1–12
49. Sattgast LH, Gallo S, Frankenfeld CL et al (2020) Nutritional intake and bone health among adults with probable undiagnosed, untreated celiac disease: what we eat in America and NHANES 2009–2014. J Am Coll Nutr 39:112–121
50. Hu FB, Satija A, Rimm EB et al (2016) Diet assessment methods in the nurses' health studies and contribution to evidence-based nutritional policies and guidelines. Am J Public Health 106:1567–1572
51. Petimar J, Smith-Warner SA, Fung TT et al (2018) Recommendation-based dietary indexes and risk of colorectal cancer in the nurses' health study and health professionals follow-up study. Am J Clin Nutr 108:1092–1103
52. Verma M, Hontecillas R, Tubau-Juni N et al (2018) Challenges in personalized nutrition and health. Front Nutr 5:117
53. Ferguson M, Capra S, Bauer J et al (1999) Development of a valid and reliable malnutrition screening tool for adult acute hospital patients. Nutrition 15:458–464
54. Olendzki B, Procter-Gray E, Magee MF et al (2017) Racial differences in misclassification of healthy eating based on food frequency questionnaire and 24-hour dietary recalls. J Nutr Health Aging 21:787–798
55. Nadkarni PM, Brandt CA (2006) The common data elements for cancer research: remarks on functions and structure. Methods Inf Med 45:594–601

Part V
Scope for Commercialization of Foods and Nutraceuticals

Chapter 21
Principles of Food Analysis and Food Laws

Ranjan Kaushik, Ankit Kumar, Rekha Phogat, Rakesh Gehlot, and Neha Rani

1 Introduction

Among various branches of analytical chemistry, food analysis is a specialized branch that deals with the research, development, and implementation of analytical procedures to distinguish the properties of foods and their constituents, food hazards, fraudulent, and traceability [1]. Understanding food and its components is crucial because it forms the foundation of the other aspects. Food is necessary for the survival of living beings other than part of culture and a source of communication. The compositions of most foods are water, lipids, protein, carbohydrates, enzymes, antioxidants, and phenols. Each component has its own physical and chemical characteristics which add to the final properties of food products [2]. It provides necessary micro and major nutrients adequately for the body's growth, maintenance, repair, and reproduction. The population is increasing day by day globally and nowadays food security is a major concern for policymakers. It is vital to feed such a huge population with good quality food without any food hazard which impacts the health of people directly. The scientists associated with the food industry including food manufacturers, ingredient suppliers, analytical service laboratories, government laboratories, and university research laboratories analyze food for various purposes to ensure its safety, quality, and nutritional value [3]. People nowadays prefer the quality of food products over quantity. The attentiveness of the consumers regarding health is leading to more specific and new food products in the market. The innovations in food are not just introducing variations and opportunities for developing new methods of processing packaging and storage, but also new risks related to pathogenic spoilage, contamination, potential health hazards, and fraud [4]. To overcome these problems, food analysis and food laws play a vital role

R. Kaushik · A. Kumar · R. Phogat (✉) · R. Gehlot · N. Rani
Centre of Food Science and Technology, Chaudhary Charan Singh Haryana Agricultural University, Hisar, Haryana, India

J. S. Yaradoddi et al. (eds.), *Frontiers in Food Biotechnology*,
https://doi.org/10.1007/978-981-97-3261-6_21

in both the fresh and processed food products industry. The advancement of today's food distribution system depends on food analysis techniques for the expansion of existing food products, quality control for food safety and authenticity, regulatory compliance, and resolving the problems in concerned food products.

To manage the aforementioned issues, food analysis must adhere to a set of critical principles. This includes ensuring that food testing laboratories follow local, national, and international regulations, seek accreditation from recognized bodies, use validated analytical methods, employ proper sampling procedures, maintain a secure chain of custody, keep well-documented records, employ trained personnel, regularly calibrate instruments, participate in quality control programs, provide accurate and transparent reporting, protect client confidentiality, and engage in continuous improvement efforts to stay current with evolving regulations. Adherence to these principles is essential for upholding food safety and quality standards, as violations or non-compliance can result in legal consequences such as fines and penalties.

2 Purpose of Food Analysis

Analysis of food is an essential part of determining the various attributes some of which are described below:

Food composition: Food contains macronutrients (carbohydrates, fats, and proteins), micronutrients (vitamins and minerals), fiber, and many bioactive compounds [5]. Food composition data is important for different sectors and users for educational purposes by researchers, consumers of food, regulatory authorities, food processing, and the agriculture industry. Analysis of food composition determination includes laboratory testing, chemical testing, and databases which are compiled from published sources and, thus, provide information about a wide variety of different characteristics of foods [6].

Nutritional value: Nowadays, nutritional security is a major concern around the world. A small nutrition causes serious health problems which include restricting the growth and development of an individual. Food analysis is a fundamental tool for determining the nutritional value of food and can solve health concerns around the globe by enlightening the people about absence/lack of essential nutritional components in food [7, 8].

Shelf life: Shelf life is the time duration in which food remains suitable for human consumption encompassing safety, nutritional value, and sensory attributes under specific storage conditions. In the food industry, there are two major issues: (1) fulfilling the food demand of a continuously growing population and (2) foods that have longer storage periods with minimal processing. To meet the food demand of a huge population, the agriculture sector is contributing a lot by boosting food production [9].

Food hazards: Hazard analysis is a systematic process used to identify, assess, and evaluate potential hazards or risks associated with food production, processing, storage, and distribution. Food hazards encompass biological (pathogens, toxins, allergens), chemical (contaminants, additives, mycotoxins), and physical (foreign objects, temperature abuse, improper handling) risks in food that have the potential to harm consumers. To encounter these problems various approaches like Hazard Analysis and Critical Control Points (HACCP), good manufacturing practices (GMPs), proper labeling, regulatory enforcement, and people consciousness play an important role. The analytical data under the approaches helps to economically produce foods that are safe, nutritious, and looked for by consumers [10].

Labeling: Labeling for food analysis is a crucial aspect of the food industry to ensure consumer safety and provide necessary information to consumers. Food labels typically include a variety of information related to the product's composition, nutritional content, and safety. These labels typically include the product name, ingredients list, allergen information, nutrition facts (including serving size and calorie count), expiration or use-by date, storage instructions, net weight or volume, manufacturer's contact details, country of origin, barcode, health claims, certifications, batch or lot numbers for traceability, and any specific instructions for use. Adherence to regulatory standards set by government agencies ensures transparency and accuracy in food labeling, safeguarding consumer health and choice [11].

Food regulations: Food regulations are a set of rules and standards established by government authorities to ensure the safety, quality, and integrity of food products. Food analysis plays a vital role in the development and enforcement of food regulations. Regular monitoring and enforcement depend on food analysis to uphold regulations, taking corrective actions when standards are violated. Furthermore, food analysis adapts to new technologies and innovations, ensuring regulations remain relevant and protective of public health and consumer interests [12].

Food traceability: Market food product traceability refers to the ability to track and trace food products throughout the entire supply chain, from the point of production to the point of consumption. Tracking data enables quick identification and removal of unsafe products, enhances supply chain efficiency, facilitates international trade, protects brand reputation, supports sustainability efforts, and provides valuable data for informed decision-making, making it an indispensable tool for navigating the complexities of the modern food industry [13].

Food adulteration: Food fraud in general comes under economically motivated adulteration, where fraudsters intentionally alter the food for economic benefits which can risk public health [14]. Food adulteration problems can be solved in two ways: a) Quantitative analysis: where the adulterant is known and b) Qualitative analysis: where the food may or may not be adulterated [15].

3 Procedure of Food Analysis

Food analysis in general follows various steps like sampling, sample preparation, analysis, data acquisition, data interpretation, and decision-making. Before following the procedure, the motive of the analysis should be clear, as the sampling and sample preparation methods vary with the purpose of the analysis [16]. There are generally four steps involved in food analysis:

1. Sampling and sample analysis
2. Analysis
3. Data acquisition
4. Data interpretation and decision-making

4 Analysis Techniques

From the beginning of the twentieth century, the analysis of food was done by so-called wet chemistry which consists of those analytical techniques where mechanical or electronic instruments are not used. The method usually relies on chemical reactions between the substances to be analyzed (the analyte) and a reagent added to the analyte. Wet techniques often depend on the formation of a product of the chemical reaction that is easily detected and measured [17], but, nowadays, it is replaced by instrumental techniques which refer to a wide range of analytical methods and tools used in scientific research, quality control, and various industries to obtain precise and objective measurements and data about the physical, chemical, and biological properties of materials and substances. These techniques rely on specialized instruments and equipment to gather information and often provide more accurate and sensitive results compared to traditional methods [18]; some of the advanced and major instrumental techniques for testing food proximate composition and nutritional analysis, sensory analysis, microbial analysis, toxins, and contamination, food integrity testing, etc., which are widely used, are explained below:

4.1 Spectroscopy

Spectroscopy operates on the principle that molecules and atoms of food constituents interact with electromagnetic radiation in unique ways, resulting in characteristic absorption and emission patterns. As atoms, molecules, or particles absorb energy from incoming radiation, electrons transition between discrete energy levels, and subsequent emissions correspond to specific wavelengths or frequencies. Spectrometers, which separate and measure radiation intensity across a range of wavelengths or frequencies, are employed to produce spectra that hold valuable information about the composition, structure, and properties of substances. These techniques find their foundation in the fundamental interaction between light and matter, leading to various phenomena such as the absorption, emission, or scattering of incident electromagnetic radiation. These interactions can be detected and analyzed through a diverse array of spectroscopic methods. These include ultraviolet-visible (UV-Vis), near-infrared (NIR), mid-infrared (MIR), far infrared (FIR), and Raman spectroscopy. Additionally, techniques using terahertz waves, microwaves, radio waves, and nuclear magnetic resonance (NMR) cover different wavelength ranges of the electromagnetic spectrum. These spectroscopic tools play a pivotal role in various analytical applications, empowering researchers to delve into molecular details, identify compounds, and accurately quantify analytes [19]. The various instrumental techniques based on the principle are discussed below:

4.2 Chromatography

Instrumental chromatography is a powerful analytical technique that operates on the principle of separating complex mixtures of compounds into their components for detailed analysis. It involves the injection of a sample into a chromatographic system, where the mixture is separated based on the varying interactions of its components with a stationary phase. The separated compounds are then detected using specialized instruments, and the resulting signals are analyzed to determine the identity and quantity of each component [20]. This versatile technique finds applications in a wide range of fields, enabling scientists to investigate the composition of substances and identify contaminants. Various examples of chromatography are gas chromatography, liquid chromatography, and high-performance liquid chromatography (HPLC).

4.3 Microscopy

It is a fundamental scientific technique that operates on the principle of magnifying and visualizing objects and structures that are too small to be seen with the naked eye. It achieves this by using lenses and/or other optical components to bend and focus light or electrons, allowing for high-resolution imaging. There are various microscopy techniques, including optical microscopy, electron microscopy, and scanning probe microscopy. Optical microscopy uses visible light to magnify and observe samples, while electron microscopy utilizes a beam of electrons for much higher resolution imaging. Scanning probe microscopy employs a physical probe to scan and map surfaces at the nanoscale. Microscopy is crucial in fields such as biology, materials science, medicine, and nanotechnology for studying and understanding the detailed structures and characteristics of various specimens and materials [21]. Various examples of microscopy are electron microscopy, and scanning electron microscopy (SEM).

4.4 Thermal Analysis

It encompasses a set of techniques used to determine the thermal properties and behavior of the substances. Differential scanning calorimetry (DSC) is employed to measure heat flow within a sample, enabling the characterization of phase transitions, melting points, and thermal stability. Thermogravimetric analysis (TGA), on the other hand, tracks changes in the mass of a sample as it undergoes temperature variations, allowing for the analysis of decomposition processes, volatilization, and thermal stability. These techniques are widely used in fields such as chemistry, materials science, and pharmaceuticals to gain insights into the thermal characteristics of substances, aiding in product development, quality control, and research endeavors [22].

Differential scanning calorimetry (DSC): It is based on the principle of comparing the heat flow into or out of a sample and a reference material as both are subjected to the same temperature program. The difference in the heat flow between the sample and reference is recorded, allowing the detection of phase transitions, such as melting, crystallization, glass transitions, and chemical reactions, which involve heat absorption or release [23].

Thermogravimetric analysis (TGA): It is a thermal analysis technique that is used to study the changes in the weight of a sample as it is heated or cooled under controlled conditions. TGA provides information about the composition, thermal stability, and decomposition kinetics of substances. The principle of TGA involves measuring the weight of a sample as a function of temperature or time, while it is subjected to a programmed temperature ramp in a controlled atmosphere (usually inert, oxidative, or reducing). As the sample is heated, any changes in its composition, such as decomposition, vaporization, or oxidation, are detected by the change in its weight [24].

4.5 Molecular Biology Techniques

Identification of microorganisms by conventional methods requires the isolation of pure cultures followed by laborious characterization experiments [25]. Method, i.e., polymerase chain reaction (PCR) amplifies DNA to detect and analyze specific genetic sequences. Polymerase chain reaction (PCR) is a widely used molecular biology technique that allows for the amplification of a specific DNA sequence. On the other hand, **DNA sequencing** determines the order of nucleotide bases (adenine [A], cytosine [C], guanine [G], and thymine [T]) in DNA molecules. These sequences, when combined in various patterns, constitute the genetic information that is fundamental to understanding and advancing our knowledge of genetics and molecular biology. Modern DNA sequencing technologies have not only empowered the characterization of bacterial genomes but have also allowed for more precise identification of the diverse microorganisms in complex microbiomes, which encompass the genetic material of microbes in various environments, from the human gut to food production facilities [26].

4.6 Sensory Analysis

These instruments include devices for evaluating taste, aroma, texture, color, sound, and more, as well as software and controlled environments for conducting sensory evaluations. It plays a crucial role in ensuring consistent quality and meeting consumer preferences in food production and quality control processes. Some of the instruments are given below:

Electronic noses and tongues: Electronic noses and tongues have emerged as instrumental techniques that can overcome the drawbacks of traditional methods [27]. These devices consist of an array of sensors capable of characterizing complex samples. Specifically, arrays of gas sensors are commonly referred to as electronic noses, enabling the detection and identification of various odors and volatile compounds. In contrast, arrays of liquid sensors are known as electronic tongues, facilitating the analysis of taste and chemical composition in liquids [28].

Texture analyzer: A texture analyzer is a laboratory instrument used to measure the mechanical properties of materials, especially the texture and consistency of food products. Texture analysis helps manufacturers ensure product quality, consistency, and consumer satisfaction [29]. For texture analysis, nowadays chewing robot has also been developed that objectively assesses food texture. The robot's precise replication of the rhythmic molar motion of human chewing makes it a valuable tool for conducting in vitro food texture assessments, ensuring its ability to mimic human chewing effectively and contributing to the evaluation of food properties [30].

These various advanced techniques have many applications in the food industry some of which are mentioned in Table 21.1.

Table 21.1 Different instrumental techniques used for the analysis of food

Technique	Application	Inferences	Reference
Gas chromatography-olfactometry (GC-O) and electronic nose (e-nose)	Odor determination	Both the techniques are better than classical organoleptic analyses. The electronic nose (e-nose) stands out as a superior choice because of its quick analysis time, high sensitivity, and excellent reproducibility. It also offers the advantage of objective odor identification and can be used for in situ analysis as a portable device	[31]
Spectroscopic methods like UV, VIS, IR and sensors	Assessing the quality of fish and other seafoods	Quality evaluation of fish freshness, microbial behavior, and spoilage patterns, as well as certain physicochemical and textural characteristics were predicted	[32]

(continued)

Table 21.1 (continued)

Technique	Application	Inferences	Reference
Spectroscopic method	Color of olive oils	Spectrophotometric data can serve as a viable alternative to measurements conducted using the Lovibond system for assessing the color of products. Spectrophotometry also allows the estimation of chlorophyll and carotenoid content in addition to color characterization	[33]
Material strength evaluation methods (compression, guillotine blade probe, and puncture test)	Assessing the texture of French fries from different fast-food restaurants	The methods offer a time-saving and efficient approach for evaluating the texture of French fries	[34]
Nuclear magnetic resonance (NMR) Microwave plasma-atomic emission spectrometry (MP-AES) Inductively coupled plasma mass spectrometry (ICP-MS) Isotope ratio mass Spectrometry (IRMS) Vibrational spectrometric techniques: Near-infrared spectroscopy (NIR) Mid-infrared spectroscopy (MIR) Fourier transform infrared spectroscopy (FT-IR) Attenuated total reflection (ATR) spectroscopy Raman spectroscopy	Quality, safety, and adulteration analysis	Used to determine milk authenticity, chemical composition of Canadian milk samples, coffee authentication, detection of genetically modified organisms/GMOs, beer spoilage bacteria, determination minerals in honey	[35]
Spectroscopy, electronic nose, and tongue	Estimate taste and aroma in beverages	The integration of chemometric tools with instrumental methods and real-time sensor data enables the continuous assessment of the overall quality of intricate food systems or matrices, including beverages, providing valuable insights for quality control and product optimization	[36]

(continued)

Table 21.1 (continued)

Technique	Application	Inferences	Reference
Texture analyzer and viscoelastograph	Cooking quality of spaghetti	High correlation between these instruments and good agreement in ranks	[37]
Texture analyzer	Sensory and instrumental study of Taralli	Perception of food quality during storage	[38]
Electronic tongues (e-tongues)	Assessment of freshness and shelf life, verification of authenticity, and characterization of food products.	While allowing for the rapid determination of multiple compounds in foodstuffs and sample classification through direct measurement, these systems are hampered by the substantial number of prior measurements needed for modeling, calibration, or learning phases	[27]
SEM analysis	Structural properties of freeze-dried guava, sapota and papaya	The presence of pectin in papaya is a likely factor causing the papaya powder to exhibit stickiness, in contrast to the smooth and free-flowing characteristics of guava and sapota freeze-dried powders	[39]
Differential scanning calorimetry (DSC) and thermogravimetric analysis (TGA)	Thermal properties of freeze-dried guava	DSC curve shows weight loss of fruits due to water loss during the temperature rise, volatilization of matter, dehydroxylation, demethoxylation, and decarboxylation reactions on pectin	[39]
Scanning electron microscopy	Morphological characteristics of spray-dried jamun juice powder	Powders produced at higher temperature showed spherical particles with some degree of shrinkage compared to the powders produced at lower temperatures showing smooth surfaces	[40]
(U) HPLC	Determination of biogenic amines in food samples	Allows excellent separations in shorter analysis time compared to those obtained by conventional HPLC	[41]
GCMS	Determination of formaldehyde level in different fish species	Results of GCMS determined formaldehyde hazard in the fish species which should be less than 0.2 mg kg^{-1} body weight, but the species had higher content	[42]

5 Purpose of Food Laws

Food safety is a critically important issue that has a profound impact on the lives and well-being of individuals. The primary objective of food laws should be to ensure the safety of food and protect public health. However, the existing legal framework often falls short of achieving this goal. These laws were formulated at a time when the food system faced different hazards and challenges compared to the present environment. One significant change in the food system is the globalization of food supply chains, which has increased the risk of people encountering common food-borne pathogens like viruses and bacteria. This has led to higher rates of disease transmission. Additionally, there is a growing understanding of the threat posed by noncommunicable diseases (NCDs), such as heart disease, diabetes, Alzheimer's disease, and cancer. The World Health Organization (WHO) predicts that NCDs will cause 35 million deaths annually [43]. These health issues cannot be solely attributed to common pathogens, as was previously thought. Instead, the modern Western diet, characterized by fast food, processed meals, and sugary drinks, is recognized as a key driver of NCDs. Surprisingly, under the current regulatory system, food products falling within this category are still deemed "safe." This implies that the current laws do not adequately address the risks associated with these harmful dietary choices. To ensure food safety and protect public health, it is crucial to update the legal framework to reflect the changing nature of food hazards and the health consequences of consuming unhealthy diets. By doing so, we can better safeguard individuals and promote a healthier society.

6 Quality Assurance and Control

6.1 Good Manufacturing Practices (GMP)

Good manufacturing practices (GMPs) encompass a set of manufacturing and administrative procedures with the aim of ensuring consistent product quality that meets defined specifications and customer requirements. These regulations apply to various consumer goods, including human and veterinary pharmaceutical products, biologically derived products, medical devices, and the manufacturing, packaging, or storage of human and processed food items [44]. In the context of food, adherence to GMP standards results in the production of safe and high-quality food products. In the United States, the Food and Drug Administration (FDA) has established GMP regulations as the minimum manufacturing requirements. Most countries have their own GMP regulations for pharmaceuticals. The initial formal GMP guidelines were introduced by the US FDA in 1963.

The core principles underlying GMP can be summarized as follows [45]:

1. GMP aligns closely with disciplines such as quality management, food safety, and food quality.
2. GMP is developed by food manufacturers for food manufacturers.
3. GMP covers the entire spectrum of food business operations, from policy establishment to implementation.
4. GMP is a proactive and hands-on approach.
5. GMP allows for exceeding customer expectations and instils confidence in the product's consistency and quality.
6. Every activity within the food business affects the final product.
7. GMP adds value to the product and fosters brand loyalty.

Benefits of GMP

GMP, or good manufacturing practices, is a dynamic program aimed at enabling organizations to institute and uphold quality enhancements. The foundational principles of GMP provide enhanced assurance concerning food safety and quality by means of comprehensive prerequisite programs. These principles underscore the importance of proactively preventing and managing processes throughout the entire food supply chain. The core objective of GMP is to introduce real-time process controls that can respond swiftly in potentially risky situations, all while continuously monitoring quality parameters. This approach allows for the timely identification of issues and fluctuations as they arise, even before the product is released to market. When combined with a well-devised HACCP (Hazard Analysis and Critical Control Points) plan, this system ensures that any safety concerns related to the processed food can be promptly detected and rectified at crucial junctures in the production process, well before the product reaches its final packaging. One notable feature of GMP is its adaptability across various industries due to its generic nature. Nevertheless, the specific management of components, production procedures, packaging, labeling, and storage and distribution may require tailored approaches depending on the specific nature of the food product [46].

6.2 Hazard Analysis Critical Control Points (HACCP)

Hazard analysis and critical control point (HACCP) techniques were initially introduced to the chemical processing industry, with a significant application in Great Britain, more than four decades ago, as noted by. This innovative system was originally developed through collaboration between the Pillsbury Company, NASA, and the U.S. Army's Research, Development and Engineering Center at Natick, with the primary objective of ensuring the safety of food for astronauts, as documented by. Over time, the HACCP system has gained universal recognition and acceptance as a fundamental method for guaranteeing food safety. The World Health Organization (WHO) has acknowledged the significance of the HACCP system in preventing foodborne diseases for over two decades, actively contributing to its advancement and promotion. A pivotal moment in the history of the HACCP system occurred when the Codex guidelines for its application were adopted by the FAO/WHO Codex Alimentarius Commission in 1993, making them mandatory for international trade [47].In the United States, large chemical industries have also embraced hazard control programs, with the US Occupational Safety and Health Administration (OSHA) introducing HACCP techniques to enhance accident reduction efforts.

Benefits of HACCP

The absence of robust food safety systems imposes significant financial burdens on the food industry, resulting in substantial annual losses attributed to waste, reprocessing, recalls, and the consequent decline in sales. It is crucial to note that foodborne diseases are no longer confined to developing nations. In the United States, according to data from the Centers for Disease Control and Prevention (CDC), foodborne illnesses are responsible for a staggering 76 million cases of illness, 325,000 hospitalizations, and 5000 deaths every year. On an international scale, it is widely acknowledged that the most cost-effective approach to ensuring food safety lies in the application of the Hazard Analysis and Critical Control Point (HACCP) technique. By embracing a comprehensive safety system founded on HACCP principles, the food industry can significantly reduce the likelihood of mishaps and guarantee the safety of food products. Importantly, HACCP is not in competition with existing processing systems, such as good manufacturing practices (GMPs) and ISO 9000 standards; rather, it complements them by introducing crucial safety elements, as emphasized by [48].

Seven Principles of HACCP

1. Analyze the hazards and risk assessment:
    (a) Hazard identification: Identifying and listing all potential hazards related to the food production process.
    (b) Hazard characterization: Evaluating the adverse effects of these hazards on human health, both quantitatively and qualitatively.
    (c) Exposure assessment: Evaluating the likely degree of exposure to these hazards, whether quantitatively or qualitatively.
    (d) Risk characterization: Integrating the above three components to estimate the potential adverse effects on the target population.

2. Identify the critical control points (CCPs):
   (a) Differentiating true CCPs from control points to avoid an unmanageable number of CCPs. For example, in smoked fish processing, only critical steps like salt penetration, smoking, and storage might be identified as CCPs (Garrett and Hudak-Ross 1990).
   (b) Two approaches for identifying CCPs include the decision tree developed by the Codex Alimentarius Committee of Food Hygiene and risk analysis [49, 50].
3. Define critical limits for each CCP:
   (a) Establishing specific criteria, such as time, temperature, physical attributes, acidity, pH, moisture content, water activity, salt concentration, and chlorine levels, to determine acceptability or rejection for each CCP.
   (b) Critical limits can be derived from experiments, regulatory requirements, codes of practice, or other valid sources [51].
4. Establish controls to monitor the CCPs:
   (a) Planning and documenting measurement or observation procedures to ensure that each CCP remains under control.
   (b) Monitoring is essential for evaluating the effectiveness of the system's operation, detecting deviations from critical limits, and providing evidence of compliance with the HACCP plan.
   (c) Monitoring can be done at defined time intervals or continuously for more reliable results [52].
5. Establish corrective actions:
   (a) Developing predetermined and documented procedures to be executed when a deviation at a CCP exceeds critical limits.
   (b) Corrective actions are necessary to resolve deviations that pose unacceptable health risks, and they include identifying the cause, implementing preventive measures, and verifying their effectiveness [53].
6. Establish verification procedures:
   (a) Applying methods, procedures, tests, and evaluations, in addition to monitoring, to determine compliance with the HACCP plan.
   (b) Verification assesses the plan's effectiveness and its compatibility with the HACCP system, and it may be performed at planned intervals, during process changes, or when deviations occur.

7. Define record-keeping and documentation requirements:
   (a) Maintaining records to document the history of the process, measurements, CCP deviations, and corrective actions taken.
   (b) An HACCP manual typically includes various sections, such as the organization's profile, policies, products, flow diagrams, hazard analysis, control schedule, consumer complaint and recall procedures, corrective action, and record and document control procedures.
   (c) Records and documents may be in various forms, including charts, written procedures, and electronic records [54].

7 Regulatory Authorities and Compliance

7.1 FDA (Food and Drug Administration) and USDA (United States Department of Agriculture)

In the United States, three executive branch agencies oversee the regulation of food and agriculture, along with other areas including drugs, cosmetics, and pollution. These agencies play pivotal roles in safeguarding public health and the environment. First, the Food and Drug Administration (FDA) operates under the US Department of Health and Human Services (DHHS) and is entrusted with the responsibility of ensuring that foods within its purview are "safe, wholesome, sanitary, and properly labelled" [55]. Essentially, the FDA's primary mandate is to determine whether the foods produced are safe for consumption. While the FDA also oversees animal feed and drug regulation, its primary legal instrument is the Federal Food, Drugs, and Cosmetics Act (FFDCA), which was enacted in 1938 [56]. Additionally, the FDA ensures the safety of food additives. The second agency responsible for food regulation in the United States is the US Department of Agriculture (USDA). The USDA's jurisdiction extends to the regulation of livestock, meat, poultry products, and certain egg products. Its mission is to "provide leadership on food, agriculture, natural resources, rural development, nutrition, and related issues based on sound public policy, the best available science, and efficient management" [57]. As an agency primarily focused on agricultural oversight, the USDA's core concern is determining whether specific crops are safe to cultivate. The third agency involved in food regulation is the US Environmental Protection Agency [58], which is charged with the mission to "protect human health and the environment." Given the inherent link between agriculture, food production, and environmental concerns, the EPA's regulatory purview extends to various aspects of food regulation. Specifically, the EPA regulates environmental factors under the National Environmental Policy Act (NEPA), biodiversity and threatened species under the Endangered Species Act [59], pesticides under the Federal Insecticide, Fungicide, and Rodenticide Act (FIFRA) [60], and water and air pollution under the Clean Water Act [58], and the Clean Air Act [61], respectively. Furthermore, the

United States has embarked on a significant overhaul of food and food safety legislation through the Food Safety Modernization Act (FSMA). This groundbreaking legislation marks a shift in focus from response to prevention in food safety efforts, equipping the FDA with enhanced tools to ensure consumer food safety [62]. These agencies, along with legislative measures such as the FFDCA and FSMA, collectively contribute to the United States' efforts to ensure the safety and quality of its food supply while addressing critical public health and environmental concerns.

Food Safety Initiatives in Canada

In Canada, the responsibility for ensuring food safety and addressing foodborne illness outbreaks is a collaborative effort among federal, provincial, and territorial governments, industry stakeholders, and consumers. The Canadian Food Inspection Agency (CFIA) is tasked with upholding the policies and standards established by Health Canada for food distributed in Canada. This encompasses the enforcement of federal food inspection programs and the verification of industry compliance with food safety regulations. The CFIA also assumes a leading role in coordinating with provincial and territorial public health authorities during investigations and responses to foodborne illness outbreaks that involve multiple regions or countries.

The regulation of food in Canada is governed by a comprehensive framework consisting of five statutes: the Food and Drugs Act, the Fish Inspection Act, the Meat Inspection Act, the Canada Agricultural Products Act, and the Consumer Packaging and Labeling Act. To address contemporary challenges in food safety, the federal government made a commitment to modernize legislation. Consequently, on June 7, 2012, the Safe Food for Canadians Act (SFCA) was tabled. Analogous to the Food Safety Modernization Act (FSMA), the SFCA substantially reforms the existing system by consolidating the food-related provisions of four statutes into a singular legislative framework. The SFCA equips the CFIA with enhanced authorities to formulate regulations pertaining to the tracing and recalling of food, providing the necessary tools to respond effectively to potentially hazardous food commodities. This includes a stringent prohibition on the sale of food commodities subject to recall. The Act further incorporates provisions necessitating the registration or licensing of importers, thereby holding them accountable for the safety of food commodities imported into the country. The overarching objective is to ensure the safety and integrity of the food supply chain within Canada [63].

7.2 EU Food Safety Authority

With more than 500 million citizens in 28 Member States, the European Union (EU) represents one of the world's largest food markets. The European Commission emphasizes that the substantial size of this common market enables the food industry to continually enhance productivity and leverage economies of scale effectively.

Consequently, gaining an understanding of European food law, regulation, and policy becomes integral to comprehending the global food law system [64]. In Europe, food manufacturers and distributors operate within a stringent regulatory framework governing food hygiene. This framework encompasses several pivotal regulations and directives, all designed to standardize food safety standards across European Member States. In 2004, three significant legislations were enacted to overhaul pan-European food hygiene regulations. Major regulation of EU along with their description has been postulated in Table 21.2.

These regulations collectively aim to ensure food hygiene is maintained throughout all stages of production and the supply chain. They mandate that food processors adhere to the principles of Hazard Analysis and Critical Control Points (HACCP), an internationally recognized food safety standard introduced by the Codex Alimentarius of the United Nations Food and Agriculture Organization. In 2008, the European Commission made substantial revisions to pan-European regulations concerning food additives, flavorings, and food enzymes through the "Food Improvement Agents Package" (FIAP). These FIAP Regulations possess direct applicability across all European Member States and have superseded previous European Directives in these domains. This package encompasses several key regulations as postulated in Table 21.3.

Table 21.2 Major regulation of EU along with their description

Regulation	Description
Regulation (EC) No. 852/2004	Focuses on food product hygiene
Regulation (EC) No. 853/2004	Outlines precise hygiene rules applicable to food of animal origin
Regulation (EC) No. 854/2004	Establishes regulations for organizing official controls on products of animal origin intended for human consumption
Regulation (EC) No. 178/2002	Lays out the general principles of food law, establishes the European Food Safety Authority, and defines procedures for food safety matters
Regulation (EC) No. 882/2004	Delineates official controls to ensure compliance with feed and food law, along with animal health and welfare rules
Directive 2002/99/EC	Governs animal health regulations concerning the production, processing, distribution, and introduction of products of animal origin for human consumption
Commission Regulation (EC) No. 2073/2005	Establishes microbiological criteria for food products
Commission Regulation (EC) No. 2074/2005	Specifies implementation measures for certain products under Regulation (EC) No. 853/2004, the organization of official controls under Regulation (EC) No. 854/2004, and Regulation (EC) No. 882/2004

Table 21.3 Pan-European regulations concerning food additives, flavorings, and food enzymes

Regulation	Description
Regulation (EC) No. 1331/2008	Establishes a common authorization procedure for food additives, enzymes, and flavorings
Regulation (EC) No. 1333/2008	Specifically addresses food additives
Regulation (EC) No. 1334/2008	Pertains to food flavorings and certain food ingredients used for flavoring in and on foods
Regulation (EC) No. 1332/2008	Deals with food enzymes

7.3 CODEX Alimentarius

In the late 1950s, the United Nations (UN), the World Health Organization (WHO), and the Food and Agriculture Organization (FAO) recognized the need for global standards to facilitate international trade while ensuring strict food safety and consumer protection. This led to the establishment of the Codex Alimentarius Commission (CAC) by WHO and FAO in 1963. The main goals of CAC were to create universally harmonized food standards, guidelines, and Codes of Practice to protect consumer health and promote fair practices in the food trade. The Commission also aimed to enhance coordination among governmental and non-governmental organizations involved in food standards. By its 50th anniversary in 2013, the Codex Alimentarius Commission had 185 member countries, including the European Union (EU), and the participation of 221 observer organizations. Covering 99% of the world's population, Codex Alimentarius holds significant influence in international food trade. Its standards serve as a crucial reference for the development of national and regional food regulations, easing the international trade of products that adhere to Codex standards. Since 1995, these standards have played a key role in resolving food-related cases brought before the World Trade Organization (WTO), arising from the Uruguay Round Agreements during 1986–1994.

Within the Codex Alimentarius framework, there are 17 active technical committees addressing various aspects of food trade. These committees cover general areas like food additives, food labeling, and contaminants in food, as well as specific food categories like fish and fish products and fats and oils. Each committee is responsible for developing standards, guidance, and codes of practice in their respective domains, with decisions made by international experts based on impartial assessments of scientific data. Consensus within the relevant committee and subsequent endorsement by the Codex Alimentarius Commission are required for the adoption of standards, guidelines, and codes of practice. It's worth noting that the Codex process, while effective in gaining international approval, can be time-consuming. Committees meet annually or biannually, and the multi-stage consultation process can lead to delays in standard development. Although Codex standards and codes of practice lack inherent legal standing in international law, many countries worldwide are increasingly adopting them, making conformity to these standards a prerequisite for accepting imports from other countries [65].

Sanitary and Phytosanitary Measures (SPS) and Technical Barriers to Trade (TBT)

From the Uruguay round within the World Trade Organization (WTO), two pivotal agreements emerged: the Technical Barriers to Trade (TBT) agreement and the Sanitary and Phytosanitary (SPS) agreement. The TBT agreement aims to prevent technical regulations and standards, including those related to packaging, labeling, and analytical procedures, from creating unjustified trade barriers. It emphasizes active participation by Member States in developing international standards, fostering broad harmonization. In the European Union (EU), the harmonization of food and agriculture-related legislation centers on common sanitary and phytosanitary (SPS) codes and veterinary measures. Echoing the WTO SPS agreement, EU agreements include measures on sanitary, veterinary, and plant quarantine adopted in 2009. The core principle is that SPS measures must be scientifically justified, prioritizing the safety and health of humans and animals. The agreements also commit to mutually recognizing SPS measures introduced by third countries, aligning with WTO principles.

SPS measures for food products involve a risk analysis of potential hazards like additives, contaminants, toxins, or microorganisms. Member states must adhere to the outlined measures but can introduce interim ones under specific conditions, such as a significant deterioration of sanitary and epidemiological safety. Any additional measures must be transparent, with comprehensive communication to other member states and third countries about new procedures and inspections. The agreement on sanitary measures specifies that member states maintain a unified list of goods subject to sanitary and epidemiological control. Products on this list must meet Uniform Sanitary Requirements, detailing hygienic standards for market goods. This comprehensive approach ensures both safety and fairness in international trade [66].

The Impact of Codex Alimentarius on Reducing Barriers to International Trade

In the late 1950s, the United Nations (UN), the World Health Organization (WHO), and the Food and Agriculture Organization (FAO) recognized the imperative of establishing international standards and controls to facilitate global trade while upholding the highest levels of food safety and consumer protection. This recognition led to the creation of the Codex Alimentarius Commission (CAC) by WHO and FAO in 1963. The primary goals of the CAC were to develop harmonized international food standards, guidelines, and codes of practice, ensuring the health and well-being of consumers and promoting equitable practices in the food trade. Additionally, the Commission took on the responsibility of coordinating all efforts related to food standards, involving both governmental and non-governmental organizations. The Codex Alimentarius encompasses 17 active technical committees, each dedicated to addressing specific aspects of the food trade. Among these committees, the "horizontal" ones, such as the Codex Committee on Food Additives, the Codex Committee on Food Labeling, and the Codex Committee on Contaminants in Food, focus on issues within their purview that impact all categories of food. This comprehensive approach underscores the commitment of Codex Alimentarius to ensuring global standards and promoting the well-being of consumers in the international food trade.

7.4 Codex India/National Codex Committee

It was established by FSSAI which reflects FSSAI's commitment to align with international food safety standards and foster collaboration with the global community to ensure the safety and quality of food products in India. The primary role of these shadow committees is to support the National Codex Committee (NCC) in the examination and deliberation of technical issues. Presently, there are a total of 12 Shadow Committees, each focusing on different subject matters related to the Codex Alimentarius. The Food Safety and Standards Authority of India (FSSAI) under the administration of Ministry of Health and Family Welfare has been designated as the nodal point for liaison with the Codex, known as National Codex Contact Point of India (NCCP) [67].

8 Indian Food Laws

For a substantial period, India faced challenges in aligning its food regulations with contemporary requirements. Consequently, in 2006, India embarked on a comprehensive reform of its food regulatory framework, encompassing standards governing food products as well as the manufacturing, distribution, sale, and import of food. The transformation was chiefly codified within the Food Safety and Standards Act (referred to as the "FSS Act" or simply "FSS") and associated rules and regulations. The phased implementation of these changes commenced in 2007. The architects of the FSS Act drew inspiration from global trends and international legal frameworks, particularly the Codex Alimentarius Commission. Their vision was for the FSS Act to align with international norms, establishing a unified policy framework to oversee the entire spectrum of food-related processes, from production and processing to handling, transportation, import, and sale. This approach aimed to replace the fragmented and multi-departmental control that previously characterized these processes. In pursuit of the goal of streamlining India's food regulatory landscape, the FSS Act led to the repeal of the Prevention of Food Adulteration Act, 1954, along with six other laws/orders related to specific food products. Concurrently, it established the Food Safety and Standards Authority of India (commonly referred to as the Food Authority).

8.1 Activities of FSSAI

Licensing and registration: Registering Petty Food Business Operators, which includes individuals such as stall owners, itinerant vendors, and home-based kitchen workers, poses notable practical challenges that require attention. Authorities primarily focus on regulating mobile vendors without fixed business locations, like bicycle-based milk sellers or those engaged in distributing homemade food. However, many of these vendors encounter difficulties in grasping the legal requirements and obtaining registration due to factors such as illiteracy and limited access to media. It's worth noting that the Licensing Regulations also encompass provisions for religious gatherings, permitting the distribution of food on special occasions. While the effort to implement registration for the sake of sanitation is commendable, there remain uncertainties regarding its enforcement for individuals providing assistance to the needy. Regarding licenses for food businesses, they are generally issued at the state level, although federal licenses are available for specific cases such as importers and multi-state operators.

Product approval: Products that do not conform to the prescribed standards as outlined in the Food Safety and Standards Act (FSS Act) or contain ingredients that lack established criteria must go through a separate approval procedure before being eligible for licensing. The intricacy of this approval process varies depending on the product category. Products devoid of plant or animal ingredients, compliant with FSS regulations, and recognized by international authorities can obtain approval relatively easily. However, products incorporating such ingredients necessitate more comprehensive information, potentially extending the approval timelines. It's noteworthy that traditional or ethnic foods are exempt from the product approval mandate. Nevertheless, challenges arise in maintaining consistency for products with established international standards.

Food standards and ingredients: The regulations pertaining to food and beverage items encompass a diverse array of approximately 150 products. These items are systematically categorized into 13 overarching groups, each characterized by comprehensive standards dictating permissible ingredients and their quantities. The regulatory scope extends broadly, encompassing both globally recognized items such as cheese and chocolates, as well as local products like ghee and jaggery.

Labeling: In India, the regulations governing food labeling primarily revolve around the Packaging and Labelling Regulations and the LMPC Rules. These regulations play a crucial role in ensuring that labels on pre-packaged food items, whether they are imported, stored, or sold in India, contain vital information, declarations, and specifications. In addition to these provisions, there are several other labeling requirements that must be adhered to, including:

- The use of either English or Hindi as the primary labeling language.
- The inclusion of a declaration for the maximum retail price.
- Clear instructions on the correct utilization of the product.
- Comprehensive address details for the manufacturer, packer, and brand owner.
- Lot/batch identification for effective tracking.
- Dates indicating the manufacture, packing, best before, and use-by dates.
- Declarations of net weight or volume, exclusive of packaging weight.

Trademarks protection: Prominent global and Indian food brands have firmly established themselves in Indian markets and households. The significance of effective branding in terms of recognition, competitiveness, and market share cannot be emphasized enough. To safeguard their intellectual property, leading players have made trademark protection a paramount concern, encompassing various elements such as words, logos, shapes, slogans, colors, and even sounds.

Genetically modified food: India's food regulations encompass genetically modified (GM) foods within their definition of food. Under the Food Safety and Standards (FSS) Act, a Scientific Panel for GM organisms and foods is established. While GM foods are permitted, the existing food laws do not currently prescribe specific regulations or labeling requirements for them. Nevertheless, GM produce has encountered significant opposition in India, leading to notable cases and developments. Initially, trials for BT cotton were allowed but were subsequently prohibited due to public protests. Similarly, the cultivation of GM BT brinjal/eggplant was banned following public outcry. Trials for GM okra have also been conducted. In 2014, a moratorium on GM food crop field trials was lifted, permitting such trials to proceed. This decision was based on the belief that sufficient regulatory mechanisms were in place. The Genetic Engineering Approval Committee (GEAC) is responsible for granting approval for GM products.

9 Food Laws in China

Chinese consumers increasingly prioritize food quality and safety, driving demand for organic products. The Chinese government has continually enhanced its food safety management system, emphasizing a "from farm to table" supply chain model involving certification, standards, legislation, and regulation. Food safety in China is influenced by two key factors: mandatory requirements, driven by laws and rules and market incentives. These factors encourage food companies to improve safety, resulting in increased customer loyalty, higher prices, and reduced internal quality costs. However, challenges remain, including coordination issues among regulators, weak law enforcement, and inefficient government oversight. Both government regulations and voluntary certifications are vital in improving food quality [68].

Developed countries like the United States and the EU have clearly defined regulatory agencies with minimal overlap. China's regulatory system is considered less effective and lacks efficient social supervision. The legislative framework in China is found to be less comprehensive and meticulous compared to developed countries. Developed countries have well-established food safety laws and regulations, including primary and secondary legislation, while China's system is characterized by the lack of coordination and uniformity due to insufficient legislation. Public participation in law formulation and revision is more pronounced in developed countries like the United States, with active involvement of government departments, stakeholders, NGOs, and individuals, resulting in more transparent and accountable processes. The enforcement methods differ, with China relying heavily on fines for non-compliance, while the United States emphasizes measures like recalls and detentions.

10 Food Laws in Russia

The Russian food safety system involves several state bodies responsible for controlling and supervising food quality and safety, including Rostekhregulirovaniye, Rospotrebnadzor, and Rosselkhoznadzor. Each of these agencies has specific functions related to regulatory compliance, consumer protection, food safety, and sanitary standards. These agencies work collectively to ensure the safety of goods and food products within the Russian Federation.

In Russia, food laws and regulations encompass various aspects of food safety and standards. These regulations include federal laws for foodstuffs, as well as numerous sanitary and regulatory documents like Sanitary Rules and Regulations (SanPiN), national standards (GOST), and other technical regulations, totaling more than 700 state laws and standards governing food production. Some important set of points have been discussed below [69]:

Hygienic requirements for food safety: The "Hygienic Requirements for Food Safety and Nutritional Value" (SanPiN 2.3.2.1078–01) sets out general food safety requirements, and Rospotrebnadzor oversees the domestic food market based on these regulations. It also regulates entities involved in food production, import, distribution, retail, public catering, and enforces hygiene standards for substances in contact with food.

Food additive regulations: Russia has specific regulations for food additives, including a list of permitted food additives, and more stringent rules for certain products, such as baby food.

Pesticides and contaminants: Pesticide and contaminant levels are regulated under SanPiN 2.3.2.1078-01, with maximum limits for antibiotics, pesticides, DDT, and radionuclides. Other regulations include standards for nitrosamines, benzopyrene, microbiological indices, and heavy metals in meat, poultry, and processed products.

Maximum residue levels (MRLs) for Pesticides and Agrochemicals: The country also enforces MRLs for pesticides and agrochemicals in crops, managed by Rospotrebnadzor and Rostekhregulirovaniye. The Ministry of Agriculture maintains an annual "State Catalogue of Pesticides and Agricultural Chemical Substances Allowed for Use in the Russian Federations," with unlisted chemicals prohibited in food production.

Mycotoxin regulations: Russia has set regulations for mycotoxins like ochratoxin A and deoxynivalenol, with limits depending on the type of food product. The EU and other countries have different standards for these mycotoxins. Additionally, Russia has comprehensive legislation for patulin, with limits conforming to international standards.

Labeling requirements: Regarding labeling, the "General Requirements for Consumer Information Regarding Foodstuffs" (GOST P 51074–2003) aligns with Codex Alimentarius International Food-Packaging Standards. It focuses on nutrition labeling and aims to ensure objective and trustworthy methods for assessing Russian food product quality and safety, preventing trade barriers.

11 An Overview of Food Laws in Other Asian Countries [70]

11.1 Cambodia

Inter-agency coordination: The Inter-Ministerial Prakas IMP868 plays a pivotal role as a legislative measure for facilitating inter-agency coordination. Such coordination is essential to ensure a cohesive approach to food safety and regulation in the country.

Food safety policy development: Cambodia is actively working on the development of a national food safety policy. The establishment of a comprehensive policy is crucial to set clear guidelines and standards for the safety of food products within Cambodia.

Consumer rights protection: Cambodia demonstrates a commitment to protecting consumer rights by embedding them in the national constitution. This legal safeguard emphasizes the importance of ensuring that consumers have access to safe and quality food products.

Laboratory accreditations: Several laboratories in Cambodia, including Industrial Laboratory Center of Combodia (ILCC), have achieved notable accreditations such as National Association of Testing Authorities (NATA), Singapore Accreditation Council (SAC), and International Organization for Standardization (ISO 17025:2005). These accreditations reflect their competence in conducting testing and quality assurance activities.

International accreditation compliance: Cambodia is working toward ensuring that its national certification system and accreditation bodies comply with international requirements and secure accreditation from recognized international bodies. This is a vital step toward aligning with global food safety standards.

Need for expert assistance: Expert assistance is required in the development of food safety standards. Collaborating with experts can help Cambodia establish effective, internationally recognized standards for food safety.

11.2 Lao PDR

Key food safety agency: The Food and Drug Department (FDD) holds a central role as the primary food safety agency in Lao PDR. Empowering this agency with resources and authority is critical for effective oversight and enforcement of food safety regulations.

Food law revision: Lao PDR is in the process of revising Food Law 2004 to include provisions related to SPS, farm-to-table food services, labeling, and GMOs. This update is a positive step in enhancing food safety and regulatory standards.

Existing national food safety policy: Lao PDR has had a National Food Safety Policy in place since 2009. The continuous review and improvement of this policy are necessary to address evolving food safety challenges.

Facility development: Lao PDR needs to focus on the development of laboratory and border checkpoint facilities for managing, inspecting, and monitoring food safety and agricultural products. Investment in infrastructure is essential to strengthen the oversight of food safety.

Expertise shortage: The shortage of human resources and expertise in various food safety areas, including food identification, diagnosis, and testing, is a significant concern. Enhancing the workforce's capabilities through training and education is crucial.

Promotion of Lao organic certification: The presence of Lao organic certification and the promotion of organic produce is a positive step towards safer and more sustainable food production.

Immediate concerns: Addressing immediate concerns, such as the lack of data on veterinary drug abuse, consumer complaints reporting, and the capacity to test or handle residues in meat products, is crucial for enhancing food safety in Lao PDR.

11.3 Myanmar

Multiple implementing agencies: Myanmar features several food safety implementing agencies coordinated through two primary national institutions, the Department of Food and Drug Administration (FDA) and the Myanmar Food and Drug Board of Authority (MFDBA). Effective coordination among these agencies is essential for consistent food safety enforcement.

Consumer protection: The Consumer Protection Law, adopted on March 14, 2014, enshrines consumer protection as a fundamental principle in Myanmar.

Ensuring the effective enforcement of consumer protection laws and regulations is crucial to safeguard the rights of consumers.

Labeling and rights: Existing laws related to labeling and rights to information should be upheld to provide consumers with necessary information about food products.

Awareness programs: The FDA's efforts to deliver programs and training using various media, including IEC materials and traditional media, to raise awareness among food producers about food safety are commendable. Continued education and outreach are essential to ensure food producers are well-informed.

Staff qualifications: It is important to ensure that government agency staff members possess sufficient experience and the appropriate qualifications to implement the Food Safety and Compliance Management (FSCM) effectively.

Food inspection and surveillance: The quality of food inspection, surveillance, and monitoring at border checkpoints needs to be strengthened to prevent the entry of unsafe food products.

Coordination and communication: Effective coordination and communication between the various authorities responsible for food safety control are essential for the consistent enforcement of food safety standards and regulations in Myanmar.

11.4 Bangladesh

Food safety act adoption: Bangladesh has adopted the Food Safety Act, and a new Food Safety Authority has been established. This legal framework provides a foundation for regulating and ensuring the safety of food products in the country.

Food safety indicators: The determination of food safety indicators is a positive step in assessing and improving food safety standards.

Quality standards: While a limited number of food safety and quality standards are in place, there is room for expansion and enhancement to cover a broader range of food products.

Food inspection challenges: Food inspections in Bangladesh face challenges due to a lack of manpower. Despite this limitation, administrative units are widely distributed, resulting in broad coverage.

Facility infrastructure: The range of food safety facilities and infrastructure in Bangladesh varies from outdated to modern. Addressing this disparity and ensuring consistency in facilities and equipment is essential. Additionally, there is an insufficiency in the number and skills of technical experts.

ISO certification: A small number of companies in Bangladesh possess ISO food safety management certification, indicating their commitment to food safety standards. Encouraging more companies to obtain such certification can elevate overall food safety standards in the country.

11.5 Malaysia

Ministerial responsibility: In Malaysia, the coordination of food imports' safety is a ministerial responsibility. Effective oversight by ministerial authorities is crucial to ensure the safety of imported food products.

Web-based platform: Since 2003, Malaysia has utilized a web-based platform called FoSIM to manage food import activities and facilitate online information sharing. This digital platform enhances transparency and efficiency in managing food imports.

Inspection and clearance process: Malaysia employs a structured approach to the inspection and clearance process, consisting of six examination levels. This system helps ensure thorough checks and verification of imported food products.

Adequate manpower: Adequate manpower is essential to maintain an effective food safety system. Sufficient staffing resources are crucial to enforce safety standards and conduct inspections effectively.

11.6 Thailand

Strategic food management: Thailand is actively planning a Strategic Framework of Food Management. Such a framework is essential for setting clear objectives and guidelines for food safety, quality, and education.

Inter-agency coordination: An umbrella institution, the National Food Committee, is being established to ensure inter-ministerial and inter-agency coordination for food security, safety, quality, and food education. This coordination is vital for maintaining consistent food standards and quality.

Consumer education: Thailand is focused on delivering consumer education through branding and recognizable food quality and safety logos. This educational approach helps raise consumer awareness and trust in the safety and quality of food products.

Strengthening food education: Efforts to strengthen food education, including basic, best practices, and output based on research and development (R&D) are commendable. Knowledge dissemination is key to ensuring the continuous improvement of food safety standards.

12 Food Laws in Australia and New Zealand

12.1 Food Supply Scenario

Australia food supply system is shaped by several factors that distinguish it from the rest of the world. Australia's relative isolation and a traditional belief in limited food importation contribute to a higher level of trust in the country's food supply [71]. Australians tend to follow the principle of "innocent until proven guilty," which sets them apart from some other regions. In contrast to Europe, where there are more negative perceptions of "foreign food," Australia has a more positive view, partially due to concerns about food safety and quality standards and past food scares in other countries [72].

Australia is considered one of the world's safest food suppliers and has not experienced significant food scares similar to those in Europe. However, there are growing concerns in Australia about the use of pesticides, food additives, and preservatives, as well as local issues such as food microbiological scares, new technological developments, and food irradiation, which challenge consumer confidence in food safety and security [73]. In Europe, food regulatory reform was driven by the need to reduce the regulatory burden on the food industry and remain competitive in the export market. In contrast, FSANZ was established in Australia and New Zealand to ensure compliance with World Trade Organization treaties governing food safety and free trade [74, 75]. This requires Australia to protect public health without restricting trade, and the approach to food safety in Australia is more preventive, defining what should be achieved rather than how it should be achieved. Australia is moving towards co-regulation of food production methods, with greater self-regulation of food hygiene by the food manufacturing industry. However, the Australian food governance system appears to be less streamlined than those in other countries, potentially resulting in weak consumer understanding and interaction with relevant bodies and stakeholders [45, 76].

12.2 Australia and New Zealand Food Safety Regulations

The cooperative arrangement among Australia, New Zealand, and the Australian states and territories is aimed at developing and implementing uniform food standards to protect public health and safety. These standards are incorporated into the Australia New Zealand Food Standards Code, which is regulated by food legislation in each Australian state, territory, and New Zealand. However, food regulation in Australia remains fragmented, with various levels of governance involved. The federal government has responsibilities that include establishing standards, importation restrictions, quarantine, and food labeling. State and territory governments are responsible for establishing food legislation, implementing and enforcing standards, and local councils manage food inspection [77].

Table 21.4 Standards outlined in Chap. 3 (Australia only) of the Food Standards Code

Standards	Inference
3.1.1	Serves as the introductory standard, explaining the terminology used in other standards and outlining the provisions required for compliance
3.2.1	Sets requirements for controlling food safety hazards during food production, manufacturing, and handling. It applies to certain high-risk industry sectors but is not mandatory for all food businesses
3.2.2	Specifies food handling controls related to the receipt, storage, processing, display, packaging, transportation, disposal, and food recall
3.2.3	Outlines requirements for food premises, fixtures, fittings, equipment, and food transport vehicles

The Food Safety Standards were designed to create more effective and nationally consistent food safety legislation in Australia and New Zealand. These standards are outlined in Chap. 3 (Australia only) of the Food Standards Code, including mandatory standards (3.2.2, 3.2.3, and 3.3.1) for all food businesses. A guide to these standards is available through FSANZ (Table 21.4).

In New Zealand, the Ministry of Primary Industries (MPI) is responsible for developing and implementing legislation to ensure the safety and suitability of food, both domestically and for export. MPI collaborates with industry and consumers to improve food safety, prevent health and environmental issues, and enhance market access for exports. Key components of this legislation include the Food Act 1981, Animal Products Act (APA) 1999, Agricultural Compounds and Veterinary Medicines (ACVM) Act 1997, and Wine Act 2003. MPI adopts a risk management framework and cooperates with international networks to regulate food. New Zealand and Australia also collaborate to enhance trade and food safety through FSANZ and other cooperative agreements, with MPI introducing a new Food Act currently under parliamentary consideration.

13 Food Safety Legislation in African Countries

In African countries, food safety legislation faces challenges that hinder its effectiveness. One key issue is the limited application of Hazard Analysis and Critical Control Points (HACCP) in the food processing industry, particularly because its implementation is not mandatory. While multinational corporations and certain food value chains, especially those catering to export markets, have embraced HACCP or even stricter systems based on customer demands, the level of adoption varies across countries. South Africa leads African countries in the adoption of third-party certification for food companies. Developing nations need to recognize that effective food safety management systems benefit their citizens and economy. However, these insights may not have been clearly communicated and understood by those responsible for food safety legislation and monitoring.

Many developing countries host multinational companies like Coca Cola, Unilever, Mars Incorporated, and Wrigley's, which maintain stringent food safety management systems. These companies often rely on their internal standards and auditing techniques, rather than government inspections. They possess extensive resources, including well-equipped laboratories, personnel, and logistics. Paradoxically, these corporations coexist with uncoordinated and unregulated street food supply chains, which are significant sources of food, especially for urban, poor, and middle-class populations. This dual-tier food safety operation in developing economies, comprising multinationals and local startups, highlights the need to address food safety comprehensively for all segments of the population [78].

13.1 Food Laws in Western Africa

In Ghana, the food production system is primarily composed of informal, small-, and medium-sized enterprises (SMEs). Hazard Analysis and Critical Control Point (HACCP) systems are rarely applied, except in response to safety incidents or consumer demands in foreign markets. Local foods and brews often do not conform to Ghana's standards, creating a gap where food safety programs cannot be effectively implemented. Neglected food supply chains, despite their significant contribution to the population's nutrition, could become sources of foodborne illnesses. There is a lack of awareness about food safety, with people perceiving it as a concern for the affluent or a different social class. Food safety is often not a top priority in developing countries, resulting in limited funding for related endeavors. Consequently, many food supply chains in rural and urban areas are unregulated, and traceability mechanisms are almost non-existent.

Ghana Food Safety Authority faces challenges related to its food testing capacity, including issues with equipment, personnel, and the centralization of the government laboratory in Accra, serving the entire country. The government of Ghana has various ministries involved in food safety legislation, with leading roles played by the Ministry of Environment and Science, Trade and Industry, Ministry of Health, and Ministry of Food and Agriculture. However, there is a need to revise and modernize food safety standards, which are currently disconnected from the bodies responsible for control and enforcement. Separation between risk analysis and advice from those tasked with managing and operationalizing food safety regulations is also prevalent. Nevertheless, Ghana has taken an important step by drafting the National Food Safety Action Plan (NFSAP) to restructure food safety, agriculture, and health institutions for improved efficiency and governance. This allows for coordinated efforts among Ministries, Departments, and Agencies, streamlining operations, reducing duplication, resolving conflicts, and enhancing the enforcement of food safety legislation [79].

Food safety in Nigeria faces numerous challenges, including inadequate application of Good Agricultural Practices (GAP), misuse of agrochemicals and pesticides, contamination by chemical substances like lead, and improper use of additives. Other issues involve toxic packaging materials, public ignorance, a lack of technical expertise, and poor laboratory facilities. Furthermore, existing legislations are inadequately enforced, and droughts may force the inclusion of contaminated food in the value chain.

Nigeria's legal framework for food safety closely mirrors that of Ghana. It encompasses legislation, policy, institutionalization, inspection, and laboratory testing services. The National Food Safety Policy, established in 1999, plays a vital role in coordinating food safety legislation in Nigeria, acknowledging the contributions of both the public and private sectors to address food safety challenges. Nigeria has a plethora of food-related legislations focusing on consumer protection, proper coordination, policy development, and priority setting for food legislation. These laws advocate Hazard Analysis and Critical Control Points (HACCP) as the basis for all food production and processing operations. Several organizations in Nigeria are responsible for food safety, including the Ministry of Health, the National Agency for Food and Drug Administration and Control, the Standards Organization of Nigeria, the Ministry of Agriculture and Research and Development and Fisheries, Plant Quarantine Service, the Consumer Protection Council, the Federal Ministry of Environment, and the Federal Ministry of Trade and Industry. Additionally, universities, research institutions, and local governments play a role in creating awareness and providing training for personnel in food safety-focused agencies. However, to achieve success in implementing food safety legislation in Nigeria, it is crucial to increase public awareness and educate consumers about the risks of neglecting food safety. Effective coordination among the various bodies responsible for food safety is essential. Capacity building through training for personnel, producers, and regulators is vital. Encouraging small and medium-sized enterprises (SMEs) to form associations can facilitate government support in terms of training and awareness programs [80].

13.2 Food Laws in Kenya

In Kenya, there are over 22 food safety-related legislations managed by different government departments and agencies. Kenya has been a member of the Codex Alimentarius Commission since 1969, actively participating in Codex committees and currently chairing the CCAFRICA region. It has adopted more than 100 Codex standards. Kenya has also developed a National Food and Nutrition Security Policy, with food safety being a key component, falling under the subtopic of "Food Safety, Standards, and Quality Control."

Kenya is in the process of establishing a National Food Safety Authority, which will play a significant role in conducting risk assessments, an area that has been relatively underrepresented in Kenya and other developing economies. The authority is expected to promote human health and improve coordination among multiple stakeholders in food safety. Kenya's food safety management model involves several bodies across different sectors with mandates to ensure safety, quality, and trade promotion for specific products. Kenya is a member of the East African Community (EAC), and the EAC has been actively working to harmonize commodity standards, with 42 standards already harmonized. The goal is to facilitate trade, remove barriers, and enable faster movement of goods across borders, thereby reducing food losses. Despite the efforts made to enhance food safety in Kenya, incidents of foodborne illnesses, outbreaks, fraud, and other challenges are still reported regularly [81].

14 Conclusion

Food analysis plays a pivotal role in upholding the safety, quality, and authenticity of food products. It is intricately linked with food laws and regulations, which are designed to safeguard consumers and maintain the integrity of the global food supply chain. Compliance with these principles and regulations is essential for both food industry stakeholders and public health.

References

1. Crew S, Authenticity F (2022) Food analysis: a practical guide. Food Sci Technol 36(3):52–57. https://doi.org/10.1002/fsat.3603_12.x
2. Zeb A (2020) Concept, mechanism, and applications of phenolic antioxidants in foods. J Food Biochem 44(9):1–22. https://doi.org/10.1111/jfbc.13394
3. Rajput H, Rehal J (2019) Methods for food analysis and quality control
4. Gizaw Z (2019) Public health risks related to food safety issues in the food market: a systematic literature review. Environ Health Prev Med 24(1):1–21. https://doi.org/10.1186/s12199-019-0825-5
5. Sulistyowati E, Rudijanto A, Soeharto S, Handayani D (2019) The identification of characteristic macro- and micronutrients and the bioactive components of Indonesian local brown rice as a functional feed in obesity nutrition therapy. Curr Nutr Food Sci 16(4):494–500. https://doi.org/10.2174/1573401315666190328223626
6. Fairulnizal M, Gunasegavan RDN, Khalid NM, Balasubramaniam V, Mustar S, Rashed AA (2020) Recent techniques in nutrient analysis for food composition database. Molecules 25(19). https://doi.org/10.3390/molecules25194567
7. Li YO, Komarek AR (2017) Dietary fibre basics: health, nutrition, analysis, and applications. Food Qual Saf 1(1):47–59. https://doi.org/10.1093/fqs/fyx007
8. Garhwal R et al (2023) Milk from Halari donkey breed: nutritional analysis, vitamins, minerals, and amino acids profiling. Food Secur 12(4):853. https://doi.org/10.3390/foods12040853
9. Corradini MG (2018) Shelf life of food products: from open labeling to real-time measurements. Annu Rev Food Sci Technol 9:251–269. https://doi.org/10.1146/annurev-food-030117-012433
10. Singh PK, Singh RP, Singh P, Singh RL (2019) Food hazards: physical, chemical, and biological. In: Food safety and human health. Elsevier Inc., pp 15–65. https://doi.org/10.1016/B978-0-12-816333-7.00002-3

11. Martini D, Menozzi D (2021) Food labeling: analysis, understanding, and perception. Nutrients 13(1):1–5. https://doi.org/10.3390/nu13010268
12. FSSAI (2015) Manual on general guidelines on sampling. Food Saf. Stand. Auth. India, Minist. Heal. Fam. Welfare, Gov. India, p. 40
13. Musa A, Dabo AAA (2016) A review of RFID in supply chain management: 2000–2015. Glob J Flex Syst Manag 17(2):189–228. https://doi.org/10.1007/s40171-016-0136-2
14. Spink J, Moyer DC (2011) Defining the public health threat of food fraud. J Food Sci 76(9). https://doi.org/10.1111/j.1750-3841.2011.02417.x
15. Callao MP, Ruisánchez I (2018) An overview of multivariate qualitative methods for food fraud detection. Food Control 86:283–293. https://doi.org/10.1016/j.foodcont.2017.11.034
16. Ramos L (2020) Basics and advances in sampling and sample preparation. In: Chemical analysis of food. Elsevier Inc., pp 3–31. https://doi.org/10.1016/b978-0-12-813266-1.00001-2
17. Braun RD (2023) Chemical analysis. In: Encyclopedia Britannica. Available: https://www.britannica.com/science/chemical-analysis
18. Cifuentes A (2012) Food analysis: present, future, and foodomics. ISRN Anal Chem 2012:1–16. https://doi.org/10.5402/2012/801607
19. Power AC, Chapman J, Chandra S, Cozzolino D (2019) Ultraviolet-visible spectroscopy for food quality analysis. In: Evaluation technologies for food quality. Elsevier Inc., pp 91–104. https://doi.org/10.1016/B978-0-12-814217-2.00006-8
20. Kumari VC et al (2022) Chromatographic techniques: types, principles, and applications. In: Analytical techniques in biosciences. Academic Press, pp 73–101
21. Maugel TK, Goodhew PJ (1976) Electron microscopy and analysis, vol 95(2). CRC Press. https://doi.org/10.2307/3225078
22. Gabbott P (2008) A practical introduction to differential scanning calorimetry. In: Principles and applications of thermal analysis, pp 1–50. https://doi.org/10.1002/9780470697702.ch1
23. Almoselhy RIM (2020) Applications of differential scanning calorimetry (dsc) in oils and fats research. a review. Am Res J Agric 6(1):1–9. https://doi.org/10.21694/2378-9018.20002
24. De Blasio C (2019) Fundamentals of biofuels engineering and technology: thermogravimetric analysis (TGA). In: Green energy and technology. Springer, pp 91–102
25. Sanz JL, Köchling T (2007) Molecular biology techniques used in wastewater treatment: an overview. Process Biochem 42(2):119–133. https://doi.org/10.1016/j.procbio.2006.10.003
26. Fanning S, Proos S, Jordan K, Srikumar S (2017) A review on the applications of next generation sequencing technologies as applied to food-related microbiome studies. Front Microbiol 8:1829. https://doi.org/10.3389/fmicb.2017.01829
27. Escuder-Gilabert L, Peris M (2010) Review: highlights in recent applications of electronic tongues in food analysis. Anal Chim Acta 665(1):15–25. https://doi.org/10.1016/j.aca.2010.03.017
28. Deisingh AK, Stone DC, Thompson M (2004) Applications of electronic noses and tongues in food analysis. Int J Food Sci Technol 39(6):587–604. https://doi.org/10.1111/j.1365-2621.2004.00821.x
29. Liu YX, Cao MJ, Liu GM (2019) Texture analyzers for food quality evaluation. In: Evaluation technologies for food quality. Woodhead Publishing, pp 441–463. https://doi.org/10.1016/B978-0-12-814217-2.00017-2
30. Chen B, Dhupia JS, Morgenstern MP, Bronlund JE, Xu W (2022) Development of a biomimetic masticating robot for food texture analysis. J Mech Robot 14(2):021012. https://doi.org/10.1115/1.4052379
31. Wardencki W, Biernacka P, Chmiel T, Dymerski T (2009) Instrumental techniques used for assessment of food quality. Proc ECOpole 3(2):273–279
32. Hassoun A, Karoui R (2017) Quality evaluation of fish and other seafood by traditional and nondestructive instrumental methods: advantages and limitations. Crit Rev Food Sci Nutr 57(9):1976–1998. https://doi.org/10.1080/10408398.2015.1047926

33. Moyano MJ, Heredia FJ, Meléndez-Martínez AJ (2010) The color of olive oils: the pigments and their likely health benefits and visual and instrumental methods of analysis. Compr Rev Food Sci Food Saf 9(3):278–291. https://doi.org/10.1111/j.1541-4337.2010.00109.x
34. Li P et al (2020) Applying sensory and instrumental techniques to evaluate the texture of French fries from fast food restaurant. J Texture Stud 51(3):521–531. https://doi.org/10.1111/jtxs.12506
35. Artavia G, Cortés-Herrera C, Granados-Chinchilla F (2021) Selected instrumental techniques applied in food and feed: quality, safety and adulteration analysis. Food Secur 10(5):1081. https://doi.org/10.3390/foods10051081
36. Smyth H, Cozzolino D (2013) Instrumental methods (spectroscopy, electronic nose, and tongue) as tools to predict taste and aroma in beverages: advantages and limitations. Chem Rev 113(3):1429–1440. https://doi.org/10.1021/cr300076c
37. Sissons M, Ames N, Egan N, Rhymer C (2008) A comparison of two instrumental techniques used to discriminate the cooking quality of spaghetti. Int J Food Sci Technol 43(7):1323–1329. https://doi.org/10.1111/j.1365-2621.2007.01644.x
38. Barbieri S, Bendini A, Balestra F, Palagano R, Rocculi P, Gallina Toschi T (2018) Sensory and instrumental study of Taralli, a typical Italian bakery product. Eur Food Res Technol 244(1):73–82. https://doi.org/10.1007/s00217-017-2937-8
39. Athmaselvi KA, Kumar C, Balasubramanian M, Roy I (2014) Thermal, structural, and physical properties of freeze dried tropical fruit powder. J Food Process 2014:1–10. https://doi.org/10.1155/2014/524705
40. Santhalakshmy S, Don Bosco SJ, Francis S, Sabeena M (2015) Effect of inlet temperature on physicochemical properties of spray-dried jamun fruit juice powder. Powder Technol 274:37–43. https://doi.org/10.1016/j.powtec.2015.01.016
41. Sentellas S, Núñez Ó, Saurina J (2016) Recent advances in the determination of biogenic amines in food samples by (U)HPLC. J Agric Food Chem 64(41):7667–7678. https://doi.org/10.1021/acs.jafc.6b02789
42. Bianchi F, Careri M, Musci M, Mangia A (2007) Fish and food safety: determination of formaldehyde in 12 fish species by SPME extraction and GC-MS analysis. Food Chem 100(3):1049–1053. https://doi.org/10.1016/j.foodchem.2005.09.089
43. Alwan A (2010) Global status report on noncommunicable diseases. World Health Organization, p 176
44. cGMP Resources (n.d.) cGMP Resources. Available: http://www.cgmp.co/resources.htm
45. FSANZ (2008) Consumer attitudes survey 2007: a benchmark survey of consumers' Attitudes to food issues. Food Stand. Aust. New Zealand, Canberra
46. Hills D (2023) Establishing waste reduction benchmarks and good manufacturing practice for open mold laminating, 2023. Available: http://www.p2pays.org/ref/01/ 00372
47. Joint FAO/WHO Codex Alimentarius Commission (1993) Codex guidelines for the application of HACCP, 20th Session
48. Motarjemi Y, Käferstein F, Moy G, Miyagawa S, Miyagishima K (1996) Importance of HACCP for public health and development the role of the World Health Organization. Food Control 7(2):77–85. https://doi.org/10.1016/0956-7135(96)00003-5
49. MFSC (Microbiology and Food Safety Committee of the National Food Processors Association) (1993) Implementation of HACCP in a food processing plant. J Food Prot 56(6):548–554
50. Brooks J, Reeves M (1995) Evaluating control points, managing food safety, Food Technology Research Centre. Department of Food Technology, Massey University, Palmerston North
51. FAO (2011) Food and Agriculture Organisation of the United Nations: Mekong Basin no. 17 April 2019. Rome, 2011
52. Norton C (2003) HACCP—Developing and verifying a flow diagram for food production. Food Manage 38(5):80. Available: http://search.ebscohost.com/login.aspx?direct=true&db=bth&AN=9938245&amp%0Alang=es&site=ehost-live
53. FAO (1998) Food and Agriculture Organisation of the United Nations, Food Quality and Safety Systems, Section 3, Rome

54. Rooney JJ, Kilkelly J (2022) On today's menu: quality. Qual Prog 35(25)
55. FDA (2014) Food Safety Modernization Act (FSMA). Available: http://www.fda.gov/Food/GuidanceRegulation/FSMA/default.htm
56. FDA (1938) Federal Food, Drugs, and Cosmetics Act (FFDCA)
57. USDA (2014) Supplemental Nutrition Assistance Program (SNAP)
58. EPA (2014) National Environmental Policy Act (NEPA). Available: http://www.epa.gov/compliance/nepa/
59. ESA (2014) National Oceanic and Atmospheric Administration Endangered Species Act 2014. Available: http://www.nmfs.noaa.gov/pr/laws/esa/
60. EPA (2014) Federal Insecticide, Fungicide, and Rodenticide Act (FIFRA)
61. EPA (2013) Clean Air Act (CAA). Available: http://www.epa.gov/air/caa/
62. Tateishi W (2018) FSMA: food safety modernization act. Nippon Shokuhin Kagaku Kogaku Kaishi 65(1):32–33. https://doi.org/10.3136/nskkk.65.32
63. Keener L, Nicholson-keener SM, Koutchma T (2013) Harmonization of legislation and regulations to achieve food safety: US and Canada perspective. https://doi.org/10.1002/jsfa.6295
64. EU (2009) Commission of the European Communities, Staff Working Document, Exploiting the Employment Potential ofICTs. Available: http://ec.europa.eu/enterprise/policies/industrialcompetitiveness/files/industry/doc/sec_2009_1111_en.pdf
65. Joint FAO/WHO Codex Alimentarius Commission (2016) CODEX Alimentarius International Food Standards: Pesticide MRLs. 2016. Available: http://www.fao.org/fao-who-codexalimentarius/standards/pestres/en/
66. Krishna P (2004) World trade organization. In: The social science encyclopedia, pp 1078–1079. https://doi.org/10.4337/9781839105609.world.trade.organization
67. FSSAI (2006) Food Safety and Standards Act, Food Safety and Standards Regulations (contaminants, toxins residues) (34 2006) Food Saf. Stand. Auth. India, vol. F.No.2-150, 2006 (Online). Available: http://www.fssai.gov.in
68. Zhang M, Hui Q, Xu W, Pu MZ, Yu ZJ, Zheng FT (2015) The third-party regulation on food safety in China: a review. J Integr Agric 14(11):2176–2188. https://doi.org/10.1016/S2095-3119(15)61114-5
69. Shamtsyan M (2017) Food legislation and its harmonization in Russia. August 2014. https://doi.org/10.1002/jsfa.6197
70. Pdr L (2013) Food safety control measures in developing asian countries
71. PMSEIC (2012) Australia and food security in a changing world. The Prime Minister's Science, Engineering and Innovation Council, Canberra
72. Lupton DA (2005) Lay discourses and beliefs related to food risks: an Australian perspective. Sociol Heal Illn 27(4):448–467. https://doi.org/10.1111/j.1467-9566.2005.00451.x
73. Buchler S, Smith K, Lawrence G (2010) Food risks, old and new: demographic characteristics and perceptions of food additives, regulation and contamination in Australia. J Sociol 46(4):353–374. https://doi.org/10.1177/1440783310384449
74. Henderson J, Ward P, Coveney J, Meyer S (2012) Trust in the Australian food supply: innocent until proven guilty. Heal Risk Soc 14(3):257–272. https://doi.org/10.1080/13698575.2012.662948
75. Henderson J, Coveney J, Ward P (2010) Who regulates food? Australians' perceptions of responsibility for food safety. Aust J Prim Health 16(4):344–351. https://doi.org/10.1071/PY10011
76. Ghosh D (2021) Food safety regulations in Australia and New Zealand Food Standards. August 2014. https://doi.org/10.1002/jsfa.6657
77. Taylor AW et al (2012) The Australian Food and Trust Survey: demographic indicators associated with food safety and quality concerns. Food Control 25(2):476–483. https://doi.org/10.1016/j.foodcont.2011.11.003

78. Leake LL (2015) Food safety in Africa: food safety and quality farm to fork, vol. 20. Available: http://www.foodqualityandsafety.com/article/out-of-africa/
79. Sefa-dedeh S (2009) Ghana: overview of food safety situation in Africa-country position. In: Go-global conference, Accra, April 2009. Available: http://www.goglobalnetwork.eu/NR/rdonlyres/A6F8EDAE-CE06-48E2-9C5B-21933D25B360/124727/OverviewofFoodSafetySituationinAfricaGhanaCountryP.pdf
80. Omojokun J (2013) Regulation and enforcement of legislation on food safety in Nigeria. In: Mycotoxin and food safety in developing countries. https://doi.org/10.5772/54423
81. Robert Kilonzo MG (2018) The Kenya Food Control System. Available: http://www.fao.org/fileadmin/user_upload/agns/news_events/Pre_CCAFRICA_KenyaEN.pdf

Chapter 22
Food Business Management and Bioentrepreneurship

Bharati S. Meti, Shreya Bhat, Shivaleela V. B., and Shilpa K. Jigajinni

1 Introduction

Lives are not complete without nutritional food. The food sector is fundamental and significant to every country. It is essential for social development, nutrition, food security, safety and public health. Major difficulties in the food sector include issues with product quality, health and hygiene/cleanliness. Food supply, production, harvesting, processing, packing, transportation, distribution, consumption and disposal are only a few of the varied operations covered by the food business management. Early in the twentieth century, the food business started to grow, to supply food for all population segments was raised in the twenty-first century due to the worldwide escalation of the food crisis, which has an impact on the processes of population reproduction. Because the availability of food is intimately related to the health of the people, ongoing study on important factors at the micro, meso and macro levels is logically necessary. Local solutions that include the geographical characteristics of economic growth are necessary for the global problem of food supply. Growing in significance is the requirement to establish the scientific underpinnings for the creation and administration of business processes in the food sector within the framework of economic transitions, taking into consideration worldwide conceptual approaches to food safety. According to the experience of the developed nations of the globe, this is only achievable if institutional support is provided for the

B. S. Meti (✉) · S. Bhat · S. K. Jigajinni
Department of Biotechnology, Basaveshwar Engineering College,
Bagalkote, Karnataka, India

Shivaleela V. B.
Pomla Natuals LLP, BEC STEP, Bagalkote, Karnataka, India

J. S. Yaradoddi et al. (eds.), *Frontiers in Food Biotechnology*,
https://doi.org/10.1007/978-981-97-3261-6_22

establishment of creative strategic structures in the food sector. Highlighting regional features for economic levers and incentives for the growth of the agri-food sector requires special consideration.

1.1 Importance of Biotechnology in Food Processing

The production, packaging and storage of food have all been revolutionized because of the myriad uses and developments made possible by biotechnology. Safety regulations are some important factors to be taken into account. To maintain the safety of the goods, particular hygiene practices must be followed throughout food processing. It is necessary to avoid cross-contamination and the admission of viable or non-viable particles into manufacturing areas.

Innovation and improvements: Genetically modified food, biopharmaceuticals, and other novel food sector goods have all been made possible through biotechnology. These developments have raised the nutritional value, nutritional quality and production effectiveness of food.

1.2 Bioentrepreneurship in the Biotechnology Sector

Identifying possibilities, coming up with creative solutions and commercializing biotechnological knowledge are all examples of bio-entrepreneurship, which are the entrepreneurial activity carried out within the biotechnology industry by people or teams.

Economic growth: The biotechnology sector, which includes food biotechnology, supports economic expansion by generating employment opportunities, luring capital and promoting innovation.

Sustainable agriculture: The development of sustainable agricultural methods, such as genetically modified crops with enhanced resilience to pests, diseases, and environmental conditions, is made possible by bio-entrepreneurship in the food industry.

Better nutrition and health: Biotechnological advancements in the food industry may result in the manufacture of functional foods with improved nutritional profiles that treat nutritional deficits and advance better health outcomes.

2 Food Business Management

The use of management concepts in the food sector is referred to as food business management. It entails planning, controlling and directing advancements in the food industry. A special combination of chemistry, economics, engineering, marketing and biology is used in food business management. The term 'food industry' refers to a large field that encompasses both for-profit and nonprofit restaurants, food production, processing, and distribution, as well as food safety and sanitation. The food business has experienced rapid expansion in recent years, demonstrating its resilience to economic upheaval. Food business management is a crucial component of the food industry and has a big influence on the global economy since it affects how businesses provide people with food.

2.1 Importance of Food Business Management

The food business is a crucial component of the overall food industry and it has a considerable impact on the world economy. Here are several justifications for the significance of food industry management:

Cost management: The day-to-day operations of restaurants and other enterprises in the food service sector must be managed. Keeping a successful business while delivering high-quality service, food and drinks is one of the largest challenges food service managers face. By training staff members about serving and preparation standards and keeping an eye on food and labour expenses, food service managers support firms in remaining profitable.

Customer contentment: In general, customer happiness brings repeat business and favourable word-of-mouth promotion to firms. To improve the customer experience, managing a food business, like managing any business, calls for a detailed and proactive strategy. Customers need to feel cared for the moment they walk into your restaurant or retail food store. Businesses may remain relevant and draw in new clients by staying up to date with seasonal changes and emerging trends in the food industry.

Compliance with rules and regulations: Several federal, state and municipal legal obligations apply to commercial activities. Managers of food services educate their workers about the regulations that apply to their particular position. They also keep an eye on workers to verify that all rules are followed to the letter. Other crucial facets of managing a food company include getting and renewing permits, keeping books and paying employees.

Business administration: Managing the day-to-day operations of the restaurant entails a wide range of duties for food service managers. Managers of food service organizations supervise workers to guarantee that diners have a positive eating experience. They manage the business administration and calculate food costs for a profitable Return On Investmant (ROI). Knowledge of resource management is needed to keep an operation profitable.

2.2 Principles of Food Business Management

The principles of food business management are essential to ensure the success and sustainability of any food-related enterprise. These principles encompass various aspects such as strategic planning, financial management, marketing, human

Fig. 22.1 Principles of a business management

resource management and customer service. By adhering to these principles, food businesses can effectively manage their operations, optimize resources, maintain profitability and meet the ever-evolving demands of consumers (Fig. 22.1).

2.3 Functions of Business Management

Management means the maintenance and creation of an internal environment of an enterprise, where individual working in groups can perform effectively and efficiently towards the attainment of a group goal. Management as a process, there are 7 functions for proper functioning of a business- planning, organizing, staffing, directing, controlling, reporting and budgeting.

Planning is to prepare plan, methods, policies and to take decisions.

Organizing is to divide the work among workers for creation of responsibility and authority.

Staffing is to employ the personnel in various positions in the industry and also to train them properly

Directing is to instruct the subordinates for proper functioning of an industry.

Controlling is to compare actual performance with predetermined standards and also to take corrective measures.

Reporting is to collect the report from all the section and to analyse it for proper functioning of the business.

Budgeting is also considered as one of the functions of business management. It is to prepare and distribution of budget. It helps for budget control.

2.4 Laws and Regulations of the Food Industry

Food safety and standards authority of India (FSSAI): In India, it is responsible authority for food business. It is to ensure food safety management and to enable food safety across distribution, supply chain and retail. It also has certified and trained food safety supervisors also food handlers in all premises [1].

Food and drug administration (FDA): The FDA Food Code is a set of regulations that suggest best practices for handling food safety in retail and dining facilities. The code addresses topics including worker health and hygiene, food preparation and storage and tools and utensils. The FDA Food Code is a set of rules that control how food is handled and handled safely in the United States. The Federal Food, Drug, and Cosmetic Act, as Amended, is the primary legislation that the US Food and Drug Administration (FDA) uses to regulate it. The Code, which incorporates requirements for many facets of food safety, is described in Title 21 of the Code of Federal Regulations (CFR) [2].

Over the past century, laws and regulations about the food business have changed. These laws seek to guarantee food safety, enable food tracing and get rid of financial fraud. To stop food product adulteration and unhygienic activities in factories particularly in packaging, federal control was established [3]. The Food Safety Modernization Act (FSMA), which focuses on food safety regulation, has made substantial changes to US food legislation [4]. For high-risk goods, the new standards mandate improved recordkeeping and traceability. However, difficulties remain in executing these laws across a variety of industries, and there are still no globally standardized criteria. To guarantee food safety and avoid deception, novel technologies like ultrasonography must adhere to legal and regulatory requirements. Overall, the goals of the laws and rules governing the food sector are to safeguard the public's health, supply healthful meals, and educate customers.

Food merchants are required by these rules to adhere to safe handling practices and maintain records of their activities.

Food safety modernization act: A federal law known as the FSMA was passed in 2011 to increase the security of the American food supply. A documented food safety plan, hazard evaluations and the implementation of preventative controls are all mandated by law for food facilities [4]. The Food Safety Modernization Act (FSMA) is a substantial reform to the laws and rules that the FDA regulates to ensure the safety of foods produced domestically and internationally. The FDA is given more authority to create rules for the food business under the FSMA, which puts a priority on avoiding foodborne diseases. The FSMA laws and regulations seek to protect US customers from unsafe food and reduce the incidence of foodborne diseases [5]. FSMA requires a paradigm shift in prevention and shared responsibility across the supply chain for all foods consumed in the United States [6]. It aims to move from a response-based system to a supply chain system with risk-based preventative strategies. For small food facilities, the FSMA's legal standards might be difficult to grasp, as can the expense of implementation and staff readiness.

Hazard Analysis and Critical Control Points (HACCP)

A food safety management method called HACCP is employed to find and eliminate possible risks during the food production process. The system entails recognizing possible risks, putting controls in place to stop or reduce the risks and checking on the efficiency of the measures. Throughout the whole food supply chain, food safety hazards are identified, assessed and controlled using a systematic process called Hazard Analysis and Critical Control Points (HACCP). It was created in the 1960s for the space programme and has now grown into a method for assuring food safety that is widely recognized. In HACCP, production processes are analysed, dangers are identified, important control points are found, control limits are set, procedures are verified, and monitoring systems are put into place. For meat products governed by the USDA and juices and seafood governed by the FDA, HACCP systems are required. Although HACCP is not required for the majority of foods, many processors have voluntarily adopted the procedures. The successful adoption of HACCP systems is supported by sanitation and other pre-requisite activities [7–10].

Good Manufacturing Practices (GMPs)

GMPs are a set of rules that specify the essential conditions for the manufacture of secure, superior food items. The rules encompass things like staff, facilities, tools and sanitation. To guarantee product safety and quality, good manufacturing procedures (GMPs) are crucial in the food industry. GMPs include a wide range of topics, including facility design, monitoring systems, documentation, staff commitment and understanding. GMP compliance is essential throughout the whole production process, including the choice of raw materials, processing, packaging, storage and distribution. For the production of safe and high-quality food items, GMP compliance is not optional but rather essential. It's crucial to include quality in food products from the start. The use of certain ideas and practices in business management, personnel training, facility provision, quality assurance, new product development, food regulation and product distribution is considered good practice in the food manufacturing industry. Using process analytical technology (PAT), such as near-infrared spectroscopy (NIRS), can help the food sector better understand and manage its manufacturing processes. The gap between the promise of NIRS and its actual application in real-time monitoring and modelling at an industrial scale has to be closed [11, 12].

Food Allergen Labelling and Consumer Protection Act (FALCPA)

According to the FALCPA, federal legislation, food producers must disclose whether any of the eight main food allergens are present in their goods. Additionally, according to the legislation, producers must identify the contents on the label in decreasing weight order. To warn customers who have food allergies and safeguard them from potentially fatal responses, the Food Allergen Labeling and Consumer Protection Act (FALCPA) was passed into law in 2004. Major food allergies must be identified on packaged goods, either on the packaging or in the ingredients list, according to the FALCPA. According to the legislation, significant dietary allergies include cow's milk, eggs, seafood, shellfish, tree nuts, peanuts, wheat, and soybeans. FALCPA, however, does not govern the use of conditional warnings such 'may conastain [allergen]', which customers may find confusing. Additionally, FALCPA does not mandate allergy labelling in restaurants and solely pertains to packaged items. Making allergens recognizable to consumers and assisting them in avoiding exposure to allergens that can cause a response are the objectives of FALCPA [13–15].

2.5 Inventory Management and Cost Control

For the food industry, inventory control and cost management are essential. To determine the best procedures, a variety of inventory methodologies have been researched. The optimal inventory techniques, including Economics Order Quantity (EOQ), Minimum Order Quantity (MOQ), First-in-First-out (FIFO/LIFO), and Safety Stock Inventory, were determined using the Analytical Hierarchy Process (AHP). To compete in an expanding market, micro, small, and medium enterprises (MSMEs) in the food and beverage sector need to increase cost-effectiveness and inventory control. In handling imported frozen food, continuous inventory systems have been proven to increase effectiveness and save cost. Dynamic inventory models are needed in the food business to handle supply and demand variations and the perishable nature of raw materials. Meeting client demands while lowering inventory costs is a difficulty for small and medium-sized food manufacturing businesses [16–20].

2.6 Food Business Management Strategies

The success of any food-related firm depends on effective management practices. These methods assist in handling the many difficulties associated with operating a food business, including demand forecasting, inventory control, order management and labour and time management, among many others. Optimizing food transportation, putting food safety management systems (FSMS) in place, adhering to food safety laws, and minimizing food waste are all part of food company management strategies. A hybrid architecture is suggested to optimize food delivery and save expenses for food-supplying businesses. Every food company participating in global supply chains must implement FSMS, and variables including managerial accountability, human resources, teamwork and food safety governance are critical to implementation success. Strategies including lowering financial stress, cost controls, customer behaviour and employee knowledge transfer are needed to comply with food safety regulations. There are several tactics that may be used to decrease food waste, such as reducing leftovers and serving waste, but there aren't many all-encompassing ones. These tactics seek to increase food safety, promote operational and financial efficiency, and support the development of food items more securely and sustainably.

There are several sorts of management strategies for the food industry, including:

Inventory management: This tactic entails keeping track of the food company's stock to make sure there is enough of it on hand to fulfil demand while reducing wastage [21].

Customer relationship management (CRM): This tactic entails controlling the interaction between a food company's clients and themselves to guarantee client satisfaction [22, 23].

Financial management: To make sure that the food industry is successful and long-lasting, this method entails controlling the industry's finances [24].

Marketing management: This tactic entails controlling the food industry's marketing to draw in and keep customers [25].

Materials requirement planning (MRP): This method involves forecasting demand for inventory and ordering the required materials in advance [21, 25].

Economic order quantity (EOQ): This method involves ordering inventory in large quantities to take advantage of bulk discounts [21, 26].

Days sales of inventory (DSI): This method involves calculating the number of days it takes to sell inventory and using this information to manage inventory levels [21, 27].

Depending on the unique requirements of the organization, effective inventory management necessitates a combination of these techniques. A company that deals in perishable commodities, for instance, may use JIT to reduce waste, whereas a company that sells seasonal goods might utilize MRP to make sure it has adequate inventory to fulfil demand during high seasons. The expense of keeping inventory, the risk of product obsolescence and the risk of stockouts are all difficulties with inventory management. Businesses can employ a variety of inventory management strategies, including demand forecasting, safety stock management and inventory optimization, to address these issues. Any firm must have effective inventory management aids in controlling inventory levels, cutting down on storage costs, and raising client satisfaction. Depending on the requirements of a firm, several types of inventory management techniques exist, each having advantages and disadvantages. Depending on the individual demands of the organization, effective inventory management necessitates a combination of these techniques. The difficulties posed by inventory management can be solved by businesses using techniques [21].

2.7 Marketing Management

The efficient administration of marketing resources and decision-making processes within the food sector is a key component of food business management. It includes a range of tasks including determining target audiences, coming up with marketing strategies, making marketing campaigns and gauging marketing effectiveness. Food businesses may make educated decisions regarding investments, marketing plans and company growth with the help of effective marketing management. Additionally, it aids in assuring compliance with rules and controlling the risks connected to marketing investments. It is impossible to emphasize the importance of marketing management in the food business. It facilitates the effective utilization of marketing resources, which is essential for the continuation and expansion of businesses. Food businesses may benefit from marketing management by identifying target audiences, developing marketing strategies, launching marketing campaigns and tracking marketing performance. Additionally, it is essential for maintaining client connections, increasing brand recognition and creating thorough marketing strategies. Target market identification is one of several tasks involved in marketing management in the food sector. This task entails determining the precise demographic groupings that are most likely to purchase a certain item or service. Another crucial task in marketing management is creating a marketing strategy. Making a strategy for how to approach and sell to the target market is part of this activity. Another crucial task in marketing management is developing marketing campaigns. This activity includes creating and carrying out marketing strategies intended to engage and reach the target market. Another crucial task in marketing management is performance measurement. In this activity, marketing data will be analysed to find trends, patterns, and other insights that might guide food firms in making decisions regarding investments, marketing plans and company growth. Customer relationship management is another crucial task in marketing management. Building and sustaining relationships with clients are necessary for this activity to guarantee their happiness and loyalty. In conclusion, marketing management is an essential component of managing a food business. It entails managing marketing resources and decision-making procedures effectively within the food sector. It includes a range of tasks including determining target audiences, coming up with marketing strategies, making marketing campaigns and gauging marketing effectiveness. For food businesses to succeed, effective marketing management is crucial because it facilitates the effective use of marketing resources, controls the risks associated with marketing investments and ensures regulatory compliance [25, 26].

3 Bioentrepreneurship

3.1 Introduction to Bioentrepreneurship

With the breakthroughs in biotechnology and entrepreneurial concepts, bioentrepreneurship has become a prominent sector in the twenty-first century. It includes the use of creative thinking and entrepreneurial abilities to design, produce and market goods and services for the biotechnology sector. As the biotechnology industry grows and develops, entrepreneurship is essential in bridging the gap between scientific discoveries and commercially viable solutions.

Recent years have seen significant advancements in the field of biotechnology, including developments in the fields of genomics, proteomics, synthetic biology and other fields. These developments have created new opportunities for the creation of novel goods, treatments and technologies that have the potential to solve pressing problems in agriculture, healthcare, and environmental sustainability. However, turning these scientific breakthroughs into useful applications in the real world necessitates a certain combination of abilities and knowledge, which is where bioentrepreneurship comes into play.

To commercialize these advances, entrepreneurship entails recognizing interesting scientific discoveries, evaluating their economic viability and carefully navigating the complicated biotechnology industrial environment. It necessitates a multidisciplinary strategy that combines corporate savvy, regulatory framework expertise and scientific knowledge. Bioentrepreneurs combine their commercial acumen with their scientific knowledge to translate scientific discoveries into marketable goods and services that may benefit society.

3.2 The Role of Bioentrepreneurship

The biotechnology sector has a wide range of applications. It acts as a catalyst for innovation, encouraging the creation of cutting-edge products and services with the potential to upend several industries. Bioentrepreneurs are essential in determining the focus of research and development initiatives by spotting unmet needs and commercial potential. They serve as go-betweens, establishing connections between academic researchers, financiers and industry players to promote cooperation and speed the commercialization of innovative ideas, in addition to being essential for economic development and employment generation. The biotechnology sector has a wide range of applications that stimulates the development of innovative goods and services that have the potential to completely alter several sectors. By identifying unmet needs and economic possibilities, bioentrepreneurs play a crucial role in deciding the direction of research and development projects. They act as middlemen, fostering collaboration and accelerating the commercialization of creative ideas by forming linkages between university researchers, investors and industrial participants. Bioentrepreneurship is vital for generating jobs and for advancing the economy. By establishing new businesses, opening up new job possibilities, and generating new income streams, the biotechnology sector has the potential to significantly increase economic value. Over 1.8 million people were employed by the biotechnology sector in the United States alone in 2021, according to a report by the Biotechnology Innovation Organization (BIO). With their capacity to spot and seize market possibilities, bioentrepreneurs support the expansion and long-term viability of the biotechnology ecosystem.

Additionally, bioentrepreneurship has the potential to tackle urgent global issues including raising agricultural output, reducing environmental impact and improving healthcare results. Bioentrepreneurs have the ability to change healthcare delivery by making it more individualized, accurate and accessible through commercializing cutting-edge biotechnologies and treatments. Additionally, they can create sustainable farming methods and responses to problems with food security. In order to combat climate change and advance sustainability, bioentrepreneurs are also at the forefront of creating environmentally friendly technology and solutions. The area of bioentrepreneurship has grown to be important and vibrant in the twenty-first century. Entrepreneurship and biotechnology developments are combined to promote innovation, economic growth and society influence. Bioentrepreneurs have a variety of roles in the biotechnology sector, from spotting promising scientific breakthroughs to negotiating the challenging commercialization environment. Bioentrepreneurs serve a crucial role in bridging the gap between scientific research and practical applications by utilizing their scientific expertise, financial survey and entrepreneurial zeal, eventually helping society as a whole [77] (Fig. 22.2).

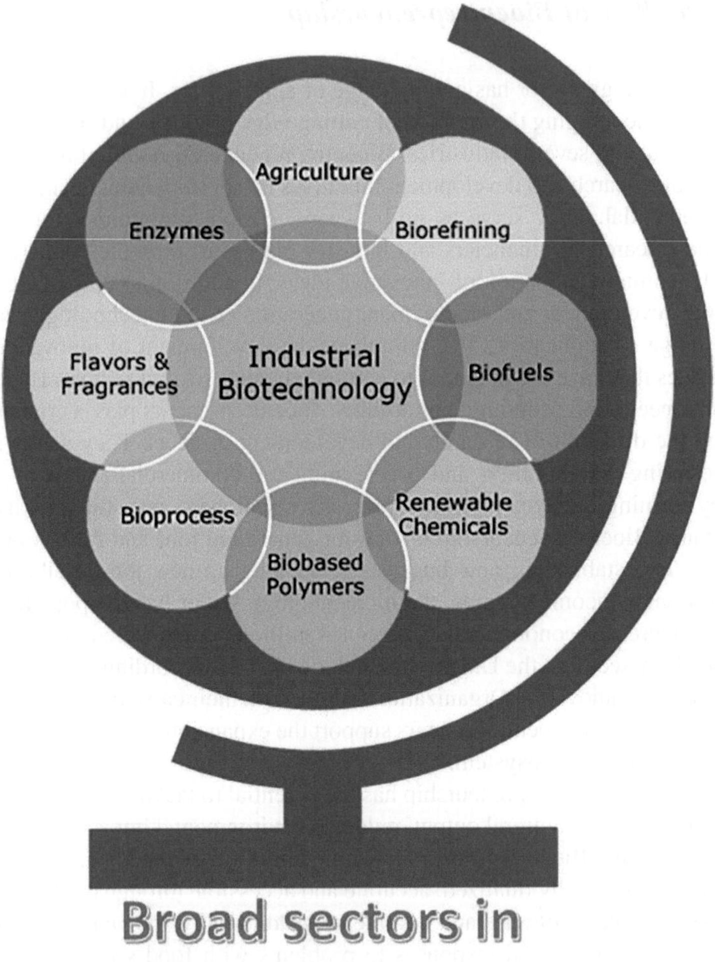

Fig. 22.2 Major areas of bioentrepreneurship

3.3 Biotechnology Dynamics in the Global State

Different nations and areas have contributed significantly to the expansion and development of the biotechnology sector, which has become a truly global business. The dynamics of biotechnology on a worldwide scale, including the participation of various nations and regions in the sector, is covered in this chapter.

India is one of the top 12 countries for biotechnology in the world and a pioneer in the manufacture of pharmaceuticals and vaccines. With several biotech businesses and research facilities, the nation places a major emphasis on biotechnology research and development. The Biotechnology Industry Partnership Programme (BIPP) and the Biotechnology Ignition Grant (BIG) Scheme are two of the policies and programmes the Indian government has put into place to aid in the expansion of the biotech sector.

With several biotech businesses and research facilities, biotechnology is a significant sector in the United States. Some of the top biotech firms in the world, including Amgen, Biogen and Gilead Sciences, are based in the nation. The US government has also adopted a number of programmes and laws to encourage the expansion of the biotechnology sector, including the Biotechnology Innovation Organization and the Small Business Innovation Research (SBIR) programme (BIO).

Europe is a significant player in the biotech business as well, with robust biotech sectors in nations like Germany, France and the United Kingdom. The European Union has put into effect a number of policies and initiatives, including the European Union Framework Programme for Research and Innovation and the European Union Biotechnology Strategy, to assist the expansion of the biotechnology sector.

Biotechnology entrepreneurship has been expanding quickly recently, particularly in South Africa and Brazil. The South African Biotechnology Strategy and the Brazilian Biotechnology Development Program are just two examples of the numerous policies and programmes that both nations have put in place to encourage the expansion of the biotech sector.

In general, a complex interaction of elements, including governmental policy, research and development, investment, and market demand, shapes the dynamics of biotechnology on a worldwide scale. New technologies and improvements are continually being developed in the biotech sector. As a result, it's critical for business owners and investors to keep abreast of the most recent advancements and trends in the sector.

3.4 *Scope of Biotechnology Venture in India*

One of the top nations in the world for biotechnology research and development is India. The government of India has enacted a variety of laws and measures to assist the development of the biotech industry, and the nation is home to several biotech businesses and research facilities. For instance, two programmes have been introduced to stimulate biotech entrepreneurship in India: the Biotechnology Industry Partnership Programme (BIPP) and the Biotechnology Ignition Grant (BIG) Scheme [27].

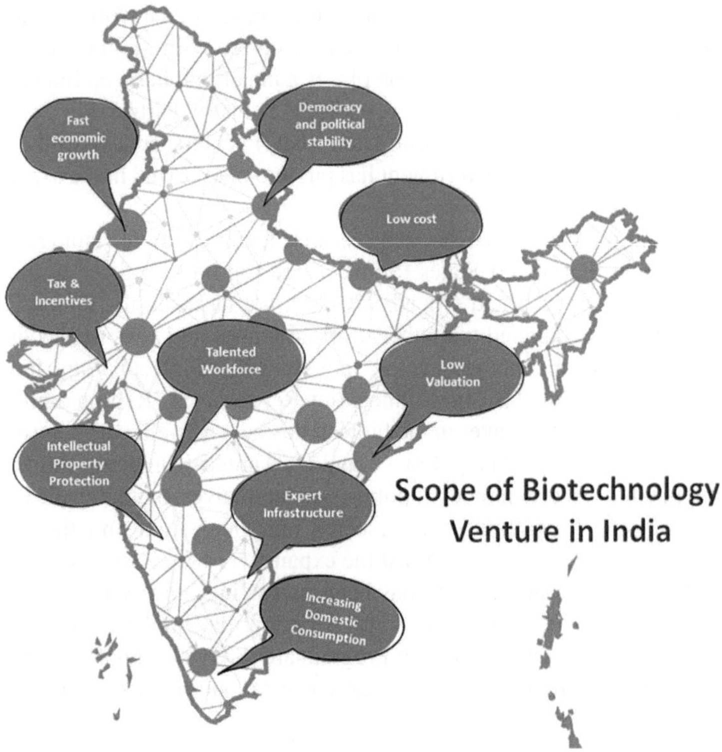

Fig. 22.3 Scope of entrepreneurship in India

With a concentration on the creation of pharmaceuticals and vaccines, the Indian biotech industry has experienced remarkable growth in recent years.

The introduction of Bt cotton is a remarkable success storey in the development of biotech crops, which has also advanced significantly in the nation. India is now one of the largest producers of Bt cotton in the world, and the technology has helped to increase yields and reduce pesticide use.

India has developed into a significant hub for clinical trials in addition to its burgeoning biotech sector [27]. Due to the enormous patient population and affordable prices in India, several international pharmaceutical firms are increasingly conducting clinical studies there. Government regulations, R&D, investment and market demand are only a few of the complex interplaying elements that influence the dynamics of biotechnology in India. With a strong emphasis on research and development and a developing biotech entrepreneurial environment, India is a major participant in the biotech sector. Entrepreneurs and investors will need to remain current on the most recent innovations and trends in the sector as it continues to develop immensely (Fig. 22.3).

3.5 Planning a Biotech Startup [28]

Planning a small-scale biotech startup in India requires careful consideration of various factors, including financing, market analysis, product development and scaling strategies. Here is an overview of the key steps involved in planning a small-scale biotech startup in India:

Market Analysis and Identification of Opportunities

1. Conduct a thorough market analysis to identify potential areas of opportunity in the biotech sector in India
2. Identify specific market needs, gaps, and challenges that your startup can address
3. Consider emerging trends, technological advancements, and regulatory landscape in the biotech industry

Business Plan Development

1. Develop a comprehensive business plan that outlines your startup's mission, vision, goals, and strategies
2. Include a detailed analysis of the target market, competitive landscape, and potential customers
3. Define your startup's unique value proposition and competitive advantage
4. Outline your marketing and sales strategies, pricing models, and distribution channels

Financing and Funding

1. Explore various sources of financing, including venture capital, angel investors, government grants and loans
2. Prepare a solid financial plan, including revenue projections, expenses, and return on investment (ROI) analysis
3. Seek support from biotech-focused incubators, accelerators, and startup programmes in India
4. Familiarize with the financing options and know-how for biotech startups in India

Product Development and Intellectual Property [29]

1. Develop a roadmap for product development, including research, prototyping, testing and regulatory compliance
2. Protect your intellectual property by filing for patents or other forms of intellectual property rights
3. Collaborations with research institutions, universities, and industry experts to enhance your product development capabilities

Regulatory Compliance

1. Familiarize yourself with the regulatory requirements and approvals necessary for biotech startups in India
2. Ensure compliance with ethical guidelines, biosafety regulations, and intellectual property laws
3. Seek guidance from regulatory experts or consultants to navigate the regulatory landscape

Scaling Strategies

1. Develop a scaling strategy that aligns with your long-term goals and market demand
2. Explore partnerships, collaborations and licensing opportunities to expand your reach
3. Continuously innovate and adapt to changing market dynamics to sustain growth
4. Learn from successful biotech companies that have successfully scaled their business and analyse their portfolio and pipeline composition

Network and Collaborate

1. Build a strong network of industry professionals, mentors, advisors and potential partners
2. Attend industry events, conferences and exhibitions to connect with key stakeholders in the biotech ecosystem
3. Collaborate with research institutions, universities and other startups to leverage expertise and resources

Risk Assessment and Mitigation

1. Identify potential risks and challenges associated with your startup and develop mitigation strategies

2. Perform a risk assessment early in the design process to address critical concerns and find creative solutions
3. Seek guidance from experienced industry experts who have successfully scaled similar processes

In conclusion, planning a small-scale biotech startup in India requires a comprehensive approach that includes market analysis, business plan development, financing, product development, regulatory compliance, scaling strategies, networking and risk assessment. By carefully considering these factors and leveraging the resources and support available in the biotech ecosystem, entrepreneurs can increase their chances of success in the dynamic and evolving biotech industry in India.

3.6 Market Data Generation Tools and Methods

Entrepreneurs may use a wide range of tools and approaches in their quest to obtain thorough market research data, each of which has its own merits. For instance, integrated market research systems unify the whole procedure, from survey preparation through data collection and analysis, streamlining the work and improving the effectiveness of decision-making. Contrarily, competitive analysis dives into comprehending the landscape of rival firms, shedding light on elements like direct and indirect rivals, the level of competition, possible entrants and the impact of suppliers and consumers on pricing strategies. This examination helps to forge sustainable income streams and carve out a distinct competitive advantage. Moreover, as Gerald Zaltman's important book 'How Customers Think' emphasizes, the importance of understanding the language of the market cannot be emphasized. To find the specific terms and phrases that connect with customers looking for products or services similar to their own, business owners may use keyword research tools like Ahrefs or Google Keyword Planner. This priceless information acts as a compass for content marketing initiatives, enabling deeper interactions with the target market. Another crucial factor for getting trustworthy market research data is selecting the right sample and study methodology. There are several options, from systematic or non-random sampling approaches like convenience, purposive, or quota sampling to complete data source use. The selection should align harmoniously with the research objectives and the population under scrutiny. Additionally, research platforms like Qualtrics CoreXM offer entrepreneurs the means to tailor their research methodologies, ensuring precision through pre-launch and soft launch survey phases, and enabling data assessment before embarking on analysis.

In certain circumstances, entrepreneurs may opt to enlist the expertise of research services from specialized consulting firms or agencies dedicated to market research. These services offer a wealth of knowledge in crafting research strategies, conducting surveys, dissecting data and extracting illuminating insights. They serve as a valuable resource for entrepreneurs who may find themselves lacking the requisite resources or expertise to undertake exhaustive market research independently.

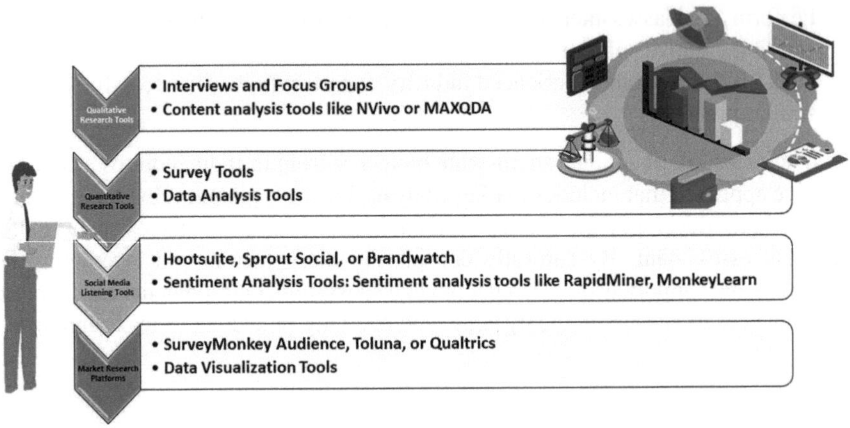

Fig. 22.4 Tools for entrepreneurs to gather market research data

Entrepreneurs may efficiently get the necessary market research data by utilizing this arsenal of strategies and tools. However, it is crucial to adapt these techniques to the specific requirements and goals of the company enterprise, assuring a focused and deliberate search for understanding of the competitive environment (Fig. 22.4).

3.7 Types of Manpower Requirements

A biotech business has a pressing need for a number of jobs and positions that are essential to its success. The management team, which is made up of people with the knowledge necessary to seek finance, make strategic choices and promote the scientific activities of the company, plays one of these crucial responsibilities. This team, which investors frequently examine, should have a history of successful fundraising and a thorough understanding of the biotech sector. The Chief Executive Officer (CEO), who is in charge of the company's overall leadership and management and requires a solid experience in biotechnology and entrepreneurship, is one of the key members. The Chief Scientific Officer (CSO) is in charge of all scientific and research activities; hence, they need to be knowledgeable in the startup's particular biotech niche. The Chief Financial Officer (CFO) is responsible for overseeing all financial matters, which calls for knowledge of the biotech sector and previous financial management expertise. The Chief Commercial Officer (CBO) manages business growth, licencing agreements and strategic alliances, which calls for excellent networking and negotiating abilities. With experience in clinical research and regulatory procedures, the Chief Medical Officer (CMO) is responsible for clinical development and regulatory affairs. A Chief Commercial Officer (CCO) is essential for commercialization strategies, market access and sales as the firm develops since they involve marketing strategy, stakeholder connections and revenue development [30] (Fig. 22.5).

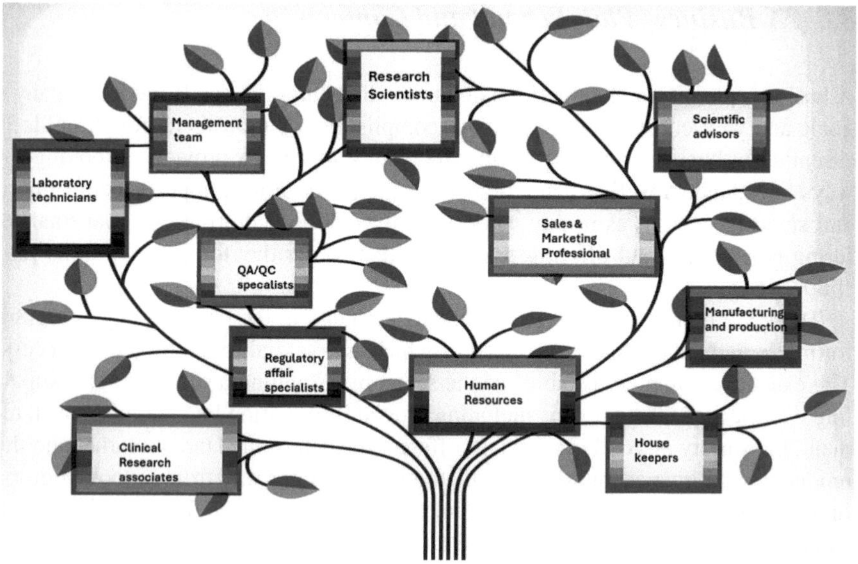

Fig. 22.5 Key components of a biotechnology startup—skilled workforce

Other human resources are necessary to support a biotech startup's ongoing operations, research, development and commercialization initiatives in addition to these crucial executive responsibilities. These positions include research scientists, who are the foundation of the firm and carry out studies and advance scientific understanding. Laboratory technicians oversee experiments, equipment and data gathering for the benefit of scientists. Quality Assurance/Quality Control (QA/QC) specialists make sure that regulatory requirements are followed as well as quality control procedures. Specialists in regulatory affairs prepare documentation and secure clearances while navigating the difficult regulatory environment. Clinical Research Associates (CRAs) manage clinical studies while assuring regulatory compliance and protocol adherence. Scaling up product production, improving procedures and managing the supply chain are all aspects of manufacturing and production. By promoting items, finding clients or partners, and boosting sales, sales and marketing specialists aid in commercialization. HR professionals oversee employee interactions, performance, onboarding, training and recruiting.

It's critical to understand that a startup's unique human resource requirements may change depending on its type, stage of growth and area of concentration. Every company should carefully analyse its particular needs to identify the key personnel needed for success. A biotech company's capacity to innovate, maintain compliance and provide biotechnological solutions to solve critical difficulties depends on these combined human resources, which also determine the success and impact of the firm in the biotech sector [31]

3.8 A Business Plan Format and Contents

A business plan is a crucial written document that carefully spells out a company's goals and well-constructed plans for accomplishing them. This thorough guide is essential for business owners, whether they are starting new projects or looking for ways to expand. A well-structured business plan is of utmost importance for micro and small businesses as it may work as a keystone in acquiring essential finance, luring new clients and offering priceless direction for day-to-day operations [29] (Fig. 22.6)

There are numerous important elements included in the business plan structure for micro and small businesses, each of which is essential to the project's success. The executive summary establishes the scene by providing a succinct yet compelling overview of the company, including crucial information like the mission statement, the variety of goods or services offered, the determined target market and the unique competitive advantages that distinguish the company from its competitors. In addition to acting as an introduction, the first part engages the reader's interest and piques their curiosity about the details of the business plan.

Beyond the executive summary, the company description offers a more thorough explanation of the firm, covering important details like the legal structure, the company's location, its history and an in-depth look at ownership. It also provides

Fig. 22.6 Key components of a business plan

information about the management team's qualifications and experience, which is crucial for stakeholders assessing the startup's chances of success.

The part on market analysis dives into the larger market and industry in which the company competes. It does a thorough investigation, analysing important elements including market size, growth prospects, prevailing trends and the competitive environment. Additionally, it describes the target market while outlining the competitive advantages that will help the company succeed in the current market environment.

The focus is on marketing and sales planning, which provides a strategy for successfully marketing and selling the goods and services. This part carefully specifies the price plan that will promote profitability, maps out the chosen marketing channels, outlines the sales strategy and provides a detailed overview of the target market. The operations sector acts as the operational roadmap, providing details on how the company will run on a daily basis. It explores the manufacturing procedures, explains supply chain management tactics and makes the workforce needs clear. Detailed financial estimates are highlighted in the financial plan section. There is a careful breakdown of the income statements, balance sheets and cash flow statements. A well-thought-out finance strategy is also offered, which is essential for maintaining sustainability and expansion. A break-even analysis is also included. By including supplemental data that gives the proposal more support, the appendix part completes the business plan's main sections. It frequently contains crucial records like the resumes of important employees, information from market research and important legal records.

It is important to stress that, while this thorough format provides a solid framework for creating a business plan tailored to micro and small enterprises, the specific requirements and nuances may change depending on the particular needs and circumstances of the company and the industry it operates in. As a result, business owners are advised to carefully evaluate their unique requirements to guarantee the effectiveness of their business strategies.

There are other requirements that increase the thoroughness of the business proposal in addition to the format itself. These might comprise firm incorporation, financial statements, bank information, performance statistics, UAM and MSME registrations and other documents according to the unique requirements of the organization. Additionally, using best practises while creating business proposals helps to increase the likelihood of success. Gaining a thorough understanding of the client's needs, remaining succinct, emphasizing the advantages of the suggested goods or services, using visuals to increase engagement, offering a clear and compelling call to action and meticulous proofreading and editing to ensure a polished and error-free proposal are some of these practises. Entrepreneurs may greatly increase their chances of not just writing a successful business proposal but also succeeding in their undertakings by implementing these best practises. A well-structured business plan is, in essence, more than just a paper; it is a dynamic instrument that may guide a micro or small firm toward its goals, guaranteeing strategic clarity and aiding informed decision-making at every crossroads (Fig. 22.7).

Fig. 22.7 Generalized contents of a business plan format of a biotechnology venture

3.9 General Layout of a Biotechnology Plant

A biotech company's layout is a key component that has a significant influence on both its operation and overall performance. It includes a number of essential elements intended to maximize space usage, guarantee regulatory compliance, improve safety and allow effective workflows. This design places a strong emphasis on laboratory rooms that are placed in key locations, outfitted with the most advanced equipment for study and testing, and supported by rigorous safety measures and designated research zones. Office spaces near laboratories let staff members, especially CEOs and regulatory professionals, communicate effectively. For sterile production and delicate operations, clean rooms are essential, and manufacturing facilities must be designed with material flow, GMP compliance and quality control checkpoints in mind. Raw materials and completed goods need enough storage space, and quality assurance and control departments should be strategically located for ongoing supervision. The thorough plan is completed by utilities, support services, staff amenities and environmental concerns. The following are important considerations: environmental awareness, safety, regulatory compliance, effectiveness, flexibility and process optimization. In essence, a biotech company's ability to conduct research, manufacture goods, and operate in a safe, effective and compliant way depends on a painstakingly designed structure (Fig. 22.8).

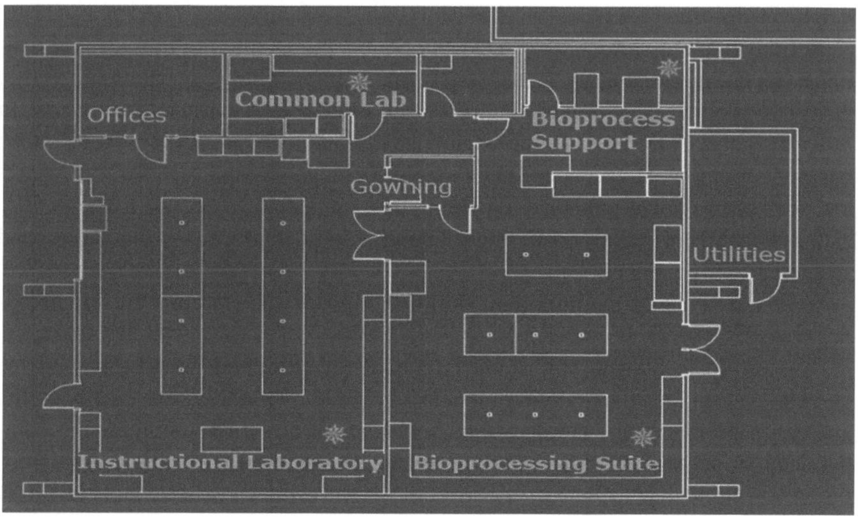

Fig. 22.8 Generalized biotechnology plant layout

3.10 Legal Requirements to Start a Venture in Biotechnology

An important pillar of a biotech startup's profitability and morality is legal compliance. These firms work in a highly regulated sector where following rules and legislation is crucial. It is impossible to exaggerate the significance of legal compliance since it protects not just the company but also patients, customers and the environment. The establishment of business entities, intellectual property protection, regulatory compliance, research ethics, the import and export of biological materials, occupational health and safety standards, employee regulations, taxation, data privacy, licences and permits, clinical trials, contracts and agreements, export controls, and financial reporting are important areas of legal compliance for biotech startups. Each of these elements is controlled by certain laws and rules that change depending on the location and the operations of the biotech firm.

Regulatory organizations like the Centers for Disease Control and Prevention (CDC), the European Medicines Agency (EMA), the U.S. Food and Drug Administration (FDA) and others are crucial in ensuring compliance in the biotech sector. To make sure that biotech businesses adhere to the appropriate norms and criteria, they establish rules, examine applications and keep an eye on activities.

For the biotech industry to achieve and maintain legal compliance, it is crucial to follow certain formats and take certain factors into account. These formats and considerations include keeping meticulous records, submitting required reports, putting audit trails in place, creating quality management systems and offering ongoing employee training [30].

Serious repercussions may result from noncompliance with these legal obligations, including regulatory penalties, legal responsibilities, reputational harm, delays in product development and even the cessation of corporate activities. In order to successfully traverse the complicated regulatory landscape of the biotech business, biotech entrepreneurs must emphasize legal compliance from the onset and seek professional guidance. By doing this, companies can ensure the security and welfare of all stakeholders while laying a solid platform for innovation and sustained growth (Fig. 22.9).

Soft skill is one of the important skill that supports the entrepreneurship activities (Fig. 22.10).

To sum up, acquiring soft skills is crucial for entrepreneurs to thrive in the quickly evolving business environment. To increase their entrepreneurial potential and company outcomes, entrepreneurs should concentrate on strengthening communication, negotiating, leadership, problem-solving and time management abilities along with strong technical, scientific and engineering knowledge and individual passion.

Legal Compliance Areas for Biotech Startups	Business Entity Formation, Intellectual Property (IP) Protection, Regulatory Compliance, Research Ethics, Biological Material Import/Export, Environmental Regulations, Occupational Health and Safety, Employee Regulations, Taxation, Data Privacy, Licensing and Permits, Clinical Trials, Contracts and Agreements, Export Controls, Financial Reporting
Regulatory Bodies and Agencies:	**FDA (U.S. Food and Drug Administration):** Regulates pharmaceuticals, medical devices, and biologics in the United States. **EMA (European Medicines Agency):** Regulates medicines in the European Union. **CDC (Centers for Disease Control and Prevention):** Sets guidelines for biosafety and bioterrorism prevention. **NIH (National Institutes of Health):** Funds and regulates biomedical research in the United States. **IRBs and Ethics Committees:** Review and approve research involving human subjects.
Essential Formats and Considerations	**Documentation:** Maintain meticulous records of research, experiments, and compliance efforts. **Reporting:** Prepare and submit required reports, such as financial statements and clinical trial results. **Audit Trails:** Implement audit trails to track changes and maintain data integrity. **Quality Management Systems:** Establish robust quality management systems to ensure product quality and safety. **Training:** Train employees on compliance requirements and ensure ongoing education.

Fig. 22.9 List of legal requirements, regulatory bodies and essential formats of a biotechnology startup

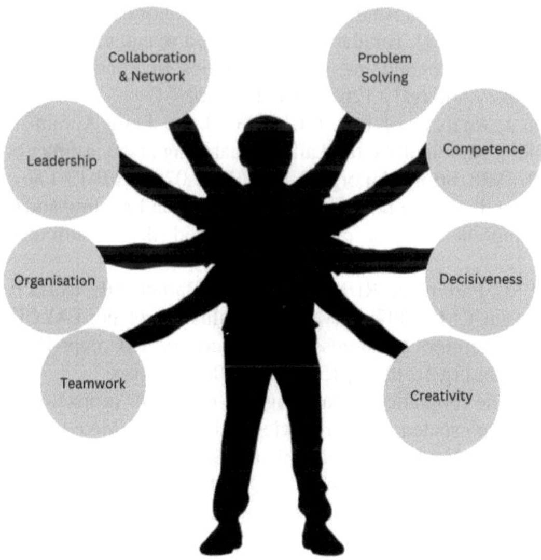

Fig. 22.10 Various skills required are shown in the image

References

1. https://www.fssai.gov.in/cms/food-safety-and-standards-regulations.php
2. https://www.fda.gov/regulatory-information/laws-enforced-fda
3. Mercier SA, Halbrook SA (2020) "Food Safety Laws," Palgrave Studies in Agricultural Economics and Food Policy. In: Agricultural Policy of the United States. Palgrave Macmillan, pp 127–142
4. Rasco B (2017) Chapter 20—Laws and regulations for novel food processing technologies. In: Bermudez-Aguirre D (ed) Ultrasound: advances for food processing and preservation. Academic Press, pp 499–524
5. DeWaal CS, Plunkett DW (2013) The Food safety modernization act—a series on what is essential for a food professional to know. Food Protect Trends:44–48
6. Crawford WM (2019) Chapter 5—The Food Safety Modernization Act (FSMA)'s role in the safety of functional foods and their ingredients. In: Bagchi D (ed) Nutraceutical and functional food regulations in the United States and around the world, 3rd edn. Academic Press, pp 61–74
7. Nummer B, Gump D, Wells S, Zimmerman S, Montalbano A (2015) Hazard analysis and critical control points (HACCP). In: Regulatory foundations for the food protection professional, pp 163–178. https://doi.org/10.1007/978-1-4939-0650-5_11
8. Hulebak KL, Schlosser W (2002) Hazard analysis and critical control point (HACCP) history and conceptual overview. Risk Anal 22(3):547–552. https://doi.org/10.1111/0272-4332.00038
9. Frestedt JL (2017) Hazard analysis and critical control points, pp 51–89. https://doi.org/10.1016/B978-0-12-805470-3.00003-X
10. Gehring KB, Kirkpatrick R (2020) Hazard analysis and critical control points (HACCP) history and conceptual overview. Risk Anal:191–204. https://doi.org/10.1007/978-3-030-42660-6_8
11. Meghwal M, Heddurshetti U, Biradar R (2017) Good manufacturing practices for food processing industries: principles and practical applications. In: Food technology. Apple Academic Press, pp 3–28. https://doi.org/10.1201/9781315365657-1
12. Gendel SM, Zhu J, Analysis of U.S. (2013) Food and Drug Administration food allergen recalls after implementation of the food allergen labeling and consumer protection act. J Food Prot 76(11):1933–1938. https://doi.org/10.4315/0362-028X.JFP-13-171
13. Roses JB (2011) Food allergen law and the Food Allergen Labeling and Consumer Protection Act of 2004: falling short of true protection for food allergy sufferers. Food Drug Law J 66(2):225–242
14. Taylor SL, Hefle SL, Farnum K, Rizk SW, Yeung J, Barnett ME, Busta F, Davis S, Newsome R, Shank FR, Bryant CM (2007) survey and evaluation of pre-FALCPA labeling practices used by food manufacturers to address allergen concerns. Compr Rev Food Sci Food Saf 6(2):36–46. https://doi.org/10.1111/j.1541-4337.2007.00016.x
15. Fortin ND (2006) The food allergen labeling and consumer protection act: the requirements enacted, challenges presented, and strategies fathomed. J Med Law 10:126. MSU Legal Studies Research Paper No. 04-13 , Available at SSRN: https://ssrn.com/abstract=948727
16. Mogbojur AO, Olanrewaju OA, Ogunleye TO (2022) Evaluation of inventory management practice in food processing industries in Lagos: analytical hierarchy process approach. Nigerian J Technol 41(2):236–246. https://doi.org/10.4314/njt.v41i2.5
17. Maulana FQ, Widjaja AW (2020) Improvement of inventory control and cost efficiency Warteg ABC. In: E3S web of conferences, vol 202, p 16004. https://doi.org/10.1051/E3SCONF/202020216004
18. Jittham S, Chaovalitwongse P (2017) Inventory management for import frozen food of food services industry. WMS J Manage 6(1):54–72. Retrieved from https://so06.tci-thaijo.org/index.php/wms/article/view/75799
19. Shin M, Lee H, Ryu K, Cho Y, Son YJ (2019) A two-phased perishable inventory model for production planning in a food industry. Comput Ind Eng 133:175–185. https://doi.org/10.1016/J.CIE.2019.05.010

20. Liang C-C, Cheng K-C, Chen C-H, Wu P-H, Cao R-Z (2012) Inventory management of food processing industry—a case study. In: International conference on engineering and business management (Part-2 Enterprise Operation and Management), pp 1146–1149
21. Golas Z, Bieniasz A (2016) Empirical analysis of the influence of inventory management on financial performance in the food industry in Poland. Eng Econ 27(3):264–275. https://doi.org/10.5755/j01.ee.27.3.5933
22. Duong LN, Wood LC, Wang WY (2018) Effects of consumer demand, product lifetime, and substitution ratio on perishable inventory management. Sustainability 10(5):1559. https://doi.org/10.3390/su10051559
23. Buttle F, Maklan S (2019) Customer relationship management concepts and technologies, 4th edn. Routledge, p 468
24. Dorfman JH (2014) Economics and management of the food industry, textbook in environmental and agricultural economics. Routledge
25. Wood B, Williams O, Nagarajan V et al (2021) Market strategies used by processed food manufacturers to increase and consolidate their power: a systematic review and document analysis. Global Health 17:1–23. https://doi.org/10.1186/s12992-021-00667-7
26. Abuselidze G, Talavyria M, Vyshnivska B, Bondarenko L, Makedon H, Kniazieva T, Salkova I (2023) The economic mechanism of marketing activity management of food enterprises. In: E3S web of conferences, vol 371. EDP Sciences, p 05002. https://doi.org/10.1051/e3sconf/202337105002
27. Sinha D et al (2021) Introduction to bioentrepreneurship. In: Agarwal S et al (eds) Bioentrepreneurship and transferring technology into product development. IGI Global, pp 1–21. https://doi.org/10.4018/978-1-7998-7411-9.ch001
28. Kolchinsky P (2018) The entrepreneur's guide to a biotech startup, 4th ed., pp. 1–96. https://www.ctsi.ucla.edu/researcher-resources/files/view/docs/EGBS4_Kolchinsky.pdf
29. Hübel A, Schmelcher T, Storz U (2012) Biopatent law: patent strategies and patent management. Springer Science & Business Media
30. Zirra D, Căruţaşu G, Garais GE, Mărgărit GL, Perju-Mitran A (2019) Start-up and management features in biotech business. In: Introduction to biotech entrepreneurship: from idea to business: a European perspective. Springer International Publishing, pp 195–213. https://doi.org/10.1007/978-3-030-22141-6_9
31. Bigliardi B, Galati F (2013) Innovation trends in the food industry: the case of functional foods. Trends Food Sci Technol 31(2):118–129

Chapter 23
Poultry and Meat Processing

Ishrat Jahan Peerzade, Tanaji Kudre, and Prakash M. Halami

1 Introduction

The consumption of meat is a ubiquitous phenomenon across the globe. This trend has been brought about by the increase in population and a consequent rise in income levels. Consequently, the production of various sources of protein, such as poultry and red meat, has been steadily growing over the course of the past half-century. One of the reasons also being its competitive prices with veggies and cereals. Human use of food products derived from animals has a detrimental effect on biodiversity and terrestrial ecosystems [1]. The consumption of meat is proportional to its production to suffice the demand; hence, the processing of meat comes into the picture. Advancements in genetics, nutrition, medicine, and engineering at the farm and processing plant levels have enabled the poultry industry to develop specialized meat breeds that grow to 2.5 kg within ~40 days and to introduce many cut-up fresh poultry and further processed products [2]. There are several challenges faced by the industry, such as the rise in myopathies and diseases spread among the animals, which require solutions from breeders, farmers, and processing plants [3]. The changes in the processing industry, including the building of large, centralized automated processing plants capable of processing 15,000 birds per hour on a single line, require higher bird uniformity [2]. Due to myopathy and spaghetti meat, poultry processing hinders deboning and further portioning [4]. There are certain interlinked parameters for the quality of meat processing, such as stability, eating quality,

I. J. Peerzade · P. M. Halami (✉)
Department of Microbiology & Fermentation Technology, CSIR-Central Food Technological Research Institute, Mysore, Karnataka, India
e-mail: prakashalami@cftri.res.in

T. Kudre
Department of Meat and Marine Science, CSIR-Central Food Technological Research Institute, Mysore, Karnataka, India

J. S. Yaradoddi et al. (eds.), *Frontiers in Food Biotechnology*,
https://doi.org/10.1007/978-981-97-3261-6_23

convenience, nutritive value, and ethical aspects. Poultry meat has taken up the meat industry to the poultry level, where over 80% of poultry is sold as cut-up parts or further processed products, especially in Western markets [2]. This shift in marketing is due to consumers having more income to spend on food and looking for convenient semi/fully prepared food products. The poultry industry has built large, dedicated processing plants to meet these demands, where line speed is continuously increasing. Demographic changes such as both parents working outside the home have also been a strong driving force in increasing the trend of buying prepared foods. Poultry and meat processing is an integral part of the food industry, ensuring that the meat products we consume are safe, nutritious, and of high quality. In conclusion, poultry and meat processing play a vital role in meeting the global demand for safe and high-quality food products.

2 Worldwide Prospective of Different Types of Meat

In comparison to fifty years ago, it can be observed that the world currently produces more than three times the amount of beef, with approximately 340 million tons being produced in 2018 [5]. While poultry output is experiencing the most significant growth, pigmeat remains the most widely consumed meat globally, leading to the annual slaughter of 80 billion animals for meat production. In 2014, the average worldwide meat consumption was 43 kg, with variations observed across different regions [6, 7]. Consumption ranged from over 100 kg in the United States and Australia to as low as 5 kg in India [1]. These differences can be attributed to varying levels of affluence worldwide, as meat consumption tends to increase with rising prosperity. Additionally, it is worth mentioning that global milk production has more than doubled in the past 50 years, currently reaching an annual volume of 800 million tons. Milk consumption per person is generally higher in wealthy nations [8]. The quantity of meat produced for a certain animal varies greatly around the world due to different production methods. Livestock farming has environmental effects, including increased greenhouse gas emissions and water and land use. The environmental impact of beef and lamb is much greater than that of pork and poultry [9].

The typical person's global meat consumption has risen by approximately 20 kg since 1961, according to observations. This average increased to almost 43 kg in 2014 [10]. The upwards direction of per capita meat consumption trends highlights the fact that the total amount of meat produced has been growing faster than the population increase.

Countries have experienced significant variations in the direction and pace of change. The most noticeable increase in meat consumption per capita has occurred in nations that have undergone significant economic transformations. For instance, in China, meat consumption has increased by approximately 15 times since 1961, while in Brazil, it has tripled (Fig. 23.1). However, India has been the main exception to this trend. The prevailing lactovegetarian preferences in the country have

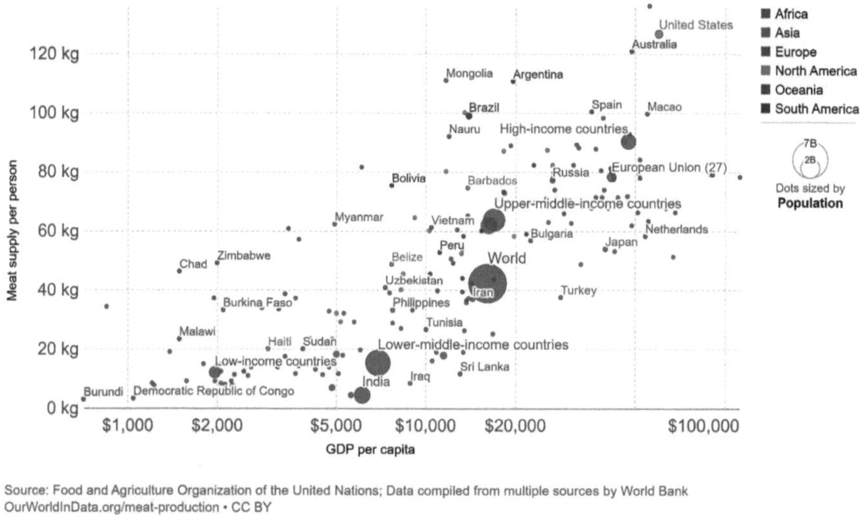

Fig. 23.1 Meat consumption vs. GDP per capita 2020—average meat consumption per capita measured in kilograms per year versus gross domestic product (GDP) per capita measured in constant international dollar. (Figure do not include fish or seafood)

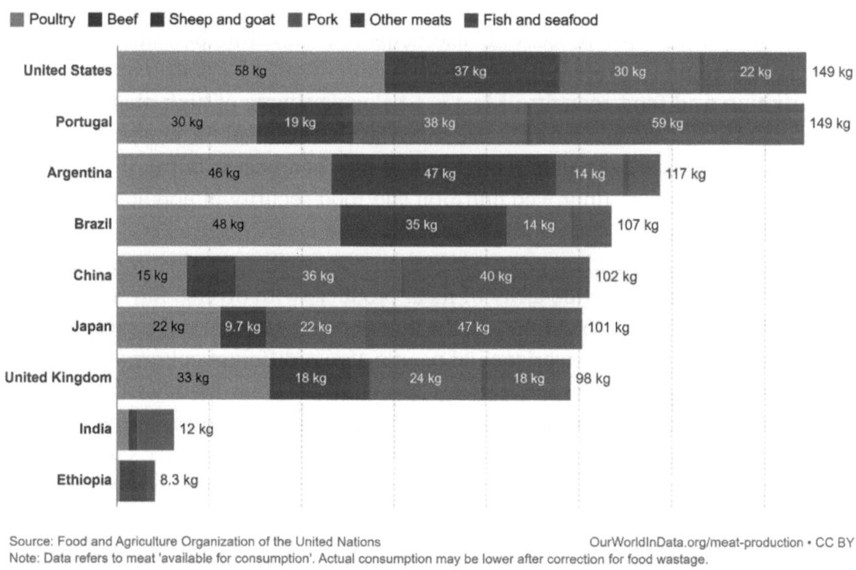

Fig. 23.2 Per capita meat consumption by type of meat; 2020

resulted in the average person's meat consumption in 2013 being less than 4 kg, nearly identical to the levels seen in 1961 [8]. The per capita consumption of meat is depicted in Fig. 23.2. The graph clearly illustrates that countries with high meat consumption, predominantly contributed by poultry, beef, and pork, contrast with countries like India where per capita consumption is significantly lower and fish and seafood are the primary sources of meat consumption.

3 Significance of Meat Processing

Meat processing holds immense significance in modern societies and has played a crucial role in human civilization throughout history. It involves various techniques and methods aimed at transforming raw meat into safe, nutritious, and flavorful food products. Some study has investigated the moderate association between meat consumption and lung carcinoma and other illness due to intake of high-temperature cooking [11]. The meat processing is explored with different techniques such as coating the meat with oil, concentrating oil in the meat, preventing loss of meat juice, sterilizing bacteria, and increasing distribution process and period [12]. The significance of meat processing can be understood from several perspectives:

- **Food safety and preservation:** One of the primary reasons for meat processing is to ensure food safety and prevent spoilage. By applying techniques such as curing, smoking, refrigeration, and canning, meat processors can extend the shelf life of meat products, reducing the risk of bacterial growth and foodborne illnesses [13]. Proper processing also helps in eliminating parasites and pathogens that might be present in raw meat, making it safer for consumption.
- **Nutrition and food security:** Meat is a valuable source of high-quality protein, essential amino acids, vitamins, and minerals that are vital for human health [14, 15]. Meat processing enhances the availability and accessibility of meat products, contributing to improved nutrition and food security in populations around the world.
- **Culinary diversity:** Meat processing techniques such as marinating, seasoning, and cooking create a wide array of meat products with diverse flavors, textures, and appearances. These culinary innovations add excitement and variety to diets, enriching the culinary experience of people across different cultures [16].
- **Economic impact:** Meat processing is a significant contributor to the economy. It generates employment opportunities, both in large-scale industrial facilities and smaller, local processing units. Additionally, the meat industry supports various related sectors, such as packaging, transportation, and equipment manufacturing, further boosting economic growth [17].
- **Waste reduction:** Proper meat processing can minimize waste by utilizing different parts of the animal for various products. This reduces the environmental impact and ensures that as much of the animal as possible is utilized, promoting sustainability in the industry.

- **International trade:** Meat processing allows countries to participate in global trade by exporting meat products to other nations. This trade promotes economic interdependence, facilitates cultural exchange, and provides access to a wider range of food products for consumers worldwide.
- **Convenience and accessibility:** Processed meat products such as sausages, deli meats, and canned meats offer convenience and portability, making them popular choices for busy individuals and those with limited cooking facilities. These products enable people to enjoy meat as a part of their diet more easily.
- **Cultural and social significance:** Meat has cultural significance in many societies and is often associated with celebrations, gatherings, and traditional ceremonies. In capitalist societies, meat is a basic substance for human nourishment and plays a crucial role in the commodity economy The availability of processed meat products has helped preserve culinary traditions and rituals, contributing to cultural identity and social cohesion [18].

  While meat processing offers numerous benefits, it is essential to balance its advantages with ethical considerations, environmental impact, and health concerns. Responsible and sustainable practices in meat processing are crucial to ensuring the long-term viability of the industry and the well-being of both consumers and animals.

4 Automated Meat Processing

Automation in meat processing has revolutionized the industry by enhancing efficiency, productivity, and safety. Recent advancements in technology have played a crucial role in this transformation. Cutting-edge technologies such as high-pressure processing, pulsed electric field processing, and hurdle technology have been shown to enhance microbial safety while preserving nutrition and sensory attributes [19]. Automated production systems, developed by companies like Scott Technology Ltd, have been instrumental in improving productivity and diversifying the industry [20]. Red meat slaughterhouse automation has become a necessity due to the growing demand for meat consumption and the need for sustainable practices. Simple, cost-efficient, and robust tooling, such as air-driven grippers, are being sought to support these automation systems [21]. Additionally, the development of smart knives and sensing technologies, such as optical methods and electromagnetic wave-based sensing, has the potential to further enhance automation in meat cutting processes [22]. These advancements in technology have significantly contributed to the efficiency, productivity, and safety of the meat processing industry. With advancements in technology, meat processing plants have increasingly adopted automated systems to handle various stages of the meat production process. Here are some key aspects of automation in meat processing.

- **Improved efficiency:** Automation streamlines various tasks in meat processing, reducing the need for manual labor and human intervention. Robots and machines can perform repetitive and labor-intensive processes, such as cutting, deboning, and packaging, at a much faster rate and with higher precision than human workers. This increased efficiency translates to higher production rates and reduced processing time [19].
- **Enhanced food safety:** Automation plays a vital role in ensuring food safety during meat processing. Automated systems can maintain consistent hygiene standards, as they are not subject to human errors or fatigue. Furthermore, automated processes minimize human contact with the meat, reducing the risk of contamination and cross-contamination of pathogens.
- **Standardization:** Automated systems can enforce standardized processing procedures, leading to consistent product quality. By eliminating human error and individual variations in handling meat, the end-product meets specific quality standards, which are especially crucial for large-scale meat processing plants and franchises.
- **Traceability:** Automation technologies often incorporate sophisticated tracking and tracing mechanisms, enabling a comprehensive record of the meat processing journey. From the sourcing of the meat to each stage of processing and packaging, automated systems can create detailed data logs, making it easier to identify the source of any potential issues and facilitate recalls if necessary.
- **Labor savings:** By automating certain tasks, meat processing plants can significantly reduce their reliance on manual labor, leading to cost savings in the long run. While automation requires an initial investment, the reduced labor costs and increased productivity can provide a favorable return on investment over time.
- **Flexibility and customization:** Modern automation systems in meat processing are designed to be flexible and adaptable to different processing requirements. They can be programmed to handle various meat types, cuts, and processing methods, allowing producers to cater to a diverse range of customer demands.
- **Workplace safety:** Meat processing can be a hazardous industry, with sharp tools and heavy machinery involved. By automating certain processes, the risk of worker injuries is reduced, contributing to a safer working environment for employees.
- **Waste reduction**: Automation enables more precise cuts and portioning of meat, minimizing waste and maximizing yield. This efficiency in processing helps reduce the overall environmental impact of meat production.

Despite the numerous benefits of automation, it is essential to strike a balance and maintain human oversight in critical areas of meat processing. Human expertise is still needed for tasks that require judgment, creativity, and decision-making. Additionally, a combination of human expertise and automation can create a more resilient and adaptive meat processing industry capable of meeting future challenges and demands.

5 HACCP Management

Hazard Analysis and Critical Control Points (HACCP) is a systematic and preventive approach to food safety that is widely utilized in the meat processing industry. HACCP management in meat processing plays a vital role in ensuring the production of safe and high-quality meat products for consumers. It aims to identify, evaluate, and control hazards that are significant for food safety, ensuring consumer health protection. HACCP is recognized as a key part of food safety management practice in the global food industry and can be applied at any stage of the food supply chain [23]. The traditional HACCP system has been improved with modern innovations, including the integration of principles, applications, steps, plans, and standards, as well as the use of modern technologies such as artificial intelligence and automation for easy detection and control of contaminants. Implementing HACCP helps food businesses identify and control a broad spectrum of food safety hazards, including biological, chemical, and physical contaminants, ensuring the safety and quality of products [24].

The HACCP system in meat processing typically follows seven principles to conduct a hazard analysis. The first step involves identifying and assessing potential biological, chemical, and physical hazards that could pose risks to food safety during meat processing. This includes evaluating the presence of bacteria such as *Escherichia coli, Staphylococcus aureus, Bacillus cereus, Listeria monocytogenes, Clostridium perfringens, Salmonella* spp., and *Shigella* spp. [24]. This also includes hazards related to raw materials, processing equipment, handling, and the processing environment.

- **Determine critical control points (CCPs):** CCPs are specific points in the meat processing flow where control measures can be applied to prevent, eliminate, or reduce identified hazards to an acceptable level. These points are critical to ensuring food safety, and their proper management is essential.
- **Establish critical limits:** Critical limits are criteria set for each CCP, specifying the maximum or minimum values to which the identified hazard must be controlled to ensure food safety. These limits are scientifically determined and serve as benchmarks for evaluating the effectiveness of the control measures.
- **Monitoring CCPs:** Regular monitoring of CCPs is crucial to verify that the critical limits are consistently met. This involves routine testing, observation, or measurements to ensure that the meat processing remains within the defined safety parameters.
- **Implement corrective actions:** If monitoring reveals a deviation from the critical limits, immediate corrective actions must be taken. These actions are predefined and aim to bring the process back under control to prevent unsafe meat production.
- **Verify and validate the HACCP system:** Regular verification ensures that the HACCP system functions effectively. This includes reviewing records, conducting audits, and validating the control measures to ensure they are scientifically sound and up-to-date.

- **Establish documentation and record-keeping:** Proper documentation is essential to keep track of the HACCP plan, including hazard analysis, CCPs, critical limits, monitoring procedures, corrective actions, and verification activities. Detailed records provide evidence of compliance with food safety regulations and serve as a valuable resource for continuous improvement.

Implementing an effective HACCP management system in meat processing requires commitment from all levels of the organization. It empowers meat processors to systematically identify and control potential hazards, thus ensuring the production of safe and high-quality meat products that meet regulatory requirements and consumer expectations. By prioritizing food safety through HACCP, the meat processing industry can maintain consumer trust and safeguard public health.

6 Microbiology and Spoilage

Human expertise and automation can create more resilient and adaptive meat processing industry human expertise, and automation can create a more resilient and adaptive meat processing industry. Microbiology and the management of spoilage are of utmost importance in the realm of meat processing in order to guarantee the production of meat products that are safe and of high quality. The utilization of natural compounds or bio-preservation technology, such as the application of lactic acid bacteria (LAB) and their by-products, has garnered significant attention in the quest to prolong the shelf life of meat and meat products [25]. The amalgamation of diverse microbial barriers, including starter cultures, nonstarter protective cultures, bacteriocins, and other antimicrobial compounds, has demonstrated favorable outcomes in impeding the proliferation of pathogenic and spoilage bacteria in meat products [26]. Furthermore, the exploration of novel antimicrobial compounds and nonthermal procedures has been undertaken to curtail microbiological hazards while simultaneously upholding product quality [27]. Moreover, the utilization of bacteriophages and their derivatives has emerged as a risk-free alternative for controlling microbial contamination in the meat processing environment (Manal et al. 2022; [13]). In conclusion, these approaches contribute to the extension of meat's shelf life, the preservation of its freshness, and the mitigation of the risk of foodborne illnesses. Some common methods and practices used in the meat processing industry for microbiology and spoilage control are as follows:

- **Hygiene and sanitation:** Maintaining a clean and hygienic processing environment is paramount to controlling microbial contamination. Proper sanitation practices, including regular cleaning and disinfection of equipment, surfaces, and facilities, help prevent the buildup and spread of harmful bacteria.
- **Temperature control:** Temperature management is crucial for controlling microbial growth. Chilling or refrigerating meat at appropriate temperatures

slows down the growth of spoilage-causing microorganisms and extends the shelf life of the product.

- **Modified atmosphere packaging (MAP):** MAP involves altering the atmosphere inside the meat packaging to extend shelf life. It typically involves reducing oxygen levels and increasing carbon dioxide or nitrogen levels to inhibit the growth of spoilage microorganisms and maintain product quality.
- **Vacuum packaging:** Vacuum packaging removes air from the meat package, creating an anaerobic environment that slows microbial growth and prevents oxidation, thus reducing spoilage.
- **High-temperature treatments:** Processes such as cooking, pasteurization, and canning involve subjecting meat products to high temperatures, effectively killing harmful pathogens and spoilage microorganisms.
- **pH control:** Monitoring and controlling the pH of meat products can inhibit the growth of certain microorganisms. Lowering the pH (acidification) through methods such as fermentation or the addition of acidic ingredients can contribute to spoilage prevention.
- **Water activity (aw) control:** Controlling the water activity level in meat products can help prevent microbial growth. Reducing the available water content by using salt, sugar, or other moisture-binding ingredients inhibits microbial proliferation.
- **Use of preservatives:** Certain natural or artificial preservatives, such as salt, nitrates, nitrites, and some food-grade antimicrobial agents, can be added to meat products to extend their shelf life and inhibit the growth of spoilage microorganisms.
- **Hurdle technology:** Employing multiple preservation methods in combination, known as hurdle technology, creates multiple barriers to microbial growth, making it harder for microorganisms to survive and spoil the meat.
- **Quality assurance and HACCP:** Quality assurance programs and Hazard Analysis and Critical Control Points (HACCP) are employed to identify potential microbiological hazards and implement appropriate control measures at critical points in the meat processing flow.

It is important to note that while these methods and practices can significantly control microbial growth and spoilage, they do not guarantee indefinite shelf life. Proper storage conditions, adherence to recommended expiration dates, and following food safety guidelines are essential for ensuring the safety and quality of meat products throughout their shelf life. Additionally, regular testing and monitoring of meat products for microbial contamination and spoilage are essential parts of quality control in the meat processing industry.

Microbial spoilage refers to the deterioration of meat due to the growth of microorganisms such as bacteria, molds, and yeasts. These microorganisms can cause changes in the appearance, texture, smell, and taste of meat, making it unsuitable for consumption. To control microbial spoilage, various methods are employed:

- **Temperature control:** Keeping meat at low temperatures, such as refrigeration or freezing, can slow the growth of microorganisms and extend the shelf life of meat.
- **Packaging:** Proper packaging, such as vacuum sealing or modified atmosphere packaging, can create a barrier that prevents the entry of microorganisms and delays spoilage.
- **Food additives:** The use of preservatives, such as salt, nitrites, and organic acids, can inhibit the growth of spoilage-causing microorganisms and extend the shelf life of meat.
- **Hygiene practices:** Maintaining good hygiene during the processing, handling, and storage of meat is crucial to prevent the introduction and spread of microorganisms.
- **Irradiation:** Meat can be treated with ionizing radiation to kill or reduce the number of spoilage-causing microorganisms, thereby extending its shelf life.
- **Use of antimicrobial agents:** Some antimicrobial agents, such as bacteriocins or natural antimicrobial compounds, can be used to inhibit the growth of spoilage microorganisms in meat.

It is important to note that while these methods can help control microbial spoilage, they do not completely eliminate the risk. Proper storage, handling, and cooking practices are also essential to ensure the safety of meat consumption.

7 Preservation Methods

Different techniques are utilized for the preservation of meat in order to prolong its storage life, uphold its quality, and prevent spoilage as well as the proliferation of harmful microorganisms. Preservation methods such as drying, salting, pickling, and smoking have been conventionally employed, while contemporary approaches like freezing, chilling, and irradiation have also been developed [25]. Recently, there has been an increasing interest in the utilization of natural compounds and bio-preservation technology for the control of spoilage in meat products. This includes the application of microorganisms or their by-products, such as lactic acid bacteria (LAB), which have demonstrated efficacy in extending the shelf life of meat and meat products. Furthermore, essential oils and plant extracts have exhibited potential as natural preservatives and antioxidants in meat products, with studies focusing on their application and effects in this domain [28]. Autochthonous microbial cultures, including LAB, have also been investigated for their role in reducing risks and enhancing safety in traditional dry fermented meat products [29]. Encapsulation techniques have been explored to stabilize and safeguard bioactive compounds within meat systems, thereby providing improved functionality and controlled release [30]. Some common methods of meat preservation include:

- **Refrigeration:** One of the simplest preservation methods involves storing meat at low temperatures, typically between 0 °C and 4 °C (32 °F to 39 °F). Refrigeration slows the growth of spoilage microorganisms and helps maintain the freshness and quality of meat for a limited period, usually a few days to a few weeks.
- **Freezing:** Freezing meat at temperatures below −18 °C (0 °F) effectively halts microbial activity and enzymatic reactions, providing long-term preservation. Frozen meat can remain safe to eat for several months, although the quality may deteriorate over time due to factors such as freezer burn.
- **Curing**: Curing involves the addition of salt, sugar, and nitrites/nitrates to meat to inhibit microbial growth, enhance flavor, and give a characteristic color (e.g., pink color in cured ham). Common cured meat products include bacon, ham, and sausages.
- **Smoking:** Smoking exposes meat to smoke from burning wood or other materials, which not only enhances flavor but also acts as a preservative by inhibiting microbial growth and preventing oxidation.
- **Drying:** Drying or dehydrating meat removes moisture, creating an environment where microorganisms cannot thrive. Dried meat products, such as beef jerky, have a long shelf life and are convenient for storage and transport.
- **Fermentation:** Fermentation is a traditional preservation method that involves the controlled growth of beneficial microorganisms to alter the meat's characteristics and inhibit the growth of harmful pathogens.
- **Canning:** Canning involves sealing meat in airtight containers and then heat-processing it to kill bacteria and other microorganisms. Canned meat products have a long shelf life and can be stored without refrigeration.
- **Modified atmosphere packaging (MAP):** MAP involves modifying the atmosphere within the meat packaging to extend shelf life. It typically includes reducing oxygen levels and increasing carbon dioxide or nitrogen levels to slow microbial growth and prevent spoilage.
- **Vacuum packaging:** Vacuum packaging removes air from the meat package, reducing the potential for microbial growth and oxidation, thereby preserving the meat's quality.
- **Use of preservatives:** Certain natural or artificial preservatives, such as salt, nitrates, nitrites, and some food-grade antimicrobial agents, can be added to meat products to extend their shelf life and inhibit the growth of spoilage microorganisms.

It is important to note that while these preservation methods can significantly extend the shelf life of meat, proper handling, storage, and adherence to food safety guidelines are essential to ensure the meat remains safe for consumption. Additionally, different preservation methods may impact the taste, texture, and nutritional profile of preserved meat products.

8 Sensory Evaluation

Sensory evaluation of meat constitutes a vital facet of quality control in the meat industry. This process entails the appraisal of meat products by individuals who have undergone specialized training or by consumers themselves, with the aim of assessing their sensory attributes, which encompass factors such as appearance, flavor, texture, and overall acceptability [31]. Within the realm of meat products, sensory evaluation has been employed to gauge the preferences and acceptability pertaining to diverse meat sources, including goat meat and products derived from insects. These investigations have demonstrated that sensory attributes, such as aroma, color, and texture, wield a significant influence over the consumer acceptance of meat products. By comprehending and addressing consumer preferences via sensory evaluation, the meat industry can enhance the quality and acceptability of its meat products, thereby fostering greater consumer satisfaction and achieving success in the market. Here are some key aspects of the sensory evaluation of meat:

- **Appearance:** The visual aspects of meat, including color, marbling, and surface texture, are essential for consumers' initial impression of the product. The color can indicate freshness, while marbling (intramuscular fat) affects tenderness and flavor. Sensory evaluation examines these visual attributes to ensure that they meet established quality standards.
- **Flavor:** The taste and aroma of meat are significant factors influencing consumer preference. Sensory evaluation allows panelists to assess the flavor profile, including attributes such as sweetness, bitterness, saltiness, and umami. Proper cooking methods, seasoning, and aging can influence the overall flavor of meat products.
- **Texture:** Texture plays a vital role in the perception of meat quality. Sensory evaluation assesses characteristics such as tenderness, juiciness, chewiness, and fibrousness. These attributes are influenced by factors such as muscle type, cooking method, and postmortem aging.
- **Mouthfeel:** Mouthfeel refers to the sensory sensations experienced while consuming meat, including aspects such as tenderness, juiciness, greasiness, and astringency. It is an essential factor in the overall eating experience.
- **Overall acceptability:** This aspect considers the overall impression and likability of the meat product. Sensory evaluation helps determine whether a product meets consumer expectations and preferences.
- **Sensory panels:** Trained sensory panels, consisting of individuals with well-developed sensory abilities, are commonly used for formal evaluations. Panelists undergo rigorous training to accurately identify and describe sensory attributes. Their evaluations are used to assess product consistency, make quality improvements, and support research and development efforts.
- **Consumer sensory testing:** In addition to trained panels, consumer sensory testing is also conducted to understand preferences and gain insights into the target market's taste preferences. These tests involve a larger number of untrained consumers who evaluate products based on their personal preferences.

- **Quality control and product development:** Sensory evaluation is a valuable tool for ensuring consistent product quality and guiding product development efforts. By understanding how various factors affect the sensory attributes of meat, producers can optimize processes and formulate new products to meet consumer demands.
- **Compliance and certification:** In some cases, sensory evaluation is an essential requirement for compliance with industry standards and certifications, such as ISO 8589 or specific certifications related to organic or premium meat products [31].

Overall, sensory evaluation of meat is an indispensable tool for the meat industry to maintain quality, meet consumer expectations, and continuously improve products to stay competitive in the market. By understanding the sensory attributes of their meat products, producers can make informed decisions to deliver high-quality and enjoyable eating experiences to consumers.

9 Quality Assessment and Grading

The process of assessing and grading meat and poultry for quality is a complex one, as it involves the consideration of multiple criteria and techniques. When evaluating poultry meat products, several factors come into play, including sensory evaluation, technological criteria, proximate chemical analysis, and histological examination [32]. Objective quality indexes, such as the ratio of fat to protein, energy value, and other indicators, can be used to classify poultry meat [33]. To accurately determine the quality grades of poultry meat samples, attribute selection and classification steps can be combined [34]. Research has shown that the use of mineral additives in poultry farming can have a positive impact on the quality of meat and poultry products, enhancing their safety, structure, and maturity [35]. In the case of pork, quality assessment involves the evaluation of various parameters, such as fat thickness, muscle score, color values, pH, and marbling score (Whitney and [36]). In its entirety, the evaluation and grading of meat and poultry require the use of objective measurements, sensory evaluation, and histological examination to ensure the quality and safety of the products. Here is a general overview of how quality assessment and grading of meat are typically performed.

- **Visual inspection:** Meat is visually inspected for color, marbling (intramuscular fat), surface texture, and any visible defects or abnormalities. Freshness indicators such as bright red color and absence of discoloration are assessed.
- **Odor assessment:** The smell of meat is checked for any signs of spoilage, off-odors, or taints, which may indicate deterioration.
- **Palpation:** The texture and firmness of meat are assessed by palpating the surface and ensuring that it meets the desired quality characteristics.
- **pH measurement:** The pH level of meat is measured, which can provide insights into meat freshness and aging.

- **Marbling score:** For certain cuts of meat (e.g., beef), marbling is scored based on the amount and distribution of intramuscular fat. Higher marbling scores often indicate higher quality and tenderness.
- **Water holding capacity:** The ability of meat to retain water during cooking is evaluated, as this can affect juiciness and tenderness.
- **Cooking tests:** Sensory evaluation through cooking and tasting the meat is also conducted to assess flavor, tenderness, and overall acceptability.
- **Grading systems:** The grading of meat is a standardized process that assigns a grade to the meat based on specific quality attributes. Different countries have different grading systems, and the criteria may differ based on factors such as meat type, age, and fat content.
- **Yield grading:** In some grading systems, yield grading is used to assess the proportion of lean meat to fat and bone in a carcass. This is particularly important for assessing the meat used in processing, as it affects the usable yield for various products.
- **Quality grading:** Quality grading assesses attributes related to tenderness, juiciness, flavor, and other sensory characteristics of the meat. It helps consumers and businesses make informed choices about the meat they purchase.
- **Official inspection:** Grading of meat is often conducted by trained inspectors from relevant governmental or regulatory bodies. These inspectors ensure that grading is performed accurately and consistently according to established standards.
- **Labelling:** Grading information is often displayed on meat packaging or labels, providing consumers with information about the quality and characteristics of the meat product.

Overall, quality assessment and grading of meat are crucial steps in ensuring the delivery of safe, high-quality, and desirable meat products to consumers. These processes enable transparency and consistency in the meat industry and empower consumers to make informed decisions based on their preferences and requirements.

10 Packaging, Handling and Transportation

Packaging, handling, and transportation conditions, along with adherence to recommended expiration dates, are of utmost importance in guaranteeing the safety and quality of meat and poultry products. The utilization of appropriate packaging systems, such as tray and overwrap, modified atmosphere packaging, and vacuum packaging, serves to augment the microbiological safety and quality of poultry meat. The implementation of Good Hygiene Practices (GHP) during the processing, packaging, storage, and distribution stages of poultry meat is indispensable in preventing bacterial proliferation and averting financial losses [38]. Furthermore, advances in food packaging technologies, including antimicrobial packaging, oxygen scavengers, and intelligent packaging concepts such as time-temperature and freshness indicators, significantly contribute to the provision of wholesome poultry products to consumers [37]. The establishment of trust among consumers in expiration date labels constitutes a pivotal factor in reducing household food wastage, and the incorporation of smart labels on meat packaging can assist in alleviating erroneous interpretations of expiration dates. In-store demonstrations and the provision of information can serve to support and foster the adoption of smart labels by consumers. Here are some key considerations for each stage:

10.1 Packaging

- **Selection of packaging materials:** Packaging materials for meat should be suitable for the specific type of meat and its intended shelf life. Common packaging options include vacuum-sealed bags, modified atmosphere packaging (MAP), trays, and shrink wrap.
- **Hygiene and sanitation:** Packaging materials and equipment must be kept clean and sanitized to prevent microbial contamination. Packaging areas should be separate from processing areas to avoid cross-contamination.
- **Leak-proof and airtight sealing:** Proper sealing of packaging prevents leakage of liquids and exposure to air, which can lead to spoilage. It also helps maintain the desired modified atmosphere for MAP products.
- **Product labelling:** Accurate and clear labelling is crucial for consumer information and regulatory compliance. Labels should include product name, expiration date, storage instructions, and any allergen information.

10.2 Handling

- **Temperature control:** Maintaining proper temperature throughout handling is essential to ensure meat safety and quality. Fresh meat should be kept at refrigerated temperatures (0–4 °C or 32–39 °F), while frozen meat should be kept at or below -18 °C (0 °F) to prevent thawing.
- **Avoiding cross-contamination:** Handlers should avoid contact between raw meat and ready-to-eat foods or surfaces that will come into contact with consumers. Separate equipment and tools should be used for handling different types of meat.
- **Hygiene practices:** Handlers should adhere to strict hygiene practices, including wearing appropriate protective clothing, washing hands regularly, and using sanitizers to prevent microbial contamination.
- **Proper lifting and stacking:** Proper lifting techniques and stacking procedures should be followed to prevent damage to the packaging and ensure that the meat remains in good condition during transportation.

11 Conclusion

The global perception of meat processing and poultry is undergoing a significant transformation driven by shifting cultural norms, environmental consciousness, health considerations, and ethical awareness. The challenges faced by these industries present an opportunity for positive change, encouraging stakeholders to adopt more sustainable practices and explore innovative alternatives. Striking a balance between meeting nutritional needs, safeguarding the environment, and respecting ethical considerations will be critical in shaping the future of the meat and poultry industries on a global scale. Collaborative efforts between governments, businesses, consumers, and the research community are necessary to address these complex challenges and ensure a secure and sustainable food future. Embracing sustainable and responsible practices in meat processing and poultry can contribute to healthier diets, reduced environmental impacts, and improved animal welfare. Furthermore, the promotion of diversity in protein sources, including plant-based alternatives, can offer inclusive solutions for global nutrition and foster a more resilient and equitable food system. Ultimately, a collective commitment to balance cultural traditions, environmental stewardship, and ethical considerations will determine the long-term success and acceptance of meat processing and poultry in a rapidly evolving world.

References

1. Machovina B, Feeley KJ, Ripple WJ (2015) Biodiversity conservation: the key is reducing meat consumption. Sci Total Environ 536:419–431
2. Barbut S, Leishman EM (2022) Quality and processability of modern poultry meat. Animals 12. https://doi.org/10.3390/ani12202766
3. Petracci M, Soglia F, Madruga M, Carvalho L, Ida E, Estévez M (2019) Wooden-breast, white striping, and spaghetti meat: causes, consequences and consumer perception of emerging broiler meat abnormalities. Compr Rev Food Sci Food Saf 18:565–583. https://doi.org/10.1111/1541-4337.12431
4. Nusairat B, Tellez-Isaias G, Qudsieh R (2021) An overview of poultry meat quality and myopathies. IntechOpen
5. Ponnampalam EN, Bekhit AE, Bruce H, Scollan ND, Muchenje V, Silva P, Jacobs JL (2019a) Production strategies and processing systems of meat: current status and future outlook for innovation—a global perspective. In: Sustainable meat production and processing. Academic Press. https://doi.org/10.1016/B978-0-12-814874-7.00002-X
6. Whitton C, Bogueva D, Marinova D, Phillips CJ (2021) Are we approaching peak meat consumption? Analysis of meat consumption from 2000 to 2019 in 35 countries and its relationship to gross domestic product. Animals 11(12):3466
7. Parlasca MC, Qaim M (2022) Meat consumption and sustainability. Ann Rev Resource Econ 14:17–41
8. Hannah Ritchie, Pablo Rosado, Max Roser (2017) Meat and dairy production. Published online at OurWorldInData.org. Retrieved from: https://ourworldindata.org/meat-production
9. Lesschen JP, van den Berg M, Westhoek HJ, Witzke HP, Oenema O (2011) Greenhouse gas emission profiles of European livestock sectors. Animal Feed Sci Technol 166:16–28

10. Allievi F, Vinnari M, Luukkanen J (2015) Meat consumption and production–analysis of efficiency, sufficiency and consistency of global trends. J Clean Prod 92:142–151
11. Tasevska N, Sinha R, Kipnis V, Subar AF, Leitzmann MF, Hollenbeck AR, Caporaso NE, Schatzkin A, Cross AJ (2009) A prospective study of meat, cooking methods, meat mutagens, heme iron, and lung cancer risks. Am J Clin Nutr. Jun; 89(6):1884-94. https://doi.org/10.3945/ajcn.2008.27272. Epub 2009 Apr 15. PMID: 19369370; PMCID: PMC2683000
12. Kim Won, Sam Jeong, Jun Ho (2019) Meat processing method
13. Wenhuan Z (2023) Reduction of risks associated with processed meats. https://doi.org/10.1016/b978-0-12-819470-6.00059-7
14. Harlina PW, Huda S, Yarlina VP (2023) Training on making healthy and safe meat processed food in Cileunyi Wetan, Bandung Regency. Commun Empowerment 8(6):889–893. https://doi.org/10.31603/ce.8635
15. Kuznetsov NB (2023) Meat and health. https://doi.org/10.1016/b978-0-323-85408-5.00009-1
16. Khalid W, Maggiolino A, Kour J, Arshad MS, Aslam N, Afzal MF, Meghwar P, Zafar KU, De Palo P, Korma SA (2023) Dynamic alterations in protein, sensory, chemical, and oxidative properties occurring in meat during thermal and non-thermal processing techniques: a comprehensive review. Front Nutr. 9:1057457. https://doi.org/10.3389/fnut.2022.1057457
17. Kulshreshtha SN (2022) Regional economic impacts of the Canadian beef processing sector. Int J Business Appl Soc Sci. https://doi.org/10.33642/ijbass.v8n12p4
18. Mroczek K, Ptasiuk W, Mroczek JR (2022) Ekologiczne i etyczno-religijne aspekty konsumpcji mięsa. Pol J Sustain Dev. https://doi.org/10.15584/pjsd.2022.26.2.6
19. Chang H (2023) Advancements in food processing technologies: enhancing safety, quality, and sustainability. Indian Sci J Res Eng Manage 7(6). https://doi.org/10.55041/ijsrem23682
20. Seaton M (2022) Lessons in automation of meat processing. Animal Front. https://doi.org/10.1093/af/vfac022
21. Hu Y (2022) A review of unilateral grippers for meat industry automation. Trends Food Sci Tech. https://doi.org/10.1016/j.tifs.2021.12.017
22. Mason A, Romanov D, Cordova-Lopez LE, Ross S, Korostynska O (2022) Smart knife: technological advances towards smart cutting tools in meat industry automation. Sensor Rev 42(1):155–163. https://doi.org/10.1108/sr-09-2021-0315
23. Awuchi C G (2023) HACCP, quality, and food safety management in food and agricultural systems. Cogent Food & Agriculture, 9(1). https://doi.org/10.1080/23311932.2023.2176280
24. Mootassim-Billah S, Van Nuffelen G, Schoentgen J (2023) Assessment of radio(chemo) therapy-related dysphagia in head and neck cancer patients based on cough-related acoustic features: a prospective phase II national clinical trial (ACCOUGH-P/A trial). Trials 24, 619. https://doi.org/10.1186/s13063-023-07660-y
25. Kaveh S, Hashemi SM, Abedi E, Amiri MJ, Conte FL (2023) Bio-preservation of meat and fermented meat products by lactic acid bacteria strains and their antibacterial metabolites. Sustainability. https://doi.org/10.3390/su151310154
26. Karbowiak M, Szymański P, Zielińska D (2023) Synergistic effect of combination of various microbial hurdles in the biopreservation of meat and meat products—systematic review. Foods. https://doi.org/10.3390/foods12071430
27. Dhara D, Biswas S, Patra G, Biswas O, Sharma G (2023) Biosensor for maintaining the quality of meat and meat products: a review. J Food Safety Hygiene. https://doi.org/10.18502/jfsh.v9i1.12687
28. Aguiar Campolina G, das Graças Cardoso M, Rodrigues-Silva-Caetano A, Lee Nelson D, Mendes Ramos E (2023) Essential oil and plant extracts as preservatives and natural antioxidants applied to meat and meat products: a review. Food Technol Biotechnol. https://doi.org/10.17113/ftb.61.02.23.7883
29. Rossi F, Tucci P, Del Matto I, Marino L, Amadoro C, Colavita G (2023) Autochthonous cultures to improve safety and standardize quality of traditional dry fermented meats. Microorganisms 11(5):1306. https://doi.org/10.3390/microorganisms11051306

30. Siddiqui SA, Farooqi MQ, Bhowmik S, Zahra Z, Mahmud MC, Assadpour E, Gan RY, Kharazmi MS, Jafari SM (2023) Application of micro/nano-fluidics for encapsulation of food bioactive compounds—principles, applications, and challenges. Trends Food Sci Technol. https://doi.org/10.1016/j.tifs.2023.03.025
31. Christopher T (2023) Sensory perspectives into indigenous fermented foods in the tropics: challenges and opportunities. Indigenous Fermented Foods for the Tropics 483-502. Academic press https://doi.org/10.1016/b978-0-323-98341-9.00023-2
32. Hosny AA, Al-Ghayat SM, Shaheen HM (2020) Quality assessment of emulsion type poultry meat products. Suez Canal Vet Med J. https://doi.org/10.21608/SCVMJ.2020.103053
33. Makhonina VN, Agafonychev VP (2017) The scientific approach of determination poultry grade and poultry products. Theory Pract Meat Process. https://doi.org/10.21323/2414-438X-2017-2-4-114-128
34. Barbon S Jr, Barbon AP, Mantovani RG, Barbin DF (2016) Comparison of SVM and REPTRee for classification of poultry quality. In: Proceedings of the modelling, simulation and identification/intelligent systems and control–2016 (MSI 2016). https://doi.org/10.2316/P.2016.840-039
35. Drozd MN, Usevich VM (2021) Assessment of the quality of meat and poultry raw materials using mineral adaptogen. Ural Agrarian Bullet. https://doi.org/10.32417/1997-4868-2021-206-03-53-66
36. Magee WE (2017) Meat quality assessment of pork fed poultry fat, flaxseed oil, and supplemented with vitamin E. Auburn University. https://doi.org/10.22175/RMC2016.067
37. Ada M B, Jessica A (2022) Food handling practices and expiration dates: Consumers' perception of smart labels. Food control. https://doi.org/10.1016/j.foodcont.2021.108615
38. Mangalassary S. (2019). Advances in Packaging of Poultry Meat Products. In: Venkitanarayanan, K, Thakur S, Ricke S. (eds) Food Safety in Poultry Meat Production. Food Microbiology and Food Safety. Springer, Cham. https://doi.org/10.1007/978-3-030-05011-5_7

Chapter 24
Potentials of Probiotics in Food

Bhavya D. C., Virupakshaiah D. B. M., and Jayachandra S. Yaradoddi

1 Introduction

Probiotics have been described as "viable preparations in foods or dietary supplements to improve the health of humans and animals" by the European Union (EU) Expert Group on Functional Food in Europe (FUFOSE) (FUFOSE working group, 1999). Probiotics were initially defined by the "Food and Agriculture Organisation" and the "World Health Organisation" council in 2001; this definition was later improved in 2014 as "live microorganisms that confer a health benefit on the host when administered in sufficient quantities," which can be taken to mean that probiotic strains must be adequately characterized, safe when used as intended, backed up by at least one successful human clinical trial carried out by generally accepted scientific standards or, when relevant, under recommendations and regulations made by local/national authorities, during the product's shelf life at an effective dose [1]. In the past, common probiotics included Lactobacilli and several Bifidobacterium species. To better handle the vast range of bacteria allocated to the genus, Lactobacillus underwent a significant redesign in 2020. Twenty-three new genera were identified, some containing well-researched probiotic species [2]. The term "probiotic" can be further divided into designer probiotics (genetically altered probiotics), direct-fed microbial (probiotics for animal usage), and probiotic medications or foods (e.g., meals, food ingredients, and dietary supplements [3]. The

Bhavya D. C.
Research Scholar, Department of Microbiology, Davangere University, Davangere, Karnataka, India

Virupakshaiah D. B. M. (✉)
Department of Microbiology, Davangere University, Davangere, Karnataka, India

J. S. Yaradoddi
Department of Biotechnology, Basaveshwar Engineering College, Bagalkote, Karnataka, India

J. S. Yaradoddi et al. (eds.), *Frontiers in Food Biotechnology*,
https://doi.org/10.1007/978-981-97-3261-6_24

human microbiota is a diverse group of bacteria that inhabit the genitourinary, respiratory, cutaneous, oral, and gastrointestinal systems. The microbiota is thought to weigh 1–2 kg [4]. Probiotics have drawn more attention in recent years due to mounting evidence of their health advantages, which include improving oral health, preventing recurring urinary tract infections in adults, and alleviating symptoms of irritable bowel syndrome in children. According to mounting evidence, Probiotics may aid in the prevention or treatment of some diseases. This idea has inspired the development of a wide range of products sold for use by both humans and animals, including specially prepared supplements and diets [5]. Researchers from all around the world are seeking a different source of probiotics. Changing the commensal microbiota can prevent and treat various gastrointestinal infections [6]. Probiotics may aid in treating gastrointestinal microbial issues, albeit it can be difficult to show these effects in clinical settings. Numerous extra functions of the gastrointestinal system essential for well-being are now being acknowledged [7]. The intestinal tracts of humans contain more than 400 different bacterial species. Its environment, which contains facultative and anaerobic microbes, is complicated [8]. Since every genus has a unique growth niche, the number of genera is constant [9]. Probiotics are crucial for treating and preventing conditions, including metabolic syndrome, which can cause diabetes, heart disease, obesity, allergies, liver problems, intestinal problems, and maladies [10]. By enhancing the mucosal barrier and regulating the human immune system, probiotics or prebiotics may enhance host defense against severe-acute-respiratory-syndrome-related coronavirus-2 (SARS-CoV-2) infection [11, 12]. By restoring the altered microbiota, the regulation process through adjuvant probiotic or prebiotic treatment may reduce the incidence of subsequent infection in these patients [13]. On the skin, in oral, intestinal, and vaginal systems, hundreds of bacteria species coexist with humans. Bacterial populations are considered 10X1013 cells throughout the human body [14]. This number is even more astounding when you realize that it outnumbers the human cells that make up the human body by a factor of 10. In adults, the digestive tract maintains a rather stable microbial environment impacted by dietary changes and other lifestyle modifications [15]. Even when taken in large quantities, probiotic bacteria do not become permanent colonizers and are rarely found in feces or intestinal samples more than a few weeks after intake. Based on a probiotic bacterium's genus or species, it is impossible to predict whether it would be useful, harmful, or have no effect on health. Research does not support the alluring hypothesis that individuals belonging to one genus or species consistently mediate particular effects [16]. On the other hand, similar beneficial effects have been shown for many strains of various genera, such as Lactobacilli, Bifidobacteria, and Enterococci, for objectives like immunological function, anticancer effects, and antidiarrheal effects [17]. Concerning their ability to maintain their availability and viability, probiotics' health advantages have always been studied [18] without affecting the host's safety, their digestibility, the provision of their health benefits [19], and the food's organoleptic characteristics after its incorporation [20]. Probiotics increase some intestinal microflora, but not all the bacteria there. The synthesis of organic acids, ethanol, hydrogen peroxide, or protein-containing substances (bacteriocins) is linked to the probiotic's antibacterial action [21]. Developing novel probiotic-rich food products and investigating specific probiotic qualities and their effects on human health are the two key areas of concentration for researchers and various food business sectors interested in the global probiotic market [22]. Probiotic goods include fermented meals with one or more probiotic bacterial species, enzymes, vitamins, capsules, and tablets. Probiotics are mostly employed in healthcare-related products. Probiotic bacteria have a protective effect on the gut flora when taken orally. To safeguard youngsters vulnerable to the third wave of coronavirus disease 2019 (COVID-19) infection, Central Food Technological Research Institute (CFTRI) recently disclosed that it is developing probiotic food and how it can shield kids from the third COVID-19 wave. Humans have a basic desire for good health [23]. Humans prioritized using natural resources to achieve positive health results [24].

2 Probiotic Microorganisms

Even within the same species, the probiotic potential of various bacterial strains varies. Different strains of the same species (Table 24.1) are always distinct, and they may have other areas of adherence (site-specific), unique immunological responses, and extra activities in a healthy versus an inflamed mucosal milieu [25]. Many microorganisms, including yeasts like Saccharomyces and bacteria from the genera *Bifidobacterium, Bacteroides, Clostridium, Fusobacterium, Propionibacterium, Streptococcus, Enterococcus, Lactococcus, Staphylococcus, Peptostreptococcus,* and *Lactobacillus*, have been reported to have probiotic properties [26]. In addition several microbes like *E. coli*, *Bacillus* species, yeast *Saccharomyces boulardii, Clostridium butyricum,* recently certified as a novel food probiotics in European Union, and they are considered as newcomers to the probiotic ranks [2]. The 25 genera that make up the new classification of the Lactobacillus genus include two host-

Table 24.1 Different strains of the same species

Lactobacillus species	Bifidobacterium species	Yeast	Others
L. acidophilus *L. rhamnosus* *L. gasseri* *L. casei* *L. reuteri* *L. delbreuckii subsp. bulgarius* *L. crispatus* *L. plantarum* *L. salivarius* *L. johnsonii* *L. gallinarum* *L. plantarum* *L. fermentum* *L. helveticus*	*B. bifidum* *B. animalis* *B. breve* *B. infantis* *B. longum* *B. lactis* *B. adolescentis*	*S. boulardii,* *S. lactis,* *S. carlsbergensis,* *Kluyveromyces marxianu,* *S. cerevisiae*	*Enterococcus faecalis* *Enterococcus faecium* *Streptococcus salivarus* subsp. *thermophilus* *Lactococcus lactis* subsp. *lactis* *Lactococcuslactis* subsp. *cremoris* *Propionibacterium freudenreichii* *Pediococcus acidilactici* *Saccharomyces boulardii* *Leuoconostoc mesenteroides*

adapted organisms like *Lactobacillus delbrueckii* and *Paralactobacillus*, and rest 23 of them were as *Holzapfelia, Amylolactobacillus, Schleiferilactobacillus, Ligilactobacillus, Lacticaseibacillus, Latilactobacillus, Dellaglioa, Bombilactobacillus, Companilactobacillus, Lapidilactobacillus, Agrilactobacillus, Ligilactobacillus, Liquorilactobacillus, Furfurilactobacillus, Lactiplantibacillus, Limosilactobacillus, Paucilactobacillus, Fructilactobacillus, Apilactobacillus, Acetilactobacillus, Secundilactobacillus, Levilactobacillus,* and *Lentilactobacillus* [27].

3 Mechanism of Probiotics

Although molecular mechanisms are rarely investigated further due to the abundance of different pathways described for probiotics, their mode of action has been roughly defined as enhancing the host's "microbial balance" or the balance of the ambient environment, in addition to their therapeutic effects. When taken orally, they defend epithelial membranes from dangerous germs by passing through the stomach and mucous membranes. Acids like lactic acid, propionic acid, and acetic acid are produced by probiotic bacteria like *Bifidobacterium* and *Lactobacillus*, which reduce pH and prevent the growth of pathogenic bacteria. Another proposed mechanism of action relates to their function in immunomodulation [28–30]. In terms of morphological, ecological, and genotypic diversity, there are 261 species of Lactobacillus (as of March 2020). Lactobacillus was divided into 25 genera, including the emended genus Lactobacillus [31]. Probiotics create antimicrobial substances that fight infections, restrict adhesion sites, compete for nutrition, eliminate toxin receptors, and alter immunity. Following is a summary of the proposed mechanism of probiotic action, avoiding dangerous bacteria's adherence to the intestinal epithelium and binding to gram-negative bacteria, stopping pathogenic germs from developing. Keeping short-chain fatty acid concentrations within normal range, colonocyte growth to restore intestinal permeability, more efficient intestinal absorption of electrolytes, enhancing the intestine's immune response, control of lipid metabolism pro-inflammatory cytokines in the intestines are inhibited [21, 29, 32–34].

3.1 Mechanisms of Action of Probiotic Bacteria [2]

1. **Lactobacillus** can,
   (a) Stimulate T cell regulatory cells (Treg cells) to produce TGF-β, interleukin-10 (IL-10), and IL-8.
   (b) Increase levels of secreted IL-6 secreted in a Toll-like receptor (TLR)-2-dependent manner, thereby inducing the clonal expansion of all IgA-producing B cells while also stimulating the expression of macrophage mannose receptor CD206.

   (c) Inhibit the expression of Janus kinase (JAK) and nuclear factor kappa-light-chain-enhancer of activated B cells (NF-κB) genes.
   (d) Increase the release of IL-12p70 and IL-4.
   (e) Reduce the TLR expression and increase CD40 and CD80 expressions, degrade the proinflammatory chemokine IFN-γ-inducible protein 10 (IP-10).
   (f) Increase the expression of TLR-9.
   (g) Favor the expression of nucleotide binding oligomeric domain-like receptor protein 3 (NLRP3), cysteine proteinase-1 (Caspase-1), and IL-18.

2. **Lacticaseibacillus** and **Limocaseibacillus** can induce β-defensins 2 and 4 and IL-8 expressions. Distinct studies reported opposing data on TLR expression.
3. **Bifidobacterium** can.

   (a) Inhibit the expression JAK and NF-κB genes.
   (b) Favor the overexpression of IL-10 and TGF-β, while, at the same time, stimulate the production of IgAs.
   (c) Favor Treg cell differentiation.
   (d) Increase the total helper (CD4+) and activated (CD25+) T lymphocytes and NK cells.
   (e) Reduce the expression of CD19 on B cells.
   (f) Induce the production of monocyte chemoattractant protein 1 (MCP-1) and tumor necrosis factor-α (TNF-α) through TLR-9 stimulation, and
   (g) Increase the number of Foxp3(+) T regulatory cells and the release of CCL20, CCL22, CXCL10, and CXCL11.

4. **Escherichia coli** can induce the expression of TLR-5 and TNF-α as well as increase the number of CD4+ cells.
5. **Bacteroidales** stimulates the release of IL-6 accompanied by the expression of mucin-2 and claudin-1.
6. **Lactobacillus, lacticaseibacillus, bifidobacterium,** and **streptococcus** can favor the release of TNF-α, IL-6, IL-1β.
7. **Streptococcus** can induce clonal expansion of B cells stimulated to release IgAs. Dash arrow: conflicting data have been reported on the effect of lactobacilli in increasing CD4+ T cell number.

4 Health Benefits of Probiotics

Lactic acid bacteria (LAB) which has been employed for thousands of years to preserve food via fermentation (Table 24.2) may also have health advantages [2]. The bacterial strains utilized in probiotic meals and supplements are crucial [29]; some examples for probiotic strains in products are shown in Table 24.3 [35, 36].

- Colonization of the intestinal, respiratory and urogenital tracts.
- Cholesterol metabolism.
- Inhibiting the carcinogenesis.

Table 24.2 Benefits of probiotic bacteria

Immunologic benefits	Nonimmunologic benefits
• Activate local macrophages to increase antigen presentation to B lymphocytes and increase secretory immunoglobulin A (IgA) production both locally and systemically • Modulate cytokine profiles • Induce hyporesponsiveness to food antigens	• Digest food and compete for nutrients with pathogens • Alter local pH to create an unfavorable local environment for pathogens • Produce bacteriocins to inhibit pathogens • Scavenge superoxide radicals • Stimulate epithelial mucin production • Enhance intestinal barrier function • Compete for adhesion with pathogens • Modify pathogen-derived toxins

Table 24.3 Examples of probiotic strains in products

Strain (alternative designations)	Brand name	Producer
Bifidobacterium animalis DN173010	Activia	Danone/Dannon
Bifidobacterium breve Yakult	Bifiene	Yakult
*Bifidobacterium infantis*35624	Align	Procter and Gamble
Bifidobacterium lactis HN019(DR10)	Howaru Bifido	Danisco
Enterococcus LABSF68	Bioflorin	Cerbios-Pharma
Escherichia coli Nissle1917	Mutaflor	Ardeypharm
Lactobacillus casei DN-114001	Actimel, DanActive	Danone/Dannon
Lactobacillus casei F19	Cultura	Arla Foods
Lactobacillus casei Shirota	Yakult	Yakult
Lactobacillus johnsonii La1(Lj1)	LC1	Nestlé
Lactobacillus plantarum 299V	Good belly, ProViva	Next Foods Probi
Lactobacillus reuteri ATTC55730	Reuteri	Bio Gaia Biologics
Lactobacillus rhamnosus ATCC53013(LGG)	Vifit and others	Valio
*Lactobacillusrhamnosus*LB21	Verum	Norrmejerier
*Saccharomyces cerevisiae (boulardii)*lyo	Diar safe, Ultralevure	Wren Laboratories, biocodex

- Stimulating immune system.
- Metabolism of lactose.
- Absorption of calcium.
- Synthesis of vitamins.
- Reduction in yeast and vaginal infections.
- Alleviation of constipation and diarrheal diseases.

5 Best Probiotic Foods

Eating more probiotic foods can increase the number of good bacteria in our body. Fermented foods in (Table 24.4) are a good source [37].

Table 24.4 List of probiotic foods

Apples 	According to research, apples contain 100 million bacteria that can interact with our gut microbiomes in a beneficial way that is superior to taking a single probiotic supplement
Apple cider vinegar 	Numerous health advantages of ACV are frequently cited. Although research is not certain on its precise probiotic properties
Bananas 	Bananas that are just a little bit under ripe are a great source of prebiotics, which promote the development of good probiotics
Balsamic vinegar 	Balsamic vinegar's primary constituent, acetic acid, has probiotic bacteria strains that can boost the immune system and digestive health
Beer 	Hoegaarden and Westmalle Tripel, two-fermented Belgian-style beers, may contain a particular probiotic yeast strain that helps destroy unwanted bacteria in the gut
Bottled probiotic drinks 	Choosing one of the many probiotic beverages available, such as juices, teas, kombucha, and smoothies, is a simple method to enhance your probiotic intake

(continued)

Table 24.4 (continued)

Cereal 	Probiotics are an ingredient in several cereal brands' offerings, like Kellogg's special K
Cheddar cheese 	The probiotic lactic acid bacteria employed as a starter for cheddar cheese may be able to withstand the cheese-making and ageing processes
Cottage cheese 	Cottage cheeses that have been fermented with live, active cultures tend to be higher in probiotics than others
Celery juice 	According to research, when enhanced with good bacteria, celery juice may be a probiotic beverage that has health benefits
Dark chocolate 	According to research, probiotics and chocolate may have an analogous impact on gut flora
Dairy alternatives 	Numerous dairy substitutes, including yoghurt and milk made from nuts, may have living cultures. To find lactobacillus or other probiotic strains, simply read the labels
Feta 	Many forms of this brine-packed salty sheep's milk cheese include probiotic bacteria
Fermented fish 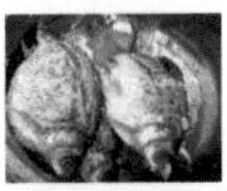	Researchers have discovered specific probiotics in a fermented fish product called utonga-kupsu
Green peas 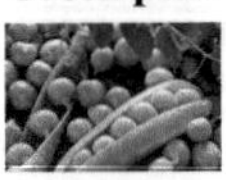	Green peas might contain the probiotic Leuconostoc mesenteroides, according to a Japanese study
Garlic 	Garlic is a flavorful and adaptable vegetable that also acts as a prebiotic food, promoting the growth of the beneficial probiotic bifidobacteria in the stomach, which may help prevent disease
Gouda 	Gouda is a probiotic-rich cheese because it can retain healthy microorganisms during the cheese-making process

(continued)

Table 24.4 (continued)

Greek yogurt 	Greek yoghurt, which is produced by straining normal yoghurt and is also a great source of probiotics, is higher in protein and lower in carbohydrates and sugars than other varieties of yoghurt
Herbal teas 	Some herbal teas are enhanced with probiotics for an additional health boost in addition to their antioxidant content
Kimchi 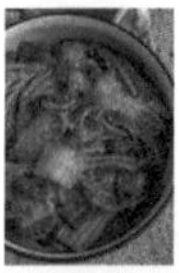	A mainstay of traditional Korean cuisine, kimchi is created by fermenting vegetables, mainly cabbage, with probiotic lactic acid bacteria. It also lowers cholesterol, fosters cognitive health, and strengthens the immune system
Kombucha 	A beverage called kombucha, which is prepared by fermenting black or green tea with sugar, yeast, and bacteria, is praised for its many health advantages, including improved digestion
Kefir 	Numerous varieties of good bacteria and yeast are present in this fermented milk beverage. It contains antibacterial and anticancer qualities in addition to being proved to help digestion
Kvass 	Kvass, a popular beverage in Eastern Europe, is a probiotic food made from malt, rye, flour, stale rye bread, and sugar
Lassi 	A lassi is a beloved beverage in India and Pakistan that is created from fermented yoghurt and fruits like mango and is high in probiotics
Miso 	Miso is a common ingredient in Japanese cuisine and is produced by fermenting soybeans with salt, koji, and other ingredients. There are numerous kinds of the paste, which is frequently used in miso soup. The vitamins B, E, K, and folic acid are also abundant in it
Mozzarella 	It has fewer calories and sodium than other types of cheese, and mozzarella still contains beneficial bacteria. This is a traditional gooey pizza topping
Microalgae 	Microalgae, which frequently comes in powder form, provides a lot of health advantages and is a rich source of probiotics

(continued)

Table 24.4 (continued)

Natto 	Natto is a probiotic-rich fermented soybean product that is a common breakfast dish in Japan. A strong source of protein and vitamin K2 that supports bone health is natto
Onions 	Onions, which are high in fiber and prebiotics, can support the development of good bacteria in the stomach
Olives 	Green, Kalamata and other kinds of olives, traditionally made via fermentation, often contain lactic acid bacteria
Provolone 	Provolone includes probiotics and is used to make most cheeses through fermentation
Parmesan 	The aged, hard Italian cheese is rich in calcium and has both prebiotic and probiotic microorganisms
Pickled beets 	Pickled and fermented beets include probiotics in addition to fiber, vitamins, iron, and other nutrients
Pickled cucumber 	Probiotics and vitamin K are abundant in cucumbers that have been allowed to naturally ferment in salted water with their own lactic acid bacteria
Pickled onions 	Probiotics are not present in every pickled onion. Look for probiotics produced by the lacto-fermentation process, which uses lactic acid bacteria in the pickle, to add more to your diet
Raw cheeses 	When creating cheese, the natural bacteria in raw, or unpasteurized, milk can survive as the cheese ferments
Skyr 	This dairy product from Iceland contains probiotic microorganisms similar to yoghurt and is produced by fermenting skim milk. It also has a low calorie and fat content while being rich in protein and other minerals

(continued)

Table 24.4 (continued)

Sauerkraut 	The salty, tart fermented cabbage is high in probiotics. Along with being abundant in fiber, vitamins, and antioxidants, it is also heavy in sodium
Smoothies 	You may make a nutritious breakfast or snack by blending your favorite fruits and veggies with probiotic-rich yoghurt
Sourdough bread 	Wild yeast and lactic acid bacteria, which are found naturally in flour, are essential components of sourdough's leavening system. Researchers also think it might have probiotic-like qualities
Sour cream 	Probably not many health advantages come to mind when you think of sour cream, but some varieties do include probiotics
Soy sauce 	Even though soy sauce is a fermented food, unless it is clearly marked as such, it may not necessarily be a probiotic. But studies suggest that it, along with other fermented foods, might also be advantageous for gastrointestinal health
Soy milk 	For added health benefits, probiotics may be added to other soy milk products as well as fermented soy milk
Spirulina 	Spirulina, a form of blue-green algae, is a well-known dietary supplement that is packed with vitamins, minerals, and probiotics
Supplements 	The majority of nutrition experts agree that the greatest probiotic sources are foods themselves, while there are also many probiotic supplements available. The probiotics lactobacillus or bifidobacterium are recommended by the Cleveland Clinic, along with products that have "one billion colony forming units" or more
Buttermilk 	Traditional buttermilk contains probiotics since it is prepared from the liquid left over after creating butter, as opposed to cultured buttermilk, which is popular in the United States but is not a probiotic food
Turhi 	This traditional middle eastern and Balkan vegetable pickle mixture, which includes carrots, celery, peppers, and other vegetables, is packed with probiotics

(continued)

Table 24.4 (continued)

Tempeh 	The protein content of tempeh makes it a well-liked meat alternative. The probiotic food and vitamin B12 source are made from fermented soybeans
Umeboshi 	These unripe ume fruit-based Japanese fermented plums are created. Umeboshi can be eaten whole, made into a paste, or kept in vinegar for storage
Yogurt 	Yoghurt is one of the best sources of probiotics since it is made from milk that has been fermented by lactic acid bacteria and bifidobacteria, both good forms of bacteria. The high calcium content is also excellent for your bones

6 Clinical Applications

For assessing clinical efficacy, meta-analyses are thought to offer the strongest degree of evidence. The heterogeneity of trial designs, probiotic interventions used, populations studied, and the relatively small numbers included in each clinical trial make it difficult to apply meta-analysis to clinical trials using probiotics. The strain-specificity of effects must be carefully taken into account in meta-analyses of probiotics, but such problems can afflict analyses of any intervention.

Below is a summary of recent findings about the clinical uses of several probiotics:

6.1 Colorectal Cancer Prevention

Although probiotics and prebiotics have been shown to improve colorectal cancer biomarkers in animal models and are thought to play a role in the development of the disease, there is a dearth of evidence demonstrating these benefits in humans.

6.2 Cardiovascular Disease

Probiotics and prebiotics have not yet been demonstrated to be effective in preventing disease and lowering the risk of cardiovascular problems [36].

6.3 Diarrhea Treatment and Prevention

Treatment of Acute Diarrhea

Some probiotic strains can help children with acute infectious diarrhea feel better faster. The duration of acute diarrheal sickness in children is reduced by about 1 day by oral dosing. According to numerous published meta-analyses of controlled clinical trials examining various probiotic strains, probiotics are probably both safe and efficient.

Prevention of Acute Diarrhea

There is evidence that, under certain circumstances, certain probiotics can be useful in preventing both adult and pediatric diarrhea. Probiotics probably have little to no effect on diarrhea lasting 48 hours or longer, according to a Cochrane meta-analysis based only on big trials with minimal risk of bias. Probiotics may consequently need to be given early on [38].

Prevention of Antibiotic-Associated Diarrhea

There is evidence of effectiveness in preventing diarrhea brought on by antibiotic medication in either adults or kids. According to meta-analyses, probiotics may have a modest benefit in reducing antibiotic-associated diarrhea in children, adults, and elderly individuals [39–41]/.

Prevention of C. Difficile Diarrhea

Probiotics are successful in reducing *C. difficile*-associated diarrhea in those taking antibiotics, according to a 2017 meta-analysis's conclusion [42]. Patients who are not immunosuppressed or seriously disabled can utilize probiotics without concern, according to research. The data suggest that patients who are at high risk of experiencing *C. difficile*-associated diarrhea would benefit from being informed of the potential advantages and disadvantages of probiotics, according to the authors, who also emphasized the need for more research.

Prevention of Radiation-Induced Diarrhea

By strengthening intestinal barrier function, enhancing innate immunity, and promoting intestinal repair mechanisms, the gut microbiota may be crucial in radiation-induced diarrhea. Probiotics may be helpful in the prevention and perhaps in the treatment of radiation-induced diarrhea, according to a 2013 meta-analysis [43].

6.4 Helicobacter pylori *Eradication*

The 2022 Maastricht VI/Florence Consensus Report on management of *H. pylori* infection reached the following conclusion: Some probiotics have been demonstrated to be useful in lowering gastrointestinal side effects produced by *H. pylori* eradication therapy and, thus, have a positive impact on the treatment. The grade of recommendation was moderate, though, and the quality of the evidence was poor [44]. Without concurrent antibiotic therapy, there is no evidence to support the idea that a probiotic would be effective on its own. Instead, probiotics appear to increase the rate of *H. pylori* eradication by lowering the side effects of eradication therapy rather than by directly affecting *H. pylori*.

6.5 Allergy

The best research shows that giving particular probiotics to expectant moms and infants up to 6 months of age can prevent atopic dermatitis. A recent clinical trial, though, did not support these findings. A few well-designed trials have shown that some probiotic strains can effectively treat a subset of individuals with atopic eczema in the therapy of allergic illness. The effectiveness of probiotics in avoiding food allergies is not well understood [30].

6.6 Hepatic Encephalopathy Prevention and Treatment

Hepatic encephalopathy is frequently prevented and treated with prebiotics, such as lactulose. There is evidence that one probiotic mixture can reverse mild hepatic encephalopathy. A 2017 Cochrane meta-analysis revealed that the evidence from three studies on the advantages of probiotics for persons with hepatic encephalopathy was of low quality [45]. The scientists came to the conclusion that probiotics may enhance recovery, quality of life, and plasma ammonia concentrations despite the fact that no difference in death rate was seen.

6.7 Immune Response

There is some evidence that oligofructose, a prebiotic, and a few probiotic strains can help boost immune function. Trials to avoid acute infectious diseases (nosocomial diarrhea in infants, influenza outbreaks in the winter) and trials to assess antibody responses to vaccines have yielded evidence that suggests improved immune responses.

Pouchitis

There is evidence that a probiotic mixture can help prevent a pouchitis attack from starting and additional relapses after antibiotic-induced remission. For adults and kids with mildly active pouchitis or as maintenance therapy for those in remission, the probiotic mixture is advised [46].

Ulcerative Colitis

Individual studies indicate that some probiotics may be equally safe and effective as standard treatment for mild to moderately active ulcerative colitis in both adult and pediatric populations. Probiotics were not shown to be useful in treating more severe ulcerative colitis, according to a 2020 Cochrane meta-analysis [47]. The analysis also found that the evidence for inducing remission in mild-to-moderate ulcerative colitis was of poor certainty.

Crohn's Disease

Crohn's disease (CD) is a chronic inflammatory disorder of the gastrointestinal tract, in which the pathogenesis is believed to be partly influenced by the gut microbiome. Probiotics can be used to manipulate the microbiome but research on probiotics with Crohn's disease has shown that there is no evidence to support their use in remission induction or maintenance.

6.8 Irritable Bowel Syndrome (IBS)

Published research consistently shows that probiotic therapies reduce stomach bloating and gas; some strains may also lessen pain and offer general relief. The research indicates that some probiotics may help people with functional abdominal pain reduce their symptoms and enhance their quality of life.

6.9 Lactose Malabsorption

*Lactobacillus delbrueckii sub*sp. *Bulgaricus* and *Streptococcus thermophilus* enhance lactose digestion and lessen lactose intolerance symptoms. This was demonstrated in several research experiments under controlled conditions including yoghurt with live cultures [48].

6.10 Necrotizing Enterocolitis

Probiotic supplementation lowers the danger of necrotizing enterocolitis in premature newborns. Although not all probiotic products examined are successful, meta-analyses of randomized controlled trials have revealed a decreased risk of death in probiotic-treated groups. Twenty patients must be treated for probiotics to be effective in preventing one death from all causes. For this particularly sensitive subset of patients, proper product quality in the probiotic is crucial [49]. Although no effect was seen on severe neurodevelopmental impairment, there was a modicum of assurance for a decrease in mortality and late-onset invasive infection [50].

6.11 Nonalcoholic Fatty Liver Disease

Numerous adult and pediatric randomized clinical trials have demonstrated the value of some probiotics as a therapeutic alternative to reduce steatohepatitis. Blood cholesterol, TNF-α, liver function tests (alanine transaminase and aspartate aminotransferase), and the results of the homeostasis model assessment (HOMA) were all improved by probiotics. To confirm long-term advantages, more research is required.

6.12 Prevention of Systemic Infections

The use of probiotics or synbiotics in critically ill adult patients in intensive care units is not backed by enough evidence. Probiotics and prebiotics have been demonstrated to influence a number of clinical outcomes that are outside the typical range of gastrointestinal disease, even though it is outside the purview of this recommendation. A relationship between these disorders and the gastrointestinal system has been established by emerging research that suggests gut microbiota may affect a number of nongastrointestinal conditions. Probiotics have been demonstrated in numerous trials to decrease bacterial vaginosis, prevent infantile atopic dermatitis, decrease oral pathogens and dental cavities, and decrease the frequency and length of common upper respiratory tract infections. The World Allergy Organization has advised probiotic usage during pregnancy, breastfeeding, and weaning in families with a high risk of allergic disease as a result of the probiotics' overall effectiveness in reducing allergic disease during the perinatal period [51].

7 Conclusion and Recommendation

The twentieth century saw the emergence of knowledge about the role of probiotics in oral and intestinal health, as well as the discovery of beneficial bacteria in the human gut. Longevity is increased by having certain gut microorganisms. The current research trends, which investigate the health-promoting qualities and mechanisms of gut microorganisms in cases of various common diseases, are based on the correlation between probiotics and long life. The experts concurred that there is sufficient scientific evidence to suggest that eating foods containing probiotics may have health benefits. Probiotics can be applied to conditions to prevent certain diseases and modulate host immunity such as gastrointestinal infections, certain bowel disorders, allergy, and urogenital infections, which affect a large portion of the world's population. There is strong evidence that certain probiotic strains can provide the host with some health advantages while remaining safe for human consumption. Currently, there is no recognized worldwide regulatory status for probiotics as a food ingredient. Only a few nations have regulations that enable probiotic products to list specific health advantages.

References

1. Hill C, Guarner F, Reid G, Gibson GR, Merenstein DJ, Pot B, Morelli L, Canani RB, Flint HJ, Salminen S (2014) Expert consensus document: The International Scientific Association for Probiotics and Prebiotics consensus statement on the scope and appropriate use of the. Nat Rev Gastroenterol Hepatol 11:506–514
2. World Gastroenterology Organisation Global Guidelines (2023) Probiotics and prebiotics
3. Venugopalan V, Shriner KA, Beringer AW (2010) US National Library of Medicine. Emerg Infect Dis 16(11):1661–1665
4. Noce A, Marrone G, Di Daniele F, Ottaviani E, Wilson Jones G, Bernini R (2019) Impact of gut microbiota composition on onset and progression of chronic noncommunicable diseases. Nutrients 11(5):1073
5. Lefebvere S (2011) Literature review- Probiotics. Banfield applied research and Knowledge team https://www.banfield.com/
6. Song HY, Zhou L, Liu DY, Yao XJ, Li Y (2018) What roles do probiotics play in the eradication of Helicobacter pylori? Current knowledge and ongoing research. Gastroenterol Res Pract 2018:937–948
7. Harish K, Varghese T (2006) Probiotics in human. A literature review on health benefits of probiotics. Calicut Med J 4(4):e3
8. Naidu AS, Bidlack WR, Clemens RA (1999) Probiotic spectra of lactic acid bacteria (LB). Crit Rev Food Sci Nutr 38(1):13–126
9. Fooks LJ, Fuller R, Gibson GR (1999) Probiotics, prebiotics and human gut microbiology. Int Diary J 9:53–61
10. Eslami M, Yousefi B, Kokhaei P, Hemati M, Nejad ZR, Arabkari V, Namdar A (2019) Importance of probiotics in the prevention and treatment of colorectal cancer. J Cell Physiol 234:17127–17143
11. Brito LP, Júnior JNS, Barros PDS, Silva EC, Calaça PRA, Soares MTCV (2021) Can postbiotics show antiviral effects against Sars-CoV-2? Res Soc Dev 10(8):e14610817259. https://doi.org/10.33448/rsd-v10i8.17259
12. Patra S, Saxena S, Sahu N, Pradhan B, Roychowdhury A (2021) Systematic network and meta-analysis on the antiviral mechanisms of probiotics: a preventive and treatment strategy to mitigate SARS-CoV-2 infection probiotics and antimicrobial. Proteins 13(4):1138–1156. https://doi.org/10.1007/s12602-021-09748-w
13. Akour A (2020) Probiotics and COVID-19: is there any link? Lett Appl Microbiol 71(3):229–234. https://doi.org/10.1111/lam.13334
14. Tannock GW (1994) More than a smell: the complexity of the normalmicroflora. In: Normal microflora. An introduction to microbes inhabiting the human body. Chapman and Hall, London, pp 1–36

15. Tannock GW, Fuller R, Smith SL, Hall MA (1990) Plasmid profiling of members of the family Enterobacteriaceae, lactobacilli, and bifidobacteria to study the transmission of bacteria from mother to infant. J Clin Microbiol 28(6):1225–1228. https://doi.org/10.1128/jcm.28.6.1225-1228.1990
16. Tannock I, Murphy K (1983) Reflections on medical oncology: an appeal for better clinical trials and improved reporting of their results. J Clin Oncol 1(1):66–70. https://doi.org/10.1200/JCO.1983.1.1.66
17. Sanders ME (1998) Overview on functional foods: emphasis on probiotic bacteria. Int Dairy J8:341–347
18. Vinderola N, Bailo JA, Reinheimer JA (2000) Survival of pro-biotic microflora in Argentinian yoghurts during refrigerated storage. Food Res Int 33:97–102
19. Ezendam J, van Loveren H (2006) Probiotics: immunomodulation and evaluation of safety and efficacy. Nutr Rev 64(1):1–14. https://doi.org/10.1111/j.1753-4887.2006.tb00168.x
20. Capela P, Hay TKC, Shah NP (2006) Effect of cryoprotectants, prebiotics and microencapsulation on survival of probiotic organisms in yoghurt and freeze-dried yoghurt. Food Res Int 39(2):203–211. https://doi.org/10.1016/j.foodres.2005.07.007
21. Ghasemian A, Eslami M, Shafiei M, Najafipour S, Rajabi A (2018) Probiotics and their increasing importance in human health and infection control. Rev Med Microbiol 29(4):153–158
22. Gupta S, Abu-Ghannam N (2012) Probiotic fermentation of plant based products: possibilities and opportunities. Crit Rev Food Sci Nutr 52:183–199
23. Guidance on Community Mental Health Services (2021) Promoting person-centred and rights-based approaches. https://www.who.int/publications/i/item/9789240025707
24. Ranjha MMAN, Irfan S, Lorenzo JM, Shafique B, Kanwal R, Pateiro M, Arshad RN, Wang L, Nayik GA, Roobab U (2021) Sonication, a potential technique for extraction of phytoconstituents: a systematic review. Processes 9:1406
25. Chibbar R, Dieleman LA (2019) The gut microbiota in celiac disease and probiotics. Nutrients 11(10):2375
26. Soccol CR, Vandenberghe LPS, Spier MR, Medeiros ABP, Yamaguishi CT, Lindner JDD, Pandey A, Soccol VT (2010) The potential of probiotics: a review. Food Technol Biotechnol 48(4):413–434
27. Brown IL, McNaught KJ, Ganly RN, Conway PL, Evans AJ, Topping DL, Wang X (2000) Carrier acts as a growth or maintenance medium for microorganisms in the large bowel or other regions of the gastrointestinal tract. US 6060050. http://www.google.com/patents/US6060050
28. Zheng J, Wittouck S, Salvetti E (2020) A taxonomic note on the genus Lactobacillus: description of 23 novel genera, emended description of the genus Lactobacillus Beijerink 1901, and union of Lactobacillaceae and Leuconostocaceae. Int J Syst Evol Microbiol 70(4):2782–2858. https://doi.org/10.1099/ijsem.0.004107
29. Ahmad A, Khalid S (2018) Therapeutic aspects of probiotics and prebiotics. In: Diet, microbiome and health. Elsevier, pp 53–91
30. Kopp-Hoolihan L (2001) Prophylactic and therapeutic uses of probiotics: a review. J Am Diet Assoc 101(2):229–241
31. Schubiger CB, Orfe LH, Sudheesh PS, Cain KD, Shah DH, Call DR (2015) Entericidin is required for a probiotic treatment (Enterobacter sp. strain C6-6) to protect trout from cold-water disease challenge. Appl Environ Microbiol 81(2):658–665
32. Zubillaga M, Weill R, Postaire E, Goldman C, Caro R, Boccio J (2001) Effect of probiotics and functional foods and their use in different diseases. Nutr Res 21(3):569–579
33. Sen M (2019) Role of probiotics in health and disease–a review. Int J Adv Life Sci Res 2(2):1–11
34. Khan AA, Khurshid M, Khan S, Alshamsan A (2013) Gut microbiota and probiotics: current status and their role in cancer therapeutics. Drug Dev Res 74(6):365–375
35. Dash SK (2001) Selection criteria for probiotic supplements. Nutr Prospect 24(4):9–11
36. World Gastroenterology Organisation (2008) Global Guidelines. Probiotics and prebiotics

37. Sweeney E (2023) Looking to give your gut a little reset? These 50 probiotic foods are a great place to start! Health and Wellness. https://parade.com/1043590/ericasweeney/best-probiotic-foods
38. Collinson S, Deans A, Padua-Zamora A, Gregorio GV, Li C, Dans LF (2020) Probiotics for treating acute infectious diarrhoea. Cochrane Database Syst Rev 12(12):CD003048
39. Guo Q, Goldenberg JZ, Humphrey C, El Dib R, Johnston BC (2019) Probiotics for the prevention of pediatric antibiotic-associated diarrhea. Cochrane Database Syst Rev 4(4):CD004827
40. Goodman C, Keating G, Georgousopoulou E, Hespe C, Levett K (2021) Probiotics for the prevention of antibiotic-associated diarrhoea: a systematic review and meta-analysis. BMJ Open 11(8):e043054
41. Zhang L, Zeng X, Guo D, Zou Y, Gan H, Huang X (2022) Early use of probiotics might prevent antibiotic associated diarrhea in elderly (>65 years): a systematic review and meta-analysis. BMC Geriatr 22(1):562
42. Goldenberg JZ, Yap C, Lytvyn L, Lo CKF, Beardsley J, Mertz D (2017) Probiotics for the prevention of Clostridium difficile-associated diarrhea in adults and children. Cochrane Database Syst Rev 12(12):CD006095
43. Hamad A, Fragkos KC, Forbes A (2013) A systematic review and meta-analysis of probiotics for the management of radiation induced bowel disease. ClinNutr 32(3):353–360
44. Malfertheiner P, Megraud F, Rokkas T, Gisbert JP, Liou JM, Schulz C (2022) Management of Helicobacter pylori infection: the Maastricht VI/Florence consensus report. Gut:327745
45. Dalal R, McGee RG, Riordan SM, Webster AC (2017) Probiotics for people with hepatic encephalopathy. Cochrane Database Syst Rev 2(2):CD008716
46. Su GL, Ko CW, Bercik P, Falck-Ytter Y, Sultan S, Weizman AV (2020) AGA clinical practice guidelines on the role of probiotics in the management of gastrointestinal disorders. Gastroenterology 159(2):697–705
47. Kaur L, Gordon M, Baines PA, Iheozor-Ejiofor Z, Sinopoulou V, Akobeng AK (2020) Probiotics for induction of remission in ulcerative colitis. Cochrane Database Syst Rev 3(3):CD005573
48. Savaiano DA, Hutkins RW (2021) Yogurt, cultured fermented milk, and health: a systematic review. Nutr Rev 79(5):599–614
49. Van den Akker CHP, Van Goudoever JB, Shamir R, Domellöf M, Embleton ND, Hojsak I (2020) Probiotics and preterm infants: a position paper by the European Society for Paediatric Gastroenterology Hepatology and Nutrition Committee on Nutrition and the European Society for Paediatric Gastroenterology Hepatology and Nutrition Working Group for Probiotics and Prebiotics. J Pediatr Gastroenterol Nutr 70(5):664–680
50. Sharif S, Meader N, Oddie SJ, Rojas-Reyes MX, McGuire W (2020) Probiotics to prevent necrotisingenterocolitis in very preterm or very low birth weight infants. Cochrane Database Syst Rev 10(10):CD005496
51. Zhang GQ, Hu HJ, Liu CY, Zhang Q, Shakya S, Li ZY (2016) Probiotics for prevention of atopy and food hypersensitivity in early childhood: a PRISMA-compliant systematic review and meta-analysis of randomized controlled trials. Medicine (Baltimore) 95(8):e2562

Chapter 25
Processing of Spices and Plantation Crops

Jagadeesh S. L., Chandan K., Umme Seema N., and Suresha G. J.

1 Processing of Spices

Spices have a significant role in the everyday diets of people all over the world. These have drawn more attention lately as flavouring, preservatives and medicinal ingredients. Spices are the dried, fragrant components of natural plants. Leaves, barks, buds, fruits, rhizomes etc. are used to make spices. Spice characteristics, such as colour and constitution, can vary depending on harvest, location, type of climate (wet or dry) and historical context [1]. The variety, harvest and postharvest circumstances, agroclimatic conditions and other elements all directly affect the physico-chemical qualitative traits of spices. Spices' flavour quality is influenced by both internal and external variables. Major extrinsic parameters are temperature of the planting environment, humidity during blossoming, soil moisture, availability of fertilizer, rainfall etc., in addition to processing techniques; major intrinsic factors are maturity and variety of spices. The primary processing steps that affect spice quality are extraction, packing, drying, grinding and sterilizing [2]. As a result, careful selection of appropriate processing methods is necessary to produce high-quality spices [3].

Jagadeesh S. L. (✉)
Regional Horticultural Research and Extension Centre, UHS Campus,
Bengaluru, Karnataka, India

Chandan K.
Department of Post Harvest Management, College of Horticulture, UHS Campus,
Sirsi, Karnataka, India
e-mail: chandan.k@uhsbagalkot.edu.in

Umme Seema N.
KSN University of Agricultural and Horticultural Sciences, Shivamogga, Karnataka, India

Suresha G. J.
Department of Post Harvest Management, College of Horticulture, UHS Campus,
Bengaluru, Karnataka, India
e-mail: suresh.gj@uhsbagalkot.edu.in

J. S. Yaradoddi et al. (eds.), *Frontiers in Food Biotechnology*,
https://doi.org/10.1007/978-981-97-3261-6_25

1.1 Drying

Drying is a common technique for preserving spices. High moisture content in spices causes microbial deterioration, which reduces their shelf life and quality. Traditional drying methods like sun drying, hot air and solar drying may achieve high drying efficiency, resulting in quality degradation by removing volatile particles, which may affect the spices' flavour and nutritional value [4–6].

Among the dehydration methods, infrared and combined infrared drying, microwave and combined microwave drying, combined solar drying and infrared and combined infrared drying are regarded to be comparatively more effective at preserving and retaining the quality of spices. These methods have several benefits over traditional methods of drying, including better nutritional value, reduced drying times and greater quality retention (better colour and appearance, flavour etc.) [7]. The common drying methods for preparing spices are compiled in Table 25.1.

Among the advanced techniques, microwave drying is an alternative for getting good quality spice products.

1.2 Microwave Drying

Electromagnetic microwaves have a frequency range of 300–300 GHz with a wavelength variation of 0.1 cm (300 GHz)–100 cm (300 MHz) [16]. When compared to traditional dehydration techniques, the microwave drying process offers the many benefits of a quick drying period, rapid rate of drying and retains the quality. Curry leaves, ginger and garlic have all been dried using this creative and promising drying technique [17, 18, 19]. However, there are several disadvantages to microwave drying alone, including inconsistent heating, the possibility of ruining the texture, the 'puffing' phenomena, a restricted degree of microwave penetration and more [19]. Thus, in order to accomplish dehydration, microwave drying is combined with additional techniques like freezing, hot air, vacuum drying etc. To achieve quicker, more efficient and more even drying.

In comparison to hot air dehydration at 60 °C and 50 °C, respectively, the combination of hot air drying and microwave-assisted convective dehydration (120 W, 50 °C) in ginger reduced the amount of time needed to produce volatile oil and oleoresin by 44 and 53% respectively [3].

The spouting, rolling state of the particles is maintained either continuously or sporadically using the microwave-induced fluidization drying process. Ginger (Zingiber officinale Roscoe) was tested under a variety of microwave fluidization drying conditions as a substitute to hot air drying at 75 °C. These conditions included continuous radiation at a variety of power density values (0.6, 1.2 and 1.8 W/g) and radiation 2 W/g with 0.4 s spouting. It was determined that pulse spouting worked well to keep the spicy flavour (gingerol) intact during the drying process and to avoid overheating [20].

Table 25.1 Comparison of spices drying techniques [8]

Drying technology	Advantages	Disadvantages	Reference
Shade drying	Simple, affordable, minimum flavour loss	Required larger area, labour-intensive, low production efficacy and easily contaminated by microbes make it unsuitable for industrial production	[9]
Hot air drying	Simple equipment with excellent performance at a low cost, quick drying, broad applicability, easy maintenance and appropriate for industrial production	Not recommended for material that are highly susceptible to heat, as it alters flavour and compromises the purity of fragrance	[8, 10, 11]
Infrared-assisted drying	Faster drying	Poor penetration strength, uneven heating, excessive use of energy, unsuitable for heat-sensitive materials and flavour-affecting due to the dryness of hard-shelled spices like cloves	[12, 13]
Freeze drying	Excellent preservation of flavour and appropriate for products that are heat-liable	Sophisticated equipment, costly maintenance and poor performance	[12, 13]
Microwave-assisted drying	Reduced drying time, easy maintenance, inexpensive equipment, appropriate for industrial production	Impacts on flavour and level of cell lysis are greater than with ultrasound-assisted drying; not recommended for materials that are sensitive to heat; uneven heating; a restricted application for spices	[8, 14]
Ultrasound-assisted drying	Reduced flavour loss, shorter drying period, non-thermal processing, appropriate for products that are heat-sensitive	Sound waves quickly attenuate and cause damage to cell structure	[14, 15]

It was discovered that myristicin's flavour and colour may be preserved by pulsed microwave-assisted hot air drying of nutmeg mace [21]. Combining the greatest aspects of hoover and microwave drying, hoover microwave drying yields higher-quality products with less energy use. The garlic slices and cloves that were dried with the 720 W vacuum oven drying process were found to be brighter and to have retained the most volatile oil [22].

Garlic slices were dehydrated employing a microwave-vacuum method of drying until their moisture content reached 10% (w.b.), following an air-drying process at 45 °C till the moisture content decreased to less than 5% (w.b.). As a result, the dried goods' quality was almost identical to the freeze-dried samples' [23]. When paired with freeze-drying, microwave drying can enhance both the quality of the spices and the velocity of drying.

The results showed that wasabi (*Eutrema yunnanense*) dried 50% faster using microwave-assisted freeze drying than using freeze drying, and that the quality (cof wasabi samples processed with pulse-spouted microwave-assisted freeze drying was comparable to freeze-dried samples [24].

1.3 Infrared Drying

Infrared (IR) rays are electromagnetic radiations having wavelengths between 0.78 and 1000 μm. Depending on the temperature of the source, IR radiation can be categorized as far-infrared (3–1000 μm), mid-infrared (1.40–3 μm) or near-infrared (0.78–1.40 μm) [25]. Infrared radiation can efficiently lower the moisture content of spices without warming the surrounding environment.

Chinese ginger (*Zingiber officinale*) was the subject of a study that examined the effects of intermittent microwave convective drying, air drying, freeze drying and infrared drying. Infrared, freeze and intermittent microwave convection techniques all resulted with increased gingerol preservation, according to the study. Also, infrared drying required less energy (12.23 kWh/g water) than freeze drying (33.7 kWh/g water) [26]. With infrared drying, the rate and duration of dehydration are significantly reduced. As the intensity of infrared radiation increased, the rates of dehydration in garlic slices subjected to airflow of 0.75 and 1.25 m/s and irradiation levels of 0.075, 0.15, 0.225 and 0.3 W/cm^2 decreased [27]. The temperature at which infrared rays dehydrate affects the quality of spices. Maximum safranal and crocin concentration was reported upon dehydration at 110 °C, according to saffron dehydrated at infrared dehydration temperatures of 60, 70, 80, 90, 100 and 110 °C [28].

When compared to the application of heat pump and far-infrared radiation, the novel drying method known as far-infrared radiation-assisted heat pump (HP-FIR) dramatically reduced the drying time for dehydrating chopped garlic [27].

1.4 Solar Drying Technique

One of the most popular alternative forms of dehydration is solar energy. Solar drying is one low-cost, environmentally friendly dehydration technique utilized in the process. In comparison to heat pump-dehumidified and tray drying methods, ginger drying on a mixed-mode solar drier (60–62.82 °C) maintained the greatest 6-gingerol content with the lowest drying time [29]. A combination of a photovoltaic-thermal solar dryer was used to dehydrate saffron at various flow rates of air mass (0.008, 0.012 and 0.016 kg/s), temperature of the air (40, 50 and 60 °C) and drier modes (with/without a heat pump system).

The solar dryer dehydrates faster and with less energy use when it employs a heat pump. At 60 °C air temperature and 0.016 kg/s flow velocity, the heat pump reached its highest drier effectiveness (72%) and specific extraction of water yield (1.16) [29].

An indirect forced convection sun dehydration system with a constant airflow velocity of 0.0636 kg/s was used to thinly dry curry leaves (Murraya koenigii). With an 8-month payback period instead of the full 15-month duration, this strategy was proven to be practical in rural settings [30]. Comparing convection solar drying to sun drying, which lowers the water content from 75.6% to 10.1% (w.b.) [31], the drying time for giant cardamom (Amomum subulatum) is reduced from 48 h to 24 h.

1.5 Cryogenic Grinding

To speed up the drying process, spices must be ground smaller. The creation of exceptional spice powders has gained popularity recently. Using an energy-intensive traditional grinding process for comminution results in a substantial amount of heat being created, which leads to a reduction in the quantity and quality of the product [32]. By using cryogenic fluids such as carbon dioxide and liquid nitrogen while operating at temperatures that are extremely low (typically below −50 °C), the cryogenic grinding process is a promising way of comminution that preserves the key qualities of spices [33]. Turmeric (*Curcuma domestica*) powder produced by the cryogenic grinding procedure was finer than that produced by the traditional grinding technique [34].

A study comparing the impact of ambient and cryogenic grinding on black pepper using a rotor mill, ball, pin and hammer found that the rotor mill performed best for the cryogenic grinding process [35]. Because of ambient grinding, grinder sieves and lids developed a mixture of powder and volatile oil coating. Grinding spices at a certain temperature affects their physical-chemical properties and the amount of volatile oil they hold. Black pepper powder was shown to have the highest volatile oil and mineral content at 120 °C and the lowest at 80, 40, 0 and 40 °C [36]. In contrast, a temperature increase of −110 to −50 °C during the grinding process does not significantly alter the volatile oil content of cloves; nevertheless, a temperature increase of 55 to 85 °C has a significant impact and does not cause any deposition on the sieve surface [37].

Because it requires less energy, has a better productivity, has fewer volatile losses, does not block the mill, is adept at controlling the size of the particles, and possesses little to no microbial survival, cryogenic grinding is consequently more advantageous.

1.6 Extraction

The majority of the compounds used to flavour spices are found as essential oils or oleoresins. The primary components of essential oils are terpenoids and their oxidative derivatives, whose particular kinds are referred to after alcohols and ketones [38]. Oleoresins are a mixture of chemicals that give spices their flavour; however, some of these compounds, including capsaicin and turmeric, are less volatile [39]. Spices generally have greater flavours and aromas with more flavouring compounds in their essential oils or oleoresins. The ratio of different flavour compounds also influences the flavour as a whole.

Traditional methods for extracting essential oils include distillation (using water, steam, or simultaneous distillation), solvent extraction (dipping, Soxhlet extraction), cold pressing and sublimation. Although being widely employed in industry, these methods have a number of drawbacks, including low extraction rates, the potential for hazardous solvent residues and the breakdown of some volatile chemicals [40]. The ultrasound extraction approach, on the other hand, is thought to be an innovative method that addresses the aforementioned drawbacks.

Traditional techniques for extracting essential oils include sublimation, cold pressing, solvent extraction (soxhlet, water and steam) and distillation (steam, water and simultaneous). Although the fact that these methods are extensively employed in industry, they have a number of drawbacks, including low extraction rates, the potential for hazardous solvent residues and the breakdown of some volatile chemicals [41]. On the contrary, the ultrasound extraction method is thought to be an innovative method that addresses the aforementioned issues. Supercritical CO_2 and ultrasound are combined to improve the flavour component extraction process. While the amounts of β-caryophyllene and acetyl in clove powder decreased by 3.77 and 16.95%, respectively, the concentrations of cinnamic aldehyde and phenols in cinnamon essential oils increased by 20.35 and 28.07%, respectively [42].

1.7 Sterilization

Sterilization of spices and their products is essential to prevent microbial degradation of flavour compounds. Sterilization methods can be classified as either thermal or non-thermal. Flavour components deteriorate due to conventional thermal techniques' imprecise temperature control. Thus, advanced thermal techniques such as ohmic heating, dielectric heating and inductive heating can be used in addition to non-thermal sterilizing techniques such as irradiation and combination irradiation techniques.

1.8 Ohmic Heating

A novel heat treatment method called ohmic heating is primarily utilized to raise materials to the required sterilizing temperature. The hot food is produced by passing an alternative charge over electrodes that have been installed. Because of the specific volume and quick heating, this enables for speedy sterilization. As opposed to 81.9% with normal heating (100 °C, 8 min), vegetative cells of different strains of Bacillus can be reduced by 99.70% by ohmic heating (50 kHz, 60 V) when pasteurizing gochujang, a conventionally fermented chilli sauce from Korea [12].

1.9 Radio Frequency Heating

Radio frequency waves are electromagnetic waves that have a frequency in the 300 kHz–300 MHz range. Direct exposure to electromagnetic radiation results in volumetric heating from molecular friction, which renders microorganisms inactive. We call this process 'radiofrequency heating.' Red pepper granules have a water activity of 0.71 and are 5-log inactivated when microwave-heated in a minute at a temperature of 70 °C [43].

1.10 Irradiation Method

As part of the non-thermal sterilization process, food items are exposed to ionizing radiations, such as gamma, X, or high-energy electron beams. The DNA of microorganisms is broken down by ionizing radiation, increasing food quality and extending the shelf life without sacrificing nutritional or organoleptic qualities.

Red pepper powder treated to X-rays, electron beams, or gamma-rays significantly decreased the activity of aerobic microbes without effecting the main qualitative qualities. Additionally, there is no noticeable change in the colour, strong flavour or capsanthin content. However, an off-flavour has been detected in all irradiation samples [44]. It has been demonstrated that the sterilizing methods indicated above work. All methods, however, have their limits.

2 Processing of Plantation Crops

2.1 Coconut

One of the major plantation crops cultivated in tropical Asia is coconut (*Cocos nucifera* L.). The world's largest producer, India, grows coconut on 2.1 million hectares of land, producing 22,000 million nuts year with an average productivity of 10,500 nuts per hectare [45].

The food sector employs coconut processing primarily for the preparation of desiccated powder, oil extraction and copra production. The processing of coconuts also includes the production of coir and coir-based products.

Kalparasa

The unopened coconut spadix contains phloem sap, which is known as kalparasa (or Neera). It is a healthy beverage that is packed with carbohydrates, protein, amino acids, minerals, antioxidants and vitamins. With ease, this sap ferments. Fresh, clean and unfermented sap is collected by the coco sap chiller that ICAR-CPCRI developed. Gathering sap in its natural state, it is a portable, lightweight HDPE-insulated device [46, 47].

Tender Coconut Water

Consumers prefer the natural flavour of tender coconut over the packaged soft drinks made from tender coconut water nowadays. When the outer shell of an 8–9-month-old nut is carefully removed, a soft white ball known as Snow Ball Tender Coconut (SBTC) is produced [48]. These nuts have the advantages of being less bulky, beautiful and having a naturally delicate watery taste. But care must be taken to prevent browning and the growth of mould in chopped nuts without compromising the tender water's purity [43].

A technique has been developed by the Defence Food Research Laboratory in Mysore to preserve soft coconut water in pouches or aluminium containers for up to 6 months. To increase their shelf life, many non-thermal food processing technologies are being tried, including ozone, cold plasma, ultraviolet, high-pressure processing, pulsed light and ultrasonography [49].

Coconut Chips

Coconut chips are a ready-to-eat non-fried healthy snack item [50]. It is prepared from 8 to 9-month-old fresh coconut kernel, which undergoes paring, blanching, slicing and osmotic treatment in glucose syrup followed by dehydration in the electric dryer at 70–80 °C for 5–6 h to less than 3% moisture content. Chips are nutritionally stable for up to 6 months of storage [51].

Coconut Cream

Partially defatted coconut milk is referred to as coconut cream. It is used widely in domestic cooking. In ice cream industries, coconut cream replaces dairy fat due to its low-fat content [52]. Also, alcoholic beverages can be blended with coconut cream.

Coconut Oil

The oil from matured coconut kernel is extracted by rotary or expeller methods. Oil is extracted by dry method in industries, but requires refinement through mechanical centrifugation. The oil recovery is around 60–65% [53]. This oil is widely used in tropical countries for culinary and other purposes.

Virgin Coconut Oil

To achieve coconut milk, the fleshy white, grated part of the coconut endosperm is crushed in a screw press. The cream and milk are separated by centrifugation. Virgin coconut oil (VCO) is extracted from the cream using a phase inversion procedure that recovers 89% of the oil. With modern method created by CFTRI, Mysore, the oil recovery is increased to 93%. The oil will have a longer shelf life, be colourless, perfectly clear, rich in nutrients and smell strongly of coconut. The oil has a lot of antioxidants and lauric acid, which has antibacterial and antiviral qualities. The oil is very good for cardiac health [54].

Desiccated Coconut

Desiccated coconut is a grated, shredded, dehydrated coconut (DC) kernel [55] with less than 3% moisture content. It is mostly utilized in India for culinary and other food preparations. Sri Lanka and East Asian countries are the major producers, whereas North America and Western Europe are the primary consumers.

Other Products

Coconut squash, a healthy drink made from coconut water, sugar and natural preservatives like lemon and ginger [56], vinegar, fermented coconut water and *Nata de coco*, a flavoured gelatinous delicacy of Philippines produced using *Acetobacter aceti* subspecies xylinum required to be commercialized throughout the world [57].

2.2 *Areca Nut*

Areca nut *(Areca catechu* L.) is an important plantation crop cultivated commercially in tropical India, East Africa, Far East Asia and South Pacific. India is the world's largest producer of areca nut, producing 710 thousand tonnes annually on 475 thousand hectares of cultivation. Globally, 1.214 million tonnes of areca nut production come from an area of 0.958 million ha [45]. In India, the areca nut industry provides the means of subsistence for around 6.50 million people. In India, it is available as whole, unhusked fruit, dehusked and dried nuts, boiled and dried whole kernels, or their slices. Roughly 25% of the total yield is eaten as ripe fruit, with the remaining portion going into processed and whole nut forms [58]. Processing areca nut into chali from a mature nut involves drying, dehusking, grading and packaging. In contrast, kalipak from a tender nut involves dehusking, boiling, with or without cutting into slices, grading and packaging.

Dehusking

The process of dehusking areca nuts requires a lot of effort since, depending on their maturity state, the nuts must be sliced into transverse or crosswise portions. The laboriousness associated with the human dehusking process is significantly reduced by the availability of rotary drum-type mechanical dehuskers, which have a capacity of 30 kg/h [59].

When it comes to chali, perfectly ripe nuts are chosen and dried in the sun for 35–40 days on a level terrain. After drying, the nuts are dehusked to obtain entire nuts. After being dehusked, the nuts in kalipak are sliced into the desired shapes and sizes. After that, the cut pieces are cooked in water or diluted extract from an earlier boiling. Barks, roots and occasionally gingelly oil are added to the syrup while it is cooking in order to provide colour. After this process, the nuts are dried.

For eight to 10 days, the boiled areca nut (kaliapak) is left out to dry in the sun. Artificial dryers are employed when the weather is cloudy. These days drying the nuts in a solar-cum-biomass dryer may lower their moisture content from 40% to 11% in just 15 h. A recently designed Automatic Areca nut Processing Unit [60] might be used to perform the full processing of raw areca nuts, including dehusking, cutting, boiling and drying, in a reliable and safe manner.

Grading

Processed areca nut is graded based on quality attributes such as size, presence of husk and dents formed on nuts. At the farm level, in the coastal districts of Karnataka, farmers classify *Chali* into '*Old Supari*' and 'new *Supari*'. In upghats, kalipak into *Hosa, Bette*, and *Gorabalu.*

Storage

Gunny sacks coated with polyethene are used to package areca nut. A nut's moisture content affects both its qualitative characteristics and shelf life. Nut deterioration due to high moisture content includes discoloration, a musty smell, loss of dry matter, tissue disintegration, loss of processing and nutritional quality and the generation of mycotoxin. Nuts that have a moisture content of less than 10% or a water activity (aw) of less than 0.71 are guaranteed to store safely [61].

Ancillary Products

The leather industries employ the liquid stain that is produced by boiling tender nuts during the kalipak preparation process. Stain solutions ranged in price from Rs. 120 to Rs. 180 per litre. Every year, there are an average of 3.5 billion areca leaves with sheaths available, which are utilized to make environmentally friendly disposable bowls and plates [62]. Farmers can make a sizable profit from the biodecomposition of husks and the vermicomposting of areca leaf trash [63]. Goods including plastics, paper boards, plyboards, hardboards, soap, toothbrushes and areca tea are becoming more and more popular. To turn these into commercially viable goods, industrial R&D must collaborate with them.

Natural dyes are environmentally beneficial dye substitutes that are non-toxic, renewable, degradable and renewable. Utilizing natural dye offers the benefit of being both cost-effective and environmentally beneficial. Research findings have demonstrated that plant-based natural colourants are appropriate for dying silk and cotton textiles, and they also validate the colour fastness and other superior qualities of naturally dyed goods. Areca nut extract dyes silk fabrics a reddish-brown colour that holds up well to washing, rubbing and perspiration. The dye silk fabric is free of azo-amine and formaldehyde chemicals. This finding demonstrates that areca nut colourants can be used to dye silk fabrics instead of synthetic ones, producing silk that is safe and fully environmentally friendly.

2.3 Cocoa

Cocoa is popularly known as the food of God. The cocoa beans, an economic part of cocoa, are derived from the matured cocoa pods and processed to get end-products such as chocolate, cocoa butter and cocoa powder. Cocoa butter is used in moisturizing creams and soaps, while cocoa powder as a chocolate flavouring agent. Cocoa is a functional food because of its richness in flavonoids, polyphenols and antioxidant properties. According to World Cocoa Foundation (WCF), there are millions of farmers throughout the world, depend on cocoa farming for their income. As a global commodity, cocoa accounts for about 70% of global exports from West African countries like Ghana and Cote d'Ivoire. The market is driven by the high demand for cocoa and chocolate in the food and beverage and confectionery industries [64].

Primary Processing

Harvesting

Traditionally, farmers harvest cocoa when the colour of the pods changes to yellow or orange using wooden mallets or poles, secateurs and sickles. The harvesting tool developed by Central Plantation Crops Research Institute, Regional Station, Vittal, Karnataka, is suitable for harvesting pods from more than two-meter height.

Fermentation

The purpose of fermentation is to get rid of the beans' mucilaginous contents. Smaller bean quantities than those needed for the conventional procedures have been used in mini box, basket, tray and heap methods of small-scale fermentation.

Based on the cut test and the pH of the beans, the effectiveness of fermenting techniques was also investigated [65, 66]. There are three ways for large-scale fermentation: heap, tray and box. Kerala Agricultural University created a cocoa bean fermenter towards the end of 2020 [67], and depending on the season, it takes 5–7 days for the fermentation to finish. Africa uses the heap approach, whereas Asia and Latin America employ the box or basket method.

Drying

After fermentation, beans are typically sun-dried for four to 12 days, depending on the weather. Nowadays, commercial cross-flow electric dryers are used to dry cocoa beans to a temperature of 70 °C. Beans that have been thoroughly dried have a moisture content of about 6–7%. Better moisture migration to the surface during the rest phase was demonstrated by interrupted drying with a brief period followed by a prolonged resting period, producing good quality [68]. The dried beans are packed in polythene-lined (150–200 gauge) gunny or jute bags after being manually or mechanically selected to remove flat, slate, shrivelled, cracked, and clumped beans as well as other extraneous items.

Secondary Processing

Secondary processing involves cleaning, roasting, kibbling/ bean breaking, winnowing and grinding to convert dried beans to chocolates and other products. Cocoa butter recovery ranges from 50% to 58% [69] and fat content in the leftover cake ranges from 11% to 25%.

Bean to Chocolate

Chocolate production with palm sugar has emerged as a healthy alternative and accords multiple health benefits due to natural sugars' low glycaemic index (GI). Bean-to-bar chocolates have high polyphenols, flavonoids, antioxidants and minerals. Home-level chocolate processing ensures women's empowerment, improves income from a unit area of land and makes farm fresh and natural chocolates available at comparatively affordable prices [70].

By-Products Utilization

The husk constitutes 70–80% of the cocoa pod weight, generally discarded after collecting beans. Pod husk is utilized for pulp and paper production [71]. Mucilage is utilized for making fruit wine. Cocoa sweating is also used for making alcohol and vinegar. Cocoa shell is another by-product of cocoa (12–25% recovery from beans). Both cocoa shells and cocoa husk contain phytochemical compounds, which are the potential to inhibit the growth of pathogenic bacteria in food products [72]. Cocoa butter extracted from discarded cocoa beans, with little or no commercial value, can produce toilet soap and body pomade.

References

1. Balasubramanian S, Roselin P, Singh KK, Zachariah J, Saxena SN (2016) Postharvest processing and benefits of black pepper, coriander, cinnamon, fenugreek, and turmeric spices. Crit Rev Food Sci Nutr 56(10):1585–1607
2. Schweiggert U, Carle R, Schieber A (2007) Conventional and alternative processes for spice production—a review. Trends Food Sci Technol 18:260–268
3. Mohanta B, Dash SK, Panda MK, Sahoo GR (2014) Standardization of process parameters for microwave assisted convective dehydration of ginger. J Food Sci Technol 51(4):673–681. https://doi.org/10.1007/s13197-011-0546-2
4. Huang LL, Zhang M, Wang LP, Mujumdar AS, Sun DF (2012) Influence of combination drying methods on composition, texture, aroma and microstructure of apple slices. LWT Food Sci Technol 47(1):183–188. https://doi.org/10.1016/j.lwt.2011.12.009
5. Mejia EG, Aguilera-Gutierrez Y, Martin-Cabrejas MA, Mejia LA (2015) Industrial processing of condiments and seasonings and its implications for micronutrient fortification. Ann N Y Acad Sci 1357(1):8–28. https://doi.org/10.1111/nyas.12869
6. Steinhaus M (2017) Confirmation of 1-Phenylethane-1-thiol as the character impact aroma compound in curry leaves and its behaviour during tissue disruption, drying, and frying. J Agric Food Chem 65(10):2141–2146. https://doi.org/10.1021/acs.jafc.7b00197
7. Zhang M, Tang J, Mujumdar AS, Wang S (2006) Trends in microwave-related drying of fruits and vegetables. Trends Food Sci Technol 17(10):354–524. https://doi.org/10.1016/j.tifs.2006.04.011
8. Wang Y, Li X, Chen X, Li B, Mao X, Miao J, Zhao C, Huang L, Gao W (2018) Effects of hot air and microwave-assisted drying on drying kinetics, physicochemical properties, and energy consumption of chrysanthemum. Chem Eng Proc Proc Intensif 129:84–94
9. Jin W, Mujumdar AS, Zhang M, Shi W (2018) Novel drying techniques for spices and herbs: a review. Food Eng Rev 10:34–45
10. Deng LZ, Yang XH, Mujumdar AS, Zhao JH, Wang D, Zhang Q, Wang J, Gao ZJ, Xiao HW (2017) Red pepper (*Capsicum annuum* L.) drying: effects of different drying methods on drying kinetics, physicochemical properties, antioxidant capacity, and microstructure. Drying Technol 36:893–907
11. Kowalski SJ, Pawłowski A, Szadzinska J, Łechtańska J, Stasiak M (2016) High power airborne ultrasound assist in combined drying of raspberries. *Innov Food Sci Emerg Technol* 34:225–233
12. Cho WI, Yi JY, Chung MS (2016) Pasteurization of fermented red pepper paste by ohmicheating. Innov Food Sci Emerg 34:180–186. https://doi.org/10.1016/j.ifset.2016.01.015
13. Nalawade SA, Sinha A, Hebbar HU (2018) Infrared based dry blanching and hybrid drying of bitter gourd slices: process efficiency evaluation. J Food Process Eng 41(4):e12672
14. Rastogi NK (2011) Opportunities and challenges in application of ultrasound in food processing. Crit Rev Food Sci Nutr 51:705–722
15. Da-Mota VM, Palau E (1999) Acoustic drying of onion drying technology. Drying Technol 17:855–867
16. Wang Y, Zhang M, Mujumdar AS, Mothibe KJ (2013) Microwave-assisted pulse-spouted bed freeze-drying ofstem lettuce slices—effect on product quality. Food Bioprocess Technol 6(12):3530–3543. https://doi.org/10.1007/s11947-012-1017-0
17. Kubra IR, Rao LJM (2012) Effect of microwave drying on the phytochemical composition of volatiles of ginger. Int J Food Sci Tech 47:53–60. https://doi.org/10.1111/j.1365-2621.2011.02806.x

18. Sharma GP, Prasad S (2006) Optimization of process parameters for microwave drying of garlic cloves. J Food Eng 75(4):441–446. https://doi.org/10.1016/j.jfoodeng.2005.04.029
19. Shivanna VB, Subban N (2013) Carotenoids retention in processed curry leaves (*Murraya koenigii*L. Spreng). Int J Food Sci Nutr 64(1):58–62. https://doi.org/10.3109/09637486.2012.694849
20. Lv W, Li S, Han Q, Zhao Y, Wu H (2016) Study of the drying process of ginger (*Zingiber officinale* roscoe) slices in microwave fluidized bed dryer. Dry Technol 34(14):1690–1699. https://doi.org/10.1080/07373937.2015.1137932
21. Meetha JN, Muhammadali P, Joy MI, Mahendran R, Santhakumaran A (2016) Pulsed microwave assisted hotair drying of nutmeg mace for better colour retention. J Spices Aromatic Crops 25(1):84–87
22. Figiel A (2009) Drying kinetics and quality of vacuum-microwave dehydrated garlic cloves and slices. J Food Eng 94(1):98–104. https://doi.org/10.1016/j.jfoodeng.2009.03.007
23. Cui ZW, Xu SY, Sun DW (2003) Dehydration of garlic slices by combined microwave-vacuum and air drying. Dry Technol 21(7):1173–1184. https://doi.org/10.1081/DRT-120023174
24. Qiu L, Min Z, Wang Y, Liu Y (2018) Physicochemical and nutritional properties of wasabi (*Eutrema yunnanense*)dried by four different drying methods. Dry Technol 37(3):363–372. https://doi.org/10.1080/07373937.2018.1458318
25. Pawar SB, Pratape VM (2015) Fundamentals of infrared heating and its application in drying of food materials: a review. J Food Process Eng 40(1):e12308. https://doi.org/10.1111/jfpe.12308
26. An K, Zhao D, Wang Z, Wu J, Xu Y, Xiao G (2016) Comparison of different drying methods on Chinese ginger (*Zingiber officinale* roscoe): changes in volatiles, chemical profile, antioxidant properties, and microstructure. Food Chem 197(PtB):1292–1300. https://doi.org/10.1016/j.foodchem.2015.11.033
27. Younis M, Abdelkarim, D. andZein El-Abdein, A. (2018) Kinetics and mathematical modeling of infrared thin-layer drying of garlic slices. Saudi J Biol Sci 25(2):332–338. https://doi.org/10.1016/j.sjbs.2017.06.011
28. Torki-Harchegani M, Ghanbarian D, Maghsoodi V, Moheb A (2017) Infrared thin layer drying of saffron (*crocus sativus*L.) stigmas: mass transfer parameters and quality assessment. Chin J Chem Eng 25(4):426–432. https://doi.org/10.1016/j.cjche.2016.09.005
29. Phoungchandang S, Nongsang S, Sanchai P (2009) The development of ginger drying using tray drying, heatpump-dehumidified drying, and mixed-mode solar drying. Dry Technol 27(10):1123–1131. https://doi.org/10.1080/07373930903221424
30. Selvaraj V, Arjunan TV, Kumar A (2017) Thin layer drying characteristics of curry leaves (*Murraya koenigii*) in an indirect solar dryer. Therm Sci 21(Supp.2):359–367. https://doi.org/10.2298/TSCI17S2359V
31. Seveda MS, Jhajharia D (2012) Design and performance evaluation of solar dryer fordrying of large cardamom. J Renew Sustain Energy 4(6):1–11. https://doi.org/10.1063/1.4769199
32. Zhang M, Zhang CJ, Shrestha S (2005) Study on the preparation technology of superfine ground powder of *Agrocybechaxingu Huang*. J Food Eng 67(3):333–337. https://doi.org/10.1016/j.jfoodeng.2004.04.036
33. Saxena SN, Barnwal P, Balasubramanian S, Yadav DN, Lal G, Singh KK (2018) Cryogenic grinding for better aroma retention and improved quality of Indian spices and herbs: a review. J Food Process Eng 41(6):e12826. https://doi.org/10.1111/jfpe.12826
34. Manohar B, Sridhar BS (2001) Size and shape characterization of conventionally and cryogenically ground turmeric (*Curcuma domestica*) particles. Powder Technol 120(3):292–297. https://doi.org/10.1016/S0032-5910(01)00284-4
35. Meghwal M, Goswami TK (2014) Comparative study on ambient and cryogenic grinding of fenugreek and blackpepper seeds using rotor, ball, hammer and pin mill. Powder Technol 267:245–255. https://doi.org/10.1016/j.powtec.2014.07.025

36. Ghodki BM, Goswami TK (2016) Effect of grinding temperatures on particle and physicochemical characteristics of black pepper powder. Powder Technol 299:168–177. https://doi.org/10.1016/j.powtec.2016.05.042
37. Singh KK, Goswami TK (2000) Cryogenic grinding of cloves. J Food Process Pres 24(1):57–71. https://doi.org/10.1111/j.1745-4549.2000.tb00405.x
38. Santos P, Aguiar AC, Barbero GF, Rezende CA, Martínez J (2015) Supercritical carbon dioxide extraction of capsaicinoids from malagueta pepper (*Capsicum frutescens* L.) assisted by ultrasound. Ultrason Sonochem 22:78–88
39. Martins FS, Borges LL, Ribeiro CSC, Reifschneider FJB, Conceiçao ECD (2017) Novel approaches to extraction methods in recovery of capsaicin from HabaneroPepper (CNPH 15.192). Pharmacogn Mag 13:375–379
40. Li DW, Bi IW, Zhao ZD (2010) Research progress on extraction and purification of carsonic-acid. Chem Ind Forest Prod 30:122–126
41. Martinez J, Rosas J, Perez J, Saavedra Z, Carranza V, Alonso P (2019) Green approach to the extraction of major capsaicinoids from habanero pepper using nearinfrared,microwave, ultrasound and Soxhlet methods: a comparative study. Nat Prod Res 33(3):447–452. https://doi.org/10.1080/14786419.2018.1455038
42. Li C, Yang RF, Fu XF (2015) Ultrasonic-assisted supercritical CO_2 extraction of volatile oil from compound recipe of clove and cinnamon. J S China U Technol 36:67–71
43. Hu S, Zhao Y, Hayouka Z, Wang D, Jiao S (2018) Inactivation kinetics for *Salmonella typhimurium* in red pepper powders treated by radio frequency heating. Food Control 85:437–442. https://doi.org/10.1016/j.foodcont.2017.10.034
44. Jung K, Song BS, Kim MJ, Moon BG, Go SM, Kim JK, Lee YJ, Park JH (2015) Effect of X-ray, gamma ray, and electron beam irradiation on the hygienic and physicochemical qualities of red pepper powder. LWT Food Sci Technol 63(2):846–851. https://doi.org/10.1016/j.lwt.2015.04.030
45. Hebbar KB, Pandiselvam R, ShameenaBeegum PP, Ramesh SV, Manikantan MR, Mathew AC (2022) Seventy five years of research in processing and product development in plantation crops—coconut, arecanut and cocoa. Intl J Innovat Hortic 11(1):103–119
46. Hebbar KB, Arivalagan M, Manikantan MR, Mathew AC, Chowdappa P (2015) Kalparasa collection and value addition. Technical bulletin no. 92, ICAR-CPCRI, Kasaragod, Kerala
47. Manikantan MR, Pandiselvam R, Beegum S, Mathew AC (2018) Harvest and postharvest technology. In: Nampoothiri KUK et al (eds) The coconut palm (*Cocos nucifera* L.). Research and development perspectives. Springer, Singapore, pp 635–722
48. Hebbar KB, Arivalagan M, Manikantan MR, Mathew AC, Thamban C, Thomas GV, Chowdappa P (2015) Coconut inflorescence sap and its value addition as sugar—collection techniques, yield, properties and market perspective. Curr Sci 109:1–7
49. Prithviraj V, Pandiselvam R, Babu AC, Kothakota A, Manikantan MR, Ramesh SV, Hebbar KB (2021) Emerging non-thermal processing techniques for preservation of tender coconut water. LWT 149:111850. https://doi.org/10.1016/j.lwt.2021.111850
50. Manikantan MR, Arivalagan M, Mathew AC, Hebbar KB (2015) Effect of process parameters on recovery of hot process virgin coconut oil and co-products utilization. J Plant Crop 43:139–144
51. Sudheer KP, Claudia KL, Sankalpa KB et al (2018) Prospects in value addition of coconut, technical bulletin no: CE-PHT/2018/22
52. Novarianto H, Tulalo MA, Mawardi S (2021) Coconut sugar production of dwarf coconut varieties. IOP Conf Ser Earth Environ Sci 800(012024):1–9
53. Subroto S, Pangawikan AD, Yarlina VP et al (2020) The extraction, purification and the recent applications of coconut oil in food products—A review. Int J Emerg Technol 11(5):234–240
54. Navin KR (2019) New technologies for value added products from coconut residues. Mater Res Forum 11:295–301

55. Jayasekara C, Gunathilake KDPP (2007) Processing technologies for virgin coconut oil and coconut based confectionaries and beverages. Proceedings of International Cococnut Summit, pp 7–11
56. Chinnamma M, Bhasker S, Binitha Hari M, Sreekumar D, Madhav H (2019) Coconut neera—a vital health beverage from coconut palms: harvesting, processing and quality analysis. Beverages 5(1):22
57. Henrietta HM, Kalaiyarasi K, Stanley Raj A (2022) Coconut tree (Cocos nucifera) products: a review of global cultivation and its benefits. J Sustain Environ Manag 1(2):257–264
58. Chowdappa P (2020) Arecanut. Today & Tomorrow's Printers and Publishers, New Delhi, p 304. ISBN-9788170196648, 8170196647
59. Kiran K, Govind AK, Bandi M, Shivasharanayya (2014) Design, development and testing of an arecanut dehusking agri-machine. Int J Eng Res Appl 7:109–115
60. Kumar Y, Louis RS, D'Souza D, Floyd S, Varghese EB, Gopalakrishna BK (2019) AIP conference proceedings 2080, 040010. Published online: 08 March 2019. https://doi.org/10.1063/1.5092928
61. Hebbar,KB, Padmanabhan S, Ramesh SV, Bhat SK, ShameenaBeegum PP, Pandiselvam R, Manikantan MR Mathew AC, (2021) Moisture content and water activity of arecanut samples: A need to revisit storage guidelines. J Plant Crop 49:136–141
62. Patil KKR, Chinnappa B, Manjunatha GR, Sowmya HS (2019) Ecofriendly arecanut leaf sheath products: An economic analysis. Indian J Ecol 46:431–436
63. Rajan A, Kurup JG, Abraham TE (2005) Biosoftening of arecanut fiber for value added products. Biochem Eng J 25:237–242
64. Beg MS, Ahmad S, Jan K, Bashir K (2017) Status, supply chain and processing of cocoa—a review. Trends Food Sci Technol 66:108–116
65. Kumaran K, Prasannakumari S, Nair PCS (1981) Experiments on small scale fermentation of cocoa beans. II. Effect of different factors for aeration on the extent of fermentation and quality of cured beans. Proc PLACROSYM III, pp 125–137
66. Premlatha T (1983) Studies on the fermentation and curing of cocoa beans. MSc. (Ag.) thesis. Kerala Agricultural University, Thrissur, Kerala, India
67. Srikanth V, Rajesh GK, Kothakota A (2019) Physico-chemical and microbial changes in cocoa beans during different methods of fermentation. Indian J Ecol 46:154–159
68. Asopa VN, Narayanan S (1990) Cocoa production and marketing in India. Oxford & IBH Publishing Co. Pvt. Ltd., pp 50–78
69. Peter KV (2002) Plantation crops. National Book Trust, New Delhi, pp 94–123
70. Suma B, Minimol JS (2018) Scope of entrepreneurial development in cocoa processing. In: Sudheer KP, Indira V (eds) Entrepreneurship and skill development in horticultural processing. New India Publishing Agency, New Delhi, pp 189–213
71. Daud Z, Kassim ASM, Aripin AM, Awang H, Hatta MZM (2013) Chemical composition and morphological of cocoa pod husks and cassava peels for pulp and paper production. Aust J Basic Appl Sci 7:406–411
72. Kayaputri L, Djali M, Sukri N, Fazaryasti RH (2020) The antimicrobial effectiveness of cacao shell and cacao husk combination on inhibition of pathogenic bacteria in food products. IOP Conference Series: Earth and Environmental Science, pp 443–451. https://doi.org/10.1088/1755-1315/443/1/012077

Chapter 26
Millets, Fermentation, and Health

Renukaradhya K. Math, Palaksha Kanive Javaregowda, Ajay S. Khandagale, Mallikarjun Goni, and Meghana Variyar

1 Introduction

Millets belong to food crops that fit into a cluster of small grain cereals and are extremely nutritious. They grow under low fertile/fragile soils conditions with low or minimal inputs of fertilizers and pesticides. These crops significantly contribute to nutritional and global food security and generally identified as nutri-cereals as they supply majority of the nutrients essential for normal healthy functioning of human system. Millets are valuable diverse and adaptive crops of resource-poor drylands worldwide. Millets had been the salvation of dry areas of Africa and Asia for food and fodder. They contribute to the diversity-richness, sustainability, and stability of agro-ecosystems. There are barely any substitutes for the potential and charactersitcs of millets to grow in a wide range of ecosystems such as in arid, semi-arid land, mountains, steppes, and tropical forest as well as their tolerability to dynamic weather and ecological conditions.

Depending on the area of cultivation and size of the grain, they are categorized as major and minor millets. Pearl millet (bajra) and sorghum millet (jowar) included under major millets. The foxtail millet, finger millet, kodo millet, brown top millet, proso millet, barnyard millet, and little millet are categorized under minor millets [1–3]. In recent decades, continuously increasing awareness on health and well-being, the demand for millet based food and food products is also increasing with their proven health benefits, in the era of "ready to kitchen" ready to eat processed foods, particularly in the urban and semi-urban societies.

R. K. Math (✉) · P. K. Javaregowda · S. K. Ajay · M. Goni · M. Variyar
SDM Research Institute for Biomedical Sciences, SDM College of Medical Sciences & Hospital Campus, Shri Dharmasthala Manjunateshwara University, Sattur, Dharwad, Karnataka, India

J. S. Yaradoddi et al. (eds.), *Frontiers in Food Biotechnology*,
https://doi.org/10.1007/978-981-97-3261-6_26

1.1 Eco-friendly and Food Security

Millets, being short duration crops (70-90 days) with drought resistance they require only a small amount of water when compared to wheat and rice. The rural livelihood is a multifaceted endeavor, further the large scale farming and sustainable prodcution is a massive challenge. Handling of unprecedented events of natural disasters, emerging environmental threats, limited resources and their sustainable utilisation require a holistic farming management strategies. There are already some key resources available in hands to meet new challenges in the form of the wide range of ignored crop species used to fulfil every household needs. Besides, research will increase crop value and availability of them would surge the livelihood choices for rural farming community.

Millets could be an answer to improve wellbeing of the people of low and middle income countries by mitigating the nutritional requirements, eradicating the hunger and thereby reducing the burden of diseases. However, being not given much importance for long period of time to millets, there is a wide gap in the knowledge exist and potential of millets. Undoubtedly/Indeed there is a tremendous scope for enhanced production and development of miller based nutritive foods and food products to overcome the nutritional deficiency related health issues. Besides, millets contributed immensely to human life in adopting sedentary life from hunter-gatherer ways, also support well-being in a good environment. Therefore, it is for us to take make a decision.

Nutrition Values of Millets

Besides ecological advantages, millets are recognized for their health and nutraceutical values compared to other cereals like wheat and rice [4]. Millets were the ancient foods known to humans; however, urbanization and industrialization have reduced cultivation and importance due to extensive cultivation of wheat and rice. As a repercussion of modern days sedentary lifestyle and unhealthy eating habits, people have become more susceptible to various lifestyle related non-communicable diseases, for example hypertension, obesity, diabetes and cardiovascular diseases. Hence, the millets based foods and food products have started returning to modern kitchen food basket as a viable option to lead a healthy life by minimizing the incidence of lifestyle related diseases.

Millets are packed with good amount of proteins, phytochemicals, vitamins, and minerals. They are rich in fiber and micronutrients compared to that of rice and wheat; however, other proximate composition remain similar [5]. Among millets, finger millets have highest carbohydrate content, whereas foxtail millet and pearl millet have high protein content. Millet proteins are also rich in critical amino acids which are required to prevent protein-energy malnutrition but have fewer crosslinked prolamins that help in improved protein digestibility [6]. With regard to lipids and fats, finger millet has lower fat content than barnyard, little foxtail, and pearl millet. However, pearl millet has high amount of palmitic linoleic and stearic acids (Table 26.1).

Table 26.1 Millets: Countries of cultivation, nutritional values and their uses

Millet	Scientific name	Grown	Major nutrients (per 100 g)	Potential use
Sorghum	*Sorghum bicolor*	Niger, India, USA, Ethiopia, Nigeria, Mexico, Sudan, Mexico, Argentina, Australia, China	Rich in carbohydrates (67.7), energy (334)	Anti-carcinogenic
Pearl millet	*Pennisetum glaucum*	India, Western & Central Africa, Eastern & Southern Africa	Rich in dietary fiber (11.5), energy (347), potassium (289), protein (11.6), total phenol (2.6)	Lignin and phytonutrients protect heart
Finger millet	*Eleusine coracana*	Rwanda, Malawi, India, Uganda, Nepal, Sri Lanka Sri Lanka, Burunda	Rich in dietary fiber (11.2), calcium (364), total phenol (10.2), flavonoids (2.4)	Low triglycerides and prevent CVD
Foxtail millet	*Setaria italica*	China, Myanmar, India, Eastern Europe	Rich in protein (12.3), energy (331), phytosterols (57)	Anti-hyperglycemic
Proso millet	*Panicum miliaceum*	India, Iran, France, Russia, Ukraine, Kazakhstan, France, South Korea, Poland, USA Belarus	Rich in carbohydrates (70.4), energy (341), Mg (153), phytosterols (26)	Improve glycemic and insulin response
Little millet	*Panicum sumatrense*	India	Rich in carbohydrate (65.5), energy (346), folic acid (36.2)	
Kodo millet	*Paspalum scrobiculatum*	India	Rich in folic acid (39.5), energy (331), total phenol (19.7), flavonoids (11.1)	
Barnyard millet	*Echinochloa crus-galli*	India, Japan, China, Malaysia	Rich in Niacin (4.2), P (280), Zn (3), carbohydrate (65.5)	Low glucose, cholesterol, triglycerides

Source: Indian Food Composition Tables, NIN—2017; Nutritive value of Indian Foods, NIN—2007

Micronutrients in millets are equivalent to other cereals like wheat and rice, with higher levels of magnesium and calcium [7, 8]. Barnyard and pearl millet have high amount of iron. However, vitamins contents are high in minor millets such as niacin, riboflavin, vitamin C and thiamine. Pearl millet has high fat-soluble vitamin E (Table 26.1). Apart from macro- and micronutrients millets also contain phytochemicals including phenolic acids, flavonoids and phytosterols, which are biological important substances [9].

Surprisingly, bioavailability of nutrients of millets is low due to the presence of antinutritional factors. Compounds like phytates, tannins, and phenol impact zinc and iron bioavailability [8]. However, procedures of food-processing like milling, cooking, soaking, and fermentation, might lessen the deleterious effect of antinutrients.

2 Fermentation of Millets

Preprocessing techniques can facilitate to get rid of antinutritional factors present in millets and help in consumption in daily diet. Results of preprocessing can bring many changes in cereal physical, nutritional, and functional characteristics. Among all the preprocessing techniques fermentation is an economical and low-energy process that can be adopted to improve food preservation, too. Fermentation process can be natural occurring due to intrinsic microorganisms or performed with selected starter cultures. The process of fermentation and the metabolites generated during the fermentation is directly dependent on the metabolic activity of an individual and/ or a group of microorganisms present in the the gut. However, each of them would release different metabolites and is well documented by Balli et al. [10]. During fermentation peptides are released in two ways: one by microbial proteolytic enzymes, and another by endogenous enzymes. Consumptions of fermented foods have increased globally including Asia and Africa; especially Europe has showed a renewed attention in fermented foods due to nutritional and sensorial properties. Fermented foods of millets not only include porridges and bakery items but also alcoholic and non-alcoholic beverages products [11, 12].

A study by Balli et al. [10] found fermentation increases mineral content in food. Millets comparatively have many advantages from a nutritional point of view than that of other common cereals. Nutrition values of a food can be amplified after fermentation; for example, an increased quantity of dietary fiber was attributed to the presence of cellulose and hemicellulose degradation in foxtail millet fermented with *Bacillus natto* [13]. Furthermore, the enhancement of starch and protein digestibility was observed in pearl millet fermented by natural fermentation and in foxtail millet fermented with lactic acid bacteria [14, 15]. Besides, fermentation of millets also releases bioactive peptides without losing essential amino acids, and these peptides are smaller than 10 kDa and are benefit to health [16]. An upsurge in total phenolic content was observed in pearl millet with a combination of Lactobacilli and yeast [10], with *Rhizopus azygosporus* [17] as well as in foxtail millet fermented by *Bacillus natto* [13]. Meanwhile, fermentation of finger millet revealed higher availability of iron, calcium, and phosphorous [12]. Therefore, millets along with the fermentation process not only enhance the nutritional value of the food but also help in protecting the environment.

Millets are not only used to produce fermented foods but also used in producing food/drinks for health use and distilled beverages. For example, the drink, "baijiu" a Chiinese national drink "preparepd from sorgham millets, isconsidered as a healthy and well-being food". Besides, in Western countries, beer production from sorghum is targeted to customers suffering celiac disease and healthy-minded drinkers who look for low calorie drinks.

3 Millets, Gut Health, and Microbiome

Millets are known for regulating digestive progression can also enhance nutrient retention and lower probability of serious gastrointestinal issues like gastric ulcer and colon cancer. Presence of high amount of fiber in millets, especially in pearl and finger millets (Table 26.1), facilitates in preventing disorders like gas, bloating cramping, and constipation.

Besides, people are suffering from a genetic disorder called celiac disease, where the immune mediated reaction (enteropathic disease) to gluten protein present in wheat and other cereals [18]. This adverse immune reaction to gluten may to lead severe abdominal pains. A diet free of gluten principally plays a key role in distressing food ingestion in the grain food community. Finding alternative gluten-free grains may help population following to gluten-free diet [19]. The most attractive property of millets is "free from gluten" which makes them not only a potential replacement for fruits and vegetables but can also be a right alternative option for peoplel suffering from celiac disease, furhter it also reduce the demand for cultivation of glugen free foods [20–23]. For example, sorghum could be a better diet for individuals who are suffering from celiac disease since it is gluten free.

The human gut harbors a dense and diverse species of microbiota, which is influenced by several internal and external factors, including diet and lifestyle. Carbohydrate polymers and dietary fiber in human gastrointestinal are neither digested nor absorbed but are subjected to microbial fermentation and consequently effect the microbial composition and its metabolic activity, including the secretion of fermentative end-products. A study by Huo et al. [24] demonstrated immune-stimulating effects of a Heat Shock Protiens-2 (HSP-2) polysaccharide present in 'Baijiu' made from fermented sorghum. Ingestion of fiber-rich millet diet promotes wide metabolic communication among microbial species living in the gastrointestine. The end-product of complex carbohydrates by bacteria in intestine is acetate, butyrate, and gases which are beneficial to colon.

4 Millets as a Source of Bioactive Peptides and Phenolic Compounds

Grains, cereals, and millets are considered as important sources in providing adequate nutrition and have displayed potential beneficial effect on lifestyle disorders. Certainly, the most studied are the bioactive peptides present in dairy, legumes, cereals, fish, and meat products. In recent times, food-derived bioactive peptides from food are increasing at significant rate in both pharmaceutical and nutraceuticals industry. The bioactive peptides, both from plants and animals typically composed of 2–20 amino acids, require a proteolytic activity to become active biomoleculre from the precursor form. Bioactive peptides are absorbed through the intestine via different modes; subsequently they exert a physiological affect on the various human systems such as endocrine, immune, digestive, cardiovascular and nervous system. Bioactive peptides derived from fermented foods of animal and plant origin with pertinent possible physiological function in human have been the subject of intensive research because of the health benefits that they may provide.

Being rich in essential nutrients and vitamins (Table 26.1), in recent years millets gained a significant importance globally to develop conventional food products as well as in scientific research towards developing consumer friendly helathy foods. In addition, millets also, contain a good amount of essential amino acids such as cysteine, methionine, etc. [25]. Peptides derived from plant source are protein fragments lesser than 10 kDa [26] and these might occur naturally or are obtained from their precursor/native protein by hydrolysis from digestive or proteolytic enzymes from microorganisms. Bioactive peptides derived from variety of biological sources come with many beneficial properties on human health by influencing on body function. For example, these exert distinct biochemical and physiological effects such as anticancer, antidiabetic, antioxidant, antimicrobial, and antihypertensive [27–30].

Polyphenols, secondary metabolites present in plants are known for their potential not only in protecting from pathogens, lifestyle diseases but also protect from ultraviolet radiations [31]. These are abundant micronutrients of our diet. Among all the millets kodo, millet has high total phenol followed by finger and pearl millet (Table 26.1).

5 Millets in Lifestyle Diseases

Millets research across the globe has significantly increased recently after the announcement of the statement that "the year 2023 is the millet year" by United Nation Organization.

Cardiovascular disease: The phytonutrients and lignin present in millet prevent heart disease by acting as antioxidants. In addition, the potassium present in pearl millet can control blood pressure by acting as an vasodilator and reduce stress to heart. Further, the consumption of millets also showed the reduced levels of cholesterol, triglycerides in serum [32] and improved high-density lipoproteins levels [33].

Cancer: In fact every millet is known to lower the risk of many cancers; this might be due to high magnesium content, phytate, tannins, and polyphenols. Furthermore, the studies have shown that antimutagenic, anticarcinogenic, and melanogenic activity of sorgham millets [34, 35].

Anti-diabetic: Millets, with high fibre content require relatively longer time to metabolise and release of glucose into blood stream, thus the millets might be an ideal foods to regulate blood sugar levels in diabetes cases. The studies have demonstrated the declining of blood glucose level following the consumption of sorghum millet foods having low glycemic index [36, 37].

Weight loss: The primary difficulty in the process of loosing weight by an individual is control over type and quantity of food intake. The virtue of slow metabolising and requiring longer time to move into intestine, the the millets based foods would be an ideal choice to individuals looking for loosing weight. Millet based foods satisfies hunger over a longer period of time and thus aids in lowering the total intake of food.

Anti-allergic: Millets are easily digestible and does not contain allergic compounds, hence their consumption is safe and Heat Shock Protiens-2 among people of all age groups and physiological status. Thus, inclusion of millets based foods in the diet plan of lactating mothers, infants, and elderly would be highly recommended.

Oxidative stress: Majority of the millets possess antioxidant property due to which it can neutralize reactive oxidative species. The free radicals generated during metabolic activity may induces oxidative stress which then lead to pathogenesis of many ailments including myocardial infraction, Alzheimer's disease, autoimmune disorders, atherosclerosis, etc. However, the foods rich in antioxidant metabolites play a key role in stopping oxidation progression, thus lowering cellular injury.

6 Millets as Savior

Fermentation process has the ability of deliberating an extra nourishing and nutraceutical levels compared to the unfermented cereal. Fermentation can be an option of making foods with improved dietary properties with this low-cost and simple method; this will help persons who already use millet as a staple food. Further, this could promote the better use of millet in nations that are not used to the ingestion of millets.

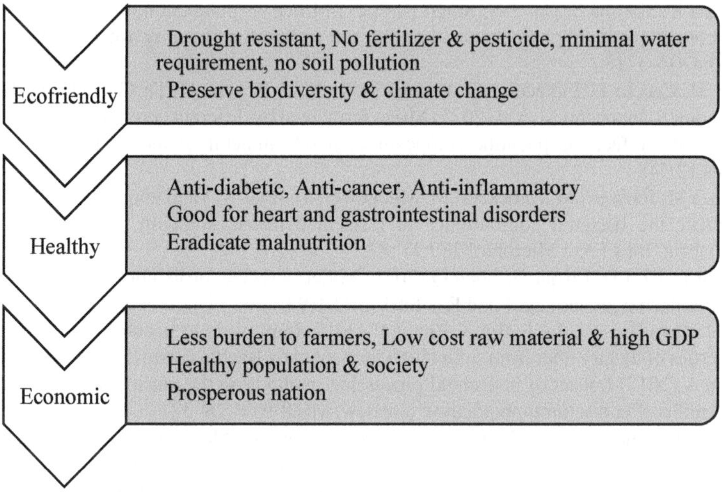

Fig. 26.1 Illustration of multi-facet connection of millets-with health and environment

Additionally, activities intended to enabling the usage of millets, which is drought resistant and identified for its agronomic benefits, could remain an exciting policy for managing through climate changes (Fig. 26.1). As of future viewpoint, millets could be considered as rescuers of the humans, environment and the Earth, no doubt combination of millet and fermentation is the answer for the sustainable life and pollution free. This is the gift which we can give to our next generation and prolong the life of human and the Earth.

References

1. Hunt HV, Vander Linden M, Liu X, Motuzaite-Matuzeviciute G, Colledge S, Jones MK (2008) Millets across Eurasia: chronology and context of early records of the genera Panicum and Setaria from archaeological sites in the old world. Veg Hist Archaeobot 17(S1):5–18
2. Bandyopadhyay T, Muthamilarasan M, Prasad M (2017) Millets for next generation climate-smart agriculture. Front Plant Sci 8:1266
3. Bhat BV, Rao BD, Tonapi AV (2018) The story of Millets. Karnataka State Department of Agriculture, Bengaluru, India with ICAR-Indian Institute of Millets Research, Hyderabad, India
4. Saleh AS, Zhang Q, Chen J, Shen Q (2013) Millet grains: nutritional quality, processing, and potential health benefits. Compr Rev Food Sci Food Saf 12:281–295
5. Muthamilarasan M, Dhaka A, Yadav R, Prasad M (2016) Exploration of millet models for developing nutrient rich graminaceous crops. Plant Sci 242:89–97
6. Sharma N, Sahu JK (2021) Postharvest processing of foxtail millet and its potential as an alternative protein source. In: Handbook of cereals, pulses, roots, and tubers. CRC Press, pp 105–114
7. Kumar M, Rani K, Ajay BC, Patel MS, Mungra KD, Patel MP (2020) Study of genetic variability, heritability and path analysis for grain micronutrients concentration, yield and component traits in pearl millet (Pennisetum glaucum (L.) R. Br.). J Pharmacog Phytochem 9(2):1402–1409
8. Hassan ZM, Sebola NA, Mabelebele M (2021) The nutritional use of millet grain for food and feed: a review. Agric Food Secur 10:16
9. Sharma R, Sharma S, Dar BN, Singh B (2021) Millets as potential nutri-cereals: a review of nutrient composition, phytochemical profile and techno-functionality. Int J Food Sci Technol 56(8):3703–3718
10. Balli D, Cecchi L, Pieraccini G, Venturi M, Galli V, Reggio M, Di Gioia D, Furlanetto S, Orlandini S, Innocenti M et al (2023) Millet fermented by different combinations of yeasts and lactobacilli: effects on phenolic composition, starch, mineral content and prebiotic activity. Foods 12:748
11. Gabaza M, Joossens M, Cnockaert M, Muchuweti M, Raes K, Vandamme P (2019) Lactococci dominate the bacterial communities of fermented maize, sorghum and millet slurries in Zimbabwe. Int J Food Microbiol 289:77–87
12. Yousaf L, Hou D, Liaqat H, Shen Q (2021) Millet: a review of its nutritional and functional changes during processing. Food Res Int 142:110197
13. Chu J, Zhao H, Lu Z, Lu F, Bie X, Zhang C (2019) Improved physicochemical and functional properties of dietary fiber from millet bran fermented by Bacillus natto. Food Chem 294:79–86
14. Urooj A (2017) Impact of household processing methods on the nutritional characteristics of pearl millet (Pennisetumtyphoideum): a review. MOJFPT 4:28–32
15. Sharma R, Sharma S (2022) Anti-nutrient & bioactive profile, in vitro nutrient digestibility, techno-functionality, molecular and structural interactions of foxtail millet (Setaria italica L.) as influenced by biological processing techniques. Food Chem 368:130815

16. Majid A, Priyadarshini PCG (2020) Millet derived bioactive peptides: a review on their functional properties and health benefits. Crit Rev Food Sci Nutr 60:3342–3351
17. Purewal S, Sandhu K, Salar RK, Kaur P (2019) Fermented pearl millet: a product with enhanced bioactive compounds and DNA damage protection activity. J Food Meas Charact 13:1479–1488
18. Catassi C, Fasano A (2008) Celiac disease. In: Gallagher E (ed) Gluten-free cereal products and beverages. Elsevier, Burlington, MA, pp 1–27
19. Thompson T (2009) The nutritional quality of gluten-free foods. In: Gallagher E (ed) Gluten-free food science and technology. John Wiley and Sons, Ltd., Oxford, UK, pp 42–51
20. Taylor JRN, Schober TJ, Bean SR (2006) Novel food and non-food uses for sorghum and millets. J Cereal Sci 44(30):252–271
21. Taylor JRN, Emmambux MN (2008) Gluten-free foods and beverages from millets. In: Gallagher E (ed) Gluten-free cereal products and beverages. Elsevier, Burlington, MA, pp 1–27
22. Chandrasekara A, Shahidi F (2011b) Antioxidant phenolics of millet control lipid peroxidation in human low density lipoproteins cholesterol and food systems. J Am Oil Chem Soc 89:275–285
23. Chandrasekara A, Shahidi F (2011c) Anti proliferative potential and DNA scission inhibitory activity of phenolics from whole millet grains. J Funct Foods 3:159–117
24. Huo J, Wu J, Huang M, Zhao M, Sun W, Sun X, Zheng F (2020) Structural characterization and immuno-stimulating activities of a novel polysaccharide from Huangshui, a byproduct of Chinese Baijiu. Food Res Int 136:109493
25. Geervani P, Eggum BO (1989) Nutrient composition and protein quality of minor millets. Plant Foods Hum Nutr 39(2):201–208
26. Farrokhi N, Whitelegge JP, Brusslan JA (2008) Plant peptides and peptidomics. Plant Biotechnol J 6(2):105–134
27. Cicero AFG, Fogacci F, Colletti A (2017) Potential role of bioactive peptides in prevention and treatment of chronic diseases: a narrative review. Br J Pharmacol 174(11):1378–1394
28. Chakrabarti S, Jahandideh F, Wu J (2014) Food-derived bioactive peptides on inflammation and oxidative stress. BioMed Res Int:1–11
29. Li S, Liu L, He G, Wu J (2018) Molecular targets and mechanisms of bioactive peptides against metabolic syndromes. Food Funct 9(1):42–52
30. Ganguly A, Sharma K, Majumder K (2019) Chapter 4: Food derived bioactive peptides and their role in ameliorating hypertension and associated cardiovascular diseases. In: Toldra F (ed) Advances in food and nutrition research, vol 89. Academic Press, New York, pp 165–207
31. Manach C, Scalbert A, Morand C, Rémésy C, Jiménez L (2004) Polyphenols: food sources and bioavailability. Am J Clin Nutr 79:727–747
32. Kumari SK, Thayumanavan B (1997) Comparative study of resistant starch from minor millets on intestinal responses, blood glucose, serum cholesterol and triglycerides in rats. J Sci Food Agric 75:296–302
33. Park KO, Ito Y, Nagasawa T, Choi MR, Nishizawa N (2008) Effects of dietary korean proso-millet protein on plasma adiponectin, HDL cholesterol, insulin levels and gene expression in obese type 2 diabetic mice. Biosci Biotechnol Biochem 72(11):2918–2925
34. Grimmer HR, Parbhoo V, McGarth RM (1992) Antimutagenicity of polyphenol-rich fractions from Sorghum bicolour grain. J Agric Food Chem 59:251–256
35. Gomez-Cordovez C, Bartolomez B, Vieira W, Viradir VM (2001) Effects of wine phenolics and sorghum tannins on tyrosinase activity and growth of melanoma cells. J Agric Food Chem 49:1620–1624
36. Shinde RKU (2004) Utility of bran products in non-insulin dependent diabetes mellitus (NIDDM) Patients. J Hum Ecol 16(3):219–222
37. Vahini J, Bhaskarachary K (2013) Effect of processing and cooking on glycemic index of Jowar varieties. Int J Food Nutr Sci 2(3):25–30

Chapter 27
Future Scope and Opportunities for the Food Biotechnologist

Manoj Girish, Manjunatha P. Talwar, Kirankumar Shivasharanappa, and Jayashree V. Hanchinalmath

1 Introduction

Food biotechnology involves the utilization of technology to genetically modify our sources of food encompassing animals, plants, and microorganisms [1]. It encompasses the application of methods to enhance both the quantity and quality of food products by enhancing the food properties like taste, aroma, and shelf life, as well as to improve their nutritional content, production capabilities, and marketing attributes [2]. Food biotechnology allows for the creation of products with desirable traits, such as plants that are resistant to diseases and droughts. It's worth noting that food biotechnology has been utilized in food production and processing for centuries with practices like fermentation and selective breeding having been employed since ancient times [3]. However, modern biotechnology offers an efficient and precise approach to achieving these enhancements by utilizing advanced techniques like recombinant DNA (rDNA) technology [4].

The significance of food biotechnology in today's world cannot be overstated, as it greatly influences food security, sustainability, and the overall food industry. Food biotechnology plays a role in tackling the challenges of global food security by enabling the production of more food on limited land. It helps minimize crop losses caused by diseases and pests thereby boosting food production. Additionally, it has

M. Girish
Department of Biotechnology, M S Ramaiah University of Applied Sciences, Bangalore, Karnataka, India

M. P. Talwar · K. Shivasharanappa
Department of Life Science, School of Sciences, Garden City University, Bangalore, Karnataka, India

J. V. Hanchinalmath (✉)
Department of Life Science, School of Sciences, Jain (Deemed to be University), Bangalore, Karnataka, India

J. S. Yaradoddi et al. (eds.), *Frontiers in Food Biotechnology*,
https://doi.org/10.1007/978-981-97-3261-6_27

the potential to reduce CO_2 emissions from farming practices and minimize the use of pesticides in food production [5]. Modern biotechnology also holds promise for enhancing the value of food products making them more beneficial for consumers. Moreover, it contributes to improving food safety standards and reducing chemical residues in our foods [6]. The advancement of food biotechnology can lead to efficient processes for producing our essential sustenance while also aiding resource management by reducing environmental impact through water conservation and lower carbon emissions [7]. These advantages are vital for ensuring the feeding of the world's population as we move forward. Though there are concerns surrounding this field such, as environmental and health risks [8], ongoing research and advancements strive to address these concerns while maximizing the benefits offered by this innovative domain.

The main goal of this chapter is to provide an overview of the current state and future potential of food biotechnology. It emphasizes how this field plays a role in addressing global challenges related to food security and sustainability. Additionally, this chapter discusses applications and advancements in food processing, engineering, packaging, nutrition, and preservation. It highlights how biotechnology can contribute to improving food quality, safety, and nutritional value. Moreover, the manuscript sheds light on the career opportunities and future prospects available in the field of food biotechnology. It emphasizes the growing demand for professionals specializing in this area. It showcases diverse job opportunities within research laboratories, processing industries, and related fields. By this, it aims to provide readers with a comprehensive understanding of food biotechnology's impact on the industry as well as potential career paths for those interested, in this dynamic and rapidly evolving field.

2 Food Biotechnology and Its Impact on the Food Industry

Technology is playing a role in transforming the global food industry by impacting various aspects of food production, distribution, and supply [9]. The integration of technology has resulted in increased efficiency, productivity, and safety to meet the changing needs of consumers and address the challenges posed by a growing world population [10]. One significant way in which technology is influencing the food sector is through automation, robotics, and advanced processing techniques [11]. These innovations have enabled food manufacturers to produce effectively to cater to the demands of an expanding global population. For example, robotics has greatly improved productivity and safety in the food industry. Countries like Sweden, Denmark, the Netherlands, and Italy have experienced a rise in robot density within their food industries [12]. Additionally, food technology has brought about changes in food production through advancements like three-dimensional (3D) printing and genetically modified organisms (GMOs) [13]. The use of 3D printing has the potential to bring about a transformation in food production especially when it comes to making functional foods that have improved nutritional value and resistance against diseases [14, 15]. This technology enables the inclusion of easily degradable bioactive compounds in food products thereby enhancing their health benefits [14]. Moreover, it opens up avenues, for creating geometric structures that offer enhanced functionality and nutritional properties that can complement traditional methods of food preparation [15], while GMOs have played a vital role, in enhancing plant disease resistance and growth in contributing to increased overall food production [16]. The field of biotechnology has made progress by incorporating genes from microorganisms that can make crops more resistant to herbicides, pests, and diseases. This has resulted in the development of high-yielding crops that are resistant to diseases [17]. Genetic modification and genome editing have proven to be tools, in managing plant diseases offering the potential to increase food production by 60% by 2050 [18].

Food technology has had an impact on the food industry going beyond just production. It has influenced parts of the food cycle, such as processing, packaging, and distribution [19]. Innovations such as methods and nanotechnology have played a significant role in improving the quality, safety, and shelf life of our food [20]. Additionally, the introduction of packaging, nondestructive inspection techniques, and automation in food processing and packaging has further strengthened these advancements (Fig. 27.1) [21]. In terms of food supply chain management technologies, like RFID (Radio Frequency Identification), IoT (Internet of Things) and

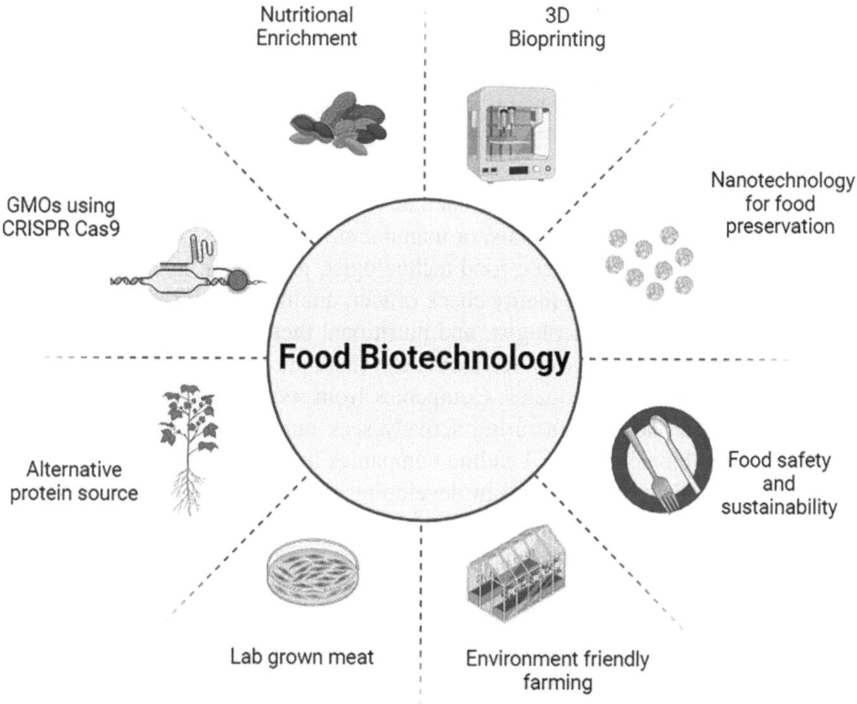

Fig. 27.1 Role of food biotechnology

blockchain have emerged as valuable tools to enhance efficiency and sustainability [22].

The food technology industry is witnessing expansion and experts predict that it will reach a market value of around $342.52 billion by 2027. This growth is being fueled by factors such as evolving family values, increased emphasis on health and nutrition standards, and the emergence of new types of consumers [23]. Furthermore, recent advancements in food processing and packaging technology including automation and advanced methods for control and monitoring are playing a role, in driving this growth [24]. This growth demonstrates the increasing importance of technology in overcoming the obstacles faced by the food production and supply sectors. By embracing technology, the food industry not only satisfies the needs of a growing population but also contributes to responsible production and consumption practices, playing a crucial role in tackling global food security and sustainability challenges [25]. Incorporating technology into the food industry has had an impact on enhancing efficiency, productivity, and safety at various stages of food production, distribution, and supply. As this industry continues to develop, technology will undoubtedly remain a driver, in meeting the demands posed by population growth while addressing issues related to food security and sustainability.

3 Career Opportunities in Food Biotechnology

The field of food biotechnology is experiencing growth providing a wide range of career prospects for both graduates and postgraduates [26]. As new products continue to evolve, and global food security becomes an increasing concern, the demand for food biotechnologists is on the rise [27]. Those with qualifications in food biotechnology can explore career paths, such as working in research laboratories, processing industries, hotels, hospitals, or manufacturing facilities. Job roles available to food biotechnologists include food technologist, production manager, purchase manager, research scientist, quality check officer, quality manager, product/process development scientist, toxicologist, and nutritional therapist [28]. The job market reflects the demand for food biotechnologists; there are opportunities available for both graduates and postgraduates. Companies from sectors like the food industry, pharmaceuticals, and manufacturing actively seek out professionals with a background in food biotechnology. Leading companies in the food and beverage sector as well as organizations involved in developing new food products or nutritional supplements are particularly interested in hiring experts from this field.

The job prospects for individuals in the field of food biotechnology are quite promising offering salaries and ample opportunities for growth and advancement (Table 27.1). According to the Institute of Food Technologists (IFT), professionals engaged in the study of food generally earn a salary of $92,000 [29]. This includes those working as food scientists/technologists, food chemists, nutritionists, and food engineers who receive compensation. With its range of opportunities in consumer research, teaching, regulatory affairs, basic research, and nutrition, the field

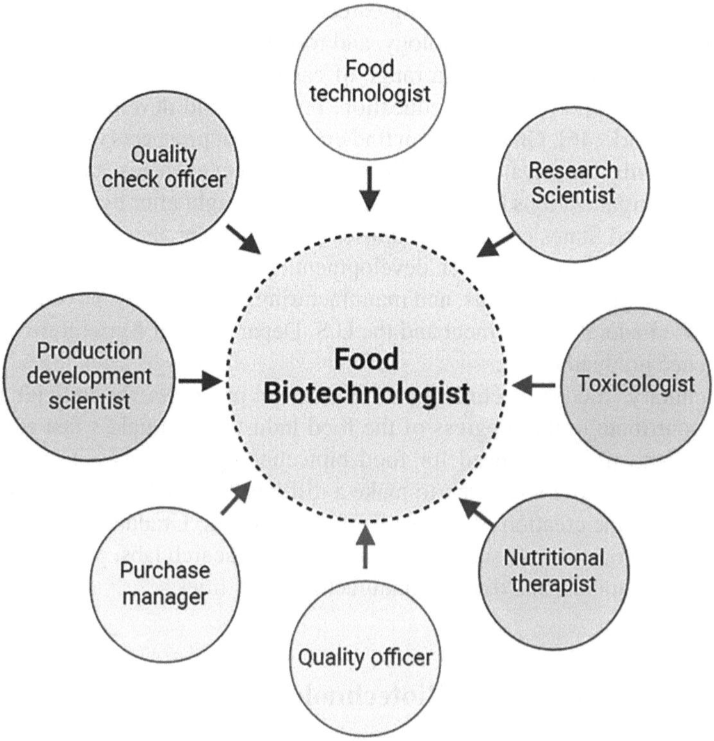

Fig. 27.2 Different job roles for food biotechnologists

Table 27.1 List of job roles and their descriptions for a graduate food biotechnologist

Job role	Description	References
Food technologist	Study food sources and determine innovative methods to create nutritious and wholesome food products	[31]
Production manager	Oversee the production process, ensuring that products are produced efficiently and meet quality standards	[32, 33]
Purchase manager	Responsible for sourcing and purchasing raw materials and equipment for food production	[34, 35]
Research scientist	Research to develop new food products, improve existing products, and test the safety and quality of food products	[36, 37]
Quality check officer	Ensure that food products meet quality standards and regulatory requirements	[38, 39]
Quality manager	Develop and implement quality control procedures to ensure that food products meet quality standards	[40]
Production/process development scientist	Develop new food products and improve existing products by modifying or enhancing the pre-existing methods of cooking or growing specific food items	[41, 42]
Toxicologist	Study the effects of toxic substances on food products and develop methods to prevent contamination	[43, 44]
Nutritional therapist	Develop and implement nutritional plans for individuals and groups based on their dietary needs and health goals	[45]

of food biotechnology is an appealing career choice for those who are passionate, about the fusion of science, technology, and food (Fig. 27.2) [30].

Food science graduates have a range of career options to choose from. These choices include opportunities in education, research, and development as well as government work [46]. Graduates can find employment prospects in various sectors such as academia, the private food industry, government agencies, and nonprofit and international organizations [47]. Food scientists are sought after by private universities in the United States and other countries [48]. They are also employed by companies involved in food product development, nutritional product development, nutraceuticals, pharmaceuticals, and manufacturing. Additionally, government entities, like the Federal Government and the U.S. Department of Agriculture, also hire food science graduates [49].

In summary, food biotechnology offers career prospects for individuals who want to contribute to the progress of the food industry. The field's fast expansion, combined with the rising need for food biotechnologists, makes it an appealing career option for those who want to make a difference in global food security, sustainability, and the creation of innovative food products. Graduates specializing in food biotechnology can pursue career paths within research labs, processing industries, hotels, hospitals, and diverse manufacturing facilities.

4 Future Scope of Food Biotechnology

4.1 Emerging Trends and Technology

Food biotechnology is an evolving field that continuously adapts to new technologies and methods. One promising trend in food biotechnology is the use of gene editing techniques like CRISPR Cas9 to develop disease-stress tolerant crops. This advancement has the potential to enhance crop yields reduce reliance on pesticides and fertilizers and promote eco-friendly agriculture [50, 51]. Another exciting area is biology, which involves designing and constructing new biological components, devices, and systems. In food biotechnology, synthetic biology can be utilized to create food products with improved nutritional profiles and unique functional properties. This technology opens up possibilities for developing foods that offer health benefits beyond basic nutrition [52]. Nanotechnology also plays a role in food processing and packaging research within the field of food biotechnology [53]. By utilizing nanomaterials as barriers against growth and oxidation, nanotechnology enhances food safety and extends shelf life [54]. Furthermore, it enables the development of food products with enhanced sensory attributes and functionalities [55]. Alternative sources of protein such as meat substitutes, lab-grown food, plant-based options, edible insects, and mycoprotein are becoming increasingly popular in the field of food biotechnology [56, 57]. These sustainable alternatives offer a range of nutrients and are produced and consumed in an efficient and environmentally friendly manner. They serve as an alternative to traditional livestock-based protein.

A growing trend in food biotechnology is the integration of 3D printing and robotics into food production processes. These technologies allow for nutrition options and the development of sustainable solutions for protein production [58]. Furthermore, incorporating robotics into food production can enhance efficiency while reducing labor costs. Another emerging trend is the digitization of food production facilities through the use of robotics, e-commerce platforms, and digital tools for managing food. This technology holds promise, for reducing food waste improving safety standards, and increasing productivity in the realm of food production [12, 59].

4.2 Potential Applications and Innovations in the Field of Food Biotechnology

In the realm of food biotechnology, there exists a profound potential to craft functional foods that extend beyond basic nutritional requirements. These specialized foods can be tailored to address specific health challenges like obesity, diabetes, and cardiovascular diseases by incorporating bioactive compounds and nutrients [60, 61]. An emerging trend in this field involves the utilization of biodegradable packaging materials, offering a sustainable alternative that reduces waste and environmental impact in contrast to traditional packaging [62, 63].

The efficiency of food production stands to benefit significantly from the advancements in food biotechnology. By developing crops that exhibit greater resilience to diseases and environmental stressors, there is a prospect of increasing crop yields while concurrently minimizing reliance on pesticides and fertilizers. Moreover, food biotechnology contributes to the ongoing efforts to diminish food waste. This is achieved through the creation of crops with extended shelf lives and improvements in the efficiency of food processing and storage. The overarching goal is to ensure the availability of food for those in need, all while lessening the environmental footprint of food production [64].

Sensory aspects play a pivotal role in consumer preferences, and food biotechnology plays a crucial role in enhancing these properties; this can be done using nanotechnology. Taste, texture, and aroma can be elevated to new heights, giving rise to innovative food products that resonate with evolving consumer tastes [65]. Furthermore, food biotechnology is at the forefront of developing novel food products, including cultured meat, lab-grown food, plant-based nutrition, edible insects, and mycoprotein [66]. These alternative protein sources boast high nutritional value and resource efficiency throughout their production and consumption cycles, positioning them as sustainable and environmentally friendly alternatives to traditional livestock-based proteins [67].

4.3 The Role of Food Biotechnology in Addressing Global Food Security and Sustainability Challenges

In the realm of addressing global food security and sustainability challenges, food biotechnology assumes a pivotal role. As the world's population is anticipated to reach 9.7 billion by 2050, there is a pressing need to augment food production while simultaneously mitigating the environmental impact of agriculture. Food biotechnology emerges as a potent solution by fostering the development of crops resilient to climate change, pests, and diseases, thereby reducing dependence on pesticides and fertilizers. Beyond bolstering crop resilience, food biotechnology contributes to the imperative of reducing food waste. Through advancements in technology, it enhances the shelf life of food products and introduces novel approaches, aligning to minimize wastage and optimize resource utilization [68].

5 Conclusion

The exciting realm of food biotechnology is buzzing with rapid growth, unveiling a plethora of potential applications and innovations across food processing, engineering, packaging, nutrition, and preservation. To highlight few, with emerging trends like gene editing techniques, synthetic biology, and nanotechnology, reflecting the vibrant evolution underway. In this landscape, food biotechnology promises breakthroughs in various domains, from crafting functional foods and developing biodegradable packaging materials to exploring alternative protein sources. Beyond these exciting applications, it plays a key role in boosting the efficiency of food production, curbing food waste, and enhancing the delightful sensory experiences of our favorite food products.

Peering into the future, the horizon of food biotechnology stretches far and wide, hinting at a potential revolution in how we produce and enjoy our food. Significantly, it steps up as a crucial player in tackling global food security and sustainability challenges. This mission is accomplished by nurturing crops with heightened resilience to climate change, pests, and diseases, all while reducing our reliance on pesticides and fertilizers. Additionally, food biotechnology lends a helping hand in the fight against food waste by extending the shelf life of products and introducing innovative preservation technologies.

In essence, food biotechnology takes center stage in a theater of transformative progress, ready to reshape the very foundations of how we grow, process, and savor our food. Its diverse contributions hold the promise of creating a more resilient, sustainable, and efficient global food system—a tantalizing prospect for the future of nourishment.

References

1. Jarvis B, Holmes AW (1982) Biotechnology in relation to the food industry. J Chem Technol Biotechnol 32(1):224–232
2. Maryam BM, Datsugwai MSS, Shehu I (2017) The role of biotechnology in food production and processing. Ind Eng 1(1):24–35
3. Fleischer J (2014) Plant biotechnology. Cambridge Int Law J

4. Roller S, Harlander S (1998) Modern food biotechnology: overview of key issues. In: Genetic modification in the food industry: a strategy for food quality improvement. Blackie Academic and Professional, London, UK, pp 3–26
5. Abah J, Ishaq MN, Wada AC (2010) The role of biotechnology in ensuring food security and sustainable agriculture. Afr J Biotechnol 9(52):8896–8900
6. Soetan KO (2011) The role of biotechnology towards attainment of a sustainable and safe global agriculture and environment–a review. Biotechnol Mol Biol Rev 6:109–117
7. Gartland KM, Gartland JS (2018) Contributions of biotechnology to meeting future food and environmental security needs. EuroBiotech J 2(1):2–9
8. Hertel TW (2015) The challenges of sustainably feeding a growing planet. Food Secur 7(2):185–198
9. Schiefer G (2003) New technologies and their impact on agriculture, environment and the food industry. In: Proceedings of the EFITA 2003 Conference (vol. 5, no. 9)
10. Ojo OO, Zigan S, Orchard J, Shah S (2019) Advanced technology integration in food manufacturing supply chain environment: pathway to sustainability and companies' prosperity. In: 2019 IEEE technology & engineering management conference (TEMSCON). IEEE, pp 1–7
11. Wallin PJ (1997) Robotics in the food industry: an update. Trends Food Sci Technol 8(6):193–198
12. Fernando Y, Mathath A, Murshid MA (2016) Improving productivity: a review of robotic applications in food industry. Int J Robot Appl Technol 4(1):43–62
13. Estrada AC, Díaz DV, Hernández CAM (2017) The role of biotechnology in agricultural production and food supply. Ciencia e investigación agraria: revista latinoamericana de ciencias de la agricultura 44(1):1–11
14. Tomašević I, Putnik P, Valjak F, Pavlić B, Šojić B, Markovinović AB, Kovačević DB (2021) 3D printing as novel tool for fruit-based functional food production. Curr Opin Food Sci 41:138–145
15. Kewuyemi YO, Kesa H, Adebo OA (2022) Trends in functional food development with three-dimensional (3D) food printing technology: prospects for value-added traditionally processed food products. Crit Rev Food Sci Nutr 62(28):7866–7904
16. Shamim M, Pandey P, Singh A, Yadav P, Bhowmick PK, Srivastava D et al (2013) Role of biotechnology in plant diseases management: an overview. J Genet Environ Resour Conserv 1(3):215–221
17. Fokunang CN (2004) Advancement in genetic modification technologies towards disease resistance and food crop production. Biotechnology (Pakistan) 3(1)
18. Van Esse HP, Reuber TL, van der Does D (2020) Genetic modification to improve disease resistance in crops. New Phytol 225(1):70–86
19. Ejye Omar O (1995) Retail influence on food technology and innovation. Int J Retail Distrib Manage 23(3):11–16
20. Kourkoutas Y, Chorianopoulos N, Nisiotou A, Valdramidis VP, Karatzas KA (2016) Application of innovative technologies for improved food quality and safety. BioMed Res Int 2016:9160375. https://doi.org/10.1155/2016/9160375
21. Mahalik NP (2014) Advances in packaging methods, processes and systems. Challenges 5(2):374–389
22. Haji M, Kerbache L, Muhammad M, Al-Ansari T (2020) Roles of technology in improving perishable food supply chains. Logistics 4(4):33
23. Chernikov YY (2021) Foodtech—a modern vector in the development of the food industry. Vestnik Universiteta (1):120–125
24. Cheruvu P, Kapa S, Mahalik NP (2008) Recent advances in food processing and packaging technology. Int J Autom Control 2(4):418–435
25. McClements DJ, Barrangou R, Hill C, Kokini JL, Lila MA, Meyer AS, Yu L (2021) Building a resilient, sustainable, and healthier food supply through innovation and technology. Annu Rev Food Sci Technol 12:1–28
26. Knorr D, Sinskey AJ (1985) Biotechnology in food production and processing. Science 229(4719):1224–1229

27. Bal'-Prilipko LV, Patyka NV, Leonova BI, Starkova ER, Brona AI (2016) Trends, achievements and prospects of biotechnology in the food industry. Mikrobiolohichnyi Zhurnal (Kiev, Ukraine: 1993) 78(3):99–111
28. Copp D (1973) Careers for biology graduates. Nature 242:387–388
29. Menrad K (2000) Economic implications of agro–food biotechnology. J Sci Food Agric 80(5):539–546
30. Oreopoulou V, Giannou V, Lakner Z, Pittia P, Mayor L, Silva CL, Costa R (2015) Career path of food science and technology professionals: entry to the world of work. Trends Food Sci Technol 42(2):183–192
31. Nehir El S, Simsek S (2012) Food technological applications for optimal nutrition: an overview of opportunities for the food industry. Compr Rev Food Sci Food Saf 11(1):2–12
32. Oakland JS, Sohal A (1989) The education, training and careers of production managers in British industry. Int J Oper Prod Manag 9(8):63–90
33. Noori H, Radford RW (1995) Production and operations management: total quality and responsiveness. McGraw-Hill
34. Bergström K, Solér C, Shanahan H (2005) Professional food purchasers' practice in using environmental information. Br Food J 107(5):306–319
35. Kamble SS, Raut RD (2019) Evaluating the factors considered for procurement of raw material in food supply chain using Delphi-AHP methodology—a case study of potato chips processing company in India. Int J Prod Qual Manage 26(2):176–189
36. Pal M, Patel AS, Bariya A, Godishala V, Kandi V (2017) A review of biotechnological applications in food processing of animal origin. Am J Food Sci Tech 5(4):143–148
37. Pyne AW (2000) Innovative new food products: technical development in the laboratory. In: Developing new food products for a changing marketplace, pp 259–275
38. Clute M (2008) Food industry quality control systems. CRC Press
39. Schoenfuss TC, Lillemo JH (2014) Food safety and quality assurance. In: Food processing: principles and applications. Wiley-Blackwell, pp 233–247
40. Criveanu I, Militaru I, Cernaianu N (2007) Quality management in food industry. Bull Univ Agric Sci Vet Med Cluj-Napoca Hortic 64:433–442
41. Bursey RG (1983) Diet modification and the development of new food products. Ann Intern Med 98(5 Part 2):872–874
42. Salmon J (1979) Developing new food products. J Hum Nutr 33(6):445–450
43. Pratt I (2000) Professional opportunities for toxicologists: the requirements of the public sector for education/teaching. Toxicol Lett 112:379–382
44. Kan CA, Meijer GAL (2007) The risk of contamination of food with toxic substances present in animal feed. Anim Feed Sci Technol 133(1–2):84–108
45. Reiter CS, Graves L (2010) Nutrition therapy for eating disorders. Nutr Clin Pract 25(2):122–136
46. Bush SD, Pelaez NJ, Rudd JA, Stevens MT, Tanner KD, Williams KS (2008) Science faculty with education specialties. Science 322(5909):1795–1796
47. McGrath M (2008) Overview of careers in the food science field. In: Careers in food science: from undergraduate to professional, pp 23–42
48. Hartel RW, Klawitter CP, Thiel A (eds) (2008) Careers in food science: from undergraduate to professional. Springer, New York, NY
49. Foegeding EA, Sathe S (2015) Federal funding for food science—challenges and opportunities. J Food Sci 80(11):iii–iv
50. Rasheed A, Gill RA, Hassan MU, Mahmood A, Qari S, Zaman QU et al (2021) A critical review: recent advancements in the use of CRISPR/Cas9 technology to enhance crops and alleviate global food crises. Curr Issues Mol Biol 43(3):1950–1976
51. Erdoğan İ, Cevher-Keskin B, Bilir Ö, Hong Y, Tör M (2023) Recent developments in CRISPR/Cas9 genome-editing technology related to plant disease resistance and abiotic stress tolerance. Biology 12(7):1037
52. Tyagi A, Kumar A, Aparna SV, Mallappa RH, Grover S, Batish VK (2016) Synthetic biology: applications in the food sector. Crit Rev Food Sci Nutr 56(11):1777–1789

53. Pradhan N, Singh S, Ojha N, Shrivastava A, Barla A, Rai V, Bose S (2015) Facets of nanotechnology as seen in food processing, packaging, and preservation industry. BioMed Res Int 2015:365672
54. Thiruvengadam M, Rajakumar G, Chung IM (2018) Nanotechnology: current uses and future applications in the food industry. 3 Biotech 8:1–13
55. Primožič M, Knez Ž, Leitgeb M (2021) (Bio)nanotechnology in food science—food packaging. Nanomaterials (Basel) 11(2):292
56. Akhtar Y, Isman MB (2018) Insects as an alternative protein source. In: Proteins in food processing. Woodhead Publishing, pp 263–288
57. Thavamani A, Sferra TJ, Sankararaman S (2020) Meet the meat alternatives: the value of alternative protein sources. Curr Nutr Rep 9:346–355
58. Lipton JI, Cutler M, Nigl F, Cohen D, Lipson H (2015) Additive manufacturing for the food industry. Trends Food Sci Technol 43(1):114–123
59. Benyam AA, Soma T, Fraser E (2021) Digital agricultural technologies for food loss and waste prevention and reduction: global trends, adoption opportunities and barriers. J Clean Prod 323:129099
60. Niba LL (2003) The relevance of biotechnology in the development of functional foods for improved nutritional and health quality in developing countries. Afr J Biotechnol 2(12):631–635
61. Sun-Waterhouse D (2011) The development of fruit-based functional foods targeting the health and wellness market: a review. Int J Food Sci Technol 46(5):899–920
62. Song JH, Murphy RJ, Narayan R, Davies GBH (2009) Biodegradable and compostable alternatives to conventional plastics. Philos Trans Royal Soc B: Biol Sci 364(1526):2127–2139
63. Guo C, Guo H (2022) Progress in the degradability of biodegradable film materials for packaging. Membranes 12(5):500
64. Smil V (2004) Improving efficiency and reducing waste in our food system. Environ Sci 1(1):17–26
65. Sun R, Lu J, Nolden A (2021) Nanostructured foods for improved sensory attributes. Trends Food Sci Technol 108:281–286
66. Hadi J, Brightwell G (2021) Safety of alternative proteins: technological, environmental and regulatory aspects of cultured meat, plant-based meat, insect protein and single-cell protein. Food Secur 10(6):1226
67. Salter AM, Lopez-Viso C (2021) Role of novel protein sources in sustainably meeting future global requirements. Proc Nutr Soc 80(2):186–194
68. Augustin MA, Riley M, Stockmann R, Bennett L, Kahl A, Lockett T et al (2016) Role of food processing in food and nutrition security. Trends Food Sci Technol 56:115–125

Correction to: Frontiers in Food Biotechnology

Jayachandra S. Yaradoddi, Bharati S. Meti, Sulochana B. Mudgulkar, and Dayanand Agsar

Correction to:
J. S. Yaradoddi et al. (eds.), *Frontiers in Food Biotechnology*, https://doi.org/10.1007/978-981-97-3261-6

The original version of the book was inadvertently published with incorrect affiliation details for editors Bharati S. Meti and Sulochana B. Mudgulkar. The complete affiliation details have been updated in the book.

The updated version of the book can be found at
https://doi.org/10.1007/978-981-97-3261-6

J. S. Yaradoddi et al. (eds.), *Frontiers in Food Biotechnology*,
https://doi.org/10.1007/978-981-97-3261-6_28

